# Vascular Plants and Paleobotany

# Vascular Plants and Paleobotany

**Editor:**

Vikas Mishra

ArclerPress

www.arclerpress.com

**Vascular Plants and Paleobotany**

*Editor: Vikas Mishra*


**708 3rd Avenue, 6th Floor**
**New York**
**NY 10017**
**United States of America**
**www.arclerpress.com**

ISBN: 978-1-68094-519-5

Library of Congress Control Number: 2016951371

# About the Editor

## Vikas Mishra

Vikas obtained his Ph.D. from King George's Medical University, India in 2010. His interests span from bio-resources, fermentation technology to medical virology and neuroscience research. He is currently working as a research associate at the University of British Columbia.

# Contents

# List of Contributors

**Dianne Edwards**
School of Earth and Ocean Sciences, Cardiff University, Cardiff CF10 3AT, UK

**Paul Kenrick**
Department of Earth Sciences, The Natural History Museum, Cromwell Road, London SW7 5BD, UK

**Aldo Piombino**
Earth Sciences Department, Florence University, 50125 Florence, Italy

**Daniele Silvestro**
Department of Biological and Environmental Sciences, University of Gothenburg, Göteborg, Sweden

**Borja Cascales-Miñana**
CNRS, UMR Botany and bioinformatics of plant architecture (AMAP), Montpellier, France

**Christine D. Bacon**
Department of Biological and Environmental Sciences, University of Gothenburg, Göteborg, Sweden

**Alexandre Antonelli**
Department of Biological and Environmental Sciences, University of Gothenburg, Göteborg, Sweden

**Emmanuel Buschiazzo**
Department of Forest Sciences, University of British Columbia, 2424 Main Mall, Vancouver, BC V6T 1Z4, Canada.
School of Natural Sciences, University of California, Merced, 5200 North Lake Road, Merced, CA 95343 USA.

**Carol Ritland**
Department of Forest Sciences, University of British Columbia, 2424 Main Mall, Vancouver, BC V6T 1Z4, Canada.

**Jörg Bohlmann**
Department of Forest Sciences, University of British Columbia, 2424 Main Mall, Vancouver, BC V6T 1Z4, Canada.
Michael Smith Laboratories, University of British Columbia, 2185 East Mall, BC V6T 1Z4, Canada.

**Kermit Ritland**
Department of Forest Sciences, University of British Columbia, 2424 Main Mall, Vancouver, BC V6T 1Z4, Canada.

**J. Gordon Burleigh**
Department of Biology, University of Florida, Gainesville, FL 32611, USA

**W. Brad Barbazuk**
Department of Biology, University of Florida, Gainesville, FL 32611, USA

**John M. Davis**
School of Forest Resources and Conservation, University of Florida, Gainesville, FL 32611, USA

**Alison M. Morse**
School of Forest Resources and Conservation, University of Florida, Gainesville, FL 32611, USA

**Pamela S. Soltis**
Florida Museum of Natural History, University of Florida, Gainesville, FL 32611, USA

**Juliana Stival Sena**
Center for Forest Research and Institute for Systems and Integrative Biology, 1030 Street of Medicine, Laval University, Quebec, QC G1V 0A6, Canada

**Isabelle Giguère**
Center for Forest Research and Institute for Systems and Integrative Biology, 1030 Street of Medicine, Laval University, Quebec, QC G1V 0A6, Canada

**Brian Boyle**
Center for Forest Research and Institute for Systems and Integrative Biology, 1030 Street of Medicine, Laval University, Quebec, QC G1V 0A6, Canada

**Philippe Rigault**
Gydle Inc., Québec, QC, Canada

**Inanc Birol**
Michael Smith Laboratories, University of British Columbia, Vancouver, BC V6T 1Z4, Canada

**Andrea Zuccolo**
Applied Genomics Institute, Udine 33100, Italy
Institute of Life Sciences, Scuola Superiore Sant'Anna, Pisa 56127, Italy

**Joerg Bohlmann**
Michael Smith Laboratories, University of British Columbia, Vancouver, BC V6T 1Z4, Canada

**Steven Jones**
Michael Smith Laboratories, University of British Columbia, Vancouver, BC V6T 1Z4, Canada

**Jean Bousquet**
Center for Forest Research and Institute for Systems and Integrative Biology, 1030 Street of Medicine, Laval University, Quebec, QC G1V 0A6, Canada
Canada Research Chair in Forest Genomics, Université Laval, Québec, QC G1V 0A6, Canada

**John Mackay**
Center for Forest Research and Institute for Systems and Integrative Biology, 1030 Street of Medicine, Laval University, Quebec, QC G1V 0A6, Canada

**Nathalie Pavy**
Canada Research Chair in Forest and Environmental Genomics, Centre for Forest Research and Institute for Systems and Integrative Biology, Université Laval, Québec, Québec G1V 0A6, Canada

**Betty Pelgas**
Canada Research Chair in Forest and Environmental Genomics, Centre for Forest Research and Institute for Systems and Integrative Biology, Université Laval, Québec, Québec G1V 0A6, Canada
Natural Resources Canada, Canadian Forest Service, Laurentian Forestry Centre, 1055 Rue du PEPS, CP 10380, Succ. Sainte-Foy, Québec, Québec G1V 4C7, Canada

**Jérôme Laroche**
Bioinformatics Platform, Institute for Systems and Integrative Biology, Université Laval, Québec, Québec G1V 0A6, Canada

**Philippe Rigault**
Canada Research Chair in Forest and Environmental Genomics, Centre for Forest Research and Institute for Systems and Integrative Biology, Université Laval, Québec, Québec G1V 0A6, Canada
Gydle Inc., 1363 Avenue Maguire, Québec, Québec G1T 1Z2, Canada

**Nathalie Isabel**
Canada Research Chair in Forest and Environmental Genomics, Centre for Forest Research and Institute for Systems and Integrative Biology, Université Laval, Québec, Québec G1V 0A6, Canada
Natural Resources Canada, Canadian Forest Service, Laurentian Forestry Centre, 1055 Rue du PEPS, CP 10380, Succ. Sainte-Foy, Québec, Québec G1V 4C7, Canada

**Jean Bousquet**
Canada Research Chair in Forest and Environmental Genomics, Centre for Forest Research and Institute for Systems and Integrative Biology, Université Laval, Québec, Québec G1V 0A6, Canada

**Elie S. M. Raherison**
Center for Forest Research and Institute for Integrative and Systems Biology, Universite Laval, Quebec, QC G1V 0A6, Canada

**Isabelle Giguère**
Center for Forest Research and Institute for Integrative and Systems Biology, Universite Laval, Quebec, QC G1V 0A6, Canada

**Sébastien Caron**
Center for Forest Research and Institute for Integrative and Systems Biology, Universite Laval, Quebec, QC G1V 0A6, Canada

**Mebarek Lamara**
Center for Forest Research and Institute for Integrative and Systems Biology, Universite Laval, Quebec, QC G1V 0A6, Canada

**John J. MacKay**
Department of Plant Sciences, University of Oxford, OX1 3RB, Oxford, UK A

**Kathleen Bauer**
Museum of Nature South Tyrol, Bindergasse 1, Bolzano/Bozen 39100, Italy
Department of Earth and Environmental Sciences, Palaeontology and Geobiology, Ludwig-Maximilians-University, Richard-Wagner-Straße 10, 80333 Munich, Germany

**Lea Grauvogel-Stamm**
4 Place du Marché aux Poissons, Strasbourg 67000, France

**Evelyn Kustatscher**
Museum of Nature South Tyrol, Bindergasse 1, Bolzano/Bozen 39100, Italy
Department of Earth and Environmental Sciences, Palaeontology and Geobiology, Ludwig-Maximilians-University, Richard-Wagner-Straße 10, 80333 Munich, Germany
Bavarian State Collection for Palaeontology and Geology, Richard-Wagner-Straße 10, 80333 Munich, Germany

**Michael Krings**
Department of Earth and Environmental Sciences, Palaeontology and Geobiology, Ludwig-Maximilians-University, Richard-Wagner-Straße 10, 80333 Munich, Germany
Bavarian State Collection for Palaeontology and Geology, Richard-Wagner-Straße 10, 80333 Munich, Germany

**Thilo C Fischer**
Department Biology I, Biocenter Botany, Ludwig-Maximilians-University Munich, Großhadernerstrasse 2-4, D-82152 Planegg-Martinsried, Germany

**Barbara Meller**
Institute of Palaeontology, Geocenter, University Vienna, Althanstrasse 14, A- 1090 Wien, Austria

**Rainer Butzmann**
Fuggerstrasse 8, D-81373 München, Germany

**Eric D Brenner**
The New York Botanical Garden, 200th St. and Kazimiroff, Bronx, NY 10458-5126, USA

**Manpreet S Katari**
New York University, Department of Biology 1009 Main Building, New York, NY 10003, USA

**Dennis W Stevenson**
The New York Botanical Garden, 200th St. and Kazimiroff, Bronx, NY 10458-5126, USA

**Stephen A Rudd**
Centre for Biotechnology, Tykistökatu 6, FIN-20521 Turku, Finland

**Andrew W Douglas**
The New York Botanical Garden, 200th St. and Kazimiroff, Bronx, NY 10458-5126, USA

**Walter N Moss**
The New York Botanical Garden, 200th St. and Kazimiroff, Bronx, NY 10458-5126, USA

**Richard W Twigg**
Biology Department, Duke University, Box 91000, Durham, North Carolina, 27708

**Suzan J Runko**
The New York Botanical Garden, 200th St. and Kazimiroff, Bronx, NY 10458-5126, USA

**Giulia M Stellari**
Department of Plant Biology, Cornell University, Ithaca NY 14850, USA

**WR McCombie**
Genome Research Center, Cold Spring Harbor Laboratory, 500 Sunnyside Blvd, Woodbury, NY 11797, USA

**Gloria M Coruzzi**
New York University, Department of Biology 1009 Main Building, New York, NY 10003, USA

**Biao Jin**
College of Horticulture and Plant Protection, Yangzhou University, Yangzhou 225009, China

**Di Wang**
College of Horticulture and Plant Protection, Yangzhou University, Yangzhou 225009, China

**Yan Lu**
College of Horticulture and Plant Protection, Yangzhou University, Yangzhou 225009, China

**Xiao Xue Jiang**
College of Horticulture and Plant Protection, Yangzhou University, Yangzhou 225009, China

**Min Zhang**
College of Horticulture and Plant Protection, Yangzhou University, Yangzhou 225009, China

**Lei Zhang**
College of Horticulture and Plant Protection, Yangzhou University, Yangzhou 225009, China

**Li Wang**
College of Horticulture and Plant Protection, Yangzhou University, Yangzhou 225009, China
Key Laboratory of Plant Molecular Physiology, Institute of Botany, Chinese Academy of Sciences, Beijing 100093, China

**Fabien L Condamine**
CNRS, UMR 7641 Applied Mathematics Centre (Ecole Polytechnique), Route de Saclay, 91128 Palaiseau, France
Department of Biological and Environmental Sciences, University of Gothenburg, Box 461, SE-405 30 Göteborg, Sweden

**Nathalie S Nagalingum**
National Herbarium of New South Wales, Royal Botanic Gardens & Domain Trust, Mrs Macquaries Road, Sydney, NSW 2000, Australia

**Charles R Marshall**
Department of Integrative Biology and Museum of Paleontology, University of California, 1101 Valley Life Sciences Building, Berkeley, CA 94720-4780, USA

**Hélène Morlon**
CNRS, UMR 8197 Institute of Biology of the Ecole Normale Supérieure, 46 rue d'Ulm, 75005 Paris, France

**B. Ravi Prasad Rao**
Biodiversity Conservation Division, Department of Botany, Sri Krishnadevaraya University,
Anantapuram, Andhra Pradesh 515003, India

**M. Chennakesavulu Naik**
Biodiversity Conservation Division, Department of Botany, Sri Krishnadevaraya University,
Anantapuram, Andhra Pradesh 515003, India

**S. Salamma**
Biodiversity Conservation Division, Department of Botany, Sri Krishnadevaraya University,
Anantapuram, Andhra Pradesh 515003, India

**M.V. Suresh Babu**
Biodiversity Conservation Division, Department of Botany, Sri Krishnadevaraya University, Anantapuram, Andhra Pradesh 515003, India

**L. Rasingam**
Botanical Survey of India, Deccan Regional Centre, Plot No. 366/1 in ZSI Campus, Attapur, Hyderguda P.O., Hyderabad, Telangana 500048, India

**Yu-Chung Chiang**
Department of Life Sciences, Pingtung University of Science and Technology, Pingtung, Taiwan, 912, Republic of China

**Kuo-Hsiang Hung**
Graduate Institute of Bioresources, Pingtung University of Science and Technology, Pingtung, Taiwan, 912, Republic of China

**Shann-Jye Moore**
Department of Life Science, Taiwan Normal University, Taipei, Taiwan, Republic of China

**Xue-Jun Ge**
South China Botanical Garden, Chinese Academy of Sciences, Guangzhou 510650, PR China

**Shong Huang**
Department of Life Science, Taiwan Normal University, Taipei, Taiwan, Republic of China

**Tsai-Wen Hsu**
Endemic Species Research Institute, Nantou, Taiwan, Republic of China

**Barbara A Schaal**
Department of Biology, Washington University, St Louis, MO 63130, USA

**TY Chiang**
Department of Life Sciences, Cheng-Kung University, Tainan, Taiwan, 701, Republic of China

**Xiuyan Feng**
Key Laboratory for Plant Diversity and Biogeography of East Asia, Kunming Institute of Botany, Chinese Academy of Sciences, Kunming 650201, Yunnan, China
University of Chinese Academy of Sciences, Beijing 100049, China

**Ying Zheng**
Key Laboratory for Plant Diversity and Biogeography of East Asia, Kunming Institute of Botany, Chinese Academy of Sciences, Kunming 650201, Yunnan, China
University of Chinese Academy of Sciences, Beijing 100049, China

**XunGong**
Key Laboratory for Plant Diversity and Biogeography of East Asia, Kunming Institute of Botany, Chinese Academy of Sciences, Kunming 650201, Yunnan, China

**Yu-Chung Chiang**
Department of Biological Sciences, National Sun Yat-sen University, Kaohsiung 80424, Taiwan

**Bing-Hong Huang**
Department of Biological Science and Technology, National Pingtung University of Science and Technology, Pingtung 91201, Taiwan

**Chun-Wen Chang**
Department of Life Science, National Taiwan Normal University, Taipei 116, Taiwan
Taiwan Forestry Research Institute, Technical Service Division, Taipei 10066, Taiwan

**Yu-Ting Wan**
Department of Biological Science and Technology, National Pingtung University of Science and Technology, Pingtung 91201, Taiwan

**Shih-Jie Lai**
Department of Biological Science and Technology, National Pingtung University of Science and Technology, Pingtung 91201, Taiwan

**Pei-Chun Liao**
Department of Biological Science and Technology, National Pingtung University of Science and Technology, Pingtung 91201, Taiwan

**Hugo Beraldi-Campesi**
Institute of Geology, UNAM, Ciudad Universitaria

**Deming Wang**
Department of Geology, Key Laboratory of Orogenic Belts and Crustal Evolution, Peking University, Beijing 100871, China

**Le Liu**
Department of Geology, Key Laboratory of Orogenic Belts and Crustal Evolution, Peking University, Beijing 100871, China

**Anne Vatén**
Department of Biology, Stanford University

**Dominique C Bergmann**
Howard Hughes Medical Institute

# Preface

Paleobotany is a field of paleontology that studies plants throughout geologic history, and is primarily concerned with the fossil record and evolutionary history of plants. Vascular plants also known as tracheophytes and also higher plants, form a large group of plants that are defined as those land plants that have lignified tissues for conducting water and minerals throughout the plant. They also have a specialized non-lignified tissue to conduct products of photosynthesis. Vascular plants include the clubmosses, horsetails, ferns, gymnosperms and angiosperms. Paleontology is the science concerned with fossils, the physical evidence of prehistoric life—including plants, animals, and microorganisms—on the earth. Paleobotany is of importance primarily because the record of fossil plants helps scientists understand the long process of plant evolution. Paleobotany focuses on plant fossils, including algae, fungi, and related organisms, as well as mosses, ferns, and seed plants. As most organisms decompose rapidly after death, their preservation in nature is a rare event. Most individuals are not represented in the fossil record, and even many species that must have existed have vanished without a trace. Normally the plant dies and it soon starts generation. This generation continues till the entire body is broken into single chemical compound. Sometimes in certain circumstances degeneration of plant body stops and part of body remains in visible form. These remains are known as fossils. Plant fossils are present in rocks composed of sediments deposited in water. These stratified rocks are super imposed upon one another. These rocks are built of sediment that accumulate in bottom of seas lakes, swamps, flooded valleys. Rock formation is co related with geological approach in the subject paleobotany. In fact fossils are the makers of geologic tume.From an evolutionary point of view, plants have ensued from filamentous green algae dwelling in shallow waters, perhaps around 500 to 600 million years ago. In around 100 to 200 million years later, land plants seems to have diversified, leading to development of vascular plants, which eventually diversified into seed plants and flowering plants, forming the vast expanse of forests on the planet earth. This book is not a comprehensive text book discussing the evolutionary extant of all plants, neither it is a complete discussion on the topic of paleobotany. It is simple effort to compile the recent research findings on the modern understanding of paleobotany and fossil records of some vascular plants. A lot can be learned about an environment from the plants. Pollen is also often used as an index fossil, so paleobotany is important for dating purposes. Knowing how plants evolved can teach us how they will evolve in the future, and plants are extremely important ant to all aspects of life.

**Editor**
**Vikas Mishra**

# INTRODUCTION

Vascular plants (from Latin vasculum: duct), also known as tracheophytes (from the equivalent Greek term trachea) and also higher plants, form a large group of plants (ca 308,312 accepted known species) that are defined as those land plants that have lignified tissues (the xylem) for conducting water and minerals throughout the plant. They also have a specialized non-lignified tissue (the phloem) to conduct products of photosynthesis. Vascular plants include the clubmosses, horsetails, ferns, gymnosperms (including conifers) and angiosperms (flowering plants). Scientific names for the group include Tracheophyta and Tracheobionta.

Paleobotany, also spelled as palaeobotany (from the Greek words paleon = old and "botany", study of plants), is the branch of paleontology or paleobiology dealing with the recovery and identification of plant remains from geological contexts, and their use for the biological reconstruction of past environments (paleogeography), and both the evolutionary history of plants, with a bearing upon the evolution of life in general. A synonym is paleophytology. Paleobotany includes the study of terrestrial plant fossils, as well as the study of prehistoric marine photoautotrophs, such as photosynthetic algae, seaweeds or kelp. A closely related field is palynology, which is the study of fossilized and extant spores and pollen.

Paleobotany is important in the reconstruction of ancient ecological systems and climate, known as paleoecology and paleoclimatology respectively; and is fundamental to the study of green plant development and evolution. Paleobotany has also become important to the field of archaeology, primarily for the use of phytoliths in relative dating and in paleoethnobotany,

### *Characteristics*

Vascular plants are distinguished by two primary characteristics:

1. Vascular plants have vascular tissues which distribute resources through the plant. This feature allows vascular plants to evolve to a larger size than non-vascular plants, which lack these specialized conducting tissues and are therefore restricted to relatively small sizes.

2. In vascular plants, the principal generation phase is the *sporophyte*, which is usually diploid with two sets of chromosomes per cell. Only the germ cells and gametophytes are haploid. By contrast, the principal generation phase in non-vascular plants is the *gametophyte*, which ishaploid with one set of chromosomes per cell. In these plants, only the spore stalk and capsule are diploid.

One possible mechanism for the presumed switch from emphasis on the haploid generation to emphasis on the diploid generation is the greater efficiency in spore dispersal with more complex diploid structures. In other words, elaboration of the spore stalk enabled the production of more spores, and enabled the development of the ability to release them higher and to broadcast them farther. Such developments may include more photosynthetic area for the spore-bearing structure, the ability to grow independent roots, woody structure for support, and more branching.

### Overview of the paleobotanical record

Macroscopic remains of true vascular plants are first found in the fossil record during the Silurian Period of the Paleozoic era. Some dispersed, fragmentary fossils of disputed affinity, primarily spores and cuticles, have been found in rocks from the

Ordovician Period in Oman, and are thought to derive from liverwort- or moss-grade fossil plants.

## An unpolished hand sample of the Lower Devonian Rhynie Chert from Scotland

An important early land plant fossil locality is the Rhynie Chert, found outside the village of Rhynie in Scotland. The Rhynie chert is an Early Devonian sinter (hot spring) deposit composed primarily of silica. It is exceptional due to its preservation of several different clades of plants, from mosses and lycopods to more unusual, problematic forms. Many fossil animals, including arthropods and arachnids, are also found in the Rhynie Chert, and it offers a unique window on the history of early terrestrial life. Plant-derived macrofossils become abundant in the Late Devonian and include tree trunks, fronds, and roots. The earliest tree was thought to be Archaeopteris, which bears simple, fern-like leaves spirally arranged on branches atop a conifer-like trunk, though it is now known to be the recently discovered Wattieza. Widespread coal swamp deposits across North America and Europe during the Carboniferous Period contain a wealth of fossils containing arborescent lycopods up to 30 meters tall, abundant seed plants, such as conifers and seed ferns, and countless smaller, herbaceous plants. Angiosperms (flowering plants) evolved during the Mesozoic, and flowering plant pollen and leaves first appear during the Early Cretaceous, approximately 130 million years ago

## Preservation of plant fossils

Ginkgoites huttonii, Middle Jurassic, Yorkshire, UK. Leaves preserved as compressions. Specimen in Munich Palaeontological Museum, Germany.

Plant fossils can be preserved in a variety of ways, each of which can give different types of information about the original parent plant. These modes of preservation are discussed in the general pages on fossils but may be summarised in a palaeobotanical context as follows.

Adpressions (compressions - impressions). These are the most commonly found type of plant fossil. They provide good morphological detail, especially of dorsiventral (flattened) plant parts such as leaves. If the cuticle is preserved, they can also yield fine anatomical detail of the epidermis. Little other detail of cellular anatomy is normally preserved.

Rhynia, Lower Devonian Rhynie Chert, Scotland, UK. Transverse section through a stem preserved as a silica petrifaction, showing preservation of cellular structure.

Petrifactions (permineralisations or anatomically preserved fossils). These provide fine detail of the cell anatomy of the plant tissue. Morphological detail can also be determined by serial sectioning, but this is both time consuming and difficult.

Moulds and casts. These only tend to preserve the more robust plant parts such as seeds or woody stems. They can provide information about the three-dimensional form of the plant, and in the case of casts of tree stumps can provide evidence of the density of the original vegetation. However, they rarely preserve any fine morphological detail or cell anatomy. A subset of such fossils are pith casts, where the centre of a stem is either hollow or has delicate pith. After death, sediment enters and forms a cast of the central cavity of the stem. The best known examples of pith casts are in the Carboniferous Sphenophyta (Calamites) and cordaites (Artisia).

Crossotheca hughesiana Kidston, Middle Pennsylvanian, Coseley, near Dudley, UK. A lyginopteridalean pollen organ preserved as an authigenic mineralization (mineralized in situ). Specimen in Sedgwick Museum, Cambridge, UK.

Authigenic mineralisations. These can provide very fine, three-dimensional morphological detail, and have proved especially important in the study of reproductive structures that can be severely distorted in adpressions. However, as they are formed in mineral nodules, such fossils can rarely be of large size.

Fusain. Fire normally destroys plant tissue but sometimes charcoalified remains can preserve fine morphological detail that is lost in other modes of preservation; some of the best evidence of early flowers has been preserved in fusain. Fusain fossils are delicate and often small, but because of their buoyancy can often drift for long distances and can thus provide evidence of vegetation away from areas of sedimentation.

## Vascular Tissue

In modern plants, we tend to differentiate between vascular and non-vascular plants. In the fossil record, we observe several pre-vascular and early vascular plants with tracheids that differ from modern plants

Grade of vascular tissue found in extinct and extant plants

- Pre-vascular plants and bryophytes: plants such as the horneophytes and some bryophytes have conducting cells that lack secondary thickenings. Sometimes these cells are referred to as hydroids.

- Basal tracheophytes: Some of the earliest vascular plants, such as rhyniophytes, had vascular conducting cells referred to as S-type tracheids.

    - S-Type tracheids: These cells had annual or helical thickening consisting of a spongy material. Prominent helical (possibly annular) thickenings show a spongy texture which is also present in the underlying and intervening lateral walls of the elements. Both thickenings and intervening lateral walls are covered by a very thin layer with numerous closely spaced holes.

### Basal eutracheophytes: e.g. Cooksonia and zosterophyllphytes

- C-type tracheids: Plants such as Cooksonia had tracheids with conventional annular and spiral tracheids, except that the imperforate lateral walls are thick when compared with the primary wall in protoxylem.
- G-Type tracheids: Names after cells found in the zosterophyllphyte Gosslingia. Secondary thickenings are annular, helical to occasionally approaching reticulate, and are connected by a sheet of resilient material which is fused to the presumed primary wall. The sheet is perforated by holes of varying size and usually rounded shape.

### Eutracheophytes: e.g. euphyllophytes and derived lycophytes

- P-Type tracheids: Named after cells found in the trimerophyte Psilophyton dawsonii. Metaxylem tracheids are long (< 3·0 mm) and markedly faceted (5–7 sides) compared with G-type. Pitting is essentially scalariform and, in specimens superbly preserved in calcium carbonate-rich nodules, pit-closing membranes are present.

### *What are Vascular Plants?*

Vascular plants have conducting, also referred to as vascular or liquefied, tissues that transport water, minerals, and photosynthetic materials throughout the plant's roots, stems, and leaves. They differ from nonvascular plants, which do not have conducting tissues, and require water for fertilization. Other names for them include tracheophytes and higher plants. They make up the majority of plants found on the Earth today, with the exception of mosses and liverworts.

There are three different systems present in the bodies of vascular plants that work together to maintain the life of the plant. These include the root system, shoot system, and vascular system. The root system consists of the roots, which anchor the plant in the ground and absorb moisture and food from the soil. The shoot system consists of the stems and leaves, which specialize in photosynthesis. The vascular system circulates water and minerals to the leaves, and the photosynthetic material from the leaves to the rest of the plant.

The plant vascular system is made up of two networks of tubes, known as the xylem and phloem, which are the water and food conducting systems. These are not present in nonvascular plants, which is the major difference between the two. The xylem transports water up from the roots and circulates it to all the other sections of the plant's body. The phloem transports food, nutrients, and photosynthetic material throughout the plant to keep it healthy and growing.

The xylem is made up of dead hollow cells known as tracheids. The phloem, however, contains living cells, which are known as sieve-tube members. They have pores to allow the passage of molecules, but lack nuclei and other organs. Their companion cells, or the cells located next to them, function to keep them alive and healthy.

Vascular plants are able to grow to a larger size than nonvascular plants, which lack woody tissue that is needed to provide support for larger growth. Nonvascular plants typically grow to around 0.39 to 0.78 inches (1 to 2 cm), while vascular ones can grow many feet (meters) high. They are also more commonly found in damp and shady areas, since they do not have vascular tissue to hydrate their bodies. They instead must absorb water on their surface, making it difficult to survive for long periods in dry areas. Vascular plants, however, can survive in nearly any climate, as long as their basic needs for water, sunlight, and appropriate temperature are met.

Some of the most common include conifers and flowering plants, although there is another type, known as seedless vascular plants. These plants contain the same vascular system described above, but do not reproduce from seed. Psilotum, also known as the whisk fern, which grows wild in the woodlands of Florida in the United States, is one such plant. Others include the Lycophyta and Sphenophyta divisions, and many types of ferns in the Pterophyta division.

### *In Plants, what is a Vascular System?*

The plant's vascular system, filled with pipe-like vascular tissue, are the crucial conduits through which water and nutrients flow in plants. Without them, plants are limited to being a few inches (centimeters) tall, as is the case with mosses, hornworts, and liverworts, the only terrestrial non-vascular plants. Most terrestrial plants alive today have a vascular system, from small shrubs to the tallest trees. The largest transport water and nutrients to the tops of trees over 379 feet (115 m) tall, such as the Giant Sequoia.

The vascular system in plants uses more than one cell type. The tissues are divided into two parts: xylem and phloem. Xylem is located more closely to the center of the plant and its main function is to transport water. The best known xylem is wood. It is mostly composed of dead cells with a tube-like structure. The phloem, however, consists mostly of living cells. Located closer to the outside of the plant, and just under the bark in trees, the phloem moves organic nutrients (known as photosynthate) to the upper reaches of the plant. The dependence of the plant upon phloem is why trees can be killed simply be removing a circular layer of bark around their circumference.

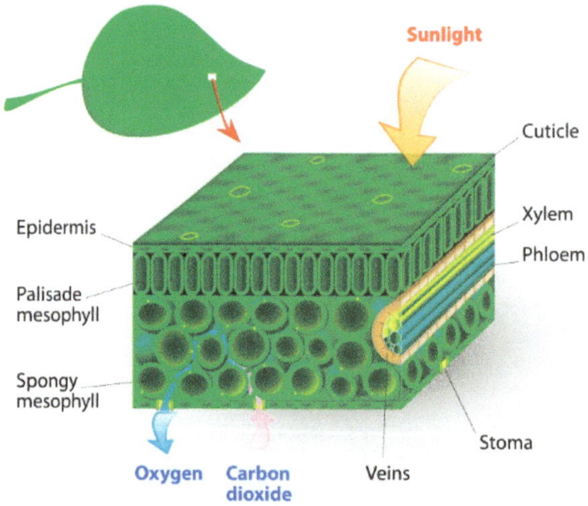


In plants, this system is found in coordinated blocks called vascular bundles, which contain xylem, phloem, and protective cells around their perimeter. As mentioned, xylem is located closer to the center, while phloem is closer to the edge. In leaves, the xylem is located on top, the phloem underneath. This is why aphids are often found on the underside of leaves: this is where they can penetrate to get sucrose from the plant's juices.

Between xylem and phloem is a layer known as the vascular cambium. It segregates the xylem and phloem from each other and is the source of the stem cells that will differentiate into both. The growth rate of cells from the vascular cambium depends on available sunlight and nutrients, and the rings found in tree trunks are indicative of the growth patterns from the vascular cambium.

### *How Many Species of Plants are there?*

There are about 375,000 species of plants, with more being discovered each year. This includes seed plants, bryophytes, ferns and relatives of ferns called fern allies. Some 297,326 species of plants have been identified, of which 258,650 are flowering and 15,000 bryophytes (liverworts, hornworts, and mosses). All plants are photosynthetic, utilizing carbon dioxide, water and light energy to produce oxygen and sugars. Within each cell are numerous organelles called chloroplasts which perform this task. Chloroplasts are thought to be ancient cyanobacteria that were absorbed by the larger eukaryotic cells which constitute plants.

Plants consist of two main groups; green algae and land plants. Other types of algae, such as red algae and brown algae, were initially identified as plants, but later classified outside the kingdom Plantae. These latter organisms are part of the same unranked group as plants, Archaeplastida, but are usually not referred to as plants except in a colloquial sense. An example of a familiar green algae is seaweed.

Most plants, of course, are the land plants, known as embryophytes. The most common are the vascular plants, meaning they have internal tubes to carry water and nutrients. The minority are species of plants called bryophytes, including liverworts, hornworts, and mosses. Without vascular tubes, plants can only grow a couple inches tall. Bryophytes first evolved during the Cambrian era, about 500 million years ago, and have remained abundant ever since.

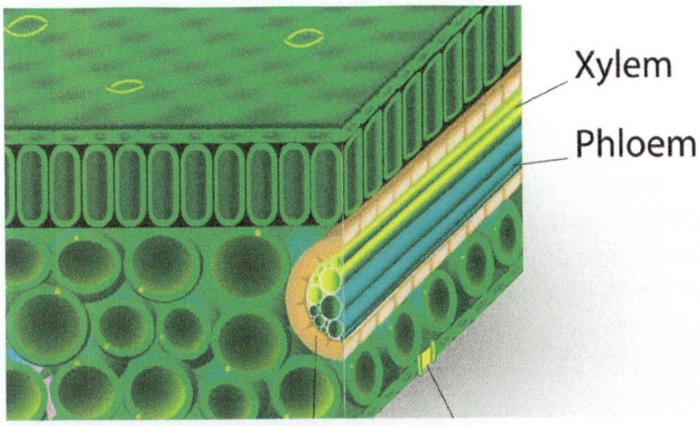


The vascular plants are a more advanced species of plant. These evolved during the Silurian era, about 420 million years ago, and went on to dominate the land, both in terms of diversity and biomass. Vascular plants are the keystone that holds up the entire terrestrial ecosystem, in the same way that cyanobacteria hold up the ecosystem in the oceans.

The first serious innovation after the initial evolution of vascular plants was that of the seed. Using seeds, plants could colonize drier areas than before, and survive seasonal drought. By comparison, plants that reproduce via spores, such as ferns, require a moist environment.

The most recent plants to evolve were the angiosperms, or flowering plants. These developed only 125 million years ago, during the Cretaceous. The point of evolving flowers was to get mobile animals, especially insects, involved in the process of exchanging pollen and therefore genetic material.

### *What is the Evolutionary History of Plants?*

The evolutionary history of plants begins a very long time ago. Algae has probably existed on moist ground for over a billion years, but plants, as in the kingdom Plantae, did not emerge until 700 million years ago at the earliest. This number is from molecular genetics analysis that suggests land plants split from the green algae around this time, although this figure is not corroborated by fossil evidence. The earliest evolutionary history of plants on land appears in the early Ordovician period, about 475 million years ago, although many paleobotanists suspect there were plants during the Cambrian period, 500 million years ago.

The first land plants were non-vascular bryophytes, represented today by mosses, hornworts, and liverworts. These plants, lacking circulatory tissues, were quite short, between 1 and 100 mm (4 in) in thickness. These bryophytes represented the basal group in the evolutionary history of plants. They could only survive in very moist areas, where all cells can easily take in water directly, and their spores could be dispersed easily. Lacking a protective coating, spores are relatively fragile, and prone to dessication (drying out). Scientists believe that the first land plants may have set the stage for the colonization of the land by animals by sequestering carbon dioxide from the atmosphere in the biopolymer lignin. This increased the portion of the atmosphere containing oxygen, making it more available to oxygen-breathing animals like the first terrestrial arthropods and mollusks.

Around 425 million years ago, the first vascular plants appeared, like the simple bifurcating, sporangia (spore-producing structure) tipped Cooksonia and the unusually advanced Baragwanathia, found in Australia. Slowly, plants grew in height, from just a couple centimeters to around 20 centimeters (8 in). At this point, plants spread mainly through vegetative growth, as spores could not be dispersed very far from the parent plant. Scientists who study the evolutionary history of plants are hard at work trying to determine which land plant was actually the first and what its ecosystem looked like.

Throughout the Devonian (416 - 360 million years ago), plants progressively grew in height to become as large as today's massive ferns. At the start of the Devonian, plants were mostly non-vascular and correspondingly diminutive, but by the end of the period, seed-bearing plants had evolved, forming huge forests. The explosion of botanical diversity during this period has been called the "Devonian explosion." Meanwhile, fish ruled the seas.

The next major innovation in the evolutionary history of plants was much later, during the Cretaceous period, when flowering plants (angiosperms) first emerged. Using flowers to attract bees, which would then go on to pollinate other plants, angiosperms were genetically diverse and a great evolutionary success. One of the latest varieties of plants are the grasses, which evolved from the angiosperms just 35 million years ago.

CHAPTER 1

# THE EARLY EVOLUTION OF LAND PLANTS, FROM FOSSILS TO GENOMICS

Dianne Edwards, Paul Kenrick

**Keywords:** Tracheophyte, bryophyte, *Cooksonia*, cryptogamic cover, Siluro-Devonian, terrestrialization

## INTRODUCTION

Plants are of fundamental importance to life on land, so developing an understanding of their origins and early evolution has been a quest for generations of botanists. Once the domain of morphology and palaeontology, recent progress in comparative genomics and molecular developmental biology brings new tools to our understanding of relationships among species and the development of their organs and tissues. At the same time, a growing understanding of the composition of early terrestrial ecosystems and the interactions among various components (e.g. plants, fungi, arthropods) is shedding light on how the evolution of life on land affected key Earth Systems (e.g. carbon cycle). Advances in these areas are driving a resurgence of interest in the earliest fossilized remains of land living organisms, whose study was initiated by W. H. Lang and published in 1937 (On the plant-remains of the Downtonian of England and Wales) [**1**].

Knowledge of modern organisms is essential to interpreting fossils, but preconceptions can also be a hindrance. Lang's paper followed pioneering earlier works on slightly younger rock sequences from Europe and Canada (**figure 1**) [**2**,**3**]. Here the fossil record revealed that prior to the origins of forest ecosystems early plants differed in notable ways from those of later floras and especially from modern species. For example, many lacked such characteristic organs as leaves and roots, making them difficult to recognize as plants and as a consequence potentially attributable to a wide range of other groups, including marine algae. They were also small—generally less than a metre in height—and their simple bifurcating stems bore novel spore-producing organs. Questions regarding the land living status of these fossils were answered beyond doubt with the discovery of specimens exquisitely preserved in silicates in the Rhynie Chert (Scotland) [**4**]. The presence of such structures as epidermis with stomata and a primitive vascular system together with the environmental context confirmed that despite the complete absence of certain categories of organ these were plants that lived on land. Drawing on this emerging body of new data and interpretation, Lang took these investigations significantly further. Geologists were unearthing organic remains in still older rocks, and these mostly defied classification in any known group of organisms. They were very challenging to interpret, because they were very small, highly fragmentary, and they varied greatly in shape and texture. Yet, they were highly abundant in certain types of sediment. Lang set out to collect and investigate this unpromising material, employing microscopy and techniques that had recently been developed to extract organics from sedimentary rocks. Cast to one side by previous generations of geologists, Lang showed that these apparently nondescript organic remains were a goldmine of data on early land ecosystems.

One of its referees, A. C. Seward FRS, the leading palaeobotanist of his generation, wrote 'This paper is exceptional in its botanical and geological importance; in it are described the remains of the oldest known British flora and considering the unpromising nature of the material, the information obtained is amazing.' Indeed, it was a model of its time in that in addition to exhaustive anatomical investigations

**Citation:** Edwards D, Kenrick P. 2015 The Early Evolution of Land Plants, From Fossils to Genomics. Phil. Trans. R. Soc. B 370: 20140343. http://dx.doi.org/10.1098/rstb.2014.0343.

of the plant fossils, he meticulously recorded the geological strata in which they were preserved and, using the nature of the rock record combined with information from animal fossils, he reconstructed depositional environments and established that the plants together with less easily categorized fossils derived from the land. It should be emphasized that all of the fossils had been transported and buried in either shallow, near shore marine sequences (Upper Silurian) or terrestrial sediments deposited by river systems (Lower Devonian) [**5**].

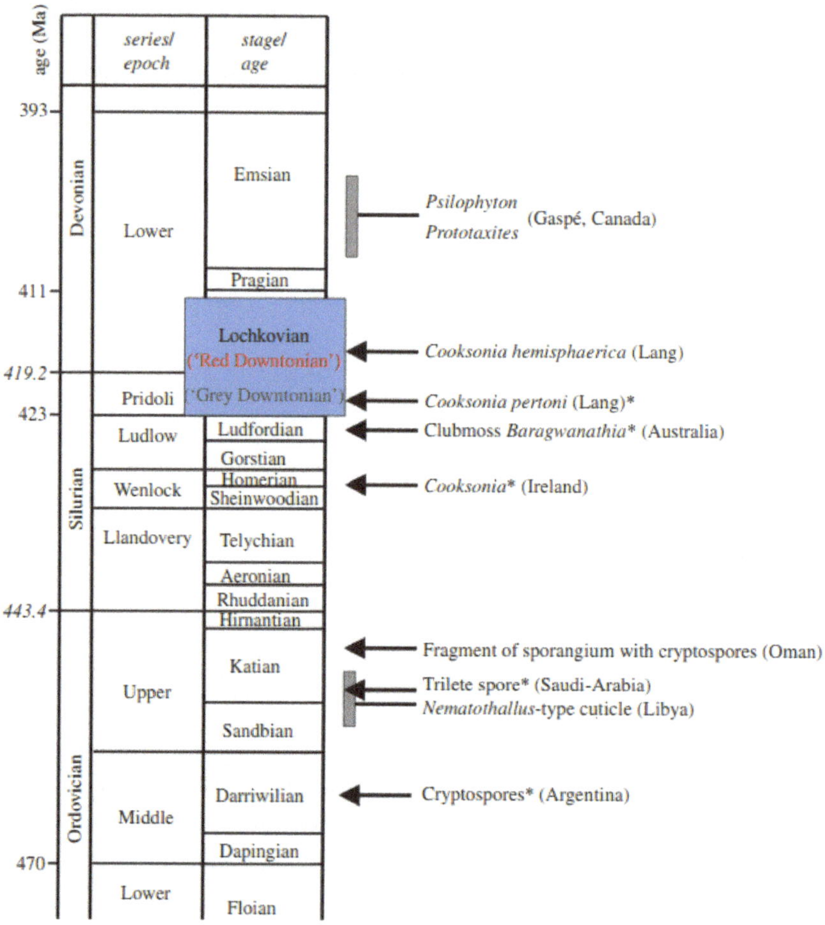


**Figure 1.** Stratigraphic occurrences of fossil plants mentioned in text. Ages in millions of years are taken from the *International Chronographic Chart of the International Commission on Stratigraphy*, 2012. All fossils mentioned are earliest occurrences (*), except for *Psilophyton* and *Prototaxites*, which indicate the age of the first specimens described by Dawson. Coloured block represents the time interval in Lang's paper: note that his 'Downtonian' encompasses the Pridoli Series (Grey Downtonian) and most of the Lochkovian Stage (Red Downtonian).

## LANG'S ACHIEVEMENTS

### Early Vascular Plants

The most enduring of Lang's achievements was his discovery and characterization of the plant he named *Cooksonia*. Fossils attributable to this genus have slender bifurcating systems, measuring from about 1 cm to no more than 6 cm in length (**figure 2***a–c*), that completely lack leaves of any kind. Each is composed of several parallel-sided axes that terminate in distinctive ellipsoidal structures (**figure 2***b*). Remarkably, Lang was able to prove that these were spore-bearing organs (sporangia) by extracting preserved spores, which measured between 25 μm and 38 μm in diameter. Technically difficult at the time, this was achieved with the application of transparent cellulose acetate, to which the spores adhered, enabling them to be removed from the sporangia and examined using light microscopy. Lang showed that the spores were smooth-walled, but that each bore a distinctive trilete (Y) mark (**figure 2***e*), indicating that they were produced by meiosis. Based on the evidence of these fragmentary remains Lang named his new plant *Cooksonia pertoni* and concluded that it was undoubtedly land dwelling. Further clarification of its affinities required crucial evidence of the plant's tissue systems, but because the fossils were preserved as thin coaly films, little of the original cellular structure remained. Several of the earliest sites (Late Silurian Period) yielded associated fragments of axis that showed the faint outlines of elongate cortical cells coupled with a distinctive central strand of thicker carbonaceous material hinting at the presence of vascular tissues. A second species, *C. hemisphaerica* (**figure 2***c,d*), was described from a slightly younger site (Early Devonian Period). Here, Lang found numerous axial fragments which he considered 'with practical certainty' belong to *C. hemisphaerica*. One such possessed a central strand composed of the degraded remains of what he interpreted were tracheids with annular thickenings (**figure 2***f*). Lang pronounced this crucial evidence as 'the most ancient piece of vascular tissue as yet demonstrated in position in a fossil plant in Britain'. Thus, by inference, he concluded that both species of *Cooksonia* were related to the vascular plants, a group that today includes among others ferns, conifers and flowering plants. Lang's discovery and meticulous observations revealed a diversity of small fossil plants of great simplicity. *Cooksonia* has since become an iconic fossil thought to demonstrate the archetypal body plan of primitive vascular plants.

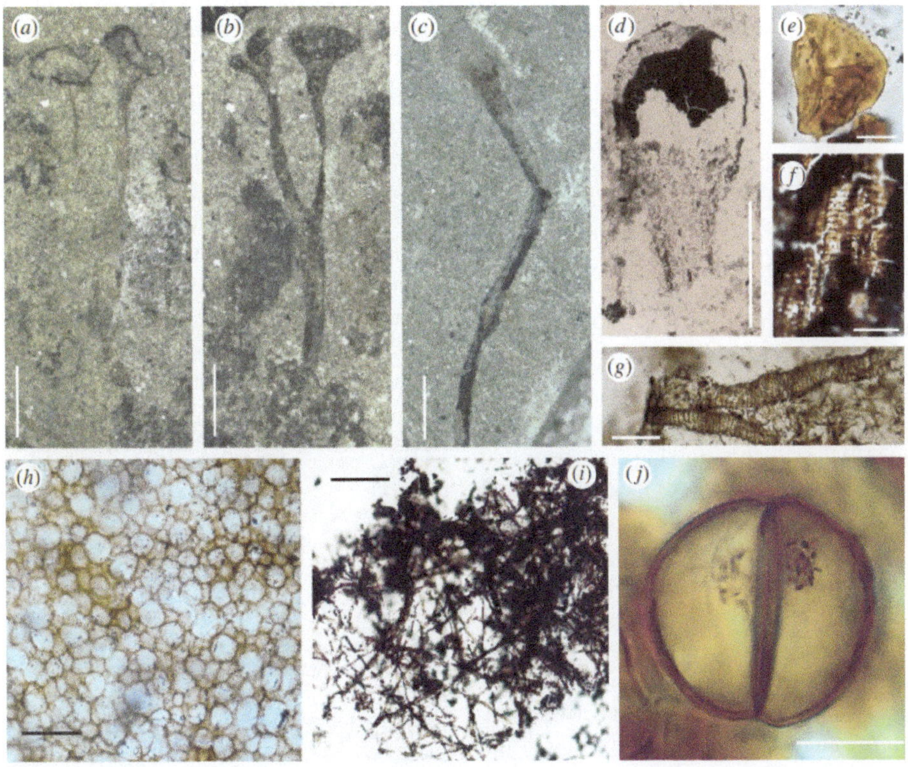


**Figure 2.** New light micrographs of material illustrated by Lang **[1]**. Original number and figure numbers in brackets. (*a*) *Cooksonia pertoni.* Lectotype. Přídolí. V58011. (124 Z b; Plate 8, fig. 8). Coalified material has been removed, possibly using cellulose acetate, for further analysis. Scale bar, 2 mm. (*b*) Counterpart of (*a*). V58010. Black patches possibly of *Nematothallus.* Scale bar, 2 mm. (*c*) *Cooksonia hemisphaerica.* Lochkovian. V58022. (186; Plate 9, fig. 33). Scale bar, 2 mm. (*d*)*C. hemisphaerica* sporangium recovered using cellulose acetate. V54592. (1062; Plate 9, fig. 34). Scale bar, 1 mm. (*e*) Spore recovered from a sporangium of *C. pertoni.* V54995. (1105: Plate 8, fig. 11). Scale bar, 10 µm. (*f*) Tracheids recovered using cellulose acetate from a sterile axis, thought by Lang to belong to *C. hemisphaerica.* Lochkovian. V55040. (1150; Plate 9, fig. 36). Scale bar, 50 µm. (*g*) Banded tubes on cellulose acetate sheet. Přídolí. (962; Plate 11, fig. 60). Scale bar, 50 µm. (*h*) *Nematothallus* 'cuticle' on a cellulose acetate sheet. Přídolí. V54697. Scale bar, 50 µm. (*i*) Typical wefts of tubes assigned to *Nematothallus.* V54851. (Not figured, but cf. plate 11). Scale bar, 160 µm. (*j*) Dyad in elongate cylindrical spore mass. Přídolí. V54654. (764; Plate 13, fig. 109). Granular material probably represents condensed cell contents. Scale bar, 20 µm.

## Spores

In addition to the individual spores that Lang recovered from the sporangia of *Cooksonia*, he also discovered within the same sediments isolated clusters or masses of spores of varying shapes and sizes. Most clusters were circular or oval, measuring 0.5–1.2 mm. One elongate specimen, measuring 4.5 mm in length and approximately 1 mm in width, contained hundreds of spores. Lang surmised that these were also derived from land plants, but that the enclosing tissue systems had decomposed whereas the spores, being more robust, had survived. The shapes of the spore masses thus reflected the shapes of the original plant's spore-bearing organs. Intriguingly, one elongate mass contained highly unusual spores each with two hemispherical cells separated by a transverse wall, which Lang described as bicellular (**figure 2***j*). This contrasts with the normal condition in land plants, in which spores develop in tetrads following meiotic cell division. Lang considered this 'unique among pre-Carboniferous spores with which I am acquainted' and entertained the possibility that they demonstrated the first stage of germination. These spores are now known to be dispersed as permanently fused dyads, and they are an abundant element of the microscopic organics extractable from sediments of the Late Ordovician through to the Early Devonian.

## Nematothallus

Arguably, the most challenging and controversial of Lang's discoveries was the organism that he named *Nematothallus pseudovasculosa*, which was abundant at many of his sites taking the form of dark 'incrustations' (**figure 2***a,b*). Lang showed that these were composed of wefts of tubes covered by a resilient film of 'cuticle' with a pseudocellular patterning (**figure 2***h*). The tubes were highly distinctive and of two size classes: narrow tubes (*ca* 2.5 µm) and wide tubes (12–40 µm; **figure 2***i*). They bifurcated occasionally. On the internal wall of the wider tubes there was evidence of narrow annular bands (**figure 2***g*). Reconstructing the three-dimensional organization in such fossils is difficult, but from consistent associations, too frequent to be coincidental, Lang considered the 'plant' to be thalloid with the wefts of fine tubes sandwiched between the uppermost pseudocellular cuticle and more deeply seated wefts of larger tubes. On decay these layers would have separated, thus accounting for their frequent recovery as isolated elements from the rock matrix. Lang also deduced that *Nematothallus* reproduced by means of spores, again based on the consistent association of tubes and spores of variable shape and surface ornament. Having worked out the general structure of *Nematothallus*, Lang was then faced with the challenge of understanding the affinities of an organism whose tubular structure defied comparison with plants, or as he put it: 'makes comparison with all plants of typical cellular structure open to serious criticism'.

Among the remains collected by Lang, there was another obscure but highly distinctive fossil that had earlier been described by Dawson [**2**] based on Canadian material, and which was widely encountered in Late Silurian and Early Devonian sediments. Named *Prototaxites*, it consisted of axes that ranged enormously in size from a few millimetres to a staggering 1 m in diameter and 8 m in length [**6**]. Like *Nematothallus*, it was constructed from profusely branched tubes of various sizes. The affinities of *Prototaxites* were vigorously debated, but starting with Carruthers [**7**] there was a growing consensus that its affinities lay with the algae, and because of its great size and tissue complexity, probably with the brown algae. An alternative view, overlooked by Lang, was the suggestion by Church [**8**] that *Prototaxites* was a giant fungus, a hypothesis recently resurrected by Hueber [**6**]. Because of similarities in their underlying structure (i.e. tissues comprising profusely branched tubes) and their frequent close association in sediments, Lang explored the possibility of a physical relationship between the two—could*Nematothallus* be the laminate appendages of the 'trunks' of *Prototaxites*? He decided to unite the two genera in a new 'class', the Nematophytales. Lang was convinced that *Nematothallus*lived on land as evidenced by its spores and cuticle. Having reviewed most of the published opinions, he concluded that nematophytes composed a unique grouping of land plants being neither algae nor vascular plants, although in possession of some characteristics of both groups. Despite major uncertainties surrounding its affinities and its functional biology, Lang himself thought of his work on *Nematothallus* as being 'most important to our botanical knowledge'.

## ADVANCES

### *Cooksonia* Today

Since the original discovery by Lang in 1937 research has focused on developing a better understanding of the diversity of plants attributable to *Cooksonia* and in particular the nature of their internal tissue systems to provide a clearer taxonomic concept [**9**], and an understanding of its functional biology [**10**] and relationship to living species [**11**]. Intensive re-collecting at a number of Lang's localities has unearthed a far more diverse assemblage of *Cooksonia*-like plants. Most variation centres on the form of the sporangia and the spores. Preserved stem anatomy is rare, but lingering doubts as to the vascular status of the genus were settled with the discovery of remarkable new fossils of *C. pertoni* preserved in charcoal (**figure 3***a,b*). These were recovered from sediments using acid extraction methods. The three-dimensional structure of the plant is faithfully preserved, including cellular level details of tracheids and stomata (**figure 3***a–e*) [**10**]. Furthermore, intensive collecting through an extended stratigraphic interval (i.e. Lang's Downtonian = Upper Silurian (Přídolí) through Lower Devonain (Lochkovian); **figure 1**) demonstrated that whereas external form is highly conserved, there are recognizable changes at the cellular level. For example, *C. pertoni* produced different kinds of spores in different geological intervals [**22**]. The spores themselves were structurally similar (i.e. equatorially thickened) but either smooth-walled or differently ornamented (**figure 3***f–j*). Such cryptic evolution in plants with simple morphology might also encompass changes in other anatomical features, particularly those relating to the vascular system (e.g. tracheids) and the epidermis (e.g. stomata). Recent research has therefore greatly expanded our knowledge of the morphology and diversity of plants attributable to *Cooksonia*, confirming and extending Lang's original concept so that *C. pertoni* is now one of the best known of the earliest stem-group vascular plants. Nevertheless, frustratingly, newly discovered types preserved in charcoal are highly fragmentary and frequently they are represented only as unique specimens.

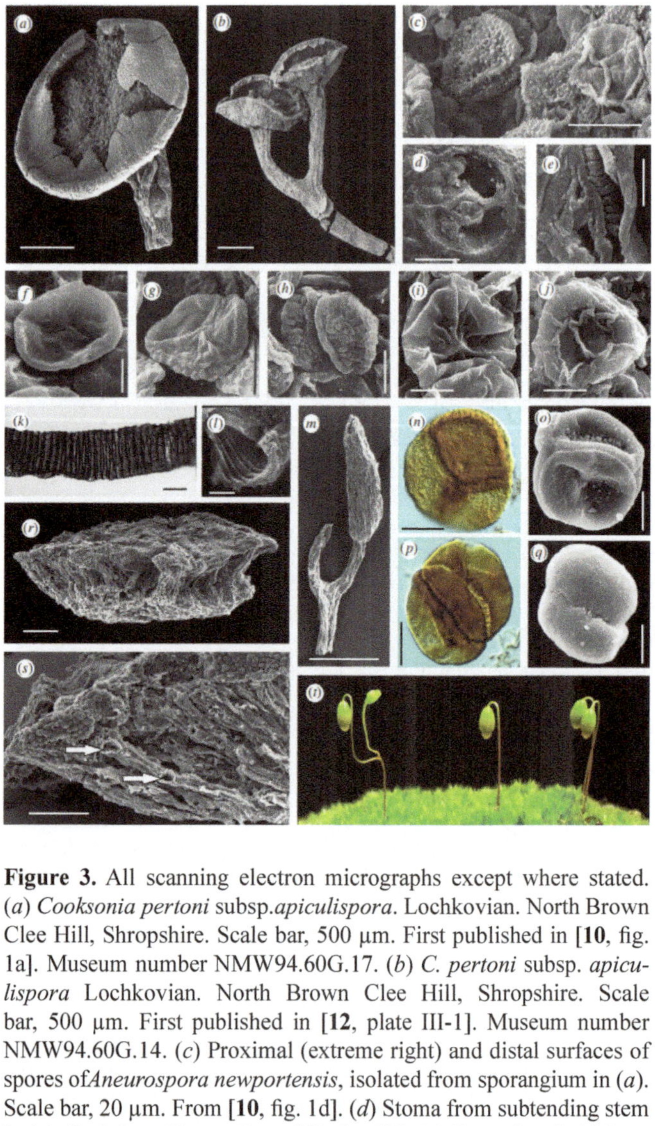

**Figure 3.** All scanning electron micrographs except where stated. (*a*) *Cooksonia pertoni* subsp.*apiculispora*. Lochkovian. North Brown Clee Hill, Shropshire. Scale bar, 500 µm. First published in [**10**, fig. 1a]. Museum number NMW94.60G.17. (*b*) *C. pertoni* subsp. *apiculispora* Lochkovian. North Brown Clee Hill, Shropshire. Scale bar, 500 µm. First published in [**12**, plate III-1]. Museum number NMW94.60G.14. (*c*) Proximal (extreme right) and distal surfaces of spores of*Aneurospora newportensis*, isolated from sporangium in (*a*). Scale bar, 20 µm. From [**10**, fig. 1d]. (*d*) Stoma from subtending stem in (*a*). Scale bar, 20 µm. From [**10**, fig. 1b]. (*e*) Cast of tracheid from stem in (a); grooves indicate position of annular thickenings. From [**10**, fig. 2a]. (*f*)–(*j*) Spores present in subspecies of *C. pertoni*. (*f*) Proximal surface of *Ambitisporites* sp. (subsp. *pertoni*), Upper Silurian. Scale bar, 10 µm. First published in [**13**, fig. 3.1b]. V.62776. (*g,h*) Proximal and distal surfaces of *Synorisporites verrucatus* (subsp. *synorispora*), Upper Silurian. Scale bar, 10 µm. First published in [**14**; plate 1, figs 3, 4]. NMW93.143G.1. (*i,j*) Proximal and distal surfaces of*Synorisporites* sp. (subsp. *reticulispora*), Lochkovian. Scale bar, 10 µm. First published in [**15**, plate VI, 3, 18]. NMW2012.29G.20

and 19. (*k*) Light micrograph of a tubular structure (?hypha/banded tube) with internal spiral thickenings isolated from late Wenlock (Silurian), Rumney, Cardiff. Scale bar, 10 µm. First published in [**16**, fig. 23]. MPK 6028. (*l*) Fractured end of banded tube, Ludlow (Upper Silurian), S. Wales. Scale bar, 5 µm. First published in [**17**, fig. 68]. NMW77.34G.33c. (*m*) *Fusitheca fanningiae* containing permanent laevigate dyads with thin envelope. Lochkovian, Shropshire. Scale bar, 500 µm. First published in [**18**, fig. 54]. NMW97.42G.4. (*n*) Light micrograph of *Velatitetras rugulata*, a permanent tetrad enclosed in a 'rucked' envelope, isolated from Wenlock rock, Shropshire. Unpublished–courtesy of Neil Burgess. Scale bar, 10 µm. (*o*) *Tetrahedraletes medinensis*, a permanent tetrad, Ordovician, Shropshire. Scale bar, 13 µm. First published in [**19**, fig. 5A]. (*p*) Light micrograph of *Artemopyra brevicosta*, a permanent dyad, Wenlock, Shropshire. Scale bar, 10 µm. First published in [**20**; plate 1, fig. 1]. (*q*) *Dyadospora murusdensa*, a permanent dyad, Ordovician, Shropshire. Scale bar, 10 µm. First published in [**19**; plate 2, fig. 11]. (*r,s*) Fragment of *Nematothallus williamii*, Lochkovian, Shropshire. First published in [**21**, fig. 1A,B]. Museum number NMW2013.39G.1. (*r*) Note three-layered thallus. Scale bar, 200 µm. (*s*) Magnification of (*r*) showing the surface patterning typical of the *Nematothallus* 'cuticle' and larger hyphae aligned perpendicular to the surface. Arrows indicate positions of lateral branches or areas in contact with a postulated photobiont. Scale bar, 100 µm. (*t*) Gametophytes and fertile sporophytes of *Funaria hygrometrica*. Note bifurcating seta in sporophyte on extreme left. (Courtesy of Jill Harrison, Yoan Coudart and Alison Reed, Cambridge University.)

## Palaeophysiological Approaches

Our growing understanding of the nature of *Cooksonia* has stimulated thinking about the origin and evolution of key land plant characteristics, and how such small plants could have functioned. One area of interest has been the extent to which early plants were capable of maintaining an internally hydrated environment under water stress (homoiohydry) as opposed to having the capacity to tolerate significant desiccation (poikilohydry). Raven addressed this question by taking an ecophysiological approach to characterizing water transport and gas exchange processes in living vascular plants (homoiohydric) and bryophytes (poikilohydric) [**23**], examining interactions between the demands of photosynthesis and water delivery as plants grew in stature. Results were used to appraise the anatomy and chemical innovations of early land plants starting with *Cooksonia*. Key elements of this analysis included the structure and position of water-conducting cells (tracheids), the synthesis of lignin (wood) and especially the function of stomata, intercellular spaces, and the cuticle system. Raven initially favoured the earlier evolution of internal conducting tissues as a key innovation in the origin of homoiohydry [**23,24**]. (i.e. tracheids must have evolved before stomata). However, later he saw ecophysiological advantages to 'stomata first' in very small sporophytes [**25**], a conclusion supported by subsequent phylogenetic analyses (see below and **figure 4**) and the order in which these features appear in the fossil record [**27**]. The small size of some cooksonias potentially restricts their ability to develop diverse tissues systems with specialized functions. Noting that Raven's analyses had concentrated on adaptations relating to transport processes in axes at the wider end of the range distribution (more than 1 mm diameter), Boyce [**28**] looked at the nature and functions of tissue systems in narrower axes and concluded that those involved in support (i.e. peripheral strengthening tissues) and water-conduction would have occupied the bulk of the plant body. In smaller species of *Cooksonia*, therefore, there would have been little room for photosynthetic tissues. In addition, *Cooksonia* had few stomata. Thus, Boyce concluded that smaller species were probably not homoiohydric and also not photosynthetically competent. As in modern bryophytes, they must have been physiologically dependent on a gametophyte, but compelling fossil evidence has yet to be found.

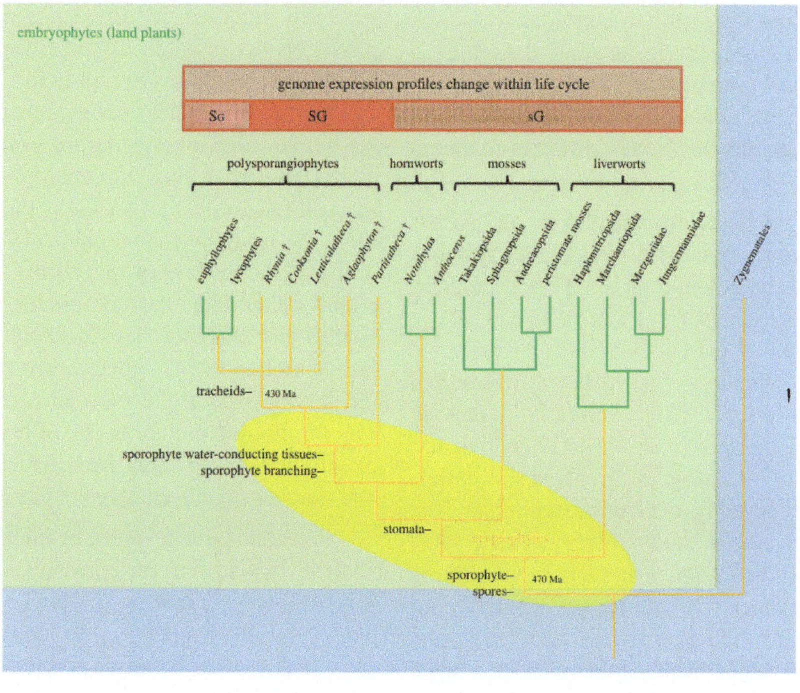


**Figure 4.** Relationships among major groups of land plants showing the hypothesized broad range of clades to which cryptophytes (extinct cryptospore-producing plants) might belong (shaded yellow oval; modified after [**5,26**]). As currently envisaged, cryptophytes is a grade of

organization potentially encompassing plants spanning the embryophyte to vascular plant stem-groups as well as stem-group hornworts, mosses and liverworts. Some key developments discussed in the text are indicated on branches. Also indicated is a major shift in life cycle among clades from gametophyte dominated (sG) to sporophyte dominated (Sg) life cycles. The intermediate isomorphic life cycle (SG) is only known in some extinct plants. Estimated divergence times in millions of years.

The challenge of interpreting the functional biology of *Cooksonia* raises questions about the function and molecular regulation of basic organs and tissue systems in plants. It has been assumed that stomata and cuticle had similar functions and controls in bryophytes, vascular plants and early fossils such as *Cooksonia*. Recently, the role played by the phytohormone abscisic acid (ABA) in stomatal physiology has been questioned, challenging conventional views. The stomata of *Lycopodium* and *Pteridium* were shown to lack a pore closure response to ABA [**29**]. Similarly, stomata in hornworts and *Sphagnum* do not respond significantly to ABA, desiccation and darkness, and there is no potassium flux between guard cells and epidermal cells [**30**]. By contrast, in the moss *Physcomitrella* and in the lycopod *Selaginella* stomatal responses to ABA appear to be similar to those in the angiosperm *Arabidopsis* [**31,32**]. The issue of when active stomatal control in plants evolved is therefore controversial and the subject of active research.

Unlike in the vascular plants, the stomata of some hornworts and mosses are known to open but then never close. Stomata in these plants may therefore have had rather a different function, perhaps aiding the desiccation of the sporangium prior to spore release [**30**]. In *C. pertoni*stomata might have played a similar role or perhaps they functioned in creating a transpiration stream for mineral nutrition of the sporangium [**27**]. There is also a functional dependency between stomata and cuticle. Some moss sporophytes share similarities in cuticle ultrastructure with vascular plants, whereas their gametophytes have thinner less well-developed cuticles [**33,34**]. This is most probably a consequence of the absence of stomata in the gametophyte, meaning that gaseous diffusion must take place through the cuticle, limiting its development. Thus, the function of stomata in early fossil land plants and in living bryophytes is very different to that in modern flowering plants.

Genomics is also beginning to provide interesting insights into the evolution of the stomata, cuticle and vascular systems. It has been shown that a homologue of the ABA regulatory protein kinase (OST1) present in *Physcomitrella* can rescue an *Arabidopsis* mutant lacking OST1 activity, and transcriptome analyses of the moss sporophyte suggest a deeply conserved mechanism of stomatal control [**31**]. Similarly, knockout mutants of moss ABCG transporter Pp-ABCG7, which has a putative orthologue in *Arabidopsis* involved in cuticle precursor trafficking, are severely deficient in cuticular wax [**34**]. PpVNS proteins have been shown to regulate the same gene families in *Physcomitrella* and *Arabidopsis*, supporting the hypothesis of a fundamentally similar genetic basis for the development of water-conducting tissues and supporting tissues in both [**35**]. So, although the physiological roles played by stomata and to a lesser extent cuticle and water-conducting cells are likely to vary across land plants, certain aspects of their development are deeply conserved.

## The Phylogeny of Early Land Plants: The Bryophyte Connection

The curious bicellular spores discovered in spore masses by Lang are now known to belong to a class of organics called cryptospores. This encompasses spores dispersed in permanently fused groups of two (i.e. dyads; **figure 3***p,q*) or four (i.e. obligate tetrads; **figure 3***n,o*) and their derivatives. Cryptospores are now known to be abundant and cosmopolitan as dispersed forms in rocks of the Ordovician through Devonian Periods [**36, 37**]. The nature and affinities of the plants that produced the cryptospores are still contentious, but knowledge of these is essential to understanding the evolution of early land floras [**38**]. Progress has come through the comparative analysis of spore morphology and ultrastructure [**39**], the recovery of spore masses from more ancient rocks [**40**] and especially through the discovery of cryptospores *in situ* [**41**]. The leading hypothesis aligns cryptospores with bryophytes. Initially, this was based on parallels between early fossil obligate tetrads and similarly configured spores known to occur in a few modern liverworts (e.g. *Riccia*, *Sphaerocarpos*, *Cryptothallus*). Studies of cell wall ultrastructure (e.g. configuration of wall lamellae) show additional similarities lending support to this hypothesis [**39,40,42,43**]. Various forms of cryptospores have also been documented within fragments of the source plants (cryptophytes) that are exceptionally well-preserved in charcoal in Silurian and Lower Devonian sediments [**5**]. So, we now know that these were minute land plants that resembled the stalk and capsule of the bryophyte sporophyte in general shape and size, but importantly they also differed in some key features, including bifurcation of the stalk (**figure 3***m*) and in their method of spore development and dispersal. Cryptospore morphology and dispersal implies a diversity of meiotic cell division in early land plants that is no longer seen in their living descendants [**5**].

Much remains unknown about early fossil cryptophytes, and because their remains are so fragmentary they have not yet been included in formal phylogenetic analyses. Modern phylogenetic research conclusively demonstrates that land plants are monophyletic [**44–46**]. The weight of evidence further indicates that bryophytes are paraphyletic to the vascular plants, and that liverworts are sister group to all other land plants (**figure 4**). However, relations among these basal clades are still uncertain, and various other topologies are plausible, including bryophyte monophyly [**46,47**]. Phylogenetic analyses group the better known species of*Cooksonia* with the vascular plants rather than the bryophytes (**figure 4**) [**9, 11**]. These studies also indicate that *Cooksonia*, as originally conceived, is not monophyletic. Species are related in various ways in the stem-groups of the vascular plant and Lycophytina clades. The developing picture is one of cryptophytes representing an extinct pool of diversity from which both modern bryophytes and vascular plants emerged (**figure 4**) [**5**].

A second issue revealed by including fossil in phylogenetic analyses is a pattern of representational bias in the geological record greatly favouring the larger and more robust vascular plants over bryophytes and bryophyte-like organisms [**11,48**]. This bias is less evident in the fossil record of dispersed spores, which tend to have similar dispersal properties and preservational characteristics across groups. A third issue relates to timetrees (i.e. calibrated molecular phylogenies) and estimates for the time of origin and initial diversification of land plants. Timetrees all indicate a lengthier geological history than the fossil evidence, but the degree to which they do so ranges from marginal to substantial [**49–51**]. The discrepancies among various lines of fossil evidence and timetrees are at least in part related to methodological issues to do with tree calibration. It is probable, however, that these discrepancies also reflect real natural phenomena including significant changes in the nature of the rock record during the Lower Palaeozoic that were operating at a global scale [**52**].

## The Branching Sporophyte and Nature of Gametophyte

One of the key innovations in plant evolution was the capacity for iterative bifurcation of the sporophyte, which laid the foundation for the development of large complex plants with multiple sporangia (polysporangiophytes) and specialized organ systems (e.g. stems, roots, leaves) [**53–55**]. Vascular plants are polysporangiate whereas bryophytes are not, and *Cooksonia* represents one of the earliest and most rudimentary stages in the evolution of this growth form. Understanding how the polysporangiophyte state first evolved and how it subsequently developed are questions that can be addressed through critical examination of the fossil record and through the methods of evolutionary developmental biology ([**56**], reviewed in [**57,58**]). Recently, the evolutionary developmental approach has been greatly enhanced though the establishment of moss model systems [**59**] and through the sequencing of the genome of *Physcomitrella patens* [**60**], enabling a comparative genomics approach spanning most of the land plant tree of life [**26,61,62**].

One emerging area of research focuses on the evolution of indeterminate growth in the spore-bearing part of the plant life cycle. The bryophyte sporophyte is not normally indeterminate, but a capacity for bifurcation in the seta has been known for some time from rare teratological specimens of both liverworts and mosses (**figure 3***t*). These are intriguing because they resemble the simple bifurcating sporophyte of Lang's *Cooksonia*. This nascent form of indeterminate growth has been induced in various ways in *P. patens*. Double sporangia were observed in *P. patens* following disruption of homologues of *Arabidopsis thaliana TEL* and *LFY*loci [**63,64**]. *TEL* encodes a putative RNA-binding protein that plays a crucial role in regulation of post-transcriptional gene expression. *PpLFY* encodes a transcription factor required for the first division of the zygote. Double sporangia were also observed following the application of auxin transport inhibitors [**65**]. Furthermore, varying the expression of the *PpCLF* gene in the gametophyte results in the apogamous development of bifurcating sporophyte-like structures [**66**]. *PpCLF* encodes part of the highly conserved Polycomb group (PcG) complex, which is involved in the epigenetic control of gene expression profiles in plants and animals. A key role for PcG in regulating the differentiation of meristematic cells in gametophyte development has also been confirmed through deletion of a *PpFIE* [**67**]. The disruption of various regulatory mechanisms in *Physcomitrella* is beginning to shed light on the origins of the polysporangiophyte state, as embodied in simple early fossils like *Cooksonia*.

## Non-vascular Plants/Cryptogamic Covers

Today, cryptogamic covers (i.e. communities of cyanobacteria, algae, fungi, lichens, bryophytes) occur on many ground and plant surfaces, where they are responsible for about 7% of the net primary productivity of terrestrial ecosystems [**68**]. There is a growing body of data indicating that comparable communities dominated the earliest terrestrial ecosystems. Mid-Palaeozoic rock sequences contain many enigmatic fossils that hint at a greater diversity of organisms accompanying and preceding the vascular plant flora. In addition to spores, assemblages frequently contain small fragments of tissue (phytodebris [**69**]) that resemble the cuticles and complexes of tubes which Lang had assigned to *Nematothallus*. Diverse affinities have been suggested [**70**]. Recently discovered fossils preserved in charcoal found at sites in the Welsh Borderland are beginning to provide a more complete picture of the *Nematothallus* organism (**figure 3***r,s*; [**21**]). The thallus was stratified with a superficial, usually uniseriate cortex. This confirms Lang's prediction about the underlying structure and the cuticle of *Nematothallus*. The cortex overlaid a palisade zone of parallel tubes with occasional septae and a basal zone of wefts of smaller, randomly interwoven and infrequently branched tubes. This general structure suggests an affinity with lichenized fungi, the tubes being interpreted as fungal hyphae. However, a photobiont has not been observed. A broadly similar type of organization, but lacking the palisade tissue, was observed in two putative lichens from the same locality [**71**]. Here, there is evidence of a photobiont, with bacterial colonies on the upper surface, and actinobacterial filaments in the medulla. Using material from Lang's fossil collections, Strother [**72**] calculated that thalloid incrustations of this type accounted for 10–50% coverage of a bedding plane, whereas axial fossils represent approximately 1%. Lichens of nematophyte affinity were probably therefore an important element of early terrestrial vegetation.

Not all coalified thalloid compressions were derived from nematophytes, and most are difficult to interpret because little cellular structure remains. Some have been interpreted as mats of cyanobacteria, whereas others were probably composed of consortia of eukaryotes [**73,74**]. Stable carbon isotope values ($\delta^{13}C_{org}$) support a terrestrial origin for much of the fossil-derived carbon in thalloid compression fossils extending back into the Late Ordovician, hinting also at a liverwort affinity [**75**]. Thus, there is a growing body of evidence for a pre-vascular plant-dominated vegetation, including cyanobacteria, algae, fungi, cryptophytes and lichens on land during the Silurian and Ordovician Periods, comparable with the cryptogamic covers that dominate certain ecosystems today [**68**].

## NEXT STEPS

The genus *Cooksonia* has now been recorded from numerous sites in Europe and the Americas (New York State, Brazil), yet the Welsh Borderland remains the most extensively studied region, yielding important new information on the biology and diversity of these early plants. The charcoal-preserved fossils are unique to this region, and they provide unparalleled insights into tissue systems at the cellular level. The research tool of choice has been the scanning electron microscope, which enables the analysis of cell structure but is limited in its capacity to build a complete three-dimensional model of the organism. X-ray synchrotron microtomography (SRXTM) provides a new and very promising alternative. This non-destructive technique has been applied to imaging the organs and tissue systems in fossil plants, yielding full three-dimensional models [**76**]. SRXTM has been used to image minute charcoalified seeds and flowers from the Cretaceous Period [77,78], which resemble *Cooksonia* in their preservation, size fraction and in their rarity. SRXTM, therefore, has the potential to yield complete reconstructions of the soft tissue systems of *Cooksonia,* shedding further light on the functional morphology of the earliest land plant.

Furthering our understanding of the cryptophytes (i.e. plants that produced the cryptospores that characterize Ordovician through Devonian rocks) and associated organisms requires both the identification of appropriate rock sequences in the field and the application of new methods. New data from Gondwana would be of great interest as there are very few early records from this continent [**79**]. The extraction and description of mesofossils (millimetre scale fossils that fall into the size range between spores and macroscopic compressions) using acid maceration techniques and *in situ* imaging will form an important part of the approach [**5**].

Evolving the capacity to maintain an internally hydrated environment under water stress (homoiohydry) was of fundamental importance in the colonization of the land. The evolution of homoiohydry involved the acquisition of a suite of features including a cutinized epidermis, stomata, gas-filled intercellular spaces (air-spaces) and vascular tissue [**80**]. Modern bryophytes possess some of these features, and their functions are now known to differ in some important respects from the better studied vascular plants [**30**]. Also, there are important differences in the chemistry of secondary compounds that X-ray spectroscopy may well distinguish. Unravelling the evolution of homoiohydry requires a better understanding in bryophytes of the nature and roles of lignin-related polyphenolics, cuticle chemistry and structure, and the function of the stomatal apparatus in relation to their distributions and the absence of intercellular spaces and water-conducting tissues.

The establishment of a moss model system and the sequencing of the genome of *P. patens* now enable a comparative genomics approach spanning most of the land plant tree of life [**59,60**]. The disruption of various regulatory mechanisms is beginning to shed light on the origins of the polysporangiophyte state [**63–66**], as embodied in simple early fossils like *Cooksonia.* Likewise, genetic and transcriptomic approaches promise to reveal mechanisms regulating the land plant life cycle [**81**]. The extension of these approaches to other key living groups, including hornworts, liverworts and charophycean algae is a necessary step in understanding the evolution of basic organs and tissue systems in plants. Also, the emerging molecular developmental models need to account for the morphology and life-cycle variants of extinct plants [**82**].

The phylogenetic tree of plants is the framework that underpins evolutionary interpretations, but several important relationships are still poorly resolved. The sister group of land plants within living charophycean algae is unclear (Charales/Zygnematales) [**46,83,84**]. This has implications for understanding the likely suite of features that characterized the common ancestor of land plants. Similarly, the relationships among the major clades of bryophytes and the vascular plants are still problematic, with low support for any one topology [**46,47**]. Whether bryophytes are monophyletic or paraphyletic to the vascular plants has important implications for our understanding of the evolution of the land plant life cycle and the evolutionary implications of early fossils.

Lang's discoveries in the Welsh Borderland brought to light hitherto unknown very small fossil plants and other enigmatic fossilized remains that opened a window on to the nature of early terrestrial ecosystems. Some of the novel organisms he described proved to be related to plants, whereas the affinities of others are still uncertain, but they are likely to be quite diverse within the eukaryotic tree of life. In modern terms, Lang's discoveries showed that early terrestrial ecosystems resembled modern soil crust or epilithic microbial communities that were dominated by cryptogamic plants. The plants were small, could not be classified comfortably in established families, and lacked many of the features associated with modern species. The iconic *Cooksonia* in particular has served as a starting point in thinking about the evolution of basic organs and tissue systems (e.g. vascular system, stem, roots, leaves). Subsequent research has clarified and extended this work. Evidence from minute fossils provides the main source of information on the composition and evolution of early terrestrial ecosystems, and new techniques and analytical approaches are helping to solve questions concerning the biology and affinities of these organisms. Molecular phylogenetics now provides a robust framework within which to appraise this evidence, and molecular developmental biology a new approach to understanding and explaining the evolution of plant form. The broader impact of early land-based life on key Earth Systems (e.g. geochemical carbon cycle) is a developing area of research [**85–87**].

Lang concluded his magnum opus with the statement that his 'present study and survey of the Downtonian flora will, it may be hoped serve as a basis of such further work'—and so it turned out to be, with much wider ramifications into areas undreamed of by this superlative botanist.

## Funding statement

We also are grateful to the Leverhulme Trust, Gatsby Charitable Foundation, NERC and Plant Origins ITN Framework 7 for funding the authors over the years.

## Acknowledgement

We are grateful to Geoff Abbott, Lindsey Axe, Liam Dolan, Jeff Duckett, Louis Emery, Jill Harrison and John Raven for their assistance in the preparation of the paper.

## REFERENCES

1. Lang WH. 1937 On the plant-remains from the Downtonian of England and Wales. Phil. Trans. R. Soc. Lond. B 227, 245– 291. (doi:10.1098/rstb. 1937.0004)
2. Dawson JW. 1859 On fossil plants from the Devonian rocks of Canada. Q. J. Geol. Soc. Lond. 15, 477– 488. (doi:10.1144/GSL.JGS.1859.015.01-02.57)
3. Kra¨usel R, Weyland H. 1930 [The flora of the German Lower Devonian]. Abh. Preussischen Geol.

Landesanstalt, Neue Folge 131, 1 – 92. (In German.)

4. Kidston R, Lang WH. 1921 On Old Red Sandstone plants showing structure, from the Rhynie Chert Bed, Aberdeenshire. Part IV. Restorations of the vascular cryptogams, and discussion on their bearing on the general morphology of the Pteridophyta and the origin of the organization of land-plants. Trans. R. Soc. Edinb. 52, 831 – 854. (doi:10.1017/S0080456800016033)
5. Edwards D, Morris JL, Richardson JB, Kenrick P. 2014 Cryptospores and cryptophytes reveal hidden diversity in early land floras. New Phytol. 202, 50 – 78. (doi:10.1111/nph.12645)
6. Hueber FM. 2001 Rotted wood – alga –fungus; the history and life of Prototaxites Dawson 1859. Rev. Palaeobot. Palynol. 116, 123 – 158. (doi:10.1016/S0034-6667(01)00058-6)
7. Carruthers W. 1872 On the history, histological structure and affinities of Nematophycus logani, Carruthers (Prototaxites logani, Dawson), an alga of Devonian age. Mon. Microsc. J. 8, 160 – 172. (doi:10.1111/j.1365-2818.1872.tb02193.x)
8. Church AH. 1919 Thalassiophyta and the subaerial transmigration. Bot. Mem. Oxf. 3, 1– 95.
9. Gonez P, Gerrienne P. 2010 A new definition and a lectotypification of the genus Cooksonia Lang 1937. Int. J. Plant Sci. 171, 199– 215. (doi:10.1086/ 648988)
10. Edwards D, Davies KL, Axe L. 1992 A vascular conducting strand in the early land plant Cooksonia. Nature 357, 683 – 685. (doi:10.1038/357683a0)
11. Kenrick P, Crane PR. 1997 The origin and early diversification of land plants: a cladistic study, p. 441. Washington, DC: Smithsonian Institution Press.
12. Edwards D. 1996 New insights into early land ecosystems: a glimpse of a Lilliputian world. Rev. Palaeobot. Palynol. 90, 159– 174. (doi:10.1016/ 0034-6667(95)00081-X)
13. Fanning U, Richardson JB, Edwards D. 1991 A review of in situ spores in Silurian land plants. In Pollen and spores (eds S Blackmore, SH Barnes), pp. 25– 47. Oxford, UK: Clarendon Press.
14. Edwards D, Davies KL, Richardson JB, Axe L. 1995 The ultrastructure of spores of Cooksonia pertoni. Palaeontology 38, 153– 168.
15. Morris JL, Edwards D, Richardson JB, Axe L, Davies KL. 2012 Further insights into trilete spore producers from the Early Devonian (Lochkovian) of the Welsh Borderland UK. Rev. Palaeobot. Palynol. 185, 35 – 63. (doi:10.1016/j.revpalbo.2012.08.001)
16. Burgess ND, Edwards D. 1991 Classification of uppermost Ordovician to Lower Devonian tubular and filamentous macerals from the Anglo-Welsh Basin. Bot. J. Linn. Soc. 106, 41 – 66. (doi:10.1111/j. 1095-8339.1991.tb02282.x)
17. Edwards D. 1982 Fragmentary non-vascular plant microfossils from the late Silurian of Wales. Bot. J. Linn. Soc. 84, 223– 256. (doi:10.1111/j.1095-8339. 1982. tb00536.x)
18. Wellman CH, Edwards D, Axe L. 1998 Permanent dyads in sporangia and spore masses from the Lower Devonian of the Welsh Borderland. Bot. J. Linn. Soc. 127, 117–147. (doi:10.1111/j.1095-8339. 1998.tb02092.x)
19. Wellman CH. 1996 Cryptospores from the type area of the Caradoc Series in southern Britain. Spec. Pap. Palaeontol. 55, 103 – 136.
20. Burgess ND, Richardson JB. 1991 Silurian cryptospores and miospores from the type Wenlock area, Shropshire, England. Palaeontology 34, 601 – 628.
21. Edwards D, Axe L, Honegger R. 2013 Contributions to the diversity in cryptogamic covers in the midPalaeozoic: Nematothallus revisited. Bot. J. Linn. Soc. 173, 505 – 534. (doi:10.1111/boj.12119)
22. Fanning U, Richardson JB, Edwards D. 1988 Cryptic evolution in an early land plant. Evol. Trends Plants 2, 13 – 24.
23. Raven JA. 1977 The evolution of vascular plants in relation to supracellular transport processes. Adv. Bot. Res. 5, 153– 219. (doi:10.1016/S0065- 2296(08)60361-4)
24. Raven JA. 1984 Physiological correlates of the morphology of early vascular plants. Bot. J. Linn. Soc. 88, 105 – 126. (doi:10.1111/j.1095-8339.1984. tb01566.x)
25. Raven JA. 2002 Selection pressures on stomatal evolution. New Phytol. 153, 371– 386. (doi:10. 1046/j.0028-646X.2001.00334.x)
26. Pires ND, Dolan L. 2012 Morphological evolution in land plants: new designs with old genes. Phil. Trans. R. Soc. B 367, 508– 518. (doi:10.1098/rstb. 2011.0252)
27. Edwards D. 2003 Xylem in early tracheophytes. Plant Cell Environ. 26, 57 – 72. (doi:10.1046/j.1365-3040.2003.00878.x)
28. Boyce CK. 2008 How green was Cooksonia? The importance of size in understanding the early evolution of physiology in the vascular plant lineage. Paleobiology 34, 179– 194. (doi:10.1666/ 0094-8373(2008)034[0179:HGWCTI]2.0.CO;2)
29. McAdam SAM, Brodribb TJ. 2012 Fern and lycophyte guard cells do not respond to endogenous abscisic acid. Plant Cell Online 24, 1510– 1521. (doi:10.1105/tpc.112.096404)
30. Pressel S, Goral T, Duckett JG. 2014 Stomatal differentiation and abnormal stomata in hornworts. J. Bryol. 36, 87 – 103. (doi:10.1179/1743282014Y. 0000000103)
31. Chater C, Kamisugi Y, Movahedi M, Fleming A, Cuming AC, Gray JE, Beerling DJ. 2011 Regulatory mechanism controlling stomatal behavior conserved across 400 million years of land plant evolution. Curr. Biol. 21, 1025– 1029. (doi:10.1016/j.cub.2011. 04.032)
32. Elizabeth M et al. 2011 Land plants acquired active stomatal control early in their evolutionary history. Curr. Biol. 21, 1030– 1035. (doi:10.1016/j.cub.2011. 04.044)
33. Budke JM, Goffinet B, Jones CS. 2011 A hundredyear-old question: is the moss calyptra covered by a cuticle? A case study of Funaria hygrometrica. Ann. Bot. 107, 1279– 1286. (doi:10.1093/ aob/mcr079)

34. Buda GJ, Barnes WJ, Fich EA, Park S, Yeats TH, Zhao L, Domozych DS, Rose JKC. 2013 An ATP binding cassette transporter is required for cuticular wax deposition and desiccation tolerance in the moss Physcomitrella patens. Plant Cell Online 25, 4000 – 4013. (doi:10.1105/tpc.113.117648)
35. Xu B et al. 2014 Contribution of NAC transcription factors to plant adaptation to land. Science 343, 1505 – 1508. (doi:10.1126/science.1248417)
36. Gray J. 1985 The microfossil record of early land plants: advances in understanding of early terrestrialization, 1970 – 1984. Phil. Trans. R. Soc. Lond. B 309, 167– 195. (doi:10.1098/rstb. 1985.0077)
37. Richardson JB. 1985 Lower Palaeozoic sporomorphs: their stratigraphical distribution and possible affinities. Phil. Trans. R. Soc. Lond. B 309, 201– 205. (doi:10.1098/rstb.1985.0079)
38. Wellman CH. 2010 The invasion of the land by plants: when and where? New Phytol. 188, 306– 309. (doi:10.1111/j.1469-8137. 2010.03471.x)
39. Taylor WA. 1995 Spores in the earliest land plants. Nature 373, 391 – 392. (doi:10.1038/373391b0)
40. Wellman CH, Osterloff PL, Mohiuddin U. 2003 Fragments of the earliest land plants. Nature 425, 282–285. (doi:10.1038/nature01884)
41. Edwards D, Duckett JG, Richardson JB. 1995 Hepatic characters in the earliest land plants. Nature 374, 635–636. (doi:10.1038/374635a0)
42. Taylor WA. 1997 Ultrastructure of lower Paleozoic dyads from southern Ohio II: Dyadospora murusattenuata, functional and evolutionary considerations. Rev. Palaeobot. Palynol. 97, 1 – 8. (doi:10.1016/S0034-6667(96)00068-1)
43. Wellman CH. 2006 The origin and early evolution of land plants: palynological evidence. Palynology 30, 230– 231.
44. Qiu YL et al. 2006 The deepest divergences in land plants inferred from phylogenomic evidence. Proc. Natl Acad. Sci. USA 103, 15 511– 15 516. (doi:10. 1073/pnas.0603335103)
45. Karol KG et al. 2010 Complete plastome sequences of Equisetum arvense and Isoetes flaccida: implications for phylogeny and plastid genome evolution of early land plant lineages. BMC Evol. Biol. 10, 16. (doi:10.1186/1471-2148-10-321)
46. Wickett NJ et al. 2014 Phylotranscriptomic analysis of the origin and early diversification of land plants. Proc. Natl Acad. Sci. USA 111, E4859 – E4868. (doi:10.1073/pnas.1323926111)
47. Cox CJ, Li B, Foster PG, Embley TM, Civa´nˇ P. 2014 Conflicting phylogenies for early land plants are caused by composition biases among synonymous substitutions. Syst. Biol. 63, 272 – 279. (doi:10. 1093/sysbio/syt109)
48. Kenrick P, Crane PR. 1997 The origin and early evolution of plants on land. Nature 389, 33 – 39. (doi:10.1038/37918)
49. Magallo´n S, Hilu KW, Quandt D. 2013 Land plant evolutionary timeline: gene effects are secondary to fossil constraints in relaxed clock estimation of age and substitution rates. Am. J. Bot. 100, 556 – 573. (doi:10.3732/ajb.1200416)
50. Smith SA, Beaulieub JM, Donoghue MJ. 2010 An uncorrelated relaxed-clock analysis suggests an earlier origin for flowering plants. Proc. Natl Acad. Sci. USA 107, 5897– 5902. (doi:10.1073/pnas. 1001225107)
51. Clarke JT, Warnock RCM, Donoghue PCJ. 2011 Establishing a timescale for plant evolution. New Phytol. 192, 266– 301. (doi:10.1111/j.1469-8137. 2011.03794.x)
52. Kenrick P, Wellman CH, Schneider H, Edgecombe GD. 2012 A timeline for terrestrialization: consequences for the carbon cycle in the Palaeozoic. Phil. Trans. R. Soc. B 367, 519–536. (doi:10.1098/rstb. 2011.0271)
53. Kenrick P. 2002 The telome theory. In Developmental genetics and plant evolution (eds QCB Cronk, RM Bateman, JA Hawkins), pp. 365– 387. London, UK: Taylor & Francis.
54. Stein WE, Boyer JS. 2006 Evolution of land plant architecture: beyond the telome theory. Paleobiology 32, 450– 482. (doi:10.1666/04036.1)
55. Rothwell GW, Wyatt SE, Tomescu AMF. 2014 Plant evolution at the interface of paleontology and developmental biology: an organism-centered paradigm. Am. J. Bot. 101, 899– 913. (doi:10.3732/ ajb.1300451)
56. Sanders H, Rothwell GW, Wyatt S. 2007 Paleontological context for the developmental mechanisms of evolution. Int. J. Plant Sci. 168, 719 – 728. (doi:10.1086/513519)
57. Shaw AJ, Szo¨ve´nyi P, Shaw B. 2011 Bryophyte diversity and evolution: windows into the early evolution of land plants. Am. J. Bot. 98, 352– 369. (doi:10.3732/ajb.1000316)
58. Tomescu AMF, Wyatt SE, Hasebe M, Rothwell GW. 2014 Early evolution of the vascular plant body plan—the missing mechanisms. Curr. Opin. Plant Biol. 17, 126 – 136. (doi:10.1016/j.pbi.2013.11.016)
59. Cove D, Bezanilla M, Harries P, Quatrano R. 2006 Mosses as model systems for the study of metabolism and development. Annu. Rev. Plant Biol. 57, 497 – 520. (doi:10.1146/annurev.arplant. 57.032905.105338)
60. Rensing SA et al. 2008 The Physcomitrella genome reveals evolutionary insights into the conquest of land by plants. Science 319, 64 – 69. (doi:10.1126/science.1150646)
61. Pires ND, Yi K, Breuninger H, Catarino B, Menand B, Dolan L. 2013 Recruitment and remodeling of an ancient gene regulatory network during land plant evolution. Proc. Natl Acad. Sci. USA 110, 9571– 9576. (doi:10.1073/pnas.1305457110)
62. Bennett T et al. 2014 Paralogous radiations of pin proteins with multiple origins of noncanonical pin structure. Mol. Biol. Evol. 31, 2042 – 2060 (doi:10.1093/molbev/msu147)
63. Vivancos J, Spinner L, Mazubert C, Charlot F, Paquet N, Thareau V, Dron M, Nogue´ F, Charon C. 2012 The function of the RNA-binding protein TEL1 in moss reveals ancient regulatory mechanisms of shoot development. Plant Mol. Biol. 78, 323– 336.

(doi:10.1007/s11103-011-9867-9)

64. Tanahashi T, Sumikawa N, Kato M, Hasebe M. 2005 Diversification of gene function: homologs of the floral regulator FLO/LFY control the first zygotic cell division in the moss Physcomitrella patens. Development 132, 1727– 1736. (doi:10.1242/ dev.01709)

65. Fujita T, Sakaguchi H, Hiwatashi Y, Wagstaff SJ, Ito M, Deguchi H, Sato T, Hasebe M. 2008 Convergent evolution of shoots in land plants: lack of auxin polar transport in moss shoots. Evol. Dev. 10, 176– 186. (doi:10.1111/j.1525-142X.2008.00225.x)

66. Okano Y, Aono N, Hiwatashi Y, Murata T, Nishiyama T, Ishikawa T, Kubo M, Hasebe M. 2009 A polycomb repressive complex 2 gene regulates apogamy and gives evolutionary insights into early land plant evolution. Proc. Natl Acad. Sci. USA 106, 16 321– 16 326. (doi:10.1073/pnas.0906997106)

67. Mosquna A, Katz A, Decker EL, Rensing SA, Reski R, Ohad N. 2009 Regulation of stem cell maintenance by the Polycomb protein FIE has been conserved during land plant evolution. Development 136, 2433– 2444. (doi:10.1242/dev.035048)

68. Elbert W, Weber B, Burrows S, Steinkamp J, Budel B, Andreae MO, Poschl U. 2012 Contribution of cryptogamic covers to the global cycles of carbon and nitrogen. Nat. Geosci. 5, 459 – 462. (doi:10. 1038/ ngeo1486)

69. Gensel PG, Johnson NG, Strother PK. 1990 Early land plant debris (Hooker's 'waifs and strays'?). Palaios 5, 520– 547. (doi:10.2307/3514860)

70. Smith MR, Butterfield NJ. 2013 A new view on Nematothallus: coralline red algae from the Silurian of Gotland. Palaeontology 56, 345 – 357. (doi:10. 1111/j.1475-4983.2012.01203.x)

71. Honegger R, Axe L, Edwards D. 2013 Bacterial epibionts and endolichenic actinobacteria and fungi in the Lower Devonian lichen Chlorolichenomycites salopensis. Fungal Biol. 117, 512– 518. (doi:10. 1016/j. funbio.2013.05.003)

72. Strother PK. 2010 Thalloid carbonaceous incrustations and the asynchronous evolution of embryophyte characters during the Early Paleozoic. Int. J. Coal Geol. 83, 154– 161. (doi:10.1016/j.coal. 2009.10.013)

73. Tomescu AMF, Rothwell GW, Honegger R. 2006 Cyanobacterial macrophytes in an Early Silurian (Llandovery) continental biota: Passage Creek, lower Massanutten Sandstone, Virginia, USA. Lethaia 39, 329– 338. (doi:10.1080/00241160600876719)

74. Tomescu AMF, Honegger R, Rothwell GW. 2008 Earliest fossil record of bacterial –cyanobacterial mat consortia: the early Silurian Passage Creek biota (440 Ma, Virginia, USA). Geobiology 6, 120 – 124. (doi:10.1111/j.1472-4669.2007.00143.x)

75. Tomescu AMF, Pratt LM, Rothwell GW, Strother PK, Nadon GC. 2009 Carbon isotopes support the presence of extensive land floras pre-dating the origin of vascular plants. Palaeogeogr. Palaeoclimatol. Palaeoecol. 283, 46 – 59. (doi:10. 1016/j.palaeo.2009.09.002)

76. Smith SY, Collinson ME, Rudall PJ, Simpson DA, Marone F, Stampanoni M. 2009 Virtual taphonomy using synchrotron tomographic microscopy reveals cryptic features and internal structure of modern and fossil plants. Proc. Natl Acad. Sci. USA 106, 12 013– 12 018. (doi:10.1073/pnas. 0901468106)

77. Friis EM, Crane PR, Pedersen KR, Bengtson S, Donoghue PCJ, Grimm GW, Stampanoni M. 2007 Phase-contrast X-ray microtomography links Cretaceous seeds with Gnetales and Bennettitales. Nature 450, 549 – 552. (doi:10.1038/nature06278)

78. Friis EM, Marone F, Pedersen KR, Crane PR, Stampanoni M. 2014 Three-dimensional visualization of fossil flowers, fruits, seeds, and other plant remains using synchrotron radiation X-ray tomographic microscopy (SRXTM): new insights into Cretaceous plant diversity. J. Paleontol. 88, 684– 701. (doi:10.1666/13-099) 79. Steemans P, Wellman CH, Gerrienne P. 2010 Palaeogeographic and palaeoclimatic considerations based on Ordovician to Lochkovian vegetation. In The terrestrialization process: modelling complex interactions at the biosphere-geosphere interface (eds M Vecoli, G Cle´ment, B Meyer-Berthaud), pp. 49– 58. London, UK: The Geological Society, Special Publications.

79. Ligrone R, Duckett JG, Renzaglia KS. 2012 The origin of the sporophyte shoot in land plants: a bryological perspective. Ann. Bot. 110, 935 – 941. (doi:10.1093/aob/ mcs176)

80. Sakakibara K, Ando S, Yip HK, Tamada Y, Hiwatashi Y, Murata T, Deguchi H, Hasebe M, Bowman JL. 2013 KNOX2 genes regulate the haploid-todiploid morphological transition in land plants. Science 339, 1067 – 1070. (doi:10.1126/science. 1230082)

81. Taylor TN, Kerp H, Hass H. 2005 Life history biology of early land plants: deciphering the gametophyte phase. Proc. Natl Acad. Sci. USA 102, 5892 – 5897. (doi:10.1073/pnas.0501985102)

82. Timme RE, Bachvaroff TR, Delwiche CF. 2012 Broad phylogenomic sampling and the sister lineage of land plants. PLos ONE 7, e0029696. (doi:10.1371/ journal. pone.0029696)

83. Bowman JL. 2013 Walkabout on the long branches of plant evolution. Curr. Opin. Plant Biol. 16, 70– 77. (doi:10.1016/j.pbi.2012.10.001)

84. Lenton TM, Crouch M, Johnson M, Pires N, Dolan L. 2012 First plants cooled the Ordovician. Nat. Geosci. 5, 86 – 89. (doi:10.1038/ngeo1390)

85. Beerling DJ, Berner RA. 2005 Feedbacks and the coevolution of plants and atmospheric $CO_2$. Proc. Natl Acad. Sci. USA 102, 1302 – 1305. (doi:10.1073/ pnas.0408724102)

86. Field KJ, Cameron DD, Leake JR, Tille S, Bidartondo MI, Beerling DJ. 2012 Contrasting arbuscular mycorrhizal responses of vascular and non-vascular plants to a simulated Palaeozoic $CO_2$ decline. Nat. Commun. 3, 835. (doi:10.1038/ ncomms1831)

CHAPTER 2

# THE HEAVY LINKS BETWEEN GEOLOGICAL EVENTS AND VASCULAR PLANTS EVOLUTION

Aldo Piombino

## NATURAL DRIVERS OF CLIMATE CHANGES AND THEIR EFFECTS ON EVOLUTION

Climate is regulated firstly by the amount of solar radiation reaching the surface; this value is mainly a function of the latitude. But there are other astronomical factors regulating it. Earth orbit variations are globally felt: the effects of orbital forcing, known as Milankovitch cycles, have been well recognized in many sedimentary series in plenty of geochemical and stratigraphic parameters; there are also changes in solar activity rate at a pluricentennial scale: for example, in Holocene colder and warmer period (such as the Medieval climatic optimum and the Little Ice Age) are linked to these variations that can produce shifts of the limits of climatic belts. All these changes are poorly reflected in life evolution because they are too brief to promote evolution. From upper Tertiary the onset of the Antarctic glaciation amplified the climatic effects of orbital forcing.

Geological events are able to modify climate and environment at all the different scales, from a global perspective to strictly delimited areas, thus affecting and driving evolution on Earth. A main process that changes climate is Plate Tectonics: the wandering continents change their latitude, thus modifying the strength of solar radiation reaching their surface; the consequent climate and environment changes constitute major evolutionary forcing. Other geological factors are orography, sea level, ocean, and landmasses distribution and their variations. Also marine currents and atmospheric winds regime are strongly influenced by landmass and ocean distribution.

The effects of the variation in landmasses distribution can be felt over long distances if they affect ocean currents. For example, in late Tertiary the northward drift of the Australian continent closed the seaway across which the warm waters from Central Pacific went to the Indian Ocean. In consequence the Indian Ocean cooled, and the East African region experienced a second drop in rain rate, after the one occurring when the mountains of the East African Rift rose [1].

The arrival of landmasses in polar area has strong consequences in their climate but also carries on global consequences: the development of ice caps drops sea level, while ice and permafrost retain variable amount of greenhouse gas like $CO_2$ and $CH_4$, triggering global temperatures falls.

After the break-up of a continent, new clades can rise from the division of formerly single populations, while the union of two continents results in a complex pattern of biotic interchanges, extinctions, and fast and spectacular radiations and differentiations.

Landmass union and intracontinental rifts have raised mountain belts that changed climate and created barriers for rains and biotic movements between the two sides of the uplifted belt.

## ATMOSPHERIC $CO_2$ AMOUNT AND LARGE IGNEOUS PROVINCES

The heavy links between life and continental drift have been well established in the second half of XX century (mainly in the 60s) while the effects of greenhouse changes (particularly

**Citation:** Aldo Piombino, "The Heavy Links between Geological Events and Vascular Plants Evolution" International Journal of Evolutionary Biology, vol. 2016, Article ID 9264357, 9 pages, 2016. doi:10.1155/2016/9264357.

$CH_4$ and, more, $CO_2$) have been poorly underscored until today, but global climate (and above all temperatures and what happens when they change) is a direct consequence of the amount of atmospheric $CO_2$, this value being influenced by many factors.

Among atmospheric factors, a main role is played by the greenhouse gas amount. Early Earth atmosphere was composed of more than 95% of $CO_2$, as the current ones of Venus and Mars. Young Sun radiation was weaker than in Phanerozoic times: only thanks to these high atmospheric pressure and greenhouse conditions Earth could retain liquid waters in Archean and Paleoproterozoic times, while Earth with the current atmosphere would be globally iced.

Volcanoes emit several hundreds of thousands of tons of $CO_2$ yearly: this "ordinary" gas flow from solid Earth to atmosphere is lower than the carbon dioxide demand of the Earth system; thus, currently, the residence time in the atmosphere of a carbon dioxide molecule is shorter than 4 years.

The best-known $CO_2$ users are plants and photosynthesis changed the world: more than 2 billion years ago, soon after the onset of this process, atmospheric carbon dioxide content dropped while the rise of oxygen occurred (the Great Oxygenation Event) because of photosynthetic unicellular algae; so these tiny organisms changed the environment from reducing to oxidizing conditions and also global temperature got down and the first global glaciation event occurred.

Sea water adsorbs $CO_2$ (the more its amount, the more the oceanic acidity); land surface and marine bottom use $CO_2$ in silicates minerals weathering and carbonates formations; another strong mineral sequestration process is the formation of hydrocarbons and coals reserves. Other $CO_2$ is trapped in marine sediments, returning in the mantle when their plate arrives at a subduction zone.

An important remover of atmospheric $CO_2$ is represented by polar areas: the arrival of landmasses in high latitude has strong consequences in their climate but also carries on global consequences: not only does the development of ice caps drop sea level, but, more, ice and permafrost retain huge amount of greenhouse gas like $CO_2$ and $CH_4$, triggering global temperatures falls. Atmospheric $CO_2$ dramatically dropped at the onset of Antarctic glaciation in Neogene and the current occurrence of ice caps is the main reason for the current very low $CO_2$ atmospheric content.

Only strong emissions from inner Earth can increase $CO_2$ atmospheric amount: this happens during the Large Igneous Provinces activity.

Large Igneous Province (from here LIP) is a concept introduced by Coffin and Eldhom in 1991 [4], joining into single definition continental flood basalts and oceanic plateaus. A LIP is an emplacement of hundreds of thousands of cubic kilometers (also millions of cubic kilometers) of mainly basaltic magmas in a very short period of time: flow-by-flow analysis of paleomagnetic directions along 10 traverses of the Western Ghats in India suggests that at the final Cretaceous a large fraction of the main sequence of the Deccan Traps was formed in a succession of less than 30 short eruptive events in a time interval lasting for less than 10 ka [5]. Similar considerations come from studies performed on Jurassic Karoo Basalts [6] and latest Permian Siberian Traps [7]. The emplacement of a LIP often results in a lot of environmental consequences at global scale, affecting atmosphere, soils, and hydrosphere because of the massive magma and volatiles production. This has been well established in Deccan Traps, the LIP that occurred at the end of Mesozoic, clearly related to dinosaur extinction (for a good examination of the topic see a recent volume of the Geological Society of America Special Publications [8]): single eruptive events of a LIP activity produce a lava volume ranging from 1,000 to 20,000 $km^3$ in few tens of years; during each of these short-lasting paroxysmal phases the releases of volatiles exceed the sequestration potential of Earth system: in consequence the accumulation in atmosphere of $CO_2$, $SO_2$, heavy metals, and other substances promotes continuous and strong environmental variations; in this way many parameters increase, like terrestrial weathering patterns, ocean and atmospheric temperatures, and water acidity in the seas and rains; many extinction events are also accompanied by severe sea level oscillations (particularly, a fast transgression preceded by a drop); in the worst cases also anoxic events with a worldwide distribution can occur. Figure 1 summarizes the climatic effects of LIP emissions in atmosphere.

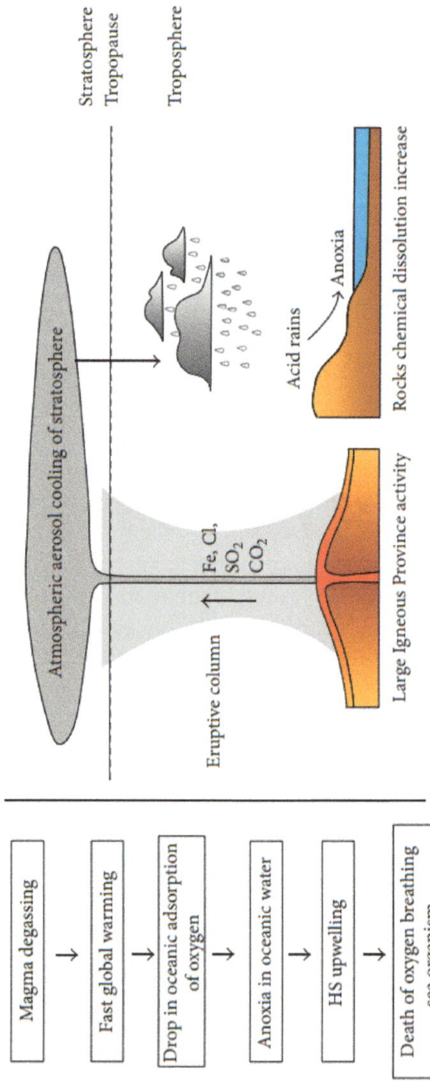


**Figure 1:** The effects on soil, atmosphere, oceanic waters, and rains of volatiles emitted by a Large Igneous Province.

Laki eruption in 1783 in Iceland can be a pale example of what happens during a LIP flow: in the island volatiles killed most of livestock and a strong famine occurred; in all Europe a sulphur cloud damaged crops and the winter 1784/84 recorded the worst mortality level in England and northern Europe of the century, especially for lung and heart diseases. And Laki produced only 16 cubic kilometers of lava in 8 months, a little quantity when compared with a single event of a LIP activity.

Environmental perturbations related to a LIP emplacement begin well before the spike of volcanic activity, probably because in the early stages the lava produces a huge amount of pyroclastics and volatiles coming from metasomatism and other reactions of the sediments in which they intrude; the higher the thickness and/or the evaporitic content of the sediments, the higher the production [9].

Figure 2, from [2], depicts the major LIPs events that occurred from early upper Permian (260 million years ago) until today. There is a clear coupling between LIP emplacements and biosphere changes, the strongest changes being the main mass extinction events. The very high speciation rate immediately before and after an extinction event reflects the continuous adaptation of the biosphere to new conditions: this is demonstrated by the biozones, which follow one another at a faster pace before and after the peak of the crisis, even if diversity generally falls because the extinction rate exceeds by far origination rate.

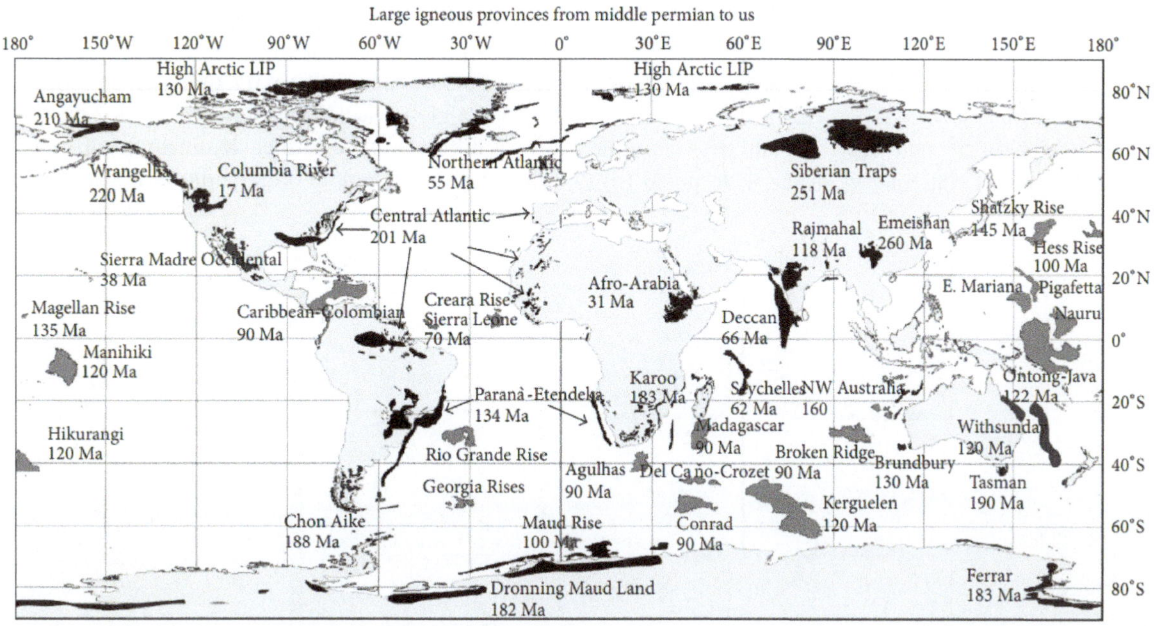


FIGURE 2: From [2], map of the Large Igneous Provinces emplaced from the Middle Permian (260 million years ago).

**Figure 2:** From [2], map of the Large Igneous Provinces emplaced from the Middle Permian (260 million years ago).

It is also noticeable that from the end of the Devonian to the beginning of the Eocene all the major chronological divisions in geological times are coeval with the occurrence of a LIP.

Figure 3, modified after [3], depicts the medium atmospheric $CO_2$ content from the Devonian to today, calculated analyzing carbon isotope composition from stomatal anatomy of fossil leaves: this method is based on the inverse relationship observed in many plants between stomatal number and $CO_2$. $CO_2$ was low (<500 ppm) during periods of long-lived and widespread continental glaciations and high (>1000 ppm) during warmer periods [10].

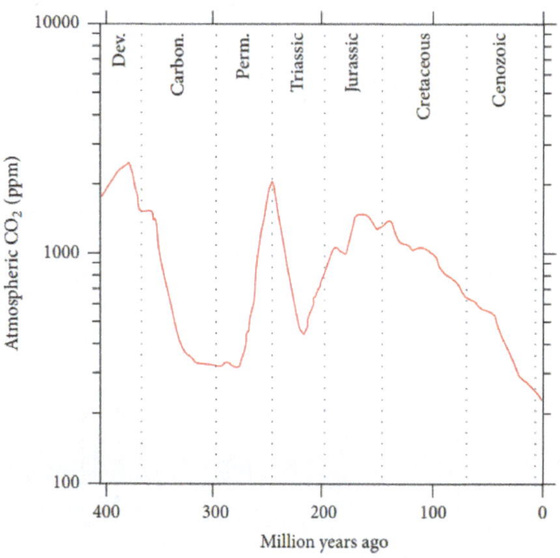


**Figure 3:** Atmospheric $CO_2$ in the last 400 Ma from [3].

The continuous drop in $CO_2$ atmospheric content in the Devonian can be related to plant colonization of the interior of continents [11]: if, during the Early to Middle Devonian, land plants were clustered in lower floodplain and coastal delta habitats, from the Middle to Late Devonian, they began to occupy more upland regions, thus increasing $CO_2$ consumption.

This trend was briefly interrupted in the Late Devonian because of the occurrence of some LIPs that injected in the atmosphere a great amount of this gas. A second process concurring with the atmospheric $CO_2$ drop started at the beginning of the Permo-Carboniferous glaciation: the arrival of landmasses in both the high latitude areas, Gondwana in the southern hemisphere and Siberia in the northern.

$CO_2$ atmospheric content grew at the end of the Permian abruptly, because of the final melting of the Permo-Carboniferous ice caps and because of the very high emissions produced by the Emeishan and Western Siberian Traps: for these reasons and for the occurrence of a large supercontinent in which moisture transport was difficult, the Early Triassic has been very hot and dry [12].

The high temperatures and $CO_2$ amount of Mesozoic atmosphere were a direct consequence of the widespread LIP magmatism of that time, related to the onset of the Gondwana break-up. So, LIP magmatism is the main dispenser of this gas and when the frequency of this activity dropped, also atmospheric $CO_2$ content did so. In Tertiary LIP activity has been weaker than in the Cretaceous (there have been only 3 smaller events on land and no one oceanic plateau has formed); contemporarily the Antarctic ice cap formation added a new, strong $CO_2$sequestration process, witnessed by the increase in the $CO_2$ drop rate from the end of the Oligocene to today. In consequence, $CO_2$ amount collapsed, as temperatures did, with plenty of environmental changes worldwide.

## MASS EXTINCTIONS AND PLANTS EVOLUTION IN THE FOSSIL RECORD

Life on Earth has been largely affected by various mass extinction episodes, defined by Jack Sepkoski as short-lasting phases showing significant extinction rate in various nonrelated clades at a global scale [13]. Before and after these events life is very different since former dominant groups vanished and new groups, with a rapid origination, assume the leadership.

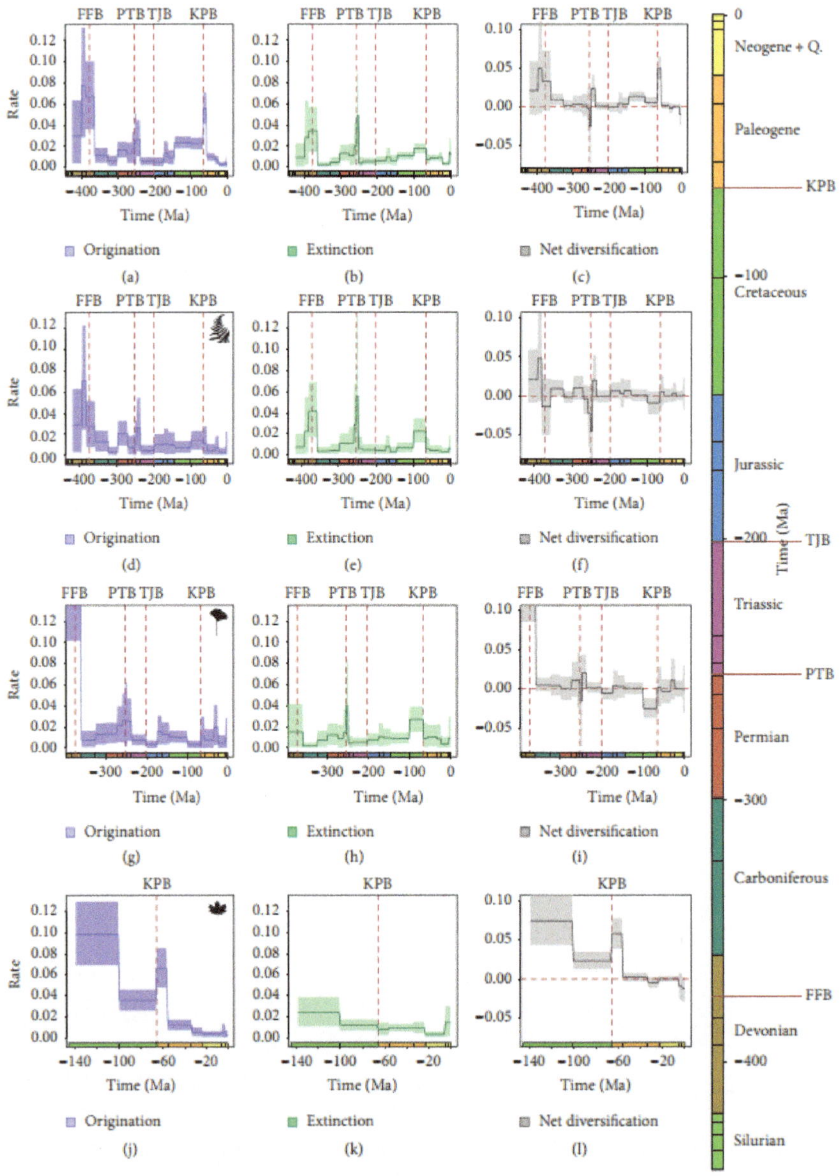


**Figure 4:** From [17], generic level diversification analysis of all vascular plants. Origination (blue) and extinction (green) rates were estimated using the Bayesian approach implemented in PyRate, within time bins defined as epochs of the geologic timescale (shown at the bottom). The

timescale (x-axis) is given in million years ago (Ma). Solid lines indicate mean posterior rates, whereas the shaded areas show 95% HPD intervals. The diversification dynamics were estimated for vascular plants as a whole (a–c), spore-bearing plants (d–f), nonflowering seed plants (g–i), and flowering seed plants (angiosperms: j–l). Net diversification rates (gray) are defined as origination minus extinction. The dashed lines indicate the major mass extinction events widely recognized: at the Frasnian-Famennian (FFB), Permian-Triassic (PTB), Triassic-Jurassic (TJB), and Cretaceous-Paleogene (KPB) boundaries. Plant silhouettes were obtained fromhttp://phylopic.org/.

In 1983 Niklas et al. described four main phases of plant biodiversity growth [14]:(1)The first origination of early terrestrial plants and the rise of the first forest in Silurian times.(2)Ferns and lycophytes of the lower Carboniferous.(3) Gymnosperm diffusion from Upper Carboniferous.(4)The rise and diffusion of the angiosperms from the lower Cretaceous.A new method, PyRate, uses a Bayesian analytical framework to origination and extinction dynamics using fossil occurrence data of plant fossil occurrences from the Paleobiology Database (PBDB;https://paleobiodb.org/cgi-bin/bridge.pl) [15]. Compared to other approaches, this method has the advantage of using all fossil occurrences in the analysis (e.g., not only first and last appearances) and of jointly estimating (1) the preservation rate, (2) times of origination and extinction of all lineages, and (3) rates of origination and extinction through time. The inference is carried out in a sound statistical framework, which has been shown to be robust under many evolutionary and sampling scenarios through extensive simulations [16]. By exploiting the properties of Bayesian modeling, PyRate generates parameter estimates provided with credible intervals while accounting for different sources of data and parameter uncertainties.

PyRate tracks vascular plant origin and diversification from the Silurian to today, dividing them into four main clades: vascular plants, spore-bearing plants, nonflowering seed plants, and flowering seed plants. The obtained data are depicted in Figure 4, from [17].The oldest uncontroversial plant fossils are cryptospores of mid-Ordovician age [18]. At these times $CO_2$atmospheric content was very high, some 8–20 times the preindustrial atmospheric level [19]. Even if climate was not as hot as expected, high $CO_2$ levels are undoubtedly a good chance for plants growing, especially with concurrent humid conditions.

## LOWER DEVONIAN EXPANSION OF VASCULAR PLANTS AND THE UPPER DEVONIAN BIOTIC CRISIS

The first vascular plants lived as early as the Early Silurian. PyRate shows a very high vascular plants origination rate in the Devonian, reaching the higher level ever in the first half of the Upper Devonian. This high level is an effect of the radiation of terrestrial plants during a huge increase in biomass amount that strongly changed continental grounds and modified hydrological cycle [20].

Nonflowering seeds plants originated in the Upper Devonian, exhibiting a high origination rate: the development of heterospory in progymnosperms with the spores transformed in seeds allowed them to reproduce also in harsher areas.

This novel reproduction system has been very useful in Devonian continents where arid or semiarid environment was widespread, raising the ability of plant to colonize larger areas than before: the fast radiation triggered a fast evolution and both origination and extinction rates have been constant up to the beginning of the Late Devonian.

In latest Devonian origination rate dropped two times, slightly before the Frasnian-Famennian border (from here FFB) and at the end of the Devonian, this second drop being huger than the first: PyRate shows an extinction rate spike some before the FFB, remaining constant throughout the border at slightly lesser levels and exhibiting a dramatic increase in the Famennian.

The Upper Devonian has been a very hard time: plant diversity modifications mirror two main faunal mass extinction events that occurred at the Frasnian-Famennian border (the Kellwasser event) and, soon after, at the Devonian-Carboniferous border (Hangenberg event). The end of Frasnian and the beginning of Famennian are characterized by fast climate changes, sea level oscillation, excursions in carbon and oxygen isotopic ratios, global anoxia, spike of atmospheric sulfide, and other phenomena.

This biotic event is one of the main Phanerozoic extinctions affecting animals, with a huge drop in biodiversity. PyRate demonstrates that also plants suffered a lot that time. Like most of mass extinctions, these events have been driven by the emplacement of Large Igneous Provinces: the Frasnian-Famennian border is clearly linked with the Viluy Traps in Yakutia [21]. The end Frasnian biotic crisis was followed by a second one at the very end of the period with widespread anoxic sediments and perturbation of the carbon isotopic ratio. Geological and geochemical observations witness that also the Hangenberg event is closely related to the emplacement of a LIP, but the position of the volcanic center has not been currently determined, Ogcheon Belt in Korea being the most candidate site today [22]. It is noticeable that the LIP activity briefly interrupted the dropping trend of atmospheric $CO_2$ content. PyRate demonstrates that the geological events of the Upper Devonian have played a major role in plants evolution, since their diversity at the beginning of the Carboniferous was well lower than at the onset of the Upper Devonian storm.

## CARBONIFEROUS PLANTS: A TWO-FACED HISTORY

In the lower Carboniferous climate has been warm, thanks to the high $CO_2$ emissions from Late Devonian LIPs, and it was also very wet in the equatorial belt, where the main plant biomass was hosted. The climate was also stable that time, so, despite the floral abundance, either origination or extinction rate was slow.

We must also note a dramatic fall of atmospheric $CO_2$ throughout all the lower Carboniferous [23]. These dryer conditions have been harder for spore-bearing and favored nonflowering seed plants. This extinction event is not related to a LIP occurrence, as occurred well later, at the Eocene-Oligocene border: it is interesting that both these biotic crises are placed near the beginning of the two last Phanerozoic major glaciation phases.

So, we can say that mid-Carboniferous rainforest collapse witnesses another main link between geology and plant evolution, even if its exact causal mechanism is still poorly known. Climate becomes colder and dryer than before throughout the Upper Carboniferous, with very low atmospheric content of $CO_2$ and methane, also because they have been partly trapped in the large areas covered by thick ice caps and permafrost and in high latitude seafloor.

Whatever the geological origin of the problems affecting vascular plant in the mid-Carboniferous, in good agreement with the drying climatic trend, PyRate detects opposite behaviors in spore-bearing and nonflowering seed plants diversity between the lower and Upper Carboniferous: in the Upper Carboniferous spore-bearing plants reduced their diversity, with a huge drop in speciation and a very little extinction rate compared to the first half of the period, while nonflowering seeds plant performed a diversity growth even if they have been affected by a stronger extinction rate than before.

It is interesting that in Upper Carboniferous reptiles performed a wide radiation that, like the Devonian nonflowering seed plants diversification, has been driven by a reproductive step: the new eggs allowed the birth of an animal quite similar to an adult with very fewer needs of water than an aquatic tadpole.

## PERMIAN AND THE END-PERMIAN GLOBAL CRISIS

Also in the Permian the lower and the upper part of the period show different trends in plant diversity, related to different climate conditions: the Early Permian marks the end of the major high latitude Gondwana glaciation [24]. The dropping $CO_2$ trend stopped. Thanks to a global temperature and moisture rise a huge increase in spore-bearing plants diversity occurred in the lower Permian, well detected by PyRate.

The warming trend can be partly ascribed to the earliest Permian Skagerrak basalt LIP emissions. The end of the Permian as well as the beginning of the Triassic has been one of the most difficult times for life on Earth: two main faunal extinction events occurred, one near the end of the Capitanian and the second at the end of the period, less than 8 Ma later. The faunal replacement has been so heavy that the end of the Permian has been the worst extinction event ever, establishing the limit between Paleozoic and Mesozoic era since the half of XIX century.

These events have been caused by the emplacement of two Large Igneous Provinces: the Emeishan Traps, now Southern China, and the Siberian Traps, east of the Ural Mountains.

Plants have been severally affected by the extinction: greenhouse gas and volatiles raised temperature and rain acidity in a drying world [25], triggering massive environmental changes that drove, either in spore-bearing plants or in nonflowering seed plants, the most dramatic diversity drop ever. But the diversity behavior shows a complex trend.

Spore-bearing plants slowed their origination rate in early upper Permian times: the onset of volcanic activity in Southern China coincided with their fall in origination rate and increase in extinction rate that began well before the faunal extinction spike of the upper Capitanian. In the latest Permian, between the two faunal extinction spikes, a lower origination rate is accompanied by a dramatic increase in extinction, prolonged toward the earliest Triassic, when it has reached its maximum level. Nonflowering seed plants show a similar pattern.

So, PyRate suggests heavy links between plant history and the environmental storm triggered by the volatiles released during the emplacement of Emeishan and Siberian basalts. It is noticeable that a spike in origination rate occurred after the second LIP activity in the earliest Triassic, but it could not compensate the even major extinction rate: thanks to the very high level of $CO_2$, partly of deglaciation origin, partly emitted during both LIPs' activity, these times have been the hottest and the driest in the Phanerozoic Eon [11], a very hostile period for life. The high plants origination and extinction rates demonstrate that environmental changes have been strong, frequent, and continuous during all the latest Permian and earliest Triassic.

## MESOZOIC ERA AND THE RISE OF FLOWERING SEED PLANTS

After the first 6 Ma of the Triassic, temperatures went down, thanks to the various $CO_2$ sequestration processes, allowing plants and animals to reconquer the equatorial belt in the early Anisian. In this stage diversity fast increased either in spore-bearing or in nonflowering seed plants. During the following part of the Triassic there are no major changes in plant biodiversity patterns.

The Triassic ends with the emplacement of the Central Atlantic Magmatic Province, the initial stage of Pangea break-up; magmatic activity triggered a severe mass extinction event in animals: many clades of Archosaurs got extinct, as most of mammalian reptiles, and a fast dinosaurs radiation and differentiation took place. This event did not affect vascular plants.

Early Jurassic warm and wet environment promoted a high differentiation in spore-bearing plants. A peak in speciation and extinction is placed in the middle Jurassic, with a low drop in diversity of spore-bearing plants while diversity in nonflowering seed plants little increased. This scenario can be related to a minor extinction event at the end of the Toarcian, corresponding to a global anoxic stage, clearly related to the environmental effects of the Karoo-Ferrar LIP, immediately before the beginning of the Gondwana break-up: a little later the blocks formed by Africa-Arabia and South America split out from the block formed by India, Antarctica, and Australia.

In Cretaceous times the main event for vascular plants has been the strong angiosperms diversification.

The origin of this group is debated because molecular data point to an older, Triassic age for this clade, when the ancient most fossil has been found in lower Cretaceous sediments, even if there are controversial claims about some Triassic fossils.

Thus, there is a gap of several tens of millions of years between the molecular and the paleontological timing (no less than 60). Many explanations have been proposed for this bias: it could be simply of paleontological origin (the lack of fossils) but also methodological issues in the analysis of genetic data must be considered: for a good and recent examination of the topic see Beaulieu et al. (2015) and references therein [26].

Could the rise of angiosperms be well older than their huge radiation?

For Doyle Triassic molecular dates may be reconciled with the fossil record if the first angiosperms were restricted to wet forest understory habitats and did not radiate outside this environment until the Cretaceous [27].

If this last reconstruction is true, something has occurred promoting the nonflowering seed plants replacement with angiosperms.

Also, in this case, a geological trigger can be considered. Climate was hot thanks to the high, even if lower than in the early Triassic, $CO_2$ content that grew (up to four times higher than today), because of the widespread LIPs activity of the lower Cretaceous, during which the Caribbean, Ontong Java-Manihiki-Hikurangi, Madagascar, Kerguelen, and other more Large Igneous Provinces had been emplaced. Climate was wetter than the early Triassic, because of the completely different arrangement of the continents: Pangaea break-up resulted in newborn oceans interposed between smaller landmasses compared to the former supercontinent; these oceanic arms allowed humidity distribution in larger areas than before, diminishing the extension of arid and semiarid lands.

Moreover, most of the boreal continent Laurasia was placed at low northern latitudes, in the tropical area.

Thanks to both of these concurrent factors, higher $CO_2$ levels and wider humidity distribution, plant biomass was higher than before, resulting in an increase of photosynthetic activity: in this way also atmospheric oxygen increased. The higher atmospheric oxygen content raised the occurrence of wildfires that became very common later, in the upper Cretaceous, witnessed by the widespread occurrence of fullerenes in the sediments. Because of their resins content, wildfires can have represented a major challenge for conifers more than for angiosperms that replaced nonflowering seed plants in many environments [28].

In the upper Cretaceous, PyRate shows a drop in all vascular plants, also in angiosperms.

This is apparently strange, because the Gondwana break-up divided into some blocks the former greater continent and this event theoretically would have triggered a diversity increase.

Why is this decline in a commonly reputed happy period for life and of splitting continents? Climate becomes newly harsher, with pronounced seasonality also at low latitudes, and an aridification of waters, rains, and soils occurred, while the high atmospheric $O_2$ and the prolonging of dry seasons increased wildfires frequency, dramatically affecting plants diversity. Also various episodes of sea level oscillation can be blamed for this diversity drop which occurred also in animals.

The Cretaceous ended with the most famous of all mass extinctions, in which not only dinosaurs have been wiped off. As written before, for this event the Yucatan meteorite impact has been blamed but this hypothesis has been challenged, and now another LIP, the Deccan Traps, is currently the best candidate for the killer role [29], because the chain of events recorded in latest Cretaceous sediments is the typical sequence of the geological, geochemical, and environmental disturbance related to a LIP emplacement.

## FROM THE PALEOCENE TO TODAY: VASCULAR PLANTS DIVERSITY IN THE GLOBAL COOLING TREND

As what happened in the other biotic restorations before mass extinctions, the Paleocene is characterized by a huge origination trend in all vascular plants, in the aftermath of K/T extinctions and environmental disturbance. From the Eocene to today, origination rates in nonflowering seeds plants and fern-bearing plants show trend opposite to that of the flowering plants: being higher in the Oligocene and Quaternary for the former, while the latter shows higher trends in the Eocene and Pliocene: crossing climatological data withPyRate-inferred plant diversity, in the Tertiary the warmer the period, the higher the origination rate in flowering plants, while colder periods show an increase in nonflowering and spore-bearing plants.

After the climax in the middle of the late Cretaceous, global temperatures began to fall: the Mesozoic greenhouse climate has been driven by the widespread occurrence of LIPs; but since their frequency lowered, Earth system has been able to soak up much more $CO_2$ than the ordinary volcanic emission amount: in consequence also temperatures trended downward after the highest point reached during the Turonian, 30 Ma before the end of the Cretaceous (90 Ma).

However Eocene climate has been still warm, accompanied by two thermal highs, at its beginning (PETM (Paleocene Eocene Thermal Maximum), synchronous with the main phase of the North Atlantic Igneous Province) and in the middle (MECO (middle Eocene climatic optimum)). Rainforest was widespread also in the areas now covered by deserts or savannas of Africa, Arabia, and SW Asia at time. But at the Eocene-Oligocene border the ongoing cooling trend triggered also a little faunal extinction event (Mongolian Remodelling orGrand Coupure).

There is probably a link between Eocene-Oligocene border cooling, Mongolian Remodelling, and the onset of Antarctic glaciation, an event for which many forcings acted:

(i)The global scale atmospheric $CO_2$ continuous fall.

(ii)The arrival of Antarctica near south pole.

(iii)The split between Antarctica and Australia and the Drake Strait opening that allowed the circumpolar current formation, preventing since then the mixing of polar and warmer waters in southern ocean [30].

From that moment permafrost and ice caps began their development in southern hemisphere (in northern hemisphere their onset is later, since North America and Eurasia arrived later than Antarctica at their current high latitudes), hence becoming a new, main, greenhouse gas sequestration process. From the beginning of the Oligocene, rainforest began to fall in higher latitudes, replaced by an open woodland environment also in Southern Africa.

It is noticeable also that these geological changes promoted in many areas C3 plants replacements with C4 plants.

## CONCLUSIONS

Vascular plants diversity dropped during various mass extinction events, especially at the two Upper Devonian events, Permian-Triassic border, Toarcian and Cretaceous-Paleogene border, while they passed the Triassic-Jurassic border event without problems. Diversity pattern shows a growth before extinction peaks, a common behavior in the frame of the biosphere recovery after a mass extinction.

Plant recovery has been slightly delayed only after the End-Permian extinction, because of the extreme climatic conditions created by the double LIP event (Emeishan Traps, Permian Traps) and by the final sudden vanishing of the last Permo-Carboniferous ice caps.

The data inferred with PyRate are in excellent agreement with geological events, thus suggesting a heavy link between plant evolution and geological events, especially the occurrence of Large Igneous Provinces.

So, the four stages detected by Niklas et al. in 1983 [14] can be interpreted with environmental changes related to geological events:

(1)The high $CO_2$ level allowed the origination of early terrestrial plants and the rise of the first forest in Silurian times.

(2)The wide occurrence in the lower Carboniferous of lands along the equatorial belt allowed the rise of the tropical rainforest and the diffusion of ferns and lycophytes.

(3)The arrival of large landmasses in polar positions and lowering in LIP occurrence rate originated the cooling and aridification trends of the Upper Carboniferous culminated with the onset of the Permo-Carboniferous glaciation, thus promoting the nonflowering seed plants diffusion.

(4)The diffusion of the angiosperms from the lower Cretaceous has been a consequence of the wide wildfires occurrence, triggered by the climatic and environmental consequences of the massive volcanic activity of that time and by the Pangaea break-up in a world in which moisture was more widely distributed than before.

## Conflict of Interests

The author declares there is no conflict of interests regarding the publication of this paper.

## Acknowledgments

The author thanks Nicola Casagli and Daniele Silvestro for the useful comments and discussion about the text.

## REFERENCES

1. M. A. Cane and P. Molnar, "Closing of the Indonesian seaway as a precursor to east African aridification around 3-4 million years ago," Nature, vol. 411, no. 6834, pp. 157–162, 2001. ·
2. S. E. Bryan and R. E. Ernst, "Revised definition of Large Igneous Provinces (LIPs)," Earth-Science Reviews, vol. 86, no. 1–4, pp. 175–202, 2008. ·
3. P. J. Franks, D. L. Royer, D. J. Beerling et al., "New constraints on atmospheric $CO_2$ concentration for the Phanerozoic," Geophysical Research Letters, vol. 41, no. 13, pp. 4685–4694, 2014.
4. M. F. Coffin and O. Eldholm, "Large igneous provinces: JOI7USSAC workshop report," Tech. Rep. 114, The University of Texas at Austin Institute for Geophysics, 1991.
5. A.-L. Chenet, V. Courtillot, F. Fluteau, et al., "Determination of rapid Deccan eruptions across the Cretaceous-Tertiary boundary using paleomagnetic secular variation: 2. Constraints from analysis of eight new sections and synthesis for a 3500-m-thick composite section," Journal of Geophysical Research B: Solid Earth, vol. 114, no. 6, pp. 1–38, 2009. ·
6. M. Moulin, F. Fluteau, V. Courtillot et al., "An attempt to constrain the age, duration, and eruptive history of the Karoo flood basalt: Naude's Nek section (South Africa)," Journal of Geophysical Research, vol. 116, no. 7, Article ID B07403, 2011.
7. V. Courtillot and F. Fluteau, "A review of the embedded time scales of flood basalt volcanism with special emphasis on dramatically short magmatic pulses," Geological Society of America Special Papers, vol. 505, Article ID 301317, pp. 301–317, 2014.
8. G. Keller and A. C. Kerr, Volcanism, Impacts, and Mass Extinctions: Causes and Effects, Geological Society of America Special Publication no. 505, 2014.
9. M. T. Jones, D. A. Jerram, H. H. Svensen, and C. Grove, "The effects of large igneous provinces on the global carbon and sulphur cycles," Palaeogeography, Palaeoclimatology, Palaeoecology, vol. 441, part 1, pp. 4–21, 2016. ·
10. D. L. Royer, R. A. Berner, I. P. Montañez, N. J. Tabor, and D. J. Beerling, "$CO_2$ as a primary driver of Phanerozoic climate," GSA Today, vol. 14, no. 3, pp. 4–10, 2004. ·
11. G. Le Hir, Y. Donnadieu, Y. Goddéris, B. Meyer-Berthaud, G. Ramstein, and R. C. Blakey, "The climate change caused by the land plant invasion in the Devonian," Earth and Planetary Science Letters, vol. 310, no. 3-4, pp. 203–212, 2011.
12. Y. Sun, M. M. Joachimski, P. B. Wignall et al., "Lethally hot temperatures during the early triassic greenhouse," Science, vol. 338, no. 6105, pp. 366–370, 2012. ·
13. J. Jack Sepkoski, "Patterns of phanerozoic extinction: a perspective from global data bases," in Global Events and Event Stratigraphy in the Phanerozoic, O. H. Walliser, Ed., pp. 35–51, Springer, Berlin, Germany, 1986. ·
14. K. J. Niklas, B. H. Tiffney, and A. H. Knoll, "Patterns in vascular land plant diversification," Nature, vol. 303, no. 5918, pp. 614–616, 1983.
15. D. Silvestro, N. Salamin, and J. Schnitzler, "PyRate: a new program to estimate speciation and extinction rates from incomplete fossil data," Methods in Ecology and Evolution, vol. 5, no. 10, pp. 1126–1131, 2014. ·
16. D. Silvestro, J. Schnitzler, L. H. Liow, A. Antonelli, and N. Salamin, "Bayesian estimation of speciation and extinction from incomplete fossil occurrence data," Systematic Biology, vol. 63, no. 3, pp. 349–367, 2014.
17. D. Silvestro, B. Cascales-Miňana, C. D. Bacon, and A. Antonelli, "Revisiting the origin and diversification of vascular plants through a comprehensive Bayesian analysis of the fossil record," New Phytologist, vol. 207, no. 2, pp. 425–436, 2015. ·

18. P. Steemans, A. Le Hérissé, J. Melvin et al., "Origin and radiation of the earliest vascular land plants,"Science, vol. 324, no. 5925, p. 353, 2009. ·

19. C. J. Yapp and H. Poths, "Ancient atmospheric $CO_2$ pressures interred from natural goethites," Nature, vol. 355, no. 6358, pp. 342–344, 1992. ·

20. T. J. Algeo and S. E. Scheckler, "Terrestrial-marine teleconnections in the devonian: links between the evolution of land plants, weathering processes, and marine anoxic events," Philosophical Transactions of the Royal Society B, vol. 353, no. 1365, pp. 113–130, 1998.

21. J. Ricci, X. Quidelleur, V. Pavlov, S. Orlov, A. Shatsillo, and V. Courtillot, "New $^{40}Ar/^{39}Ar$ and K-Ar ages of the Viluy traps (Eastern Siberia): further evidence for a relationship with the Frasnian-Famennian mass extinction," Palaeogeography, Palaeoclimatology, Palaeoecology, vol. 386, pp. 531–540, 2013. · ·

22. V. A. Kravchinsky, "Paleozoic large igneous provinces of Northern Eurasia: correlation with mass extinction events," Global and Planetary Change, vol. 86-87, pp. 31–36, 2012.

23. R. A. Berner and Z. Kothavala, "Geocarb III: a revised model of atmospheric $CO_2$ over phanerozoic time," American Journal of Science, vol. 301, no. 2, pp. 182–204, 2001. ·

24. C. R. Fielding, T. D. Frank, J. L. Isbell, and C. R. Fielding, "The late Paleozoic ice age—a review of current understanding and synthesis of global climate patterns," Geological Society of America Special Paper, vol. 441, pp. 343–354, 2008. ·

25. M. J. Benton and A. J. Newell, "Impacts of global warming on Permo-Triassic terrestrial ecosystems,"Gondwana Research, vol. 25, no. 4, pp. 1308–1337, 2014. · ·

26. J. M. Beaulieu, B. C. O'Meara, P. Crane, and M. J. Donoghue, "Heterogeneous rates of molecular evolution and diversification could explain the Triassic age estimate for angiosperms," Systematic Biology, vol. 64, no. 5, pp. 869–878, 2015. ·

27. J. A. Doyle, "Molecular and fossil evidence on the origin of angiosperms," Annual Review of Earth and Planetary Sciences, vol. 40, pp. 301–326, 2012.

28. S. A. E. Brown, A. C. Scott, I. J. Glasspool, and M. E. Collinson, "Cretaceous wildfires and their impact on the Earth system," Cretaceous Research, vol. 36, pp. 162–190, 2012. · ·

29. G. Keller and A. C. Kerr, "Foreword," in Volcanism, Impacts, and Mass Extinctions: Causes and Effects, Geological Society of America Special Paper 505, pp. 5–9, Geological Society of America, Boulder, Colo, USA, 2015. ·

30. M. E. Katz, B. S. Cramer, J. R. Toggweiler et al., "Impact of antarctic circumpolar current development on late paleogene ocean structure," Science, vol. 332, no. 6033, pp. 1076–1079, 2011. ··

CHAPTER 3

# REVISITING THE ORIGIN AND DIVERSIFICATION OF VASCULAR PLANTS THROUGH A COMPREHENSIVE BAYESIAN ANALYSIS OF THE FOSSIL RECORD

Daniele Silvestro, Borja Cascales-Miñana, Christine D. Bacon, Alexandre Antonelli

**Keywords:** Biodiversity changes, diversification, floristic turnover, mass extinction, plant fossils, PyRate

## INTRODUCTION

Understanding the tempo and mode of species diversification – that is, the interplay between speciation and extinction – has engaged naturalists for centuries (e.g. Darwin, **1859**). In particular, the evolution of vascular plants (tracheophytes) still remains an area of active research (e.g. Boulter *et al.*, **1988**; Raubeson & Jansen, **1992**; Kenrick & Crane, **1997**; Pryer *et al.*, **2001**). In the pioneering work by Niklas *et al.* (**1983**) four time periods were identified as phases of increased land plant diversification: the Devonian, Carboniferous–Permian, Triassic–Jurassic and Early Cretaceous. Land plants also went through four of the five major extinction events (the 'Big Five'; Jablonski, **2005**) identified from marine organisms at the Frasnian–Famennian, the Permian–Triassic, the Triassic–Jurassic and the Cretaceous–Paleogene boundaries (Raup & Sepkoski, **1982**).

The four phases of plant radiation described by Niklas *et al.* (**1983**) began with the Silurian–Devonian diversification and expansion of early land plants (embryophytes) (Edwards & Wellman, **2001**; Edwards & Richardson, **2004**; Wellman, **2014**) and the evolution of early forests (Meyer-Berthaud *et al.*, **2010**). The second phase was characterized by the dominance of lycophytes and Carboniferous tropical wetlands (Cleal, **2008a,b,c**; Cleal *et al.*, **2012**) together with the diversification of early 'gymnosperms' (a nonmonophyletic group including several lineages, e.g. Cordaitales) and seed ferns (Anderson *et al.*, **2007**). The third phase encompasses the recovery from the Late Permian mass extinction event (Niklas, **1997**; Rees, **2002**) and the establishment of 'gymnosperms' as the dominant plant group (Anderson *et al.*, **2007**; Bonis & Kürschner, **2012**). The last phase (Early Cretaceous) witnessed the radiation of angiosperms, which are today the world's most dominant plants in species richness and ecological differentiation (Crepet & Niklas, **2009**; Friis*et al.*, **2011**). The broad patterns of vascular plant history can also be explained in terms of turnover of five major floras, also known as the Rhyniophytic (= Eotrachyophytic), Eophytic, Paleophytic, Mesophytic and Cenophytic Floras (Cleal & Cascales-Miñana, **2014**). Following this scheme, the first two represent the early evolutionary trends of plant life, whereas the Paleophytic, Mesophytic and Cenophytic Floras represent the heyday of spore-bearing plants (i.e. lycophytes and ferns), 'gymnosperms' and angiosperms, respectively (Cleal & Cascales-Miñana, **2014**).

Except for the origin of angiosperms, the majority of biological events that have characterized plant diversification are rooted in the Paleozoic. Some of the most crucial events include the invasion of land by early plants (the 'terrestrialization process'; Kenrick & Crane, **1997**; Vecoli *et al.*, **2010**), the heyday of Carboniferous tropical wetlands (Cleal *et al.*, **2011**), the reduction of arborescent lycophyte forests in the Pennsylvanian (Dimitrova *et al.*, **2010**; Cleal *et al.*, **2012**), and the major extinction of plant diversity at the Permian–Triassic boundary (Rees, **2002**; McElwain & Punyasena, **2010**; Cascales-Miñana & Cleal, **2014**). Considerable advances have

**Citation:** Daniele Silvestro, Borja Cascales-Miñana, Christine D. Bacon, Alexandre Antonelli, Revisiting the origin and diversification of vascular plants through a comprehensive Bayesian analysis of the fossil record, doi: 10.1111/nph.13247.

been made recently in our understanding of these key events, and their impact on past ecosystem dynamics. For instance, the fossil record from South China suggests that plant diversity dynamics during the Permian–Triassic were characterized by a gradual floral turnover, rather than a mass extinction like those in the coeval marine faunas (Xiong & Wang, **2011**). Furthermore, fossil evidence suggests that diversity dynamics of land plants were considerably different from those of marine faunas and land vertebrates during the Devonian, but primary diversity changes affected all terrestrial organisms synchronously at the Devonian–Carboniferous boundary (Xiong *et al.*, **2013** and references therein). Despite these recent advances, few studies have attempted a quantitative assessment of the origin and diversification dynamics of vascular plants following uniform criteria regarding taxonomic, temporal and spatial scales. This lack of formal, data-driven approaches calls for a re-evaluation of the unequivocal events of increased plant origination and extinction, which are supported by statistical significance. Ideally, such analysis should be done at the finest reliable taxonomic scale, under a global perspective, and using recent advances in paleobotany, as well as the most sophisticated and statistically robust analytical methods.

## Estimating the Origin and Diversification of Plant Groups Using the Fossil Record

The plant fossil record has been primarily explored using range-based counts (Cleal, **1993**; Anderson *et al.*, **2007**), rather than occurrence-based data. Age ranges include the first and last appearance of a taxon, whereas taxonomic occurrence data include finer detail of fossil individuals, such as abundance and geographic distribution. Range-based approaches are commonly applied with standard datasets above the generic level to infer biodiversity and diversification trends across broad time intervals (e.g. for Phanerozoic trends; Sepkoski *et al.*, **1981**). By contrast, targeting finer-scale diversification events (e.g. the Late Permian biotic crisis and its corresponding recovery period) requires the analytical implementation of sampling parameters from fossil occurrence data (e.g. Alroy *et al.*, **2001**). Both approaches to exploring the fossil record are valid, but the use of occurrence data opens the door to a battery of methods to analyze data and test hypotheses in a more robust statistical framework.

A new method, PyRate, was developed to infer the rates of speciation (or origination for higher taxonomic ranks) and extinction and their variation through time, within a Bayesian probabilistic framework (Silvestro *et al.*, **2014b**). The method is capable of handling fossil occurrence data (i.e. not only first and last appearances) and dealing with the inherent incompleteness of the fossil record. PyRate simultaneously estimates rates of origination, extinction and preservation, as well as the lifespan of each taxa in the dataset. The method thus takes into account the uncertainties associated with the fossil preservation process.

A major goal in plant evolutionary research is to determine the temporal origin of major groups. Usually the oldest known fossil evidence of a given lineage is assumed to closely correspond to its true time of origin. However, this can be controversial due to the lack of consensus on the most reliable type of fossil (e.g. pollen data vs fossil leaves; Kenrick *et al.*, **2012**), or because of differences in the results obtained from fossil data and from dated molecular phylogenies (Doyle & Endress, **2010**). In fact, dated phylogenies of extant taxa rely heavily on the selection of fossil calibrations (Sauquet *et al.*, **2012**) and the associated prior distributions (Ho & Phillips, **2009**; Heath, **2012**). PyRate is able to produce statistically robust estimates of origination times, because it uses the complete fossil record (i.e. not only the oldest fossil) and incorporates a preservation process. Therefore, it can be used to reduce the potential biases deriving from arbitrarily defined prior distributions.

Here, we characterize the main radiations and biotic crises of plants within the Bayesian framework implemented in PyRate. From this, we (1) infer the fluctuations of origination and extinction rates through time for vascular plants (as a whole as well as subdivided into three main vegetation groups), (2) evaluate the effect of the largest widely recognized mass extinction events on both origination and extinction rates, and (3) derive the age of origin of vascular plants, seed plants and angiosperms, providing posterior distribution curves for their ages rather than point estimates, which can be directly used as prior calibrations in molecular dating analyses.

# MATERIALS AND METHODS

## Data Compilation

This study is based on plant fossil occurrences downloaded from the Paleobiology Database (PBDB; http://paleobiodb.org/cgi-bin/bridge.pl) on 6 March 2014. We performed multiple search queries in order to obtain the most comprehensive dataset of the plant fossils, on a global level. The approach that captured the maximum number of records involved querying the database for all described plant classes and orders individually. With this approach we obtained 35 666 entries, which spanned most of the Phanerozoic (the oldest occurrences dating back to *c.* 424 million yr ago (Ma)). We included all occurrences that were identified to the species level as well as those identified only at the genus level but without specific assignation (e.g. '*Psilophyton* sp.'). We did not revise fossil identifications provided by the PBDB. Several culling procedures are normally used to avoid including incompletely identified taxa, such as searching and excluding entries identified with aff., cf., sp. or other modifications to binomial species names (Janevski & Baumiller, **2009**). We initially explored the effect of excluding such records from the analyses, but because we found no differences in genus-level diversity patterns we opted for using all available records. The dataset thus includes records from all lithologies, environments and continents that were freely available from the PBDB, including their estimated ages in millions of years (Myr). Raw data are available as supplementary materials.

It is well known that the plant fossil record is highly fragmentary, not least for plants (Cleal & Thomas, **2010**). This is further complicated by the incomplete nature of fossilized plant organs, varying quality of preservation, and our incomplete knowledge on morphological evolution. These factors lead to difficulties in confidently assessing taxonomic affinity, and the existence of many fossils of uncertain taxonomic placement or '*incertae sedis*'. To minimize such problems we analyzed occurrence data at the genus level assuming that this level best captures the long-term and global-level dynamics of plant diversification (Rees, **2002**; Wang *et al.*, **2010**; Xiong & Wang, **2011**; Xing *et al.*, **2014**). An additional problem in the plant fossil record is that a single genus may be represented in several 'fossil-genera' described from different plant remains or organs (e.g. in arborescent lycophytes; Cleal *et al.*, **2012**). This might have the effect of overestimating plant diversity from some time intervals (e.g. the Carboniferous). Although our dataset includes several different types of plant remains (e.g. leaves, fertile and sterile axes, seed, fruits and roots), leaf samples represent > 70% of all the occurrences. Therefore, we considered the discrepancies between genera and fossil-genera as unlikely to alter the major fluctuations of plant diversity.

There are several practical reasons for avoiding species-level analyses. First, excluding all fossils without assignation to the species level would reduce the full dataset by 47%. Second, > 50% of the species we compiled are known from a single occurrence, which makes it difficult to estimate accurate preservation rates regardless of the methodological approach used. Lastly, species-level data should arguably be more sensitive to floristic variation due to regional endemism and fossil preservation biases. The interpretation and analytical incorporation of regional variations of species diversity is often problematic, as it could either derive from true biodiversity differences or reflect the fact that regional morphological differences are often described as new species in the absence of more thorough evidence. Thus, we consider that genus-level analyses constitute a robust way of exploring the diversification dynamics of the plant fossil record.

Plant macrofossils (e.g. whole flowers, fruits, seeds, leaves) are commonly assumed to be the most reliable source of data to study plant evolution, because they usually contain many taxonomically useful characters and can therefore be more easily and reliably assigned to different plant groups (Wang *et al.*, **2010**). By contrast, microfossils (pollen and spores) often lack a sufficient amount of diagnostic synapomorphies that facilitate taxonomic identification at the generic level. We therefore subsampled our original dataset to include only macrofossils (i.e. > 60% of the original dataset) in the analyses.

Analyses of diversification should ideally be performed on monophyletic groups, that is, clades in a phylogenetic sense. However, although monophyly assessment is straightforward in phylogenetic analyses based on molecular data, many questions remain on the phylogenetic placement of several fossil taxa. In addition, restricting the analyses to clades would require splitting the already fragmentary fossil data into smaller subsets, reducing analytical power and precision. We therefore categorized vascular plant fossil-genera into three major groups, distinguished by their functional ecology, morphology and relationships (following Kenrick & Crane, **1997**; Tomescu, **2009**; Friis*et al.*, **2011**; Hao & Xue, **2013**): (1) spore-bearing plants (i.e. early tracheophytes, rhyniophytes, lycophytes, sphenophytes, early euphyllophytes, progymnosperms, ferns and related plants, but excluding bryophytes); (2) nonflowering seed plants (i.e. 'gymnosperms' including Cycadidae, Ginkgoidae, Gnetidae, Pinidae, Bennettitales, *Caytonia* and seed ferns); and (3) flowering seed plants (i.e. angiosperms). Fossil occurrences were assigned to the different categories based on the supra-generic taxonomic information provided by the PBDB and, when available, using the *Index Nominum Genericorum* (Smithsonian Institution, http://botany.si.edu/ing/). We matched the names of fossil-genera in our dataset with those listed at *The Plant List* (http://www.theplantlist.org/) to identify which taxa are extant.

The final dataset of macrofossils included 22 415 occurrences assigned to 443 genera, of which 263 are extinct and 180 are extant. The fossil occurrences were evenly distributed across the three plant groups described above: spore-bearing plants (34%), nonflowering seed plants (34%) and angiosperms (32%). Angiosperms encompassed 58% of the diversity (with 257 genera) whereas the remaining genera were evenly distributed among the other two plant groups. The large majority of fossil occurrences included temporal ranges (minimum and maximum ages), which typically derive from the upper and lower temporal boundaries of the stratigraphic units to which fossils were assigned. In our dataset, temporal ranges spanned on average 12.7 Myr with a standard deviation of 12.6 Myr. In order to account for these dating uncertainties in the analyses of origination and extinction rates we considered the temporal ranges as uniform distributions and, from these, we randomly resampled the ages of each fossil occurrence following Silvestro *et al.* (**2014b**). We generated 100 datasets using this randomization approach and ran the analyses on all replicates.

## Diversification Rate Analyses

We carried out the analyses of plant diversification using a hierarchical Bayesian model to infer the temporal dynamics of origination and extinction (Silvestro *et al.*, **2014b**. This approach, implemented in the program PyRate (http://sourceforge.net/projects/pyrate/; Silvestro *et al.*, **2014a**), uses as input data all fossil occurrences that can be assigned to a taxon, in this case fossil-genera, to jointly model the preservation and diversification processes. The preservation process infers the individual origination and extinction times of each taxon based on all fossil occurrences and on an estimated preservation rate (expressed as expected occurrences per taxon per Myr). This process is important because the first and last appearances of a taxon are likely to underestimate the true extent of its lifespan (Liow & Stenseth, **2007**). The origination and extinction rates are inferred from the estimated lifespans of taxa based on a birth–death process, which models the diversification dynamics underlying diversity changes and constitutes the prior over the origination and extinction times (Silvestro *et al.*, **2014b**). In summary, the main parameters jointly estimated by this approach are: preservation rate and its degree of heterogeneity; origination and extinction times; and origination and extinction rates, which can vary through time. All of these parameters are sampled from their posterior distributions using a Markov Chain Monte Carlo (MCMC) algorithm, yielding parameter estimates and credible intervals, here calculated as 95% highest posterior density (HPD) intervals.

Compared to the original model described by Silvestro *et al.* (**2014a,b**), we introduced some modifications that better comply with the type of data and the aims of this study, which is mainly focused on variation in origination and extinction at global scale and large temporal ranges. We used here an homogeneous Poisson process (HPP) of preservation instead of the original 'hat'-shaped nonhomogeneous process

(NHPP; Liow *et al.*, **2010**; Silvestro *et al.*, **2014b**). Although we believe the NHPP robustly describes fossil abundance in well-preserved marine or mammalian species, we consider the HPP as a more appropriate assumption in the case of genus-level plant fossil data due to potentially sparse sampling. We also accounted for varying preservation rates across taxa using the Gamma model, that is, with gamma-distributed rate heterogeneity (Silvestro*et al.*, **2014b**). Because of the large number of occurrences analyzed and of the vast timescale considered, we used here eight rate categories to discretize the gamma distribution, instead of the default four, to accommodate the potential for more variability of preservation rates across taxa.

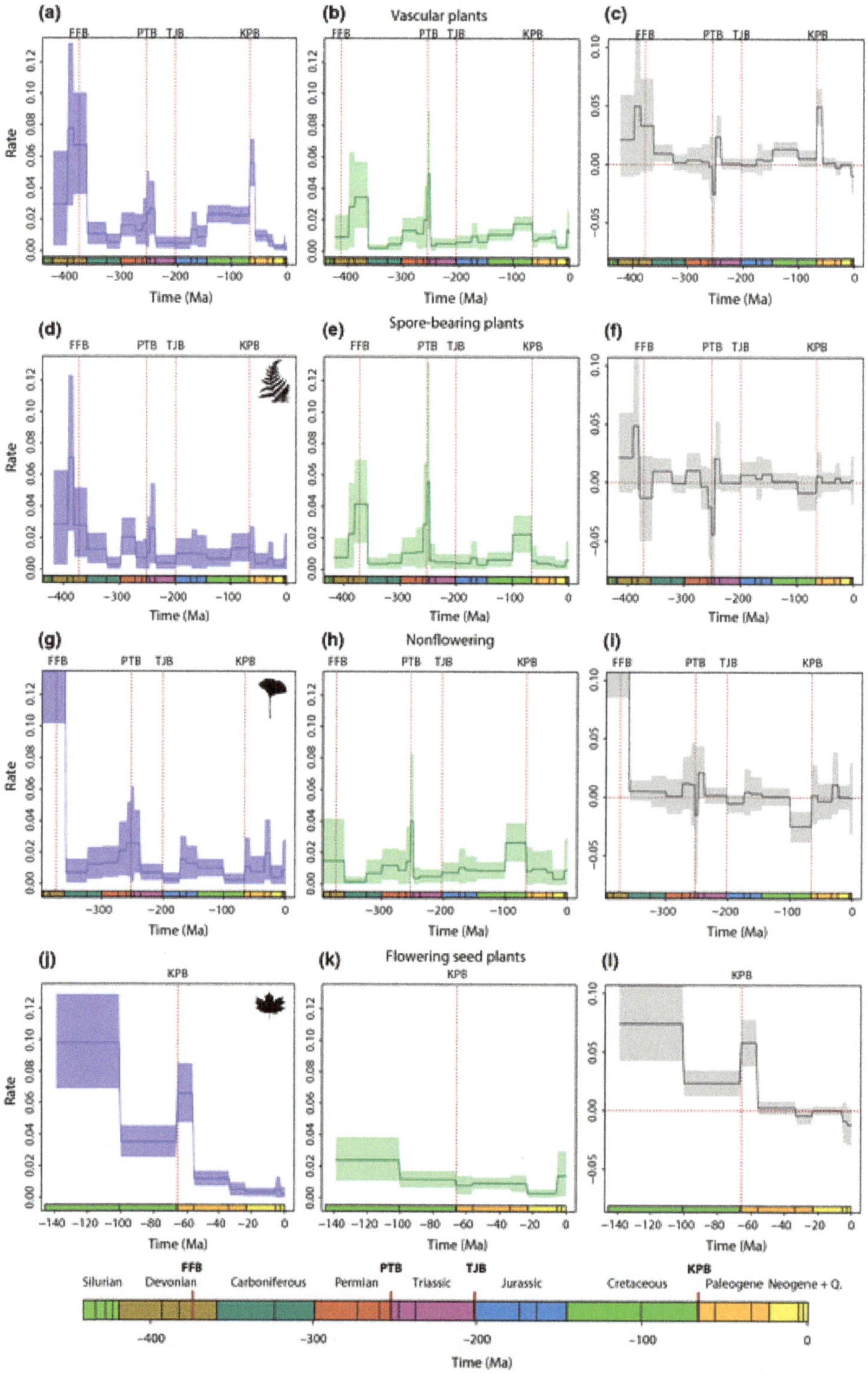


**Figure 1.** Generic-level diversification analysis of all vascular plants. Origination (blue) and extinction (green) rates were estimated using the Bayesian approach implemented in PyRate (Silvestro *et al.*, **2014a**,**b**) within time bins defined as epochs of the geologic timescale (shown at the bottom). The timescale (*x*-axis) is given in million years ago (Ma). Solid lines indicate mean posterior rates, whereas the shaded areas show 95% HPD intervals. The diversification dynamics were estimated for vascular plants as a whole (a–c), spore-bearing plants (d–f), nonflowering seed plants (g–i) and flowering seed plants (angiosperms; j–l). Net diversification rates (gray) are defined as origination minus extinction. The dashed lines indicate the major mass extinction events widely recognized: at the Frasnian–Famennian (FFB), Permian–Triassic (PTB), Triassic–Jurassic (TJB) and Cretaceous–Paleogene (KPB) boundaries. Plant silhouettes were obtained fromhttp://phylopic.org.

Furthermore, we broke down the birth–death process into segments of time, defined by the epochs of the stratigraphic geological timescale (Cohen *et al.*, **2013**) and estimated origination and extinction rates within these intervals. We adopted this solution as an alternative to the algorithms implemented in the original PyRate release that jointly estimate the number of rate shifts and the times at which origination and extinction undergo a shift. The estimation of origination and extinction rates within fixed time intervals improved the mixing of the MCMC and allowed us to obtain an overview of the general trends of rate variation throughout a long timescale, that is, almost the entire the Phanerozoic, within reasonable computational effort (a few days on a computer cluster). We emphasize that both the preservation and the birth–death processes are still modeled in continuous time and are not based on boundary crossings. Thus, the origination and extinction rates are measured as the expected number of origination and extinction events per lineage per Myr. One potential problem in fixing *a priori* the number of rate shifts is overparameterization. We overcame this issue by assuming that the rates of origination and extinction are part of two families of parameters following a common prior distribution, with parameters estimated from the data using hyper-priors (Gelman, **2004**). Thus, we used Cauchy prior distributions centered at 0 and with scale (hyper-) parameters estimated from the data in the MCMC. The use of hyper-priors is a convenient way to reduce the subjectivity in defining the priors and to reduce the effective parameter space, thus limiting the risks of overparameterization (Gelman, **2004**).

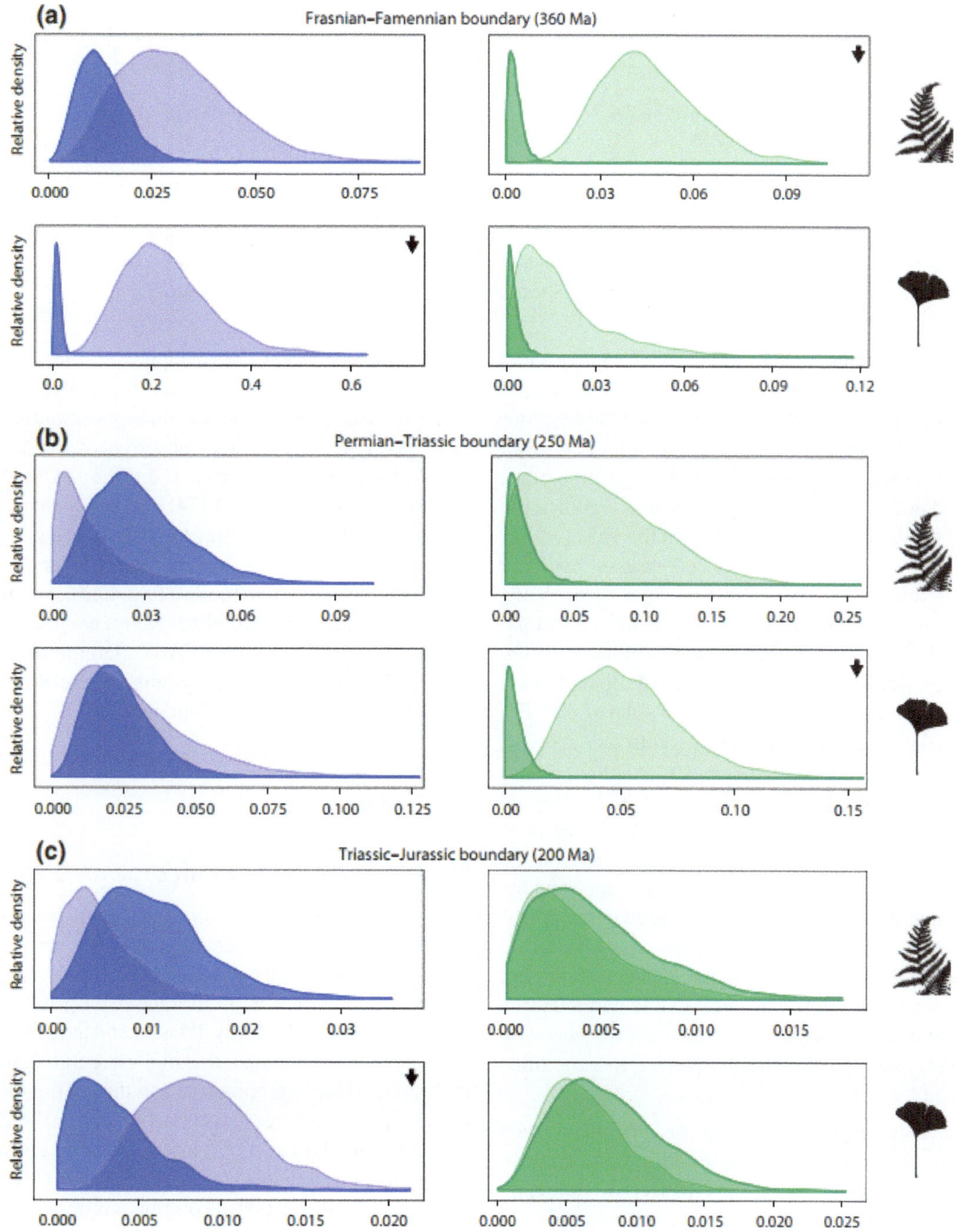

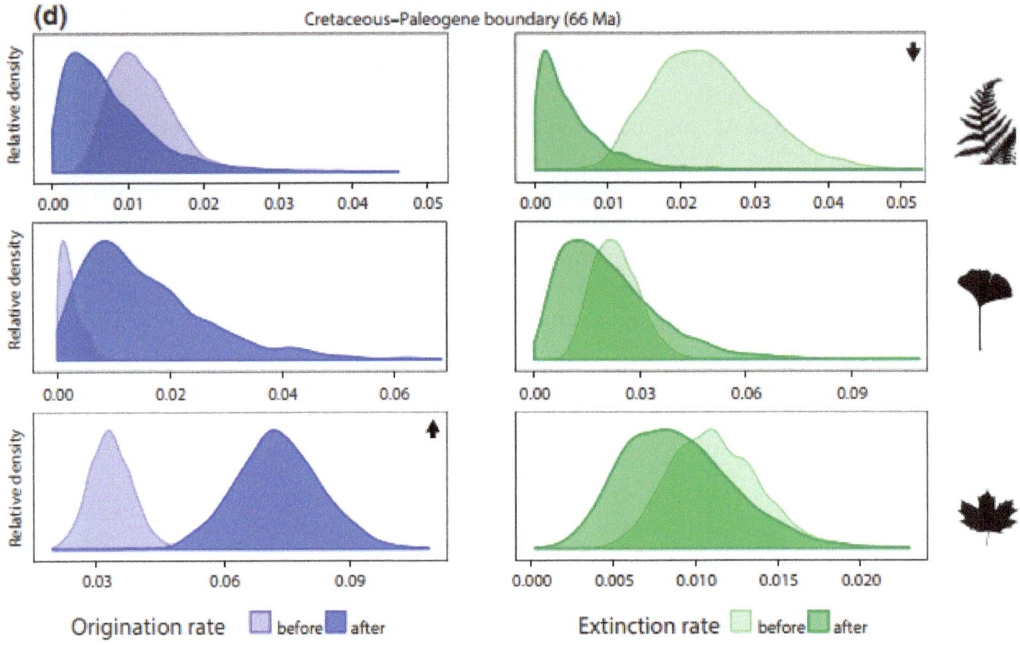


**Figure 2.** Origination and extinction rates of plants across major events of global biodiversity turnover, the Frasnian–Famennian (a), Permian–Triassic (b), Triassic–Jurassic (c) and Cretaceous–Paleogene (d) boundaries. The time of the events is provided in million years ago (Ma). These plots show the sampled posterior distributions of origination (blue) and extinction (green) rates that were estimated before and after each event. Posterior distributions were standardized to relative densities for graphical purposes. Arrows on the top-right corner of panels indicate significant rate increases (pointing up) or decreases (pointing down) based on the 95% HPD of the two distributions.

We ran PyRate for 5000 000 MCMC generations on each of the 100 randomly replicated datasets. After excluding the first 20% of the samples as burnin phase, we combined the posterior estimates of the origination and extinction rates across all replicates and used them to generate rates-through-time plots. Rates of two adjacent intervals were considered significantly different when the mean of one lay outside of the 95% HPD of the other, and vice versa. We looked at the marginal posterior distributions of origination and extinction rates through the largest extinction events documented in geological history, the so-called 'Big Five' mass extinction events in life's history (Jablonski, **2005**; McElwain & Punyasena, **2010**). In particular, we examined the diversification dynamics at four of the 'Big Five' events (the Frasnian–Famennian, Permian–Triassic, Triassic–Jurassic and Cretaceous–Paleogene boundaries; see the 'Introduction' section; Figs 1, 2). The Ordovician–Silurian extinction event occurred before the appearance of vascular plants and was therefore not considered. We focused on the magnitude of rate changes, their statistical significance and the uncertainty around those estimates. We measured extinction as the rate of disappearance of fossil-genera, which are described based on morphology and might not always represent phylogenetic lineages. Thus, it was not possible with our data to discern potential events of pseudo-extinction (i.e. when two fossil-genera describe a single lineage evolving to a modified form).

We estimated the origination times of vascular plants, seed plants and flowering plants from the posterior distributions of the origination times of the groups analyzed, following Silvestro *et al.* (**2014b**). Based on the posterior samples of the ages of origin derived from all replicates, we identified the best-fitting distribution by testing several distributions (uniform, normal, log-normal, gamma and Laplace) under maximum likelihood. We emphasize that these estimates are derived after explicitly modeling the fossil preservation process, and accounting for the uncertainties around the estimated ages of the fossil occurrences. After choosing the best-fitting distribution based on the Akaike Information Criterion (AIC; Akaike, **1974**), we computed the estimated parameter values, which may be directly used for other studies as calibrations to date molecular phylogenetic trees using relaxed molecular clock models (e.g. Lepage *et al.*, **2007**). The estimated parameters and offset values are given following the parameterization used in BEAST (Drummond*et al.*, **2012**), one of the most popular programs for molecular dating. We implemented these fitting functions in the latest version of PyRate.

## RESULTS

### Diversification Dynamics

Our results show that the diversification history of vascular plants as a whole was characterized by several changes in global origination and extinction rates through time (Fig. 1a–c). The origination rates showed three main peaks, at the Middle Devonian, during the Early Triassic and at the early Paleogene (Fig. 1a). We estimated the highest extinction rates at the Paleozoic–Mesozoic boundary and during the Late Devonian. Overall, plants suffered less extinction during the Mesozoic than the Paleozoic; however, we inferred a progressive increase in extinction rates towards the Late Cretaceous (Fig. 1b). In terms of net diversification rates (origination minus extinction), we inferred strongly negative diversification at the earliest Triassic followed by a peak of positive rates in the Middle Triassic (Fig. 1c). The highest diversification values were detected in the Late Devonian and early Cenozoic. Diversification rates declined to negative values towards the present day.

Results of spore-bearing plant diversification history (Fig. 1d–f) revealed three main episodes of origination at the Middle Devonian, the Early Permian and the Early–Middle Triassic. Origination rates declined progressively through the Late Devonian and Carboniferous. During the Mesozoic and Cenozoic, spore-bearing plant diversity witnessed quite stable levels of origination with slightly higher rates in the Late Cretaceous (Fig. 1d). Spore-bearing plants experienced phases of high extinction during the Late Devonian, Late Permian and Late Cretaceous (Fig. 1f). We detected a progressive increase in extinction through the Carboniferous and Permian, and the highest value was inferred at the Paleozoic–Mesozoic boundary (Fig. 1e). Consequently, the main fluctuations in net diversification rates were estimated in the Late Devonian and during the Permian–Triassic and Cretaceous–Paleogene transitions (Fig. 1f).

The diversification dynamics of nonflowering seed plants (Fig. 1g–i) showed very high origination rates during the Devonian, followed by a drastic drop at the Mississippian (Early Cretaceous). We then observed high turnover with a progressive increase in both origination and extinction values culminating at the late Paleozoic (Fig. 1g,h). Another origination peak was at the middle Jurassic (Fig. 1g), whereas a second maximum of extinction was detected at the Late Cretaceous (Fig. 1h). This was followed by diversity recovery during the Cenozoic with diversification rates increasing to non-negative values (Fig. 1i).

Angiosperm diversification (Fig. 1j–l) was characterized by very high initial origination rates, progressively decreasing towards the present (Fig. 1j). The clade diversified under a high origination and comparatively low extinction rate, contributing to significantly positive net diversification lasting from its origin to the Paleocene. We found a maximum value of origination in the Early Cretaceous (Fig. 1j) and a second origination peak during the early Paleogene. Overall, extinction rates appear quite low and stable through time (Fig. 1k). Angiosperm net diversification rates show negative values at the Paleogene and at more recent times (Fig. 1l).

## Shifts in origination and extinction rates

We found a significant decrease (> 20-fold) in the extinction rate of spore-bearing plants following the Frasnian–Famennian boundary, which did not take place in similar magnitude among seed plants (Fig. 2a). Although for both spore-bearing plants and nonflowering seed plants extinction rates strongly decreased (> 10-fold) after the Permian–Triassic boundary, such decrease was found to be significant only in nonflowering seed plants (Fig. 2b), potentially as a result of a high degree of uncertainty around the posterior rate estimates (as reflected by the relatively large shaded area under their curves). Across the Triassic–Jurassic boundary we did not detect any notable changes of extinction rates in spore-bearing and seed plants, but the origination rates in the latter group experienced a significant 3.6-fold decrease (Fig. 2c). This suggests that the Frasnian–Famennian boundary event had a stronger impact on the extinction of spore-bearing plants than on the extinction of other plant groups, whereas the Permian–Triassic boundary event mostly affected the extinction of seed plants. Likewise, the Frasnian–Famennian and the Triassic–Jurassic boundaries are preceded by periods of high origination, mainly for seed plant diversity. Interestingly, the transition between the Mesozoic and the Cenozoic (the Cretaceous–Paleogene boundary) involved significant changes in extinction rates only in spore-bearing plants, in which the extinction rates decreased 10-fold (Fig. 2d). The Cretaceous–Paleogene boundary was followed by a significant (2.2-fold) increase in origination rates in angiosperms (Fig. 2e).

## Origin of Major Plant Clades

The estimated times of origin of vascular plants, seed plants and angiosperms were best fitted by log-normal, gamma and Laplace distributions, respectively (Fig. 3;). Our results indicate that the diversification of vascular plants started during pre- or Early Silurian times (mean = 433.5; 95% HPD 449.0–424.0 Ma; Fig. 3a). The origins of seed plants and angiosperms were estimated to the Middle Devonian (mean = 373.8, 382.3–367.2 Ma; Fig. 3b) and Late Jurassic–Early Cretaceous (mean = 143.8, 95% HPD: 151.8–133.0 Ma; Fig. 3c), respectively. These ages predate those of the oldest fossil occurrences included in this study for each group, because they reflect the use of all available occurrence data to model the preservation process while incorporating the uncertainties around the fossil ages.

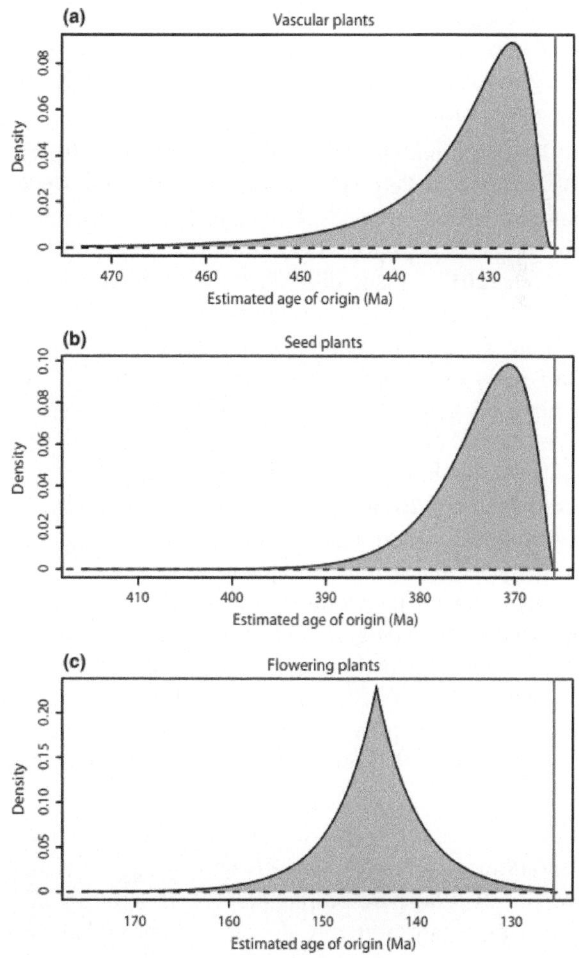


**Figure 3.** Estimated times of origin (crown ages) for vascular plants, seed plants and angiosperms. The distributions were obtained from modeling the preservation process in PyRate and after combining 100 randomization analyses to incorporate the uncertainties around the ages of the fossil occurrences (see the 'Materials and Methods' sec-

tion). The vertical red lines indicate the age of the oldest occurrence included in this study for each group. The time of origin of vascular plants (a) was best fitted by a log-normal distribution with log mean = 2.06, log standard deviation = 0.76, offset = 423.22 million yr ago (Ma; estimated mean age: 433.5). For seed plants (b) the best fit was given by a gamma distribution with shape = 2.44, scale = 3.19, offset: 366 Ma (estimated mean age: 373.8). The time of origin of flowering plants (c) was best modeled by a Laplace distribution with mean = 144.26 Ma, scale 4.36, truncated at 125.65 Ma (estimated mean age: 143.7). These distributions can be directly implemented as calibration priors for dating the phylogenies of these clades using Bayesian relaxed molecular clocks. The parameterization provided here follows that implemented in the popular molecular dating program BEAST (Drummond*et al.*, **2012**).

## DISCUSSION

### The Dynamics of Plant Diversification

The balance between origination and extinction is the main factor driving biodiversity dynamics through time (Sepkoski, **1978**, **1979**; Sepkoski *et al.*, **1981**), and the only one to be considered in global-level analyses (migration may play a role at regional scales). This relationship is thus critical to understanding current patterns of plant diversity. Our analyses have shown the main trends of origination and extinction of plant evolutionary history based on global fossil occurrences. From these results, at least two noteworthy observations arise: major mass extinction events had differential impact among plant groups; and strongly negative rates of plant diversification were never sustained through longer time intervals.

There is strong evidence of high extinction rates in terrestrial ecosystems at the Permian–Triassic and Cretaceous–Paleogene boundaries from tetrapods, birds and insects (Benton, **1985**; Labandeira & Sepkoski, **1993**; Longrich *et al.*, **2011**, **2012**; Fröbisch, **2013**). Plant diversification, however, appears to have followed different dynamics. Our results show that the Late Cretaceous biotic crisis, one of the best studied of all mass extinction events, had less effect than expected on plant diversity, despite a supposedly great impact on the configuration of terrestrial habitats due to drastic changes in the world's biota (Krug *et al.*, **2009**; Schulte *et al.*, **2010**; Meredith *et al.*, **2011**). This result supports a recent family-level analysis of extinction rates for plants, which did not find evidence for significant rate variation at the Cretaceous–Paleogene boundary (Cascales-Miñana & Cleal, **2014**). This boundary led to ecological and functional replacement of life forms, including a total collapse of Paleophytic diversity and a sustained decline of a Mesophytic flora that was progressively replaced by Cenophytic vegetation during the Cretaceous (Cleal & Cascales-Miñana, **2014**).

Interestingly, our results show that the Late Cretaceous is characterized by the lowest net diversification rates in nonflowering seed plants (Fig. 1i), and by negative diversification in spore-bearing plants (Fig. 1f). In both cases, these patterns are primarily driven by high extinction rates, although in nonflowering seed plants this is further exacerbated by decreased origination rates. The significant increase in extinction rate observed in spore-bearing plants at the end-Cretaceous (Fig. 1e) is associated with a collapse of fern diversity from the Paleophytic flora (see Fig. 5 from Cleal & Cascales-Miñana, **2014**). By contrast, positive net diversification rates are reconstructed throughout the Cretaceous for flowering plants (Fig. 1l), which rapidly accumulated generic diversity since the Late Cretaceous (Xing *et al.*, **2014**). As a result, plant diversity has been dominated by angiosperms since the early Cenozoic, with some contributions from pteropsid ferns and 'gymnosperms', mainly pinopsids (Collinson, **1996**; Anderson *et al.*, **2007**; Friis *et al.*, **2011**).

The sudden increase of extinction rates for nonflowering seed plants in the Late Cretaceous (Fig. 1h) occurs synchronically with a major loss of seed fern diversity. As noted by McLoughlin*et al.* (**2008**), seed ferns were diverse and abundant during the Triassic–Jurassic time interval, and declined in the Cretaceous synchronically with the radiation of angiosperms. These suggestions are supported by our results (Fig. 1f,l). The almost complete demise of seed ferns (only the Corystospermales briefly survived the Cretaceous–Paleogene event; McLoughlin *et al.*, **2008**) and the loss of primitive forms of pinopsids (i.e. Cheirolepidiaceae and Voltziaceae) and ginkgoopsids (i.e. Umaltolepidaceae, Leptostrobaceae, Caytoniaceae; Anderson *et al.*, **2007**), unequivocally caused a major taxonomic shift in floristic diversity. This change, together with the environmental changes following the Chicxulub impact event (Schulte *et al.*, **2010**), appear to be tightly linked to the success of flowering plants, which immediately after the end-Cretaceous extinction event underwent a second burst of diversification (Figs 1l, 2d).

The Permian–Triassic mass extinction event is associated with a much more pronounced reduction in plant diversification than observed at the Cretaceous–Paleogene boundary. In fact, the estimated extinction rates following the Permian–Triassic boundary are the highest estimated for all plant groups surveyed (Fig. 1). These results provide further evidence that a substantial depletion of plant diversity took place at the Permian–Triassic boundary (McElwain & Punyasena, **2010**), in which, according to Rees (**2002**), > 50% of all plant species were exterminated. As pointed out by Retallack (**2013**), the Late Permian mass extinction involved the demise of glossopterids, gigantopterids, tree lycopsids and cordaites. This substantial loss of plant diversity was also recently documented at the family level (Cascales-Miñana & Cleal, **2014**). Moreover, our results show a progressive increase of extinction rates since the middle Carboniferous. This result supports the idea of a gradual floral reorganization, as suggested for South China by Xiong & Wang (**2011**).

In comparison with the Permian–Triassic and the Cretaceous–Paleogene crises, our results suggest that the other widely recognized mass extinction events (at the Frasnian–Famennian and Triassic–Jurassic boundary) had less detectable effects on plant diversity. However, although Cascales-Minana & Cleal (**2015**) found no significant increase in extinction rates at the Frasnian–Famennian boundary at the family level, here we have detected important changes at the generic level. The high extinction rates estimated for the Late Devonian (Fig. 1b) were paired with high origination rates (Fig. 1a), suggesting high rates of taxonomic turnover. Indeed, during this period we observe the disappearance of several spore-bearing plant groups representing early diverging lineages, such as Zosterophylls (Cascales-Miñana & Meyer-Berthaud, **2014**). This period is also characterized by the diversification of arborescent lycopsids, early seed ferns and ligniophytes

(Wang *et al.*, **2005**; Galtier & Meyer-Berthaud, **2006**; Meyer-Berthaud *et al.*, **2010**; Decombeix *et al.*, **2011**). The Late Triassic is well documented as the heyday of 'gymnosperms' (Anderson *et al.*, **2007**). However, our results indicate that their highest net diversification may have taken place slightly earlier, in the Middle Triassic (Fig. 1i), followed by significant changes of origination rates across the Triassic–Jurassic boundary (Fig. 2c).

## Origin of Major Plant Clades

The origin of land plants can be established from the Dapingian (middle Ordovician) based on the earliest known cryptospores, whereas the first sporangia is dated from the Katian (late Ordovician) and first robust evidence of macroflora is found at the Wenlock Series (i.e. *Cooksonia*; Wellman, **2014**). In comparison, our estimated interval for the origin of vascular plants ranges from 449 Ma (late Ordovician) to 424 Ma (Ludlow, late Silurian) (Fig. 3a). Despite the fact that our inferences are based exclusively on macrofossil evidence, this estimated interval falls within the expected range considering the entire fossil record (i.e. including microfossils) of early land plants. This is consistent with the idea that, especially for early land plants, the spore fossil record is more abundant and less biased than the plant megafossil record because: these early plants produced vast numbers of spores; spores are thought to have primarily accumulated in near-shore marine environments suitable for sediment preservation; and spores have high fossilization potential due to the fact that they contain sporopollenin, one of the hardest biological materials (Wellman *et al.*, **2013**). By contrast, the presence of rich assemblages of plant megafossils from early floras is linked mainly to terrestrial environments (Hao & Xue, **2013**), which represents a considerable deposition bias. The concordance between our estimates of the origin of vascular plants based on macrofossils and previous estimates based on microfossils highlights the importance of modeling the preservation process when assessing the age of clades (Silvestro *et al.*, **2014b**). Importantly, due to the delayed appearance of characteristic vascular plant traits in the earliest land plants, the dating presented here suggests that their origin could be even earlier than revealed by the oldest cryptospores.

We placed the origin of seed plants between the Givetian (late Middle Devonian), and the Famennian (Late Devonian). Our estimation slightly predates that derived by Anderson *et al.* (**2007**), which could be expected considering the inherent constraints of identifying fossil-genera. A critical element here is the controversial assignment of fern-like foliage into spermatophytes. For instance, the genus *Pecopteris* has mainly been used for ferns. Only one seed fern has been placed in that genus, but is now referred to as *Dicksonites* (Callistophytaceae; Galtier & Bethoux, **2002**). Indeed, it was recently documented that the type of *Pecopteris* is a fern of the extinct family Tedelaceae, and the generic name should be restricted to that family (Cleal, **2015**). Similarly,*Sphenopteris* has been regularly used for both ferns and seed ferns. However, the type is a Mississippian age seed fern belonging to the Lyginopteridales (C. J. Cleal, pers. comm., 2 May 2014).

Darwin described the appearance of early angiosperms as an abominable mystery due to their 'recent' appearance in the Cretaceous fossil record, their apparent speed to adopt 'modern looking' leaves, and their current ecological dominance (Friis *et al.*, **2010**). Although our knowledge about angiosperm radiation has significantly advanced since the 19th Century, there is still considerable controversy about the timing of angiosperm origin. The root of this debate is derived from different scenarios obtained from using macro vs microfossil evidence or paleontological vs molecular data. From the fossil record, Friis *et al.* (**2011**) argue that there is no unequivocal evidence of angiosperms before the Cretaceous. Nevertheless, other authors have suggested an early Jurassic (201.3–174.1 Ma) origin based on macrofossil evidence (Wang *et al.*, **2005**, **2010**). An even earlier origin of angiosperms has been proposed based on pollen data, dating back to the Anisian (Middle Triassic, 247.2–237 Ma) (Hochuli & Feist-Burkhardt, **2013**). In summary, pre-Cretaceous fossil evidence of angiosperms has been considered either equivocal or plagued with problems of geological dating, with no consensus reached to date.

We find further discrepancies in the timing of angiosperm origins based on molecular evidence through phylogenetic dating analyses. As summarized by Doyle & Endress (**2010**), Bell *et al.* (**2010**) estimated that angiosperms originated *c.* 183–147 Ma (early to late Jurassic), whereas Smith *et al.* (**2010**) proposed older estimates ranging between 228 and 217 Ma (Late Triassic). A very wide interval is provided by Magallón *et al.* (**2013**) who, based on different analyses, estimated that angiosperms originated between 258 Ma (Late Permian) and 158 Ma (Late Jurassic; see also Magallón, **2010**; Magallón *et al.*, **2015**). Our results suggest an angiosperm origin between 152 and 133 Ma, centered at 143.8 Ma (Fig. 3c), thus assigning the highest probabilities towards the boundary between the Late Jurassic and the Early Cretaceous. Our estimation derives from an analysis of the entire set of macrofossil occurrences and accounts for the preservation process.

Although the direct incorporation of fossil taxa in molecular dating is possible and promising (Ronquist *et al.*, **2012**; Heath *et al.*, **2014**), these methods either require large morphological datasets or are computationally intensive and potentially difficult to implement on large clades. Thus, many studies still rely on fossil prior calibrations for molecular dating. The methodology used in this study constitutes a novel approach to objectively define calibration priors, in a process-based and data-driven analysis that is able to handle the inevitably incomplete nature of the fossil record. The age estimates derived here for the three major plant clades (Fig. 3) could thus be directly used as prior crown age distributions of those clades in molecular dating analyses.

## CONCLUSIONS

Our results have provided new insights into the origin and diversification of vascular plants. This represents the first Bayesian analysis of their diversification dynamics, based on publicly available plant macrofossils at a global level, and separating the three major groups of plants recognized today. We found that plants underwent important phases of extinction across three of the 'Big Five' extinction events (at the Frasnian–Famennian, Permian–Triassic and Cretaceous–Paleogene boundaries), which represented key points of floral turnover and contributed to major diversity changes of plant life. However, there was a differential effect of those events on each plant group analyzed, and substantial differences in the impact of each event on diversification dynamics. Although many uncertainties remain regarding the age of origin of major plant clades, our analyses have provided new estimates that help in reconciling current fossil evidence with a robust probabilistic approach that takes into account a complex and variable preservation process.

## ACKNOWLEDGEMENTS

We thank Susana Magallón, Erika Edwards and Xavier Meyer for discussions, and the organizers of the Plant Radiation Meeting (Zurich, Switzerland, June 2014) Colin Hughes, Peter Linder and Reto Nyffeler. Analyses were run at the high-performance computing center Vital-IT of the Swiss Institute of Bioinformatics (Lausanne, Switzerland). We are thankful to Allister Rees, Hallie Sims, Bruce Tiffney, Peter Wilf, Cindy Looy, Rick Lupia, Bill Stein and John Alroy, the major contributors of paleobotanical data to Paleobiology Database. We acknowledge financial support from Carl Tryggers stiftelse and Wenner-Gren foundation to D.S., Project ANR-2010-BLAN-607-02 'TERRES' to B.C-M., and from the Swedish Research Council (B0569601) and the European Research Council under the European Union's Seventh Framework Programme (FP/2007-2013, ERC Grant Agreement n. 331024) to C.D.B. and A.A. This is Paleobiology Database manuscript number 216.

## REFERENCES

1. Akaike H. 1974. A new look at the statistical model identification. *IEEE Transactions on Automatic Control* **19**:716–723.
2. Alroy J, Marshall CR, Bambach RK, Bezusko K, Foote M, Fursich FT, Hansen TA, Holland SM, Ivany LC, Jablonski D *et al.* 2001. Effects of sampling standardization on estimates of Phanerozoic marine diversification.*Proceedings of the National Academy of Sciences, USA* **98**: 6261–6266.
3. Anderson JM, Anderson HM, Cleal CJ. 2007. Brief history of the gymnosperms: classification, biodiversity, phytogeography and ecology. Pretoria, South Africa: South African National Biodiversity Institute.
4. Bell CD, Soltis DE, Soltis P. 2010. The age and diversification of angiosperms re-revisited. *American Journal of Botany* **97**: 1296–1303.
5. Benton MJ. 1985. Mass extinction among non-marine tetrapods. *Nature* **316**: 811–814.
6. Bonis RB, Kürschner WM. 2012. Vegetation history, diversity patterns, and climate change across the Triassic/Jurassic boundary. *Paleobiology* **38**: 240–264.
7. Boulter MC, Spicer RA, Thomas BA. 1988. Patterns of plant extinction from some palaeobotanical evidence. In:Larwood GP, ed. Extinction and survival in the fossil record. Oxford, UK: Oxford University Press, 1–36.
8. Cascales-Miñana B, Cleal CJ. 2014. The plant fossil record reflects just two great extinction events. *Terra Nova***26**: 195–200.
9. Cascales-Miñana B, Meyer-Berthaud B. 2014. Diversity dynamics of Zosterophyllopsida. *Lethaia* **47**: 205–215.
10. Cleal CJ. 1993. Plants. In: Benton MJ, ed. The fossil record. London, UK: Chapman and Hall, 779–794.
11. Cleal CJ. 2008a. Palaeofloristics of Middle Pennsylvanian lyginopteridaleans in variscan Euramerica.*Palaeogeography, Palaeoclimatology, Palaeoecology* **261**: 1–14.
12. Cleal CJ. 2008b. Palaeofloristics of Middle Pennsylvanian medullosaleans in variscan Euramerica.*Palaeogeography, Palaeoclimatology, Palaeoecology* **268**: 164–180.
13. Cleal CJ. 2008c. Westphalian-Stephanian macrofloras of the southern Pennine Basin, UK. *Studia Geologica Polonica* **129**: 25–41.
14. Cleal CJ. 2015. The generic taxonomy of Pennsylvanian age marattialean fern frond adpressions. *Palaeontographica Abteilung B* **292**: parts 1–3.
15. Cleal CJ, Cascales-Miñana B. 2014. Composition and dynamics of the great Phanerozoic evolutionary floras. *Lethaia* **47**: 469–484.
16. Cleal CJ, Oplustil S, Thomas BA, Tenchov Y. 2011. Pennsylvanian vegetation and climate in tropical variscan Euramerica. *Episodes* **34**: 3–12.
17. Cleal CJ, Thomas BA. 2010. Botanical nomenclature and plant fossils. *Taxon* **59**: 261–268.
18. Cleal CJ, Uhl D, Cascales-Miñana B, Thomas BA, Bashforth AR, King SC, Zodrow EL. 2012. Plant biodiversity changes in Carboniferous tropical wetlands. *Earth-Science Reviews* **114**: 124–155.
19. Cohen K, Finney S, Gibbard P, Fan J. 2013. The ICS international chronostratigraphic chart. *Episodes* **36**:199–204.
20. Collinson ME. 1996. "What use are fossil ferns?" – 20 years on: with a review of the fossil history of extant pteridophyte families and genera. In: Camus MG, Johns J, eds. Pteridology in perspective. Richmond, UK: Royal Botanic Gardens, Kew, 349–394.
21. Crepet WL, Niklas KJ. 2009. Darwin's second "abominable mystery": why are there so many angiosperm species? *American Journal of Botany* **96**: 366–381.
22. Darwin C. 1859. On the origin of species by means of natural selection, or the preservation of favoured races in the struggle for life. London, UK: John Murray.
23. Decombeix A, Meyer-Berthaud B, Galtier J. 2011. Transitional changes in arborescent ligniophytes at the Devonian-Carboniferous boundary. *Journal of the Geological Society* **168**: 547–557.
24. Dimitrova TK, Zodrow EL, Cleal CJ, Thomas BA. 2010. Palynological evidence for Pennsylvanian (Late Carboniferous) vegetation change in the Sydney Coalfield, eastern Canada. *Geological Journal* **45**: 388–396.
25. Doyle JA, Endress PK. 2010. Integrating Early Cretaceous fossils into the phylogeny of living angiosperms: Magnoliidae and eudicots. *Journal of Systematics and Evolution* **48**: 1–35.
26. Drummond AJ, Suchard MA, Xie D, Rambaut

A. 2012. Bayesian phylogenetic with BEAUti and the BEAST 1.7.*Molecular Biology and Evolution* **29**: 1969–1973.

27. Edwards D, Richardson JB. 2004. Silurian and Lower Devonian plant assemblages from the Anglo-Welsh Basin: a palaeobotanical and palynological synthesis. *Geological Journal* **39**: 375–402.
28. Edwards D, Wellman CH. 2001. Embryophytes on land: the Ordovician to Lochkovian (Lower Devonian) record. In: Gensel PG, Edwards D, eds. Plants invade the land. Evolutionary and environmental perspectives. New York, NY, USA: Columbia University Press, 3–28.
29. Friis EM, Crane PR, Pedersen KR. 2011. Early flowers and angiosperm evolution. Cambridge, UK: Cambridge University Press.
30. Friis EM, Pedersen KR, Crane PR. 2010. Diversity in obscurity: fossil flowers and the early history of angiosperms. *Philosophical Transactions of the Royal Society B* **365**: 369–382.
31. Fröbisch J. 2013. Vertebrate diversity across the end-Permian mass extinction: separating biological and geological signals. *Palaeogeography, Palaeoclimatology, Palaeoecology* **372**: 50–61.
32. Galtier J, Bethoux O. 2002. Morphology and growth habit of *Dicksonites pluckenetii* from the Upper Carboniferous of Graissessac (France). *Geobios* **35**: 525–535.
33. Galtier J, Meyer-Berthaud B. 2006. The diversification of early arborescent seed ferns. *The Journal of the Torrey Botanical Society* **133**: 7–19.
34. Gelman A. 2004. Bayesian data analysis. Boca Raton, FL, USA: Chapman & Hall/CRC.
35. Hao SG, Xue JZ. 2013. Earliest record of megaphylls and leafy structures, and their initial diversification. *Chinese Science Bulletin* **58**: 2784–2793.
36. Heath TA. 2012. A hierarchical Bayesian model for calibrating estimates of species divergence times. *Systematic Biology* **61**: 793–809.
37. Heath TA, Hulsenbeck JP, Stadler T. 2014. The fossilized birth-death process for coherent calibration of divergence-time estimates. *Proceedings of the National Academy of Sciences, USA* **111**: 2957–2966.
38. Ho SYW, Phillips MJ. 2009. Accounting for calibration uncertainty in phylogenetic estimation of evolutionary divergence times. *Systematic Biology* **58**: 367–380.
39. Hochuli PA, Feist-Burkhardt S. 2013. Angiosperm-like pollen and Afropollis from the Middle Triassic (Anisian) of the Germanic Basin (Northern Switzerland). *Frontiers in Plant Science* **4**: 1–14.
40. Jablonski D. 2005. Mass extinctions and macroevolution. *Paleobiology* **31**: 192–210.
41. Janevski GA, Baumiller TK. 2009. Testing for extinction selectivity in the fossil record of Phanerozoic marine invertebrates. *Paleobiology* **35**: 553–564.
42. Kenrick P, Crane PR. 1997. The origin and early evolution of plants on land. *Nature* **389**: 33–39.
43. Kenrick P, Wellman CH, Schneider H, Edgecombe GD. 2012. A timeline for terrestrialization: consequences for the carbon cycle in the Paleozoic. *Proceedings of the Royal Society B* **367**: 519–536.
44. Krug AZ, Jablonski D, Valentine JW. 2009. Signature of the end-Cretaceous mass extinction in the modern biota. *Science* **323**: 767–771.
45. Labandeira CC, Sepkoski JJ. 1993. Insect diversity in the fossil record. *Science* **261**: 310–315.
46. Lepage T, Bryant D, Phillippe H, Lartillot N. 2007. A general comparison of relaxed molecular clock models. *Molecular Biology and Evolution* **24**: 2669–2680.
47. Liow L, Skaug H, Ergon T, Schweder T. 2010. Global occurrence trajectories of microfossils: environmental volatility and the rises and falls of individual species. *Paleobiology* **36**: 224–252.
48. Liow LH, Stenseth NC. 2007. The rise and fall of species: implications for macroevolutionary and macroecological studies. *Proceedings of the Royal Society B* **274**: 2745–2752.
49. Longrich NR, Bhullar BAS, Gauthier JA. 2012. Mass extinction of lizards and snakes at the Cretaceous-Paleogene boundary. *Proceedings of the National Academy of Sciences, USA* **109**: 21 396–21 401.
50. Longrich NR, Tokaryk T, Field DJ. 2011. Mass extinction of birds at the Cretaceous-Paleogene (K-PG) boundary. *Proceedings of the National Academy of Sciences, USA* **108**: 15 253–15 257.
51. Magallón S, Gómez-Acevedo S, Sánchez-Reyes LL, Hernández-Hernández T. 2015. A metacalibrated time-tree documents the early rise of flowering plant phylogenetic diversity. *New Phytologist* **207**: 437–453.
52. Magallón S. 2010. Using fossils to break long branches in molecular dating: a comparison of relaxed clocks applied to the origin of Angiosperms. *Systematic Biology* **59**: 384–399.
53. Magallón S, Hilu KW, Quandt D. 2013. Land plant evolutionary timeline: gene effects are secondary to fossil constraints in relaxed clock estimation of age and substitution rates. *American Journal of Botany* **100**: 556–573.
54. McElwain JC, Punyasena SW. 2010. Mass extinction events and the plant fossil record. *TRENDS in Ecology and Evolution* **22**: 548–557.
55. McLoughlin S, Carpenter RJ, Jordan GJ, Hill RS. 2008. Seed ferns survived the end-Cretaceous mass extinction in Tasmania. *American Journal of Botany* **95**: 465–471.
56. Meredith RW, Janecka JE, Gatesy J, Ryder OA, Fisher CA, Teeling EC, Goodbla A, Eizirik E, Simao TLL, Stadler T*et al*. 2011. Impacts of the Cretaceous terrestrial revolution and KPg extinction on mammal diversification.*Science* **334**: 521–524.
57. Meyer-Berthaud B, Soria A, Decombeix A-L. 2010. The land plant cover in the Devonian a reassessment of the evolution of the tree habitat. *Geological Society Special Publication* **339**: 59–70.
58. Niklas K. 1997. The evolutionary biology of plants. Chicago, IL, USA: The University of Chicago Press.

59. Niklas KJ, Tiffney BH, Knoll AH. 1983. Patterns in vascular land plant diversification. *Nature* **303**: 614–616.

60. Pryer KM, Schneider H, Smith AR, Cranfill R, Wolf PG, Hunt JS, Sipes SD. 2001. Horsetails and ferns are a monophyletic group and the closest living relatives to seed plants. *Nature* **409**: 618–622.

61. Raubeson LA, Jansen RK. 1992. Chloroplast DNA evidence on the ancient evolutionary split in vascular land plants. *Science* **255**: 1697–1699.

62. Raup DM, Sepkoski JJ. 1982. Mass extinctions in the marine fossil record. *Science* **215**: 1501–1503.

63. Retallack G. 2013. Permian and Triassic greenhouse crises. *Gondwana Research* **24**: 90–103.

64. Rees PM. 2002. Land-plant diversity and the end-Permian mass extinction. *Geology* **30**: 827–830.

65. Ronquist F, Klopfstein S, Vilhelmsen L, Schulmeister S, Murray DL, Rasnitsyn AP. 2012. A total-evidence approach to dating with fossils, applied to the early radiation of the Hymenoptera. *Systematic Biology* **61**:973–999.

66. Sauquet H, Ho SYW, Gandolfo MA, Jordan GJ, Wilf P, Cantrill DJ, Bayly MJ, Bromham L, Brown GK, Carpenter RJ*et al.* 2012. Testing the impact of calibration on molecular divergence times using a fossil-rich group: the case of *Nothofagus* (Fagales). *Systematic Biology* **61**: 289–313.

67. Schulte P, Alegret L, Arenillas I, Arz JA, Barton PJ, Bown PR, Bralower TJ, Christeson GL, Claeys P, Cockell CS*et al.* 2010. The Chicxulub asteroid impact and mass extinction at the Cretaceous-Paleogene boundary. *Science* **327**: 1214–1218.

68. Sepkoski JJ. 1978. A kinetic-model of phanerozoic taxonomic diversity I. Analysis of marine orders. *Paleobiology***4**: 223–251.

69. Sepkoski JJ. 1979. A kinetic-model of phanerozoic taxonomic diversity II. Early phanerozoic families and multiple equilibria. *Paleobiology* **5**: 222–251.

70. Sepkoski JJ, Bambach RK, Raup DM, Valentine JW. 1981. Phanerozoic marine diversity and the fossil record.*Nature* **293**: 435–437.

71. Silvestro D, Salamin N, Schnitzler J. 2014a. PyRate: a new program to estimate speciation and extinction rates from incomplete fossil data. *Methods in Ecology and Evolution* **5**: 1126–1131.

72. Silvestro D, Schnitzler J, Liow LH, Antonelli A, Salamin N. 2014b. Bayesian estimation of speciation and extinction from incomplete fossil occurrence data. *Systematic Biology* **63**: 349–367.

73. Smith SA, Beaulieu JM, Donoghue MJ. 2010. An uncorrelated relaxed-clock analysis suggests an earlier origin for flowering plants. *Proceedings of the National Academy of Sciences, USA* **107**: 5897–5902.

74. Tomescu AMF. 2009. Megaphylls, microphylls and the evolution of leaf development. *Trends in Plant Science***14**: 5–12.

75. Vecoli M, Meyer-Berthaud B, Clement G. 2010. Introduction. In: Vecoli M, Clement G, eds. The terrestrialization process: modeling complex interactions at the biosphere–geosphere interface. London, UK: Geological Society, 1–3.

76. Wang Q, Ferguson DK, Guang-Ping F, Ablaev AG, Wang Y-F, Yang J, Li Y-L, Li C-S. 2010. Climatic change during the Palaeocene to Eocene based on fossil plants from Fushun China. *Palaeogeography, Palaeoclimatology, Palaeoecology* **295**: 323–331.

77. Wang Q, Geng BY, Dilcher DL. 2005. New perspective on the architecture of the late devonian arborescent lycopsid *Leptophloeum rhombicum* (Leptophloeaceae). *American Journal of Botany* **92**: 83–91.

78. Wellman C, Steemans P, Vecoli M. 2013. Palaeophytogeography of Ordovician-Silurian land plants. In: Harper DAT, Servais T, eds. Early palaeozoic biogeography and palaeogeography. London, UK: Geological Society,461–476.

79. Wellman CH. 2014. The nature and evolutionary relationships of the earliest land plants. *New Phytologist* **202**:1–3.

80. Xing Y, Onstein RE, Carter RJ, Stadler T, Linder HP. 2014. Fossils and a large molecular phylogeny show that the evolution of species richness, generic diversity and turnover rates are disconnected. *Evolution* **68**: 2821–2832.

81. Xiong C, Wang Q. 2011. Permian-Triassic land-plant diversity in south China: was there a mass extinction at the Permian/Triassic boundary? *Paleobiology* **37**: 157–167.

82. Xiong C, Wang DM, Benton MJ, Xue JZ, Meng M, Zhao Q, Zhang J. 2013. Diversity dynamics of silurian-early carboniferous land plants in south China. *PLoS ONE* **8**: e75706.

CHAPTER 4

# SLOW BUT NOT LOW: GENOMIC COMPARISONS REVEAL SLOWER EVOLUTIONARY RATE AND HIGHER DN/DS IN CONIFERS COMPARED TO ANGIOSPERMS

Emmanuel Buschiazzo, Carol Ritland, Jörg Bohlmann and Kermit Ritland

## BACKGROUND

Determining the mutational and the selective forces responsible for evolution has overarching implications in biology, e.g. in understanding what makes species unique and how organisms respond to biotic and abiotic challenges. Identifying the rate of evolution and the patterns of nucleotide substitution underlying DNA evolution has thus become a fundamental goal of molecular genomics [1, 2]. Key to the central dogma of molecular biology, protein-coding sequences (hereafter referred to as genes) have classically been regarded as a major unit of evolution. Substitutions at synonymous (silent) and non-synonymous (replacement) sites are commonly distinguished to differentiate between neutral (or at least weak) and active selective forces acting on genes, respectively. In pairwise comparisons of orthologous genes, the ratio of non-synonymous distance (i.e. number of substitutions per non-synonymous site; dN) over synonymous distance (dS) gives a general but conservative indication of the mode and strength of selection [1, 2]. An excess of non-synonymous substitutions (dN/dS > 1) suggests adaptive or diversifying selection, while an excess of synonymous mutations (dN/dS < 1) indicates purifying selection, and no difference between synonymous and non-synonymous mutation rates (dN/dS = 1) is taken as evidence for neutrality [3]. Large-scale sequence datasets now exist, allowing comparisons to be made for thousands of genes in all domains of life. Synonymous and non-synonymous substitution rates have been found to vary widely within and between taxa [4, 5, 6, 7]. From early studies based on a limited number of species and genes to the era of genomics and systems biology [8, 9], a complex blend of non-mutually exclusive biological, biochemical and demographic mechanisms emerged to explain these variations. While intraspecies differences are believed to be influenced by selection on protein structure and function (reviewed in [10, 11, 12, 13, 14]), interspecies differences are influenced by (i) the efficacy of the DNA repair machinery, (ii) life history traits (e.g. generation time), (iii) metabolic rate, (iv) effective population size (random genetic drift), (v) purifying (background) selection and (vi) reproductive strategy. Some factors (i - iii) influence the way mutations appear, while others (iv - vi) influence their fixation over generations (reviewed in [9, 13, 14]).

Among plants, most of the attention in comparative evolutionary studies has been focused on flowering plants [4, 5, 14, 15], and interest is now growing for other plant taxa as more sequence data is produced. Gymnosperms are separated from angiosperms by ~300 million years of evolution [16]. Expectedly, many biological features of gymnosperms and angiosperms differ greatly, including seed morphology, life span, diversification rate, pollination processes, environmental requirements and response to environmental stresses. With ~600 extant species, conifers make up about two thirds of all gymnosperm species, and are the dominant plants in most temperate and boreal ecosystems. Conifers have an immense ecological and economical value such as practical forestry economics, immediate ecological value of forest ecosystems

**Citation:** Buschiazzo et al.: Slow but not low: genomic comparisons reveal slower evolutionary rate and higher dN/dS in conifers compared to angiosperms. DOI: 10.1186/1471-2148-12-8.

and in the long term, large capacity for carbon sequestration. Biological differences between angiosperms and conifers and the need for long-lived conifer species to cope with challenges such as insect pests and environmental changes, underscore the importance of understanding the molecular and functional evolution of conifer genomes.

The genetic architecture of conifers has been addressed by a wide variety of studies, mainly in pine (*Pinus*[17]) and spruce (*Picea* [18]). Approaches include quantitative trait locus mapping [19, 20, 21], candidate gene approaches [22, 23], association mapping [24, 25], BAC sequencing [26, 27], transcriptome analysis [28, 29], characterization of gene families [30] and proteome analyses [31], and combinations thereof [32]. Missing from past endeavors, however, are large-scale comparative comparisons that investigate both evolutionary rates and the selective forces acting on conifer genes.

In this study, we take advantage of the existing large and high-quality sequence data in two conifer species, Sitka spruce (*Picea sitchensis*) and loblolly pine (*Pinus taeda*), consisting of a collection of *bona fide* full-length cDNA sequences (FL-cDNAs) [33, 34] and UniGenes constructed from several EST libraries, respectively. Together with whole-genome gene sets available for two angiosperms, *Arabidopsis thaliana* and *Populus trichocarpa*; a rich data set exists to identify rates and patterns of evolution between conifer species and between conifer and angiosperm species. We find evidence for significantly slower evolutionary rates in conifers. In stark contrast, we find a significantly higher dN/dS ratio in conifers as compared to angiosperms, indicating perhaps higher adaptation. We also investigate these patterns across functional categories of genes.

## METHODS

## Protein-Coding Sequences for Conifers and Angiosperms

### Conifer Sequences

Clustered ESTs from loblolly pine were downloaded from NCBI UniGene (build 10, which had 18,921 clusters). Sitka spruce FLcDNAs came from the Treenomix II project [35]; as of Nov. 10 2009, this collection comprised 10,665 FLcDNAs, of which 3,218 clustered in contigs. We used all individual FLcDNAs because our approach ultimately removes any redundant or duplicated sequences.

### Open Reading Frame (ORF) Search in Conifer Genes

All possible ORFs (from start to stop codons) found in spruce FLcDNAs were queried against the plant UniProtKB SwissProt and trEMBL datasets [36], with predicted proteins from Sitka spruce [33] removed from the trEMBL dataset. Only ORFs from the 5,680 spruce FLcDNAs that had no hit against the SwissProt dataset were queried against the trEMBL dataset. ORFs from 3,296 spruce FLcDNAs had no homology with either of the plant UniProtKB datasets; for those, the longest ORF was arbitrarily selected for further analysis. A single FLcDNA with no ORF structure in its sequence was discarded. We did not use the same strategy for loblolly pine because the pine UniGene set may contain only a truncated portion of the actual coding sequence. For conifers, we looked in each member of the UniGene set for the ORF among all possible ORFs with the same frame as the longest overlapping sequence with the best-scoring BLAST query against the spruce ORFs. Of 18,921 pine UniGenes, we found 7,627 ORFs in the same frame as spruce ORFs.

### Orthology of Conifer Genes

We used the reciprocal best hit (RBH) approach [37, 38] to infer putative 1:1 orthologues between spruce FLcDNAs and pine UniGene sequences, using BLAST with -e threshold = $10^{-20}$. We found a total of 4,774 RBHs, of which 4,250 contained a complete ORF in pine.

### Angiosperm Orthologues

*A. thaliana* was chosen because it represents the best characterized plant genome. Poplar was included in the analyses as the first completely sequenced tree genome. *A. thaliana* coding sequences were downloaded from the TAIR9 annotation release [39]. Poplar coding sequences (annotation 1.1) were downloaded from the JGI Genome Portal [40]. We used Ensembl Compara predictions through the BioMart server [41] to select a list of orthologous genes from *Arabidopsis* and poplar. Only 1:1 and apparent 1:1 orthologous coding sequences were retained for analysis, finalizing a set of 5,108 orthologues.

### Alignment

Gymnosperm (spruce-pine) and angiosperm (*Arabidopsis*-poplar) orthologous coding sequences were aligned using DIALIGN-TX [42] with highest sensitivity (-L option). Gaps in the alignments and gap-free regions > 7 bp, interpreted as non-homologous by DIALIGN-TX, were excluded from the analysis. Finally, alignments shorter than 30 amino acids were discarded. The RBH conifer orthologue set contained 3,883 alignments and the angiosperm gene set totaled 5,073 successfully aligned 1:1 orthologues.

## Data Analysis

### Substitution Rates

Pairwise distances at non-synonymous (dN), synonymous (dS) and 4-fold degenerate (4D) sites (d4) were estimated for individual genes in both gymnosperm and angiosperm alignment sets using codeml (PAML 4.0) [43, 44], with settings seqtype = 1, CodonFreq = 2, Runmode = -2, and transition-transversion ratio ($\kappa$) estimated from the data. Genes showing signs of saturated divergence were excluded because codeml results are reliable for moderate ranges of sequence divergence. For conifers, we discarded 42 orthologues with dN/dS = 98.99 and 118 with dS > 0.5, and for angiosperms, we discarded two genes with dN > 5 and 996 genes with dS > 4. Threshold dS values were determined by plotting dN as a function of dS and excluding outliers from the main distribution. Final RBH orthologue sets contained 3,723 conifer genes (average gap-free length = 510 bp) and 4,080 1:1 angiosperm genes (average gap-free length = 387 bp). 95% confidence intervals for evolutionary estimates were calculated based on 1,000 bootstrap replicates using R [45]. Absolute rates of substitution at coding sites (μ) in pairwise comparisons were inferred using the formula:

$$\mu = \frac{d}{2T}$$

with d the distance at synonymous (dS), non-synonymous (dN) or 4D (d4) sites; T divergence time between spruce and pine, or between *Arabidopsis* and poplar. Divergence times are documented from fossil records, between ~120 and ~160 MYA for conifers [46, 47, 48, 49, 50, 51], and between ~105 and ~115 MYA for*Arabidopsis* and poplar [52]. Unless mentioned otherwise, we used 140 MYA and 110 MYA, respectively, as working divergence times.

## Analyses of Functional Categories

Functions of conifer orthologues were inferred using analogy with *Arabidopsis* proteins for GO annotations, and with plant proteins for descriptive annotation. In detail, spruce ORFs were queried against the TAIR9 protein-coding genes and the plant UniprotKB database using BLASTX (-e threshold = $10^{-5}$). Of the 3,983 best hits against *Arabidopsis*, 1,230 contained an ORF that successfully aligned to loblolly pine ORFs and were assigned the GO annotation corresponding to that of the best *Arabidopsis* hits, when available.

For statistical comparisons among conifer genes, we used gene set enrichment analysis tools in the Babelomics platform [53], a web application that implements threshold-independent statistics (FatiScan and logistic regression) to investigate asymmetrical distributions of GO terms, KEGG pathways and InterPro domains within our list of annotated genes ranked by dN/dS. Fatiscan uses a Fisher exact test over a collection of partitions of the ranked list of genes, while the logistic model is used to find association of each functional block with the high or low values of the ranked list; under- and over-represented functional terms are then extracted. Prior to these analyses, we removed 43 genes that showed no non-synonymous substitution. For other functional analyses, we used the 'GO Slim' classification system provided by TAIR database [54].

# RESULTS

## Substitution Rates in Conifer Protein-Coding genes

We aligned the sequences of 3,723 spruce-pine orthologous genes and inferred the number of pairwise synonymous (dS) and non-synonymous (dN) substitutions per site (see Table 1). Mean dS was 0.191 (95% confidence interval [CI] = 0.188, 0.193), meaning that on average, one mutation occurred about every five sites along both lineages since the common ancestor. Mean dN was lower than dS (0.049; CI = 0.048, 0.050), reflecting the expected elevated mutational constraint on non-synonymous sites.

Mean genetic distances at synonymous (dS), 4-fold degenerate (d4) and non-synonymous (dN) sites are expressed as a number of substitutions per site. Absolute substitution rates are expressed in substitutions per synonymous ($\mu_S$), four-fold degenerate ($\mu_{4D}$) and non-synonymous ($\mu_N$) site per year. Species-specific rates for angiosperms were estimated based on the 1:6 difference in evolutionary rate between poplar and *Arabidopsis* [57, 58].

**Table 1:** Substitution rates in conifer protein-coding genes compared to angiosperm genes

| Pairwise comparison | Gene number | dS | d4 | dN | $\mu_S$(×10$^{-9}$) | $\mu_{4D}$(×10-$^{9}$) | $\mu_N$(×10$^{-9}$) | dN/dS |
|---|---|---|---|---|---|---|---|---|
| Gymnosperms: | 3,723 | 0.1908 | 0.1769 | 0.0492 | 0.68 | 0.64 | 0.18 | 0.3137 |
| Sitka spruce | | | | | | | | |
| Loblolly pine | | | | | | | | |
| Angiosperms: | 4,080 | 2.1846 | 2.0057 | 0.2019 | 9.93 | 9.12 | 0.92 | 0.0924 |
| *Arabidopsis* | | | | | 17.02 | 15.63 | 1.57 | |
| Poplar | | | | | 2.84 | 2.61 | 0.26 | |
| **Fold-change** | | | | | | | | |
| Angiosperm:conifers | | 11.4:1 | 11.4:1 | 4.1:1 | 14.6:1 | 14.4:1 | 5.2:1 | 1:3.4 |
| *Arabidopsis*:conifers | | | | | 25.0:1 | 24.7:1 | 9.0:1 | |
| Poplar:conifers | | | | | 4.2:1 | 4.1:1 | 1.5:1 | |

Based on fossil records, the *Pinus-Picea* divergence occurred between 120 and 160 MYA [46, 47, 48, 49, 50, 51]. Assuming an average divergence time of 140 MYA and that rates were equivalent along both lineages, we inferred an average rate of 0.68 × $10^{-9}$ (95% CI = 0.67 × $10^{-9}$, 0.69 × $10^{-9}$) substitutions per site per year at synonymous sites ($\mu_S$, see Table 1). However, to fully account for the uncertainty of divergence time between pine and spruce, we also consider that this time is between 120 and 160 MYA, giving the actual estimate of $\mu_S$as lying between 0.60 × $10^{-9}$ and 0.80 × $10^{-9}$. The neutral theory of molecular evolution predicts that the evolutionary rate at neutral sites corresponds to the actual mutation rate in an organism [55]. Because neutrality at synonymous sites is disputed [56], distance in a subset of synonymous sites known as 4-fold degenerate ($\mu_{4D}$) sites (i.e. sites where a change to any of the four nucleotides will not alter the amino acid during translation) stands as a better proxy to estimate the mutation rate. From our comparison in conifers, we inferred distance at $\mu_{4D}$ sites (d4) at 0.177 (95% CI = 0.174, 0.179), which translates into a substitution rate of 0.64 × $10^{-9}$ per 4D site per year ($\mu_4$, see Table 1), and a range of 0.55 ×

$10^{-9}$ and $0.74 \times 10^{-9}$ using the extreme estimates of divergence time between spruce and pine.

## *dN/dS in Conifer Protein-Coding Genes*

Ideally, dN/dS should be estimated at every site to find evidence of selection (which is only possible when comparing more than two species in a phylogenetic context) and not averaged over the entire gene. However, an over-representation of non-synonymous substitutions can be used as a crude indication of either adaptive evolution or at least relaxed constraint in protein-coding genes. Mean dN/dS in conifer genes was 0.314 (95% CI = 0.299, 0.329). Of the 3,723 pairwise comparisons, 100 (2.68%) had a dN/dS > 1. We note the presence of genes that are involved in abiotic and biotic stress response; some examples are protein kinases, protein phosphatases, heat shock proteins, leucine-rich repeat proteins, histone modification proteins, glycosyltransferases, and transcription factors (see Table 2).

**Table 2:** Conifer genes involved in defense, resistance and response against insects with dN/dS > 1

| Spruce clone ID | Pine UniGene ID | dN/dS | UniProt ID | Species | Putative function |
|---|---|---|---|---|---|
| WS02821_B21 | DT625383 | 7.3061 | A7P5L0 | *Vitis vinifera* | Protein phosphatase/Serine/ threonine phosphatases |
| WS0297_D22 | CX645632 | 7.0185 | A7QNM9 | *Vitis vinifera* | leucine-rich repeat family protein/binding protein |
| WS02725_C02 | DR097823 | 6.5839 | A7P656 | *Vitis vinifera* | Protein phosphatase 2C/ hydrolase/metal-binding |
| WS02757_H19 | DR165429 | 4.4902 | Q9SE11 | *Funaria hygrometrica* | Chloroplast-localized small heat shock protein (HSP20) family |
| WS02758_N18 | DR160912 | 4.4589 | Q0DTD2 | *Oryza sativa* subsp. *japonica* | Heat shock protein DnaJ |
| WS02741_E07 | DT634060 | 3.3785 | Q588B8 | *Cryptomeria japonica* | Glycoside Hydrolase Family 17 |
| WS02761 N01 | CO365391 | 3.0817 | A7PWA7 | *Vitis vinifera* | Heat shock protein DnaJ |
| WS02817_M06 | DR093347 | 2.9656 | A7NWZ2 | *Vitis vinifera* | serine/threonine-specific protein kinase |
| WS0272_J12 | DR015390 | 2.3311 | A7QFY4 | *Vitis vinifera* | Heat shock protein DnaJ |
| WS0454_E20 | DR049906 | 2.2728 | A0MMD5 | *Litchi chinensis* | Xyloglucan endotransglycosylase (Glycoside hydrolase family) |
| WS02774_M01 | DR060506 | 2.1005 | Q6VAA9 | *Stevia rebaudiana* | UDP-glycosyltransferase |
| WS02749_F04 | CO164226 | 1.7421 | Q9MA24 | *Arabidopsis thaliana* | Glycosyltransferase |
| WS0288_C08 | DR022129 | 1.5822 | A7QTB5 | *Vitis vinifera* | Glycoside hydrolase |
| WS0292_O15 | DR681862 | 1.294 | A7P0R3 | *Vitis vinifera* | heat shock protein (hsp70) |
| WS02729_N15 | DR689530 | 1.1356 | Q8LHS7 | *Oryza sativa* subsp. *japonica* | Histone deacetylase |
| WS02716_E18 | AI784893 | 1.1314 | A5AWM3 | *Vitis vinifera* | Pathogenesis-related transcriptional activator PTI6 |
| WS0298_F15 | DT638459 | 1.0692 | A7QTU5 | *Vitis vinifera* | Glycosyltransferase |
| WS02725_E03 | U39301 | 1.0481 | A0ERF9 | *Cathaya argyrophylla* | Caffeic acid ortho-methyltransferase |

Genes with dN/dS lower than 1 can in fact be under positive selection at specific sites [3] and dN/dS measured over the whole gene length is thus considered too conservative to identify genes or groups of genes putatively under positive selection. Hence, we also applied a segmentation test and a logistic regression test to look for functional groups of genes that are significantly and coordinately associated to high and/ or low values of dN/dS. Based on 1,230 GO-annotated conifer genes, we found that heat shock proteins, genes involved in signal transduction and regulation of transcription and nucleic acids seem more likely to evolve under reduced constraint; whereas genes involved in translation, protein assembly, chlorophyll biosynthesis and cellular organization are under strong selective constraint.

## Comparison between Gymnosperms and Angiosperms

We compared evolutionary distances between two representative conifer taxa, Sitka spruce and loblolly pine, and two representative angiosperm taxa, *Arabidopsis* and poplar (see Table 1). Mean dN in 4,080 *Arabidopsis*-poplar orthologous genes was 0.202 (95% CI = 0.199, 0.205), mean dS was 2.184 (95% CI = 2.164, 2.206), and mean d4 was 2.006 (95% CI = 1.985, 2.026). Based on a relatively confident divergence time of ~110 million years [52], we inferred an average synonymous mutation rate $\mu_S$ of $9.93 \times 10^{-9}$ substitutions per year along the lineages separating *Arabidopsis* and poplar (CI = $9.84 \times 10^{-9}$, $10.03 \times 10^{-9}$). This is 15-fold higher than the average mutation rate found in conifer orthologues (see Table 1). Even using the lowest estimate of divergence time between spruce and pine, $\mu_S$ is more than 10-fold higher in angiosperms. Absolute rates of substitution are calculated assuming equal rates on the poplar and the *Arabidospis* lineages, but it has been suggested that the evolutionary rate in the poplar branch is one-sixth that of the *Arabidopsis* branch since divergence [57, 58]. Using this factor, we obtained $\mu_S$ estimates of $2.84 \times 10^{-9}$ in the poplar lineage and $1.70 \times 10^{-8}$ in the *Arabidopsis* lineage, which compares well with $1.50 \times 10^{-8}$, a previously known rate in Arabideae [59]. However, this rate has since been revised to $7.5 \times 10^{-9}$ with the recent finding that the divergence time between *A. thaliana* and *A. lyrata* is about twice the previously known time, i.e. ~10 MYA instead of ~5 MYA [60].

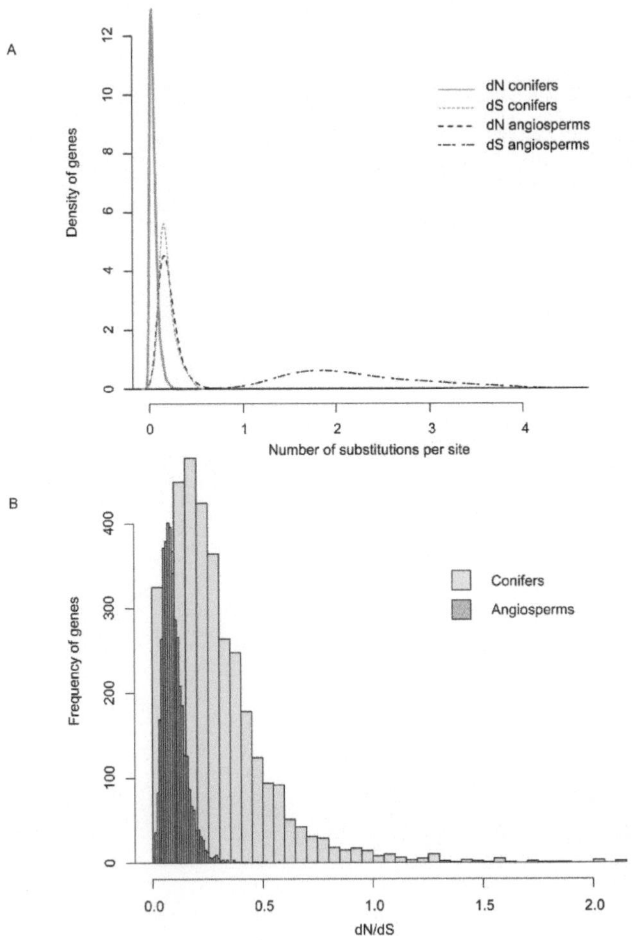


**Figure 1:** Distribution of evolutionary estimates for conifer and angiosperm protein-coding genes. A. Smoothed density plots of dS and dN estimates. B. Histogram plots of dN/dS estimates. Conifer genes with dN/dS > 2 (n = 38) are not shown. Breaks = 200.

We also found a difference in $\mu_N$ between gymnosperms and angiosperms ($0.18 \times 10^{-9}$ and $0.92 \times 10^{-9}$mutations per year, respectively), representing a five-fold difference. If we account for the differential rate between the two angiosperm species, the difference for $\mu_N$ is 9-fold and 1.5-fold with *Arabidopsis* and poplar, respectively (see Table 1). Figure 1.A illustrates the difference in dS and dN distributions between conifers and angiosperms, in particular the strikingly low dS estimates for conifers.

Overall, our results indicate a relative over-representation of non-synonymous mutations versus synonymous mutations in conifer species compared to angiosperm species. Consequently, mean dN/dS is higher in conifers than in angiosperms, i.e. 0.3137 and 0.0924, respectively, on average, and the distribution of dN/dS values for conifers extends towards and over unity (Figure 1.B). While we found 100 conifer genes with dN/dS > 1 out of 3,723 orthologues, there was a single *Arabidopsis*-poplar orthologue out of 4,080 orthologues that showed signs of positive selection over the entire alignment (dN/dS = 1.8565). This gene (*ORF25*; TAIR ID: ATMG00640; UniProt ID: Q04613) encodes a plant b subunit of mitochondrial ATP synthase.

We compared dN/dS between functional categories in conifers and gymnosperms, and consistently found higher dN/dS in conifers in most functional GO Slim categories (Figure 2; Mann-Whitney test, $P < 0.05$). However, 'DNA/RNA metabolism' (biological processes; $P = 0.37$), and 'chloroplast' and 'ribosome' (cellular component; $P = 0.46$ and $P = 0.62$, respectively) showed no significant difference.

If synonymous mutations, and even more so mutations at 4D sites, follow a neutral mode of evolution, we would expect no significant difference in average $\mu_S$ between functional categories. However, there were significant disparities among some of the functional categories, even when considering the 'more neutral' mutations at 4D sites (Kruskal-Wallis test; $H = 52.831$, $P < 0.001$), a surprising finding because it goes against the neutral expectancy. Interestingly, a recent study in birds has found evidence for selective constraints at 4D sites in the avian genome [61], and completes previous evidence accumulated in mammals [56]. Taken together, these results should call for careful attention when using dS as an estimate of neutral mutation rate, especially when inferring positive selection from dN/dS estimates or when applying molecular clocks. The present study does not claim positive selection but merely reports evolutionary trends; our results are therefore not significantly affected by the assumed neutrality of dS.

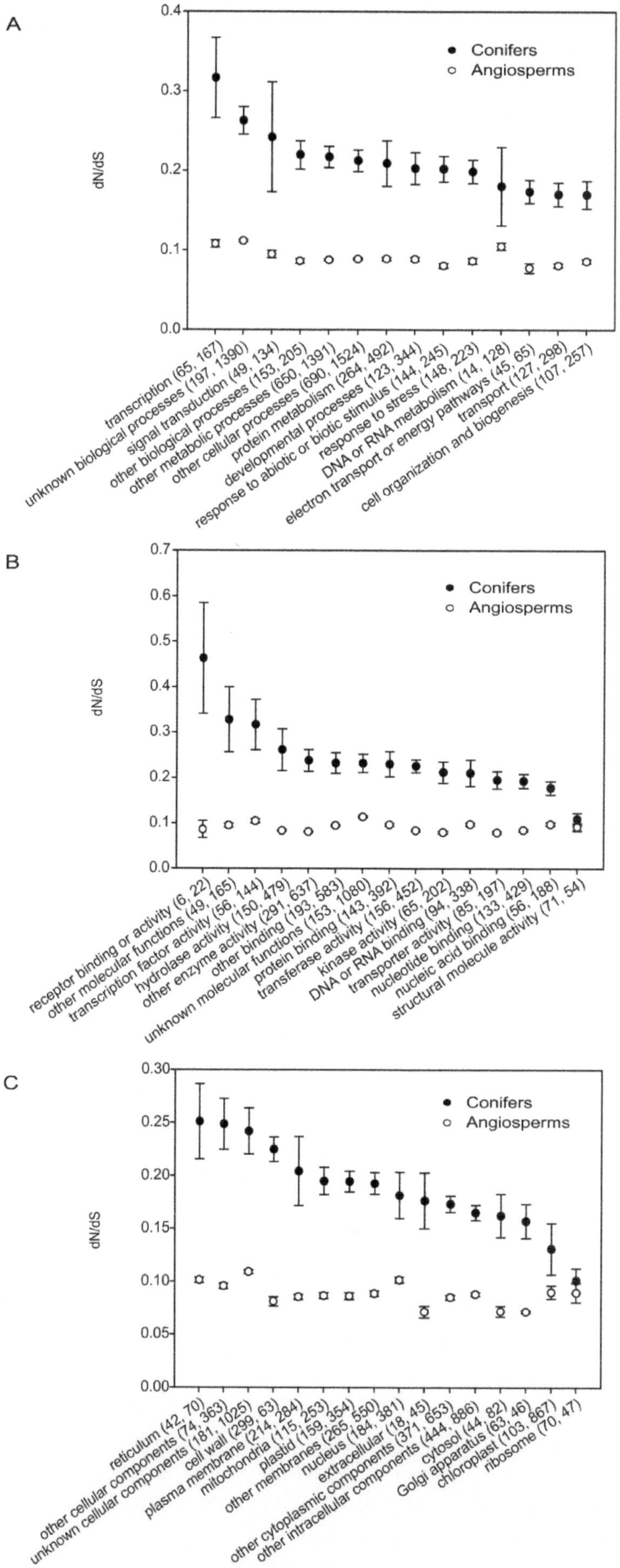


**Figure 2:** dN/dS estimates in conifer and angiosperm genes across *Arabidopsis'* GO slim functional categories. Mean dN/dS values for conifer (full circle) and angiosperm (open circle) protein-coding genes. Conifer genes were BLASTed against *Arabidopsis* gene transcripts, whose GO

Slim annotations were used for homologous conifer genes. Brackets represent the standard error of the mean. A: Biological processes; B: Molecular functions; C: Cellular component.

## DISCUSSION

Our findings, based upon large-scale sampling rather than a small set of genes, are of significance for understanding the differences in patterns of evolution between conifers and angiosperms. First, we found that evolutionary rates are dramatically lower in conifers than in angiosperms. Second, we find that such differences vary across functional categories of genes.

Classically, interspecific studies of protein-coding genes in conifers have involved very few loci. Kusumi *et al.*[62] studied evolutionary rates of 11 genes in the Cupressacea. Bouillé and Bousquet [63] compared polymorphisms of three nuclear genes in *Picea*. More recently, Palmé *et al.* [64] scrutinized patterns of selection in 21 nuclear genes in a pine phylogeny while Chen *et al.* [65] carried out similar analyses for 10 genes in four spruce species. Large-scale comparative approaches are needed to grasp global evolutionary trends representative of conifer genomes.

Genome-scale sequencing of conifer genomes is coming of age [26, 27], in particular for two economically and environmentally important species of the Pinaceae: Sitka spruce and loblolly pine. EST datasets for these species have previously been used in a comparative framework to find conifer-specific genes [66] and studying the evolution of gene families [67] and of xylem-specific genes [68] in vascular plants. Here, we carried out the first comparative study of substitution rates and mutational patterns in a sizable fraction of the conifer gene set - or that of any gymnosperm.

## Lower Rates of Evolution in Conifers As Compared To Angiosperms

### *Are evolutionary rates slower in conifers and gymnosperms than in angiosperms?*

We estimated evolutionary measures at 3,723 conifer orthologues and 4,080 angiosperm orthologues. As in any partial list of ESTs (i.e. not genome-wide), there might have been an unintentional selection of particular functional categories of genes, but we believe that our gene set is large enough to be representative of the genome as a whole. We found a much smaller dS in conifers than in angiosperms (0.1908 and 2.1846, respectively; see Table 1). A practical consequence of this difference is that we discarded almost 10 times as many angiosperm genes before final analysis; these genes showed a significant level of genetic saturation compared to conifer genes. Genetic saturation artificially reduces sequence divergence because multiple mutations at any given site of a particularly fast-evolving gene cannot be ruled out. All considered, not discarding these genes would only increase the difference in dS between conifers and angiosperms. Estimates of dN were also lower in conifers than in angiosperms (0.0492 and 0.2019, respectively), but the difference was not as dramatic as for dS (see Table 1, Figure 1.A), suggesting that substitutions at synonymous sites are particularly constrained - or that those at non-synonymous sites are less constrained, at equal mutation rate, in conifers as compared to angiosperms. Although the causes for this pattern of substitutions in conifer genes are unclear, the answer resides in what seems a unique picture of mutational processes and/or selective influences that affect conifer genes (see below).

Using published divergence times, we inferred an average synonymous mutation rate of $0.68 \times 10^{-9}$substitutions per site per year in conifer genes (see Table 1); this is 15 times less than the average rate in 4,080*Arabidopsis*-poplar orthologues ($\mu_S = 9.93 \times 10^{-9}$). If we account for the lower (1:6) rate in the poplar lineage [57], the difference is 25 times less in conifers than in *Arabidopsis* ($\mu_S = 17.02 \times 10^{-9}$), and four times less than poplar ($\mu_S = 2.84 \times 10^{-9}$). We compiled a list of substitution rates that have been published for gymnosperms and angiosperms, and our findings fall well into the range of rates reported for the two seed plant groups. For example, two phytochrome genes were shown to evolve at a synonymous rate of $0.48 \times 10^{-9}$per year in *Pinus sylvestris* and *Picea abies* [69]. For angiosperms, a rate of $1.5 \times 10^{-8}$ per year was commonly accepted for *Arabidopsis* [59] and the resulting 1:6 rate in poplar (2.5 $\times 10^{-9}$ per year) is also very similar to our results (Table 1). However, with a divergence time between *A. thaliana* and *A. lyrata* recently revised at ~10 MY [60], the current estimate of the mutation rate in *Arabidopsis* has doubled. Although it is unclear how this relates to our results, it is important to acknowledge the uncertainty that exists in our results, in the 1:6 poplar:*Arabidopsis* ratio and in timing divergence, even when relaxed molecular clocks are used.

Interestingly, at the population level, conifers also exhibit lower nucleotide diversity despite high gene flow and low population structure [65, 70, 71]. In addition, low substitution rate and low nucleotide diversity in conifers are paralleled with reports of relatively low evolutionary rates above the nucleotide level. For example, angiosperms are highly diversified while gymnosperms have experienced a very low speciation rate [72]. At least in birds, diversification has been shown to be positively correlated with mutation rate [73]. At the chromosome level, not only is there little variation in the number of haploid conifer chromosomes (n = 11-13) with only scarce evidence of whole genome duplication and polyploidy [74] but comparative genome maps also suggest that macrosynteny is conserved; making it possible to easily navigate across genomes [75] and suggesting that conifer chromosomes are 'fossilized'. There is on the contrary, a high rate of chromosome evolution in angiosperms [72], as well as frequent polyploidy and genome duplication events. Finally, Jaramillo-Correa *et al.* [76] found that recombination, which has been correlated with levels of genetic diversity, is lower in conifers compared to angiosperms.

There are only a few known exceptions to this general trend of lower evolutionary rates in gymnosperms. Conifers have larger genomes than angiosperms [74], partly due to larger gene families and abundance of pseudogenes and partly due to a very high content in repetitive DNA such as transposable elements [27, 74]. Possible elevated rates of gene duplication and transposition could have occurred along the gymnosperm lineage to cause this genome expansion, with evidence to date suggesting that these events were ancient [77]. Despite these exceptions, conifers exhibit dramatically slower evolutionary rates compared to angiosperms, in particular substitution rates in protein-coding genes, suggesting the existence of conifer-specific evolutionary mechanisms.

## What are the causes for the slow substitution rates in conifer genomes?

Substitution rates vary depending on rates at which mutations appear in individuals and are fixed in the population [9, 13].

First, the rate at which mutations appear is affected by the efficacy of the DNA repair machinery, generation time, and metabolic rate. In animals, mitochondrial genes evolve ten times faster than nuclear genes, but the inverse situation is found in plants [4]. This difference may at least in part originate from the presence of the DNA repair gene *recA* in plant mitochondrial genomes, and its absence in those of animals [78]. To our knowledge, there is no information on the efficiency of the conifer DNA repair system compared to that of angiosperm species. Life history traits such as generation time or total life span are factors that are commonly called forth to explain differences in evolutionary rates detected between species, e.g. in mammals [79], in invertebrates [80] and in plants [81]. In angiosperms, rates of evolution are higher in annuals than in perennials [15]. Our data supports this finding as *Arabidopsis* (an annual) has higher rates than trees. This accords with the germline theory of mutations [82]. However, generation time effects will be unknown until we can reconcile the difference between cell lineage division time and generation time in plants [14]. Conifers exhibit lower values of nucleotide diversity at the population level despite high gene flow and low population structure [65, 70, 71] suggesting that trees accumulate fewer mutations per unit of time than other plants and thus generation time is not sufficient to explain the annual-perennial difference in mutation rates. Finally, the low metabolite rate of conifer trees, with their large body size and temperate to boreal habitats [83], as well as reduced recombination rates [76], could generate fewer nucleotide substitutions in their genomes.

Second, the fixation rate of new mutations depends on the interplay between random genetic drift (i.e. effective population size and population structure), purifying (background) selection and reproductive strategy. Large population sizes and extensive gene flow are often suggested as the causes of low synonymous polymorphism found in conifer populations [58]. Both empirically and theoretically, grey areas remain about the effect of effective population size ($N_e$), population subdivision and selection on the pattern of nucleotide divergence between species [84, 85, 86]. Our results however support the inverse relationship between $N_e$ and neutral substitution rate that is expected by the "nearly neutral theory of molecular evolution" [87]. In addition, with low diversification rate in conifers [72], there have been fewer speciation-associated bottleneck events than in angiosperms, thus continuous low diversity between populations. That conifers are mainly outcrossing (selfing is generally avoided through high early inbreeding depression) is only adding to the homogenization of populations. Indeed, studies have shown that there is weak population structure in Sitka spruce [88] and loblolly pine [89]. Finally, the influence of background selection and other selective forces such as hitchhiking on the genomic reduction of substitution rate in conifers is mostly unknown, although selective sweeps following bottlenecks have been reported for several loci [22, 23, 90]. Teasing out the evolutionary mechanisms controlling the rate of evolution in any organism is a daunting task. When comprehensive data are available across several conifer and other gymnosperm species, comparative analyses will help elucidate if, in what manner and to what extent typical conifer features such as low metabolite rate, long generation time, large effective population and low genetic structure affect substitution rates [91, 92].

## Is the evolutionary slow-down similar between conifer and angiosperm trees?

Conifers have high levels of genetic diversity within population but experience low nucleotide substitution rates and low speciation rates. Strikingly, the same trend can be seen in angiosperm trees and all trees (angiosperm and gymnosperm) share common attributes that may explain this similarity such as perenniality, outcrossed mating system and large population sizes [58, 82]. However, vast evidences point at a more pronounced slow-down in conifers compared to angiosperm trees, for example: recombination rate [76], nucleotide diversity [58] and substitution rates. In this study, we found that conifers have a lower substitution rate at both synonymous and non-synonymous sites than poplar (see Table 1). The existence of conifer-specific factors that explain this difference is therefore likely; gymnosperms have evolved separately from angiosperms for about 300 MY. However, the exact nature and influence of these factors are still to be determined.

## High Adaptability of Conifers to Their Environment

We found that mean dN/dS was about three times higher in conifers than in angiosperms (0.3137 *vs.* 0.0924, respectively; see Table 1) despite much lower substitution rates in conifer protein-coding genes, and that this trend was found throughout almost all functional categories. Higher dN/dS in conifers could be due to a general low mutation rate and a high selective constraint on synonymous mutations, which seems at odds with the neutral expectancy but cannot be completely ruled out, or a general very low mutation rate but a proportionally lower constraint (relative to angiosperm genes) at non-synonymous sites. Assuming a relatively high rate of amino acid change in conifer proteins, high average estimates of dN/dS in conifers have important evolutionary implications, especially in light of the distinctive biology of conifer trees.

## Characteristics of Fast-Evolving Genes and Functional Gene Categories

Among 100 conifer genes with dN/dS > 1, we found a large fraction of genes involved in abiotic and biotic stress response. For example, we found two protein phosphatases with dN/dS > 6, and one protein kinase with dN/dS~3 (see Table 2). Protein phosphatases and kinases act in tandem to regulate signaling pathways for plant stress tolerance or avoidance [93]. Four heat shock proteins, one leucine-rich repeat protein, one histone modification protein, two glycosyltransferases, four glycoside hydrolases, and seven transcription factors are also gene products involved in defense, resistance and/or stress response. Other genes with dN/dS > 1 were involved in cell signaling, development and growth, vesicle trafficking and DNA/RNA binding. These single-gene results were paralleled by a gene set analysis on 1,230 annotated genes ranked by dN/dS, where functional categories involved with heat shock proteins, signal transduction and in the regulation of transcription and nucleic acids were more likely to contain genes with high dN/dS. Conifers, like other long-lived sessile plants, require responsiveness and plasticity to defend themselves against various herbivores and pathogens, as well as abiotic stresses (e.g. temperature and drought). This plasticity can for example be obtained by regulating transcription and DNA/RNA binding proteins, which could explain why these groups of genes seem to have experienced adaptive selection in conifer lineages. In contrast, categories of genes involved in translation, protein assembly, cellular organization and chlorophyll biosynthesis are under strong selective constraint (low dN/dS) because these processes are highly conserved across either the tree of Life, or across photosynthetic organisms (i.e. chlorophyll biosynthesis).

## Adaptability of Conifers

The conifer divergence was dramatically slower at synonymous sites than at non-synonymous sites (11-fold vs. 4-fold), suggesting that more adaptive mutations (and deleterious mutations, but see below) are fixed in conifers than in angiosperms. Indeed, there was a single *Arabidopsis*-poplar orthologue gene with a dN/dS > 1 while values for other orthologues were below 0.6. Conversely, we found a distribution of conifer dN/dS ratios significantly deviated near unity (Figure 1.B), with 100 genes showing values suggesting positive selection (dN/dS > 1). In addition, all GO Slim functional categories showed a significantly higher dN/dS in conifers than in angiosperms, with the exception of DNA/RNA metabolism and translation, which are evolutionary stable processes (Figure 2).

A threshold of unity is usually applied to determine if a gene shows signs of adaptive evolution, but this threshold is overly conservative in the case of pairwise comparisons over the whole length of the alignment. Algorithms exist to identify adaptive mutations at specific sites and/or on specific branches of a species tree, even when dN/dS < 1 over the entire gene, but there is an implicit requirement for comparisons of at least three species [3]. At the time of this study, loblolly pine and Sitka spruce had significantly more publicly available sequences than any other conifer, and we chose to restrain our study to two species and several thousands of genes, rather than opting for additional species but a few hundreds of genes. With more sequences becoming available for conifer species [94], it will be possible to test for positive selection using models of evolution across a tree composed of three or more species.

An overarching goal of modern biology is to uncover the genetic architecture of biological adaptations. Our study suggests that there is a substantial amount of adaptive substitutions in two conifer species and we expect that this finding will be generalized to other conifer taxa, especially in environments where conifers compete in extreme ecological niches. For example, the Vietnamese pine has evolved broad leaves, i.e. flattened needles, to compete for light with evergreen angiosperm trees in tropical forests [95]. In Western North America, lodgepole pine has evolved large and thick-scaled cones where squirrels are absent but crossbills are present, while crossbills evolve larger beaks [96]. An arms race between conifers and herbivorous insects, such as bark beetles, results in the diversification of constitutive defense and stress-induced genes in conifers [97]. Sitka spruce and loblolly pine, like most conifers in their natural environment, have been confronted by various endemic herbivorous pests, which we speculate could be reflected by high dN/dS estimates at genes involved in defense and stress response.

## Why do conifers show more signs of adaptive evolution than most plant lineages?

Our results show that the low mutational rate seen in conifer genes is congruent with higher dN/dS, i.e. higher adaptability at the amino acid level, compared to angiosperm genes. At first, this relationship might seem contradictory and counter-intuitive; it is accepted that mutations are the foundation for adaptation. In conifers, a combination of factors seems to have promoted a staggering high rate of fixation for non-synonymous mutations, despite a generalized low mutation rate.

Little evidence has been found for adaptive evolution in angiosperm genes. In *Arabidopsis thaliana* and *A. lyrata*, purifying selection is the determinant force acting on amino acid substitutions [98]. In addition, Gossmann et al. [99] found little or no signal of adaptation in nine pairs of angiosperm species, except in sunflowers. Other exceptions to this rule are European aspen [100] and the crucifer *Capsella grandiflora* [101], where 30% and 40% of amino acid substitutions have been fixed by natural selection, respectively. What differentiates sunflowers and *C. grandiflora* from the other studied angiosperms are low population genetic structure and especially large effective population size ($N_e$ > 500,000). European aspen has a lower reported $N_e$ (118,000) but it has been argued that 500,000 individuals may not be unrealistic [100]. Strasburg et al. [102] compared different species of sunflowers, and found a positive correlation between $N_e$ and levels of adaptive divergence. Sunflowers, European aspen and *C. grandiflora* are also outcrossing species but an excess of non-synonymous mutations was found in the outcrossing *A. lyrata* [98], so mating system may only have limited effect on selective pressure compared to demographic factors. Lastly, selfing *A. thaliana* appears to have rare adaptive substitutions, likely due to consequent population subdivision and reduced $N_e$ through different bottleneck episodes [98, 103, 104].

In conifers, investigations of sequence divergence at the genome level have not been performed yet. Resequencing and comparative data have already provided a large body of evidence that several individual genes in conifers species have evolved under positive selection [58, 64, 89]. In addition, there are various examples of local adaptation in conifer species, whereby a specific population within the range of the species has expressed a phenotype adapted to an environmental constraint [105, 106, 107]. Concurrent with our results, the overall picture from the study of molecular evolution of conifer genes is that ecology, demography, life history and genome stability of conifers are favorable for the fixation of non-synonymous mutations. While fixation of deleterious mutations is reduced by outcrossing and large effective population size, most non-synonymous mutations are likely beneficial mutations in the conifer phyla. In addition, although deleterious mutations could be fixed through bottlenecks and selective sweeps, it has been shown that the time to establishment of complex adaptations is minimized in species with a large effective population size, even in the advent of deleterious intermediate steps [108].

## CONCLUSIONS

Large-scale and genomewide comparative approaches go beyond comparisons of small groups of candidate genes and provide global evolutionary trends. In this study, we found that there was a dramatic slow-down in the overall mutation rate of conifer orthologues compared to angiosperm orthologues. This finding is compatible with an increase in the fixation of non-synonymous mutations, which can be beneficial for adaptation. Large effective population size is likely the main factor that contributes to this trend, along with low population structure, low recombination and outcrossing mating system.

Several genome sequencing projects in conifer species are now funded including for loblolly pine, Douglas fir, sugar pine, white spruce and Norway spruce. These data will allow phylogenetic comparisons of much greater power then we currently employ. Not only should the present approach be expanded to a phylogenetic context, but future studies may also apply comparative methods to tease out the evolutionary processes under various demographic and ecological scenarios [91, 92]. Finally, resequencing large numbers of candidate genes, once a reference genome sequence is established, will further identify the mode and strength of selection in conifer genomes.

### List of Abbreviations

BAC: Bacterial Artificial Chromosome; cDNA: complementary DNA; EST: Expressed Sequence Tag; FLcDNA: Full-length cDNA; GO: Gene Ontology; MYA: Million Years Ago; ORF: Open Reading Frame; RBH: Reciprocal Best Hit; 4D: 4 fold degenerate.

## DECLARATIONS

### Acknowledgements and Funding

This work was supported by Genome British Columbia, Genome Canada, and the Province of British Columbia (Treenomix II/Conifer Forest Health grant to KR and JB). We thank Stephen Ralph for the production of the Sitka spruce FL-cDNA, Nancy Liao at the Michael Smith Genome Sciences Centre for bioinformatics work, and Elizabeth Flavall for editing the manuscript.

## REFERENCES

1. Nielsen R: Molecular signatures of natural selection. Annu Rev Genet. 2005, 39: 197-218. 10.1146/annurev.genet.39.073003.112420.
2. Hurst LD: Genetics and the understanding of selection. Nat Rev Genet. 2009, 10: 83-93.
3. Yang Z, Nielsen R: Codon-substitution models for detecting molecular adaptation at individual sites along specific lineages. Mol Biol Evol. 2002, 19: 908-917. 10.1093/oxfordjournals.molbev.a004148.
4. Wolfe KH, Li WH, Sharp PM: Rates of nucleotide substitution vary greatly among plant mitochondrial, chloroplast, and nuclear DNAs. Proc Natl Acad Sci USA. 1987, 84: 9054-9058. 10.1073/pnas.84.24.9054.
5. Drouin G, Daoud H, Xia J: Relative rates of synonymous substitutions in the mitochondrial, chloroplast and nuclear genomes of seed plants. Mol Phylogenet Evol. 2008, 49: 827-831. 10.1016/j.ympev.2008.09.009.
6. Britten RJ: Rates of DNA sequence evolution differ between taxonomic groups. Science. 1986, 231: 1393-1398. 10.1126/science.3082006.
7. Kumar S, Subramanian S: Mutation rates in mammalian genomes. Proc Natl Acad Sci USA. 2002, 99: 803-808. 10.1073/pnas.022629899.
8. Nishant KT, Singh ND, Alani E: Genomic mutation rates: what high-throughput methods can tell us. Bioessays. 2009, 31: 912-920. 10.1002/bies.200900017.
9. Lanfear R, Welch JJ, Bromham L: Watching the clock: Studying variation in rates of molecular evolution between species. Trends Ecol Evol. 2010, 25: 495-503. 10.1016/j.tree.2010.06.007.
10. Rocha EPC: The quest for the universals of protein evolution. Trends Genet. 2006, 22: 412-416. 10.1016/j.tig.2006.06.004.
11. Pál C, Papp B, Lercher MJ: An integrated view of protein evolution. Nat Rev Genet. 2006, 7: 337-348. 10.1038/nrg1838.
12. Warnecke T, Weber CC, Hurst LD: Why there is more to protein evolution than protein function: splicing, nucleosomes and dual-coding sequence. Biochem Soc Trans. 2009, 37: 756-761. 10.1042/BST0370756.
13. Baer CF, Miyamoto MM, Denver DR: Mutation rate variation in multicellular eukaryotes: causes and consequences. Nat Rev Genet. 2007, 8: 619-631.
14. Gaut B, Yang L, Takuno S, Eguiarte LE: The patterns and causes of variation in plant nucleotide substitution rates. Annual Review of Ecology, Evolution, and Systematics. 2011, 42: 245-266. 10.1146/annurev-ecolsys-102710-145119.
15. Yue JX, Li J, Wang D, Araki H, Tian D, Yang S: Genome-wide investigation reveals high evolutionary rates in annual model plants. BMC Plant Biol. 2010, 10: 242-10.1186/1471-2229-10-242.
16. Hedges SB, Dudley J, Kumar S: TimeTree: a public knowledge-base of divergence times among organisms. Bioinformatics. 2006, 22: 2971-2972. 10.1093/bioinformatics/btl505.
17. Plomion C, Chagné D, Pot D, Kumar S, Wilcox P, Burdon R, Prat D, Peterson D, Paiva J, Chaumeil P, et al: Pines. Forest Trees. Edited by: Kole C. 2007, Berlin Heidelberg: Springer-Verlag, 29-92.
18. Bousquet J, Isabel N, Pelgas B, Cottrell J, Rungis D, Ritland K: Spruce. Forest Trees. Edited by: Kole C. 2007, Berlin Heidelberg: Springer-Verlag, 93-114.
19. Hurme P, Sillanpaa MJ, Arjas E, Repo T, Savolainen O: Genetic basis of climatic adaptation in Scots pine by Bayesian quantitative trait locus analysis. Genetics. 2000, 156: 1309-1322.
20. Ukrainetz N, Ritland K, Mansfield S: Identification of

quantitative trait loci for wood quality and growth across eight full-sib coastal Douglas-fir families. Tree Genet Genom. 2008, 4: 159-170. 10.1007/s11295-007-0097-x.

21. Pelgas B, Bousquet J, Meirmans P, Ritland K, Isabel N: QTL mapping in white spruce: gene maps and genomic regions underlying adaptive traits across pedigrees, years and environments. BMC Genomics. 2011, 12: 145-10.1186/1471-2164-12-145.

22. Eckert AJ, Wegrzyn JL, Pande B, Jermstad KD, Lee JM, Liechty JD, Tearse BR, Krutovsky KV, Neale DB: Multilocus patterns of nucleotide diversity and divergence reveal positive selection at candidate genes related to cold-hardiness in coastal Douglas-fir (*Pseudotsuga menziesii* var. *menziesii*). Genetics. 2009, 183: 289-298. 10.1534/genetics.109.103895.

23. Ersoz ES, Wright MH, González-Martínez SC, Langley CH, Neale DB: Evolution of disease response genes in loblolly pine: insights from candidate genes. PLoS ONE. 2010, 5: e14234-10.1371/journal.pone.0014234.

24. Quesada T, Gopal V, Cumbie WP, Eckert AJ, Wegrzyn JL, Neale DB, Goldfarb B, Huber DA, Casella G, Davis JM: Association mapping of quantitative disease resistance in a natural population of loblolly pine (*Pinus taeda* L.). Genetics. 2010, 186: 677-686. 10.1534/genetics.110.117549.

25. Holliday JA, Ritland K, Aitken SN: Widespread, ecologically relevant genetic markers developed from association mapping of climate-related traits in Sitka spruce (*Picea sitchensis*). New Phytol. 2010, 188: 501-514. 10.1111/j.1469-8137.2010.03380.x.

26. Hamberger B, Hall D, Yuen M, Oddy C, Hamberger B, Keeling CI, Ritland C, Ritland K, Bohlmann J: Targeted isolation, sequence assembly and characterization of two white spruce (*Picea glauca*) BAC clones for terpenoid synthase and cytochrome P450 genes involved in conifer defence reveal insights into a conifer genome. BMC Plant Biol. 2009, 9: 106-10.1186/1471-2229-9-106.

27. Kovach A, Wegrzyn J, Parra G, Holt C, Bruening G, Loopstra C, Hartigan J, Yandell M, Langley C, Korf I, Neale D: The *Pinus taeda* genome is characterized by diverse and highly diverged repetitive sequences. BMC Genomics. 2010, 11: 420-10.1186/1471-2164-11-420.

28. Pavy N, Boyle B, Nelson C, Paule C, Giguère I, Caron S, Parsons LS, Dallaire N, Bedon F, Bérubé H, et al: Identification of conserved core xylem gene sets: conifer cDNA microarray development, transcript profiling and computational analyses. New Phytol. 2008, 180: 766-786. 10.1111/j.1469-8137.2008.02615.x.

29. Verne S, Jaquish B, White R, Ritland C, Ritland K: Global transcriptome analysis of constitutive resistance to the white pine weevil in spruce. Genome Biology and Evolution. 2011

30. Keeling C, Weisshaar S, Ralph S, Jancsik S, Hamberger B, Dullat H, Bohlmann J: Transcriptome mining, functional characterization, and phylogeny of a large terpene synthase gene family in spruce (Picea spp.). BMC Plant Biol. 2011, 11: 43-10.1186/1471-2229-11-43.

31. Lippert DN, Ralph SG, Phillips M, White R, Smith D, Hardie D, Gershenzon J, Ritland K, Borchers CH, Bohlmann J: Quantitative iTRAQ proteome and comparative transcriptome analysis of elicitor-induced Norway spruce (*Picea abies*) cells reveals elements of calcium signaling in the early conifer defense response. Proteomics. 2009, 9:

32. Hall DE, Robert JA, Keeling CI, Domanski D, Quesada AL, Jancsik S, Kuzyk MA, Hamberger B, Borchers CH, Bohlmann J: An integrated genomic, proteomic and biochemical analysis of (+)-3-carene biosynthesis in Sitka spruce (*Picea sitchensis*) genotypes that are resistant or susceptible to white pine weevil. The Plant Journal. 2011, 65: 936-948. 10.1111/j.1365-313X.2010.04478.x.

33. Ralph SG, Chun HJ, Kolosova N, Cooper D, Oddy C, Ritland CE, Kirkpatrick R, Moore R, Barber S, Holt RA, et al: A conifer genomics resource of 200,000 spruce (*Picea* spp.) ESTs and 6,464 high-quality, sequence-finished full-length cDNAs for Sitka spruce (*Picea sitchensis*). BMC Genomics. 2008, 9: 484-10.1186/1471-2164-9-484.

34. Lippert D, Yuen M, Bohlmann J: Spruce proteome DB: a resource for conifer proteomics research. Tree Genet Genom. 2009, 5: 723-727. 10.1007/s11295-009-0220-2.

35. Treenomix - Conifer Forest Health. [http://www.treenomix.ca/]

36. Schneider M, Lane L, Boutet E, Lieberherr D, Tognolli M, Bougueleret L, Bairoch A: The UniProtKB/Swiss-Prot knowledgebase and its Plant Proteome Annotation Program. J Proteomics. 2009, 72: 567-573. 10.1016/j.jprot.2008.11.010.

37. Hirsh AE, Fraser HB: Protein dispensability and rate of evolution. Nature. 2001, 411: 1046-1049. 10.1038/35082561.

38. Jordan IK, Rogozin IB, Wolf YI, Koonin EV: Essential genes are more evolutionarily conserved than are nonessential genes in bacteria. Genome Res. 2002, 12: 962-968.

39. The Arabidopsis Information Resource. [http://www.arabidopsis.org/]

40. JGI Genome Portal. [http://genomeportal.jgi-psf.org/]

41. Smedley D, Haider S, Ballester B, Holland R, London D, Thorisson G, Kasprzyk A: BioMart - biological queries made easy. BMC Genomics. 2009, 10: 22-10.1186/1471-2164-10-22.

42. Subramanian A, Kaufmann M, Morgenstern B: DIALIGN-TX: greedy and progressive approaches for segment-based multiple sequence alignment. Algorithms Mol Biol. 2008, 3: 6-10.1186/1748-7188-3-6.

43. Yang Z: PAML: a program package for phylogenetic analysis by maximum likelihood. Comput Appl Biosci. 1997, 13: 555-556.

44. Yang Z: PAML 4: Phylogenetic Analysis by Maximum Likelihood. Mol Biol Evol. 2007, 24: 1586-1591. 10.1093/molbev/msm088.

45. R Development Core Team: R: A language and environment for statistical computing. R Foundation for Statistical Computing, Vienna, Austria. ISBN 3-900051-07-0, URL. 2011, [http://www.R-project.org/]

46. Savard L, Li P, Strauss SH, Chase MW, Michaud M, Bousquet J: Chloroplast and nuclear gene sequences indicate late Pennsylvanian time for the last common ancestor of extant seed plants. Proc Natl Acad Sci USA. 1994, 91: 5163-5167. 10.1073/pnas.91.11.5163.
47. Wang XQ, Tank DC, Sang T: Phylogeny and divergence times in Pinaceae: Evidence from three genomes. Mol Biol Evol. 2000, 17: 773-781.
48. Miller C: Mesozoic conifers. Bot Rev. 1977, 43: 217-280. 10.1007/BF02860718.
49. Lin C-P, Huang J-P, Wu C-S, Hsu C-Y, Chaw S-M: Comparative chloroplast genomics reveals the evolution of Pinaceae genera and subfamilies. Genome Biol Evol. 2010, 2: 504-517. 10.1093/gbe/evq036.
50. Alvin KL: Further conifers of the Pinaceae from the Wealden formation of Belgium. 1960, Bruxelles: Institut Royal des Sciences Naturelles
51. Gernandt DS, Magallón S, Geada López G, Zerón Flores O, Willyard A, Liston A: Use of simultaneous analyses to guide fossil-based calibrations of Pinaceae phylogeny. Int J Plant Sci. 2008, 169: 1086-1099. 10.1086/590472.
52. Bell CD, Soltis DE, Soltis PS: The age and diversification of the angiosperms re-revisited. Am J Bot. 2010, 97: 1296-1303. 10.3732/ajb.0900346.
53. Medina I, Carbonell J, Pulido L, Madeira SC, Goetz S, Conesa A, Tarraga J, Pascual-Montano A, Nogales-Cadenas R, Santoyo J, et al: Babelomics: an integrative platform for the analysis of transcriptomics, proteomics and genomic data with advanced functional profiling. Nucl Acids Res. 2010, 38: W210-213. 10.1093/nar/gkq388.
54. Swarbreck D, Wilks C, Lamesch P, Berardini TZ, Garcia-Hernandez M, Foerster H, Li D, Meyer T, Muller R, Ploetz L, et al: The Arabidopsis Information Resource (TAIR): gene structure and function annotation. Nucleic Acids Res. 2008, 36: D1009-1014.
55. Kimura M: The Neutral Theory of Molecular Evolution. 1983, Cambridge: Cambridge University Press
56. Chamary JV, Parmley JL, Hurst LD: Hearing silence: non-neutral evolution at synonymous sites in mammals. Nat Rev Genet. 2006, 7: 98-108. 10.1038/nrg1770.
57. Tuskan GA, Difazio S, Jansson S, Bohlmann J, Grigoriev I, Hellsten U, Putnam N, Ralph S, Rombauts S, Salamov A, et al: The genome of black cottonwood, *Populus trichocarpa* (Torr. & Gray). Science. 2006, 313: 1596-1604. 10.1126/science.1128691.
58. Savolainen O, Pyhäjärvi T: Genomic diversity in forest trees. Curr Opin Plant Biol. 2007, 10: 162-167. 10.1016/j.pbi.2007.01.011.
59. Koch MA, Haubold B, Mitchell-Olds T: Comparative evolutionary analysis of chalcone synthase and alcohol dehydrogenase loci in *Arabidopsis*, *Arabis*, and related genera (Brassicaceae). Mol Biol Evol. 2000, 17: 1483-1498.
60. Beilstein MA, Nagalingum NS, Clements MD, Manchester SR, Mathews S: Dated molecular phylogenies indicate a Miocene origin for Arabidopsis thaliana. Proc Natl Acad Sci USA. 2010, 107: 18724-18728. 10.1073/pnas.0909766107.
61. Künstner A, Nabholz B, Ellegren H: Significant selective constraint at 4-fold degenerate sites in the avian genome and its consequence for detection of positive selection. Gen Biol Evol. 2011
62. Kusumi J, Tsumura Y, Yoshimaru H, Tachida H: Molecular evolution of nuclear genes in Cupressacea, a group of conifer trees. Mol Biol Evol. 2002, 19: 736-747. 10.1093/oxfordjournals.molbev.a004132.
63. Bouillé M, Bousquet J: Trans-species shared polymorphisms at orthologous nuclear gene loci among distant species in the conifer *Picea* (Pinaceae): implications for the long-term maintenance of genetic diversity in trees. Am J Bot. 2005, 92: 63-73. 10.3732/ajb.92.1.63.
64. Palmé A, Pyhäjärvi T, Wachowiak W, Savolainen O: Selection on nuclear genes in a *Pinus* phylogeny. Mol Biol Evol. 2009, 26: 893-905. 10.1093/molbev/msp010.
65. Chen J, Kallman T, Gyllenstrand N, Lascoux M: New insights on the speciation history and nucleotide diversity of three boreal spruce species and a Tertiary relict. Heredity. 2010, 104: 3-14. 10.1038/hdy.2009.88.
66. Ujino-Ihara T, Tsumura Y: Screening for genes specific to coniferous species. Tree Physiology. 2008, 28: 1325-1330. 10.1093/treephys/28.9.1325.
67. Volokita M, Rosilio-Brami T, Rivkin N, Zik M: Combining comparative sequence and genomic data to ascertain phylogenetic relationships and explore the evolution of the large GDSL-lipase family in land-plants. Mol Biol Evol. 2010
68. Li X, Wu H, Southerton S: Comparative genomics reveals conservative evolution of the xylem transcriptome in vascular plants. BMC Evol Biol. 2010, 10: 190-10.1186/1471-2148-10-190.
69. García-Gil MR, Mikkonen M, Savolainen O: Nucleotide diversity at two phytochrome loci along a latitudinal cline in *Pinus sylvestris*. Mol Ecol. 2003, 12: 1195-1206. 10.1046/j.1365-294X.2003.01826.x.
70. Heuertz M, De Paoli E, Kallman T, Larsson H, Jurman I, Morgante M, Lascoux M, Gyllenstrand N: Multilocus patterns of nucleotide diversity, linkage disequilibrium and demographic history of Norway spruce [*Picea abies* (L.) Karst]. Genetics. 2006, 174: 2095-2105. 10.1534/genetics.106.065102.
71. Pyhäjärvi T, Garcia-Gil MR, Knurr T, Mikkonen M, Wachowiak W, Savolainen O: Demographic history has influenced nucleotide diversity in European *Pinus sylvestris* populations. Genetics. 2007, 177: 1713-1724. 10.1534/genetics.107.077099.
72. Levin DA, Wilson AC: Rates of evolution in seed plants: Net increase in diversity of chromosome numbers and species numbers through time. Proc Natl Acad Sci USA. 1976, 73: 2086-2090. 10.1073/pnas.73.6.2086.
73. Lanfear R, Ho SYW, Love D, Bromham L: Mutation rate is linked to diversification in birds. Proc Natl Acad Sci USA. 2010, 107: 20423-20428. 10.1073/pnas.1007888107.
74. Ahuja MR, Neale DB: Evolution of genome size in

conifers. Silvae Genet. 2005, 54: 126-137.

75. Pelgas B, Beauseigle S, Achere V, Jeandroz S, Bousquet J, Isabel N: Comparative genome mapping among *Picea glauca, P. mariana* × *P. rubens* and *P. abies*, and correspondence with other Pinaceae. Theor Appl Genet. 2006, 113: 1371-1393. 10.1007/s00122-006-0354-7.

76. Jaramillo-Correa J, Verdu M, Gonzalez-Martinez S: The contribution of recombination to heterozygosity differs among plant evolutionary lineages and life-forms. BMC Evol Biol. 2010, 10: 22-10.1186/1471-2148-10-22.

77. Friesen N, Brandes A, Heslop-Harrison JS: Diversity, origin, and distribution of retrotransposons (*gypsy* and *copia*) in conifers. Mol Biol Evol. 2001, 18: 1176-1188. 10.1093/oxfordjournals.molbev.a003905.

78. Lin Z, Kong H, Nei M, Ma H: Origins and evolution of the *recA/RAD51* gene family: Evidence for ancient gene duplication and endosymbiotic gene transfer. Proc Natl Acad Sci USA. 2006, 103: 10328-10333. 10.1073/pnas.0604232103.

79. Welch J, Bininda-Emonds O, Bromham L: Correlates of substitution rate variation in mammalian protein-coding sequences. BMC Evol Biol. 2008, 8: 53-10.1186/1471-2148-8-53.

80. Thomas JA, Welch JJ, Lanfear R, Bromham L: A generation time effect on the rate of molecular evolution in invertebrates. Mol Biol Evol. 2010, 27: 1173-1180. 10.1093/molbev/msq009.

81. Smith SA, Donoghue MJ: Rates of molecular evolution are linked to life history in flowering plants. Science. 2008, 322: 86-89. 10.1126/science.1163197.

82. Petit RJ, Hampe A: Some evolutionary consequences of being a tree. Annu Rev Ecol Evol Syst. 2006, 37: 187-214. 10.1146/annurev.ecolsys.37.091305.110215.

83. Gillooly JF, Allen AP, West GB, Brown JH: The rate of DNA evolution: Effects of body size and temperature on the molecular clock. Proc Natl Acad Sci USA. 2005, 102: 140-145. 10.1073/pnas.0407735101.

84. Whitlock MC: Fixation probability and time in subdivided populations. Genetics. 2003, 164: 767-779.

85. Woolfit M: Effective population size and the rate and pattern of nucleotide substitutions. Biol Lett. 2009, 5: 417-420. 10.1098/rsbl.2009.0155.

86. Li J, Li H, Jakobsson M, Li SEN, SjÖDin PER, Lascoux M: Joint analysis of demography and selection in population genetics: where do we stand and where could we go?. Mol Ecol. 2011

87. Ohta T: The nearly neutral theory of molecular evolution. Annu Rev Ecol Syst. 1992, 23: 263-286. 10.1146/annurev.es.23.110192.001403.

88. Gapare WJ, Aitken SN: Strong spatial genetic structure in peripheral but not core populations of Sitka spruce [*Picea sitchensis* (Bong.) Carr.]. Mol Ecol. 2005, 14: 2659-2667. 10.1111/j.1365-294X.2005.02633.x.

89. Eckert AJ, van Heervaarden J, Wegrzyn JL, Nelson CD, Ross-Ibarra J, Gonzalez-Martinez SC, Neale DB: Patterns of population structure and environmental associations to aridity across the range of Loblolly pine (*Pinus taeda* L., Pinaceae). Genetics. 2010, 185: 969-982. 10.1534/genetics.110.115543.

90. Palmé AE, Wright M, Savolainen O: Patterns of divergence among conifer ESTs and polymorphism in *Pinus sylvestris*identify putative selective sweeps. Mol Biol Evol. 2008, 25: 2567-2577. 10.1093/molbev/msn194.

91. Mayrose I, Otto SP: A likelihood method for detecting trait-dependent shifts in the rate of molecular evolution. Mol Biol Evol. 2011, 28: 759-770. 10.1093/molbev/msq263.

92. Lanfear R: The local-clock permutation test: a simple test to compare rates of molecular evolution on phylogenetic trees. Evolution. 2011, 65: 606-611. 10.1111/j.1558-5646.2010.01160.x.

93. Chae L, Pandey GK, Luan S, Cheong YH, Kim KN: Protein kinases and phosphatases for stress signal transduction in plants. Abiotic Stress Adaptation in Plants. Edited by: Pareek A, Sopory SK. 2010, Bohnert HJ: Springer Netherlands, 123-163.

94. Rigault P, Boyle B, Lepage P, Cooke JEK, Bousquet J, MacKay JJ: A white spruce gene catalog for conifer genome analyses. Plant Physiol. 2011, 157: 14-28. 10.1104/pp.111.179663.

95. Brodribb TJ, Feild TS: Evolutionary significance of a flat-leaved *Pinus* in Vietnamese rainforest. New Phytol. 2008, 178: 201-209. 10.1111/j.1469-8137.2007.02338.x.

96. Benkman C: Diversifying coevolution between crossbills and conifers. Evo Edu Outreach. 2010, 3: 47-53. 10.1007/s12052-009-0190-8.

97. Raffa KF, Berryman AA: Interacting selective pressures in conifer-bark beetle systems: A basis for reciprocal adaptations?. Amer Nat. 1987, 129: 234-262. 10.1086/284633.

98. Foxe JP, Dar VU, Zheng H, Nordborg M, Gaut BS, Wright SI: Selection on amino acid substitutions in *Arabidopsis*. Mol Biol Evol. 2008, 25: 1375-1383. 10.1093/molbev/msn079.

99. Gossmann TI, Song B-H, Windsor AJ, Mitchell-Olds T, Dixon CJ, Kapralov MV, Filatov DA, Eyre-Walker A: Genome wide analyses reveal little evidence for adaptive evolution in many plant species. Mol Biol Evol. 2010, 27: 1822-1832. 10.1093/molbev/msq079.

100. Ingvarsson PK: Natural selection on synonymous and nonsynonymous mutations shapes patterns of polymorphism in*Populus tremula*. Mol Biol Evol. 2010, 27: 650-660. 10.1093/molbev/msp255.

101. Slotte T, Foxe JP, Hazzouri KM, Wright SI: Genome-wide evidence for efficient positive and purifying selection in *Capsella grandiflora*, a plant species with a large effective population size. Mol Biol Evol. 2010, 27: 1813-1821. 10.1093/molbev/msq062.

102. Strasburg JL, Kane NC, Raduski AR, Bonin A, Michelmore R, Rieseberg LH: Effective population size is positively correlated with levels of adaptive divergence among annual sunflowers. Mol Biol Evol. 2011, 28: 1569-1580. 10.1093/molbev/msq270.

103. Nordborg M, Hu TT, Ishino Y, Jhaveri J, Toomajian C, Zheng H, Bakker E, Calabrese P, Gladstone J, Goyal

R, et al: The pattern of polymorphism in *Arabidopsis thaliana*. PLoS Biol. 2005, 3: e196-10.1371/journal.pbio.0030196.

104. Cao J, Schneeberger K, Ossowski S, Gunther T, Bender S, Fitz J, Koenig D, Lanz C, Stegle O, Lippert C, et al: Whole-genome sequencing of multiple *Arabidopsis thaliana* populations. Nat Genet. 2011, 43: 956-963. 10.1038/ng.911.

105. King JN, Alfaro RI, Cartwright C: Genetic resistance of Sitka spruce (*Picea sitchensis*) populations to the white pine weevil (*Pissodes strobi*): distribution of resistance. Forestry. 2004, 77: 269-278. 10.1093/forestry/77.4.269.

106. Mimura M, Aitken SN: Local adaptation at the range peripheries of Sitka spruce. J Evol Biol. 2010, 23: 249-258. 10.1111/j.1420-9101.2009.01910.x.

107. Grivet D, Sebastiani F, Alia R, Bataillon T, Torre S, Zabal-Aguirre M, Vendramin GG, Gonzalez-Martinez SC: Molecular footprints of local adaptation in two Mediterranean conifers. Mol Biol Evol. 2010, 28: 101-116.

108. Lynch M, Abegg A: The rate of establishment of complex adaptations. Mol Biol Evol. 2010, 27: 1404-1414. 10.1093/molbev/msq020.

CHAPTER 5

# EXPLORING DIVERSIFICATION AND GENOME SIZE EVOLUTION IN EXTANT GYMNOSPERMS THROUGH PHYLOGENETIC SYNTHESIS

J. Gordon Burleigh, W. Brad Barbazuk, John M. Davis, Alison M. Morse, and Pamela S. Soltis

## INTRODUCTION

Recent advances in sequencing technology offer the possibility of identifying the genetic mechanisms that influence evolutionarily important characters and ultimately drive diversification. Within angiosperms, large-scale phylogenetic analyses have identified complex patterns of diversification (e.g., [1–3]), and numerous genomes are at least partially sequenced. Yet the other major clade of seed plants, the gymnosperms, have received far less attention, with few comprehensive studies of diversification and no sequenced genomes. Note that throughout this paper "gymnosperms" specifies only the approximately 1000 extant species within cycads,Ginkgo, Gnetales, and conifers. These comprise the Acrogymnospermae clade described by Cantino et al. [4].

Many gymnosperms have exceptionally large genomes (e.g., [5–7]), and this has hindered whole-genome sequencing projects, especially among economically important Pinus species. This large genome size is interesting because one suggested mechanism for rapid increases in genome size, polyploidy, is rare among gymnosperms [8]. Recent sequencing efforts have elucidated some of genomic characteristics associated with the large genome size in Pinus. Morse et al. [9] identified a large retrotransposon family in Pinus, that, with other retrotransposon families, accounts for much of the genomic complexity. Similarly, recent sequencing of 10 BAC (bacterial artificial chromosome) clones from Pinus taeda identified many conifer-specific LTR (long terminal repeat) retroelements [10]. These studies suggest that the large genome size may be caused by rapid expansion of retrotransposons and may be limited to conifers, Pinaceae, or Pinus. Other studies have quantified patterns of genome size among gymnosperms, especially within Pinus and the other Pinaceae [6, 7, 11–14]. These studies have largely focused on finding morphological, biogeographic, or life history correlates of genome size, but the rates and patterns of genome size evolution in gymnosperms are largely unknown.

This study first synthesizes the available phylogenetically informative sequences to build a phylogenetic hypothesis of gymnosperms that reflects the recent advances in sequencing and computational phylogenetics. The resulting tree provides a starting point for large-scale evolutionary and ecological analyses of gymnosperms and will hopefully be a resource to promote and guide future phylogenetic and comparative studies. We use the tree to examine large-scale patterns of diversification of the extant gymnosperm lineages and also to examine rates of genome size evolution.

## METHODS AND MATERIALS

### Supermatrix Phylogenetic Inference

We constructed a phylogenetic hypothesis of gymnosperms from available, phylogenetically informative sequence data in GenBank that was available on June 30, 2009. We first downloaded from GenBank all core nucleotide

**Citation:** J. Gordon Burleigh, W. Brad Barbazuk, John M. Davis, Alison M. Morse, and Pamela S. Soltis, "Exploring Diversification and Genome Size Evolution in Extant Gymnosperms through Phylogenetic Synthesis," Journal of Botany, vol. 2012, Article ID 292857, 6 pages, 2012. doi:10.1155/2012/292857.

sequence data from gymnosperms (Coniferophyta, Cycadophyta, Ginkgophyta, and Gnetophyta). Additionally, we downloaded sequences from the "basal angiosperm" lineages (e.g., Amborella, Nymphaeales, Chloranthaceae, and Austrobaileyales) to represent the angiosperms and a diverse sampling of Moniliformopses taxa (including species from Equisetum, Psilotum, Ophioglossum, Botrychium, Angiopteris, and Adiantum) to use as outgroups.

To identify sets of homologous sequences from the GenBank data, we clustered sequences less than 10,000 bp in length based on results from an all-by-all pairwise BLAST analysis. The all-by-all blastn search was done with blastall using the default parameters [15]. Significant BLAST hits had a maximum e-value of $1.0e^{-10}$ and at least 50% overlap of both the target and query sequences. A Perl script identified the largest clusters of sequences in which each sequence has a significant BLAST hit against at least one other sequence in the cluster. We only considered clusters containing loci that had been used previously for phylogenetic analyses. This included plastid and mitochondrial loci as well as nuclear 18S rDNA, 26S rDNA, and internal transcribed spacer (ITS) sequences. Among these clusters, those containing sequences from at least 15 taxa were aligned using Muscle [16], and the resulting alignments were manually checked and adjusted. The resulting alignments were edited for inclusion in the supermatrix by removing hybrid taxa and those that lacked a specific epithet and also keeping only a single sequence per species. The final cluster alignments were then concatenated to make a single phylogenetic supermatrix (e.g., [17]).

## Phylogenetic and Dating Analysis

To estimate the optimal topology and molecular branch lengths for the gymnosperms, we performed maximum likelihood (ML) phylogenetic analysis on the full supermatrix alignment using RAxML-VI-HPC version 7.0.4 [18]. All ML analyses used the general time reversible (GTR) nucleotide substitution model with the default settings for the optimization of individual per-site substitution rates and classification of these rates into rate categories. To assess uncertainty in the topology and branch length estimates, we ran 100 nonparametric bootstrap replicates on the original data set [19].

We transformed the optimal and bootstrap trees to chronograms, ultrametric trees in which the branch lengths represent time, using penalized likelihood [20] implemented in r8s version 1.71 [21]. We used a smoothing parameter of 10000, which was chosen based on cross-validation of the fossil constraints. For the r8s analysis, we used the same time constraints on seed plant clades used by Won and Renner [22]. The most recent common ancestor of seed plants was constrained to a maximum age of 385 million years ago (mya). The most recent common ancestor of the extant gymnosperms was fixed at 315 mya and Gnetum at 110 mya. The following clades were given minimum age constraints: angisperms: 125 mya, cycads: 270 mya, Cupressaceae: 90 mya, Araucariaceae: 160 mya, Gnetales + Pinaceae: 225 mya, Pinaceae: 90 mya, and Gnetales: 125 mya.

## Diversification Analysis

To examine the general patterns of diversification through time among the extant gymnosperm lineages, we first made lineage through time plots using the R package APE [23]. To account for uncertainty in the dating estimates, we plotted each bootstrap tree after it had been transformed into a chronogram and all nongymnosperm taxa were removed.

Since there appears to be much variance in the divergence time estimates among trees, and branch length estimates are often unreliable, especially when estimated from such a sparse, heterogeneous sequence matrix, we used a test for changes in diversification rate that relies on tree shape, not branch lengths. Specifically, we used the whole-tree, topology-based test described by Moore et al. [24] to detect nodes associated with significant shifts in diversification rate based on the $\Delta_1$ statistic. The analyses were performed using the apTreeshape R package [25]. We used only the optimal tree estimate and again, pruned all non-gymnosperm taxa from the tree prior to analysis.

## Rates of Genome Size Evolution

We first assembled a set of mean genome size data for all gymnosperms present in the phylogenetic tree (in pg DNA) from the Kew C-value database [26]. This includes data from the studies of Murray [6] and Grotkopp et al. [14]. When there were multiple estimates available from a single species, we used the mean of the estimates. We tested for shifts in rates of genome size evolution using Brownie v. 2.1.2 [27]. We used the censored rate test, which tests for differences in rates of evolution of a continuous character (genome size) in one clade versus another clade or paraphyletic group based on a Brownian motion model. We made the following comparisons: conifers + Gnetales versus cycads + Ginkgo, Pinaceae versus other conifers + Gnetales, non-Pinaceae conifers + Gnetales versus cycads + Ginkgo, Pinus versus other Pinaceae, the non-Pinus Pinceae versus the other conifers + Gnetales, and Pinus subgenus Strobus subgenus versus Pinus subgenus Pinus. To account for topological and branch length uncertainty, we performed all hypothesis tests in Brownie on each bootstrap tree and weighted the results across replicates. The penalized likelihood analysis in r8s collapsed some branch lengths to 0, and Brownie does not work with 0 branch lengths in the tree. Thus, prior to the Brownie analysis, all 0 branch lengths were changed to 0.1.

# RESULTS

## Phylogenetic Data

The alignment from the complete supermatrix contains sequences from 950 taxa (739 gymnosperms, 108 angiosperms, and 103 nonseed plant outgroups) and is 74,105 characters in length. The 739 gymnosperm taxa include at least one representative from every family as well as from 88 genera. In total, the matrix contains 4,511,144 nucleotides and 93.6% missing data. The number of nucleotides per taxon ranges from 252 to 33,138 (average = 4,749; median = 3,355).

## Phylogenetic Inference

In the 950-taxon trees, 63.3% (601) of the nodes have at least 50% bootstrap support, 41.7% (396) have at least 70% support, 25.8% (245) have at least 90% support, and 9.7% (92) have

100% support. The seed plants have 100% support, and the angiosperms are sister to a clade of all gymnosperms (Figure 1). Within gymnosperms, a clade of Ginkgo + cycads (bootstrap support (BS) = 66%) is sister to a clade consisting of conifers + Gnetales (BS = 96%). Gnetales are sister to Pinaceae within the conifers, although the "Gne-Pine" clade has only 57% support. Within the major groups of gymnosperms (conifers, Gnetales, and cycads), family-level and generic relationships generally are congruent with those inferred in other analyses (Figure 1). Of the 54 gymnosperm genera represented by more than one species in the tree, 47 have at least 50% bootstrap support, 36 have at least 90% bootstrap support, and 26 have 100% bootstrap support. A full version of the bootstrap consensus tree is available as Supplemental Material.

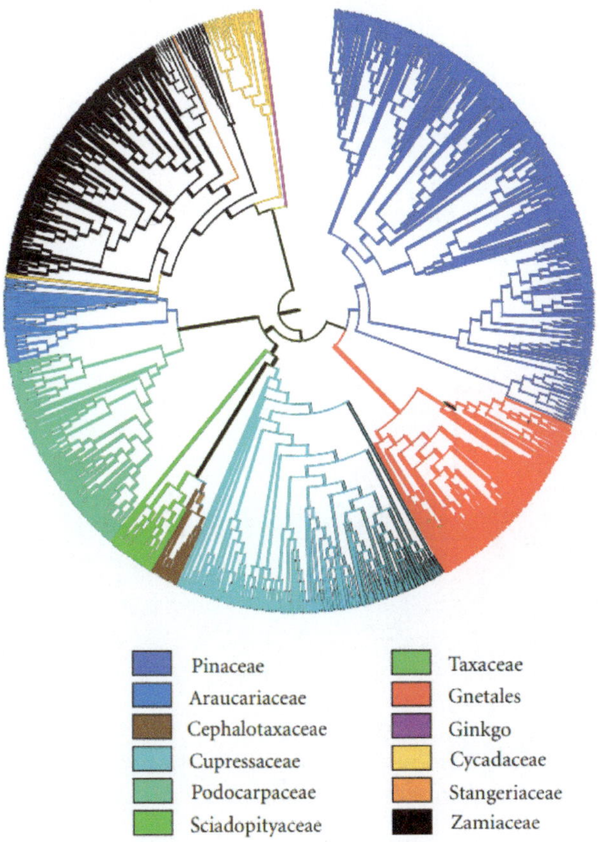


**Figure 1:** Overview of the ML tree of 739 gymnosperm taxa; angiosperms and outgroups have been removed. Colors represent the different families of conifers (Pinaceae, Araucariaceae, Cephalotaxaceae, Cupressaceae, Podocarpaceae, Sciadopityaceae, and Taxaceae), Gnetales, Ginkgo, and the families of cycads (Cycadaceae, Stangeriaceae, and Zamiaceae). A full bootstrap consensus tree is available as Supplementary Material available online at doi:10.1155/2011/292857.

## Diversification

Although the lineage through time plots display much variation among bootstrap replicates (Figure 2), the general trend among the bootstrap trees is similar, with what appears to be high and possibly increasing diversification over the last 100 million years. Still, lineage through time plots are imprecise and difficult to interpret. If this trend of high recent diversification were true, we would expect to find evidence of increased rates of diversification in some relatively young clades.

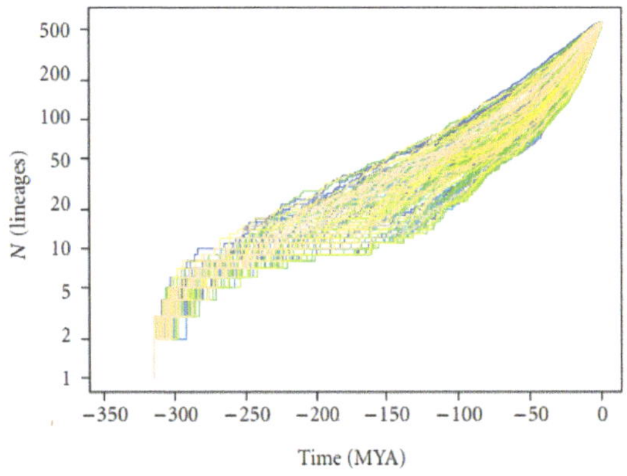


**Figure 2:** Lineage through time plot for the gymnosperm species. All bootstrap trees, with ultrametric branch lengths from r8s, were pruned to include only the gymnosperm taxa. Each line represents a single ML bootstrap tree. The graph shows the pattern of diversification of the gymnosperm taxa in the tree through time, as the tree grew from a single lineage at the root to the current sampling of 739 species.

The $\Delta_1$ statistic indicated a significant shift in the rates of diversification at 10 nodes in the tree. Several are within the cycads. This includes the node dividing Cycas and Epicycas species from the other cycads (P = 0.0474) and its daughter node separating Cycas, Epicycas, and Dioon from the other cycads (P = 0.157). Also, two basal-most nodes of Zamia show significant shifts in diversification rates (P = 0.014 and 0.316). Within conifers, there is a significant shift in diversification at the most recent common ancestor of Podocarpus (P = 0.017). Also, there are significant shifts in diversification at the two basal nodes of Cupressaceae (P = 0.0326 and 0.0366) and within Cupressaceae, at the most recent common ancestor of Callitris, Neocallitropsis, Actinostrobus, Widdringtonia, Fitzroya, Diselma, and Austrocedrus (P = 0.0387). Finally, there is a significant shift in two of the basal nodes of Picea (P = 0.0166, P = 0.0029).

## Genome Size Evolution

Based on the large size of genomes of Pinus species, we hypothesized that there may be an increase in the rate of genome size evolution (Figure 3). We performed a series of likelihood ratio tests to examine the patterns of rate variation throughout gymnosperms, with a focus on testing for rate variation associated with conifers, Pinaceae, and Pinus (Table 1).

**Table 1:** Rate estimates from the two rate parameter models from Brownie. *Indicate that the single rate model was rejected based on the Chi-squared *P* value (***P*<0.005; ****P*<0.0005). Significance was also assessed using AIC.

| Comparison | Rates of Genome Evolution |
|---|---|
| Conifers + Gnetales | 1.878*** |
| Cycads + *Ginkgo* | 0.095 |
| Pinaceae | 2.715*** |
| Other conifers + Gnetales | 0.178 |
| Other conifers + Gnetales | 0.178 |
| Cycads + *Ginkgo* | 0.095 |
| *Pinus* | 3.234** |
| Other Pinaceae | 0.431 |
| Other Pinaceae | 0.431 |
| Other conifers + Gnetales | 0.178 |
| *Pinus* Strobus subgenus | 2.66 |
| *Pinus* Pinus subgenus | 3.56 |

In all comparisons in which Pinus (or a group containing Pinus) was compared to another group, the group with Pinus showed significantly elevated rates of genome size evolution (Table 1). We detected no significant shifts in rates of evolution between any groups that did not contain Pinus, and there was no significant difference in rates of evolution between the two Pinus subgenera (Pinus and Strobus; Table 1).

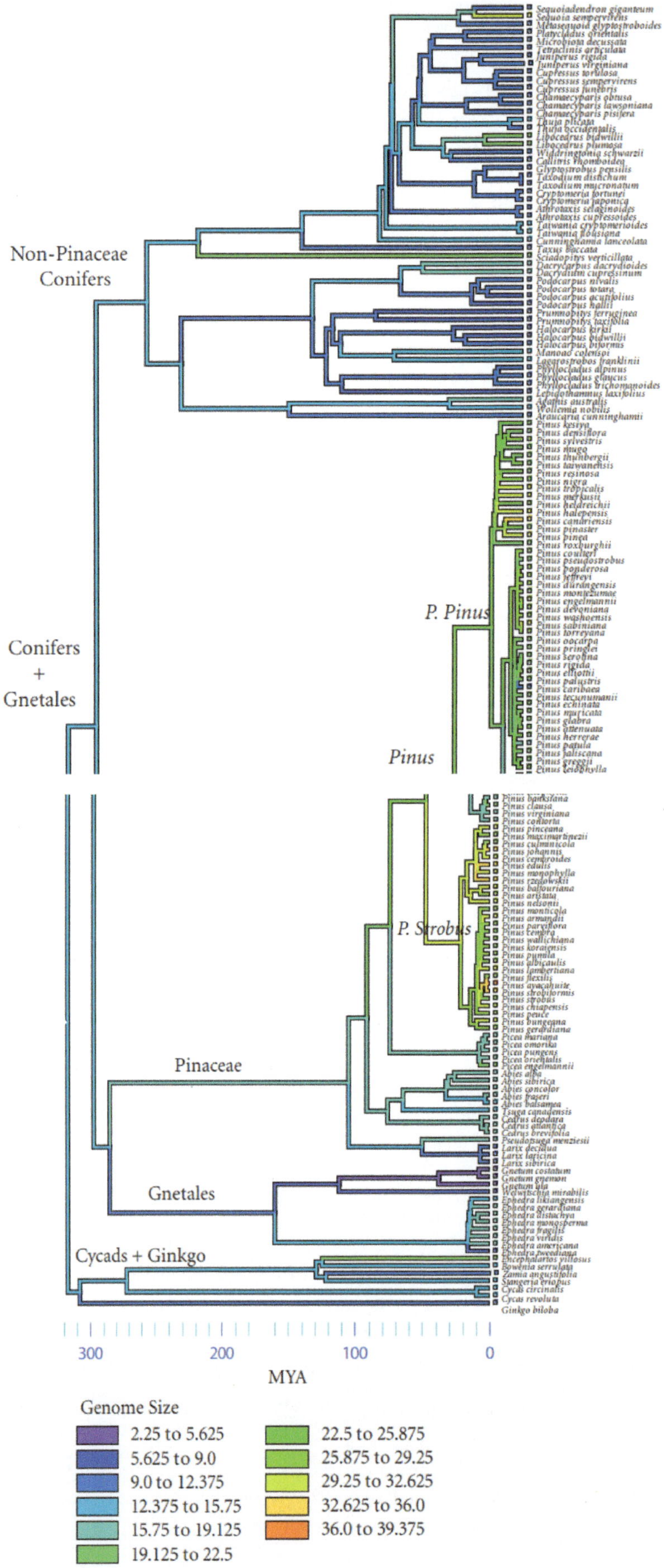


**Figure 3:** Ancestral state reconstruction of genome size (in pg DNA) on a chronogram 165 gymnosperm taxa. Different genome sizes are repre-

sented by different colors, with the ancestral genome sizes estimated with squared change parsimony.

## DISCUSSION

The analyses of gymnosperm diversification and genome size evolution demonstrate the dynamic evolutionary processes of the extant gymnosperms, which sharply contrasts with their reputation as ancient, relictual species. The lineage through time plots are consistent with high, and possible growing, rates of diversification within the last 100 million years, concurrent with major radiations of angiosperms (e.g., [1, 2, 28]) and extant ferns [29]. There is evidence of numerous significant shifts in diversification within both cycads and conifers, and there is strong evidence for a recent, large increase in the rate of genome size evolution in Pinus. Although Pinus is a species-rich genus, we find no links between increased rates of diversification and shifts in rates of genome size evolution.

Advances in sequencing technology and computational biology over the past decade enable phylogenetic estimates comprising large sections the plant diversity. This study demonstrates that it is possible to construct credible phylogenetic hypotheses including nearly three quarters of the extant gymnosperm species. Unlike supertree approaches (e.g., [14]), the supermatrix methods easily incorporate branch length estimates and estimates of topological and branch length uncertainty. Still, until there is far more data per taxon, estimates of the gymnosperm phylogeny will continue to improve, and thus, it is important to consider error and uncertainty in phylogenetic estimates when using these trees to infer evolutionary processes. There are other reasons to interpret this gymnosperm tree with caution. For example, both heterogeneity in the patterns of molecular evolution and missing data can lead to erroneous estimates of trees and branch lengths in ML phylogenetic analyses (e.g., [30, 31]). Furthermore, our analysis does not attempt to incorporate evolutionary processes, such as incomplete lineage sorting or gene duplication and loss or reticulation, that may cause incongruence between the gene trees and the species phylogeny (e.g., [32]). Although this study used thousands of sequences, it does not incorporate the evolutionary perspective of low-copy nuclear genes.

Still, in many cases, evolutionary or ecological analyses that use phylogenetic trees may be robust to topological and branch length error (e.g., [33]), and the large tree of gymnosperms enables sophisticated and comprehensive tests of evolutionary and ecological hypotheses. We demonstrate this with our diversification analysis, the results of which emphasize numerous, independent shifts in diversification rate throughout gymnosperms and apparently recent, high rates of diversification (Figure 2). Estimates of diversification may be affected by taxonomic sampling and inaccurate branch length estimates. However, we might expect that adding the remaining species, which would likely fit near the tips of the tree, would result in increased estimates of recent diversification. Thus, our analyses suggest the intriguing perspective that the extant gymnosperms are a vibrant, growing clade, and not simply the sole survivors of ancient diversity. Greater sampling and a more robust tree will provide a more complete view of gymnosperm diversification. With better branch length estimates, it will be possible to use more powerful likelihood-based approaches to identify clades with increasing and decreasing diversification rates [34]. With more complete taxon sampling, it may be possible to identify characters associated with changing speciation and extinction rates ([35], but see [36]).

One of the great challenges of evolutionary genomics is to identify the mechanisms of genome evolution that drive diversification. Some of the mechanisms that cause changes in genome size, such as whole-genome duplications or activity of retrotransposons, can have implications on diversification rates. Our analysis of the rates of genome size evolution demonstrate that Pinus is unique among gymnosperms. That is, the highly elevated rates of change in genome size appear to be limited to Pinus. However, in gymnosperms, we find no evidence of increases in diversification associated with Pinus, which displays a significantly elevated rate of genome size evolution. Furthermore, we find no obvious evidence for increase in rates of genome size evolution in clades associated with shifts in diversification. While our analysis failed to link genome size and diversification, this comparative approach for identifying shifts in genome size can inform our search for the specific drivers of the increased genomic complexity in Pinus, and this ultimately can help inform strategies for sequencing and assembling the first Pinus genomes.

## ACKNOWLEDGMENT

This work was funded by a University of Florida Research Opportunity Fund Seed Grant.

## REFERENCES

1. S. Magallón and A. Castillo, “Angiosperm diversification through time,” American Journal of Botany, vol. 96, no. 1, pp. 349–365, 2009.
2. C. D. Bell, D. E. Soltis, and P. S. Soltis, “The age and diversification of the angiosperms re-revisited,” American Journal of Botany, vol. 97, no. 8, pp. 1296–1303, 2010.
3. S. A. Smith, J. M. Beaulieu, A. Stamatakis, and M. J. Donoghue, “Understanding angiosperm diversification using small and large phylogenetic trees,” American Journal of Botany, vol. 98, no. 3, pp. 404–414, 2011.
4. P. D. Cantino, J. A. Doyle, S. W. Graham et al., “Towards a phylogenetic nomenclature of Tracheophyta,” Taxon, vol. 56, no. 3, pp. 822–846, 2007.
5. D. Ohri and T. N. Khoshoo, “Genome size in gymnosperms,” Plant Systematics and Evolution, vol. 153, no. 1-2, pp. 119–132, 1986.
6. B. G. Murray, “Nuclear DNA amounts in gymnosperms,” Annals of Botany, vol. 82, pp. 3–15, 1998.
7. M. R. Ahuja and D. B. Neale, “Evolution of genome size in conifers,” Silvae Genetica, vol. 54, no. 3, pp. 126–137, 2005.
8. T. N. Khoshoo, “Polyploidy in gymnosperms,” Evolution, vol. 13, no. 1, pp. 24–39, 1958.
9. A. M. Morse, D. G. Peterson, M. N. Islam-Faridi et al., “Evolution of genome size and complexity inPinus,” PLoS ONE, vol. 4, no. 2, Article ID e4332, 2009.
10. A. Kovach, J. L. Wegrzyn, G. Parra et al., “The Pinus

taeda genome is characterized by diverse and highly diverged repetitive sequences," BMC Genomics, vol. 11, no. 1, article 420, 2010.

11. K. L. Joyner, X.-R. Wang, J. S. Johnston, H. J. Price, and C. G. Williams, "DNA content for Asian pines parallels new world relatives," Canadian Journal of Botany, vol. 79, no. 2, pp. 192–196, 2001.

12. S. E. Hall, W. S. Dvorak, J. S. Johnston, H. J. Price, and C. G. Williams, "Flow cytometric analysis of DNA content for tropical and temperate new world pines," Annals of Botany, vol. 86, no. 6, pp. 1081–1086, 2000.

13. I. Wakamiya, R. J. Newton, J. S. Johnston, and H. J. Price, "Genome size and environmental factors in the genus Pinus," American Journal of Botany, vol. 80, no. 11, pp. 1235–1241, 1993. ·

14. E. Grotkopp, M. Rejmánek, M. J. Sanderson, and T. L. Rost, "Evolution of genome size in pines (Pinus) and its life-history correlates: supertree analyses," Evolution, vol. 58, no. 8, pp. 1705–1729, 2004. ·

15. S. F. Altschul, W. Gish, W. Miller, E. W. Myers, and D. J. Lipman, "Basic local alignment search tool,"Journal of Molecular Biology, vol. 215, no. 3, pp. 403–410, 1990.

16. J. D. Thompson, D. G. Higgins, and T. J. Gibson, "CLUSTAL W: improving the sensitivity of progressive multiple sequence alignment through sequence weighting, position-specific gap penalties and weight matrix choice," Nucleic Acids Research, vol. 22, no. 22, pp. 4673–4680, 1994. ·

17. A. de Queiroz and J. Gatesy, "The supermatrix approach to systematics," Trends in Ecology and Evolution, vol. 22, no. 1, pp. 34–41, 2007.

18. A. Stamatakis, "RAxML-VI-HPC: maximum likelihood-based phylogenetic analyses with thousands of taxa and mixed models," Bioinformatics, vol. 22, no. 21, pp. 2688–2690, 2006.

19. J. Felsenstein, "Confidence limits on phylogenies: an approach using the bootstrap," Evolution, vol. 39, no. 4, pp. 783–791, 1985. ·

20. M. J. Sanderson, "Estimating absolute rates of molecular evolution and divergence times: a penalized likelihood approach," Molecular Biology and Evolution, vol. 19, no. 1, pp. 101–109, 2002. ·

21. M. J. Sanderson, "R8s: inferring absolute rates of molecular evolution and divergence times in the absence of a molecular clock," Bioinformatics, vol. 19, no. 2, pp. 301–302, 2003.

22. H. Won and S. S. Renner, "Dating dispersal and radiation in the gymnosperm Gnetum (Gnetales)—clock calibration when outgroup relationships are uncertain," Systematic Biology, vol. 55, no. 4, pp. 610–622, 2006.

23. E. Paradis, J. Claude, and K. Strimmer, "APE: analyses of phylogenetics and evolution in R language,"Bioinformatics, vol. 20, no. 2, pp. 289–290, 2004.

24. B. R. Moore, K. M. A. Chan, and M. J. Donoghue, "Detecting diversification rate variation in supertrees," in Phylogenetic Supertrees: Combining Information to Reveal the Tree of Life, O. R. P. Bininda-Emonds, Ed., pp. 487–533, Kluwer Academic, Dodrecht, The Netherlands, 2004.

25. N. Bortolussi, E. Durand, M. G. B. Blum, and O. François, "Aptreeshape: statistical analysis of phylogenetic treeshape," Bioinformatics, vol. 22, no. 3, pp. 363–364, 2006.

26. M. D. Bennett and I. J. Leitch, "Plant DNA C-values database," 2005, http://data.kew.org/cvalues/.

27. B. C. O'Meara, C. Ané, M. J. Sanderson, and P. C. Wainwright, "Testing for different rates of continuous trait evolution using likelihood," Evolution, vol. 60, no. 5, pp. 922–933, 2006.

28. H. Wang, M. J. Moore, P. S. Soltis et al., "Rosid radiation and the rapid rise of angiosperm-dominated forests," Proceedings of the National Academy of Sciences of the United States of America, vol. 106, no. 10, pp. 3853–3858, 2009.

29. H. Schneider, E. Schuettpelz, K. M. Pryer, R. Cranfill, S. Magallón, and R. Lupia, "Ferns diversified in the shadow of angiosperms," Nature, vol. 428, no. 6982, pp. 553–557, 2004.

30. B. Kolaczkowski and J. W. Thornton, "Performance of maximum parsimony and likelihood phylogenetics when evolution is heterogenous," Nature, vol. 431, no. 7011, pp. 980–984, 2004.

31. A. R. Lemmon, J. M. Brown, K. Stanger-Hall, and E. M. Lemmon, "The effect of ambiguous data on phylogenetic estimates obtained by maximum likelihood and bayesian inference," Systematic Biology, vol. 58, no. 1, pp. 130–145, 2009.

32. W. P. Maddison, "Gene trees in species trees," Systematic Biology, vol. 46, no. 3, pp. 523–536, 1997. ·

33. E. A. Stone, "Why the phylogenetic regression appears robust to tree misspecification," Systematic Biology, vol. 60, no. 3, pp. 245–260, 2011. ·

34. D. L. Rabosky, "LASER: a maximum likelihood toolkit for detecting temporal shifts in diversification rates," Evolutionary Bioinformatics, vol. 2, pp. 247–250, 2006.

35. W. P. Maddison, P. E. Midford, and S. P. Otto, "Estimating a binary character's effect on speciation and extinction," Systematic Biology, vol. 56, no. 5, pp. 701–710, 2007.

36. D. L. Rabosky, "Extinction rates should not be estimated from molecular phylogenies," Evolution, vol. 64, no. 6, pp. 1816–1824, 2010.

CHAPTER 6

# EVOLUTION OF GENE STRUCTURE IN THE CONIFER PICEA GLAUCA: A COMPARATIVE ANALYSIS OF THE IMPACT OF INTRON SIZE

Juliana Stival Sena, Isabelle Giguère, Brian Boyle, Philippe Rigault, Inanc Birol, Andrea Zuccolo, Kermit Ritland, Carol Ritland, Joerg Bohlmann, Steven Jones, Jean Bousquet and John Mackay

**Keywords:** Genome size, *Pinus taeda,* BAC, Repeat elements, Gymnosperms, Gene expression

## BACKGROUND

Many factors related to genome size, recombination rate, expression level, and effective population size, among others, have been proposed to affect the evolution of gene structure [1, 2, 3, 4]. At the molecular level, genome size variations may result from mobile or transposable elements (TEs), whole genome duplication events, and polyploidization events, among others. Comparative studies have shown that intron lengths and the abundance of mobile elements directly correlate with genome size, such that large genomes have longer introns and a higher proportion of mobile elements [1]. Mobile elements also impact gene structure and function as they can insert into genes, including introns and exons, and thus contribute to the evolution of genes.

Conifer trees have very large genomes ranging from 18 to 35 Gbp [5] that are composed of a large fraction of repetitive sequences [6, 7]. New insight into plant genome evolution are expected from the unique structure and history of conifer genomes [8], which may contribute to a broader understanding of the relationships between gene structure and genome architecture. Draft genome assemblies were recently reported for the European *Picea abies* (Norway spruce) [9] as well as the North American species *Picea glauca* (white spruce) [10] and *Pinus taeda* (loblolly pine) [11, 12]. Nystedt et al. [9] reported that Norway spruce and other conifers accumulate long introns and showed that some introns can be very long (>10 Kbp) compared to other plant species.

A positive relationship between genome size and intron length has been observed in broad phylogenetic studies [2, 13, 14] including between recently diverged *Drosophila* species harboring considerable difference in genome size, where *D. viliris* had longer introns than *D. melanogaster*[15]. In plants, a few studies investigated this question within angiosperms, indicating that genome size is not necessarily a good predictor of intron length [16, 17] although a general trend is observed. For instance, *Arabidopsis thaliana, Populus trichocarpa, Zea mays* have well characterized genomes that range in size from 125 Mbp to 2.3 Gbp; their average exons sizes are between 250 and 259, whereas their introns sizes are 168 bp, 380 bp and 607 bp on average, respectively [18, 19, 20]. The length of introns may depend upon gene function and expression level; however, there is considerable debate surrounding this issue when it comes to plant genomes. In *Oryza sativa*and *A. thaliana* it was found that highly expressed genes contained more and longer introns than genes expressed at a low level [21], which is in contrast to findings in *Caenorhabditis elegans* and *Homo sapiens*[4].

Transposable elements are among the factors that may influence the evolution of intron size, as they represent the major component of plant genomes [22]. In *Vitis vinifera*, transposable elements comprise 80% of long introns [17]. In many plants, LTR-RT represent a large fraction of the genome but are more abundant in gene poor regions of the genome; therefore, their impact on the evolution of gene structure may actually be lesser than other classes of transposable elements

**Citation:** Stival Sena et al.: Evolution of gene structure in the conifer Picea glauca: a comparative analysis of the impact of intron size. DOI: 10.1186/1471-2229-14-95.

such as MITEs [23] and helitrons, both of which are known to insert into or close to genes [24].

To date, studies related to genome size and the evolution of plant introns have primarily involved angiosperms (flowering plants), many of which have genomes under 1 Gbp. More recently, the *Picea abies* and*Pinus taeda* genomes were shown to have among the largest average introns size [9, 12]. We aimed to develop an understanding of the gene structure in conifers through a detailed analysis of individual genes with a particular emphasis on the potential impact of long introns on gene structure trough comparative analyses. An underlying question relates to potential impacts on gene expression; therefore, our analyses took into account their expression profiles. Gene structure was analyzed in two conifers (*P. glauca* and *P. taeda*) and three angiosperms. We explored three main hypothesis: (1) Intron length is the major type of variation affecting gene structure in conifers compared to other plant species; (2) there is a positive relationship between genome size and intron length in *P. glauca* compared to *A. thaliana*, *Z. mays* and *P. trichocarpa*; (3) *P. glauca* and *P. taeda* present a conserved gene structure despite the fact that they diverged over 100 MYA in keeping with their low rate of genome evolution [8].

We present a detailed analysis of gene structure for 35 genes from the conifer *Picea glauca* obtained from BAC sequencing and genome assembly and comparative analyses with *A. thaliana*, *P. trichocarpa* and *Z. mays*. Our study also included the analysis of nearly 6000 gene sequences obtained from sequence capture aiming to explore the potential impact of repetitive sequences on intron size in *P. glauca*. Our findings show that intron size and the position of long introns within genes is variable between plant lineages but highly conserved in conifers.

## RESULTS

### Genomic Sequences

Genomic sequences were analyzed for several *P. glauca* genes. The sequences were obtained either by targeted BAC isolations, from an early assembly of the *P. glauca* genome [10], or from a sequence capture experiment (for details, see Methods).

A total of 21 BAC clones were isolated each containing a different single copy gene associated with secondary cell-wall formation or with nitrogen metabolism. Following shotgun sequencing by GS-FLX and assembly with the Newbler software, the integrity and identity of each gene was verified. Estimated size of BAC clones was 131 Kb on average and coverage was 144× (for Summary statistics, see Additional file 1: Table S6). Twenty of the 21 targeted genes were complete as determined by sequence alignment indicating full coverage of FL cDNA sequences from spruces and pines (*P. glauca, P. sitchensis, P. taeda* and *Pinus sylvestris*) [25, 26, 27, 28] (Additional file 1: Table S7). Nearly all genes were contained within a single contig, except the LIM gene which lacked one exon, and the Susy gene which was complete cDNA sequence but spanned two contigs. None of BACs contained other genes as determined by BLAST searches against the *P. glauca* gene catalog [29] and the Swiss Prot database.

Sequences were also isolated from a whole genome shotgun assembly of *P. glauca*[10]. Sequences with ubiquitous expression were targeted in order to complement the set of more specialized genes which had been selected for BAC isolation. The *P. glauca* genome shotgun assembly was screened with the complete CDS derived from cDNA sequences (according to Rigault et al. [29]) that were highly expressed in most tissues (according to Raherison et al. [30]). A total of 18 genomic sequences were randomly selected among those that spanned the entire coding region of the targeted gene.

### Gene Expression Profiles

Transcript accumulation profiles from eight different tissues were obtained from the PiceaGenExpress database [30] for each of the gene sequences described above (Figure 1). The transcript data indicated that the group of highly expressed genes was detected in all tissues and with average abundance class above 9.7 (out of 10) across all tissues (Figure 1, top). In contrast, the genes associated with wood formation and nitrogen metabolism nearly all had tissue preferential expression patterns; they were detected in six tissues on average (range of two to eight tissues) and had an average transcript abundance class of 5.8 in those tissues where the genes were expressed (Figure 1, bottom).

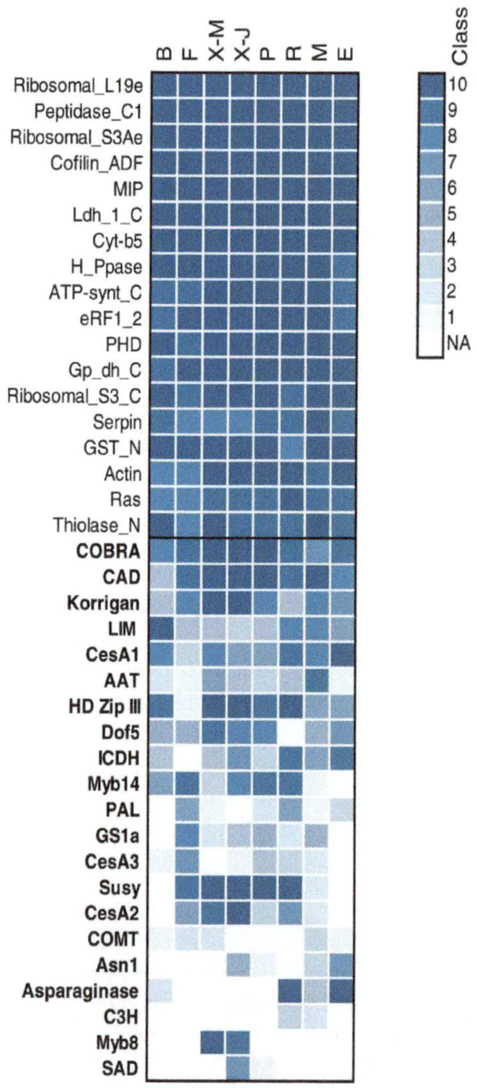


**Figure 1: Transcript accumulation profiles from the PiceaGen-Express database (Raherison et al. [[30]]) of the *P. glauca*genes.**

The transcript abundance data are classified from 1 to 10, from lowest to highest microarray hybridization intensities detected within a given tissue. The profiles of highly expressed genes (top) (according to Raherison et al. [30]; class 8 to 10) are contrasted with most of the genes associated with secondary cell wall formation and nitrogen metabolism (bottom, names in bold). NA: Not detected. Tissues: B (Vegetative buds), F (Foliage), X-M (Xylem – from mature trees), X-J (Xylem –juvenile trees), P (Phelloderm), R (Adventitious roots), M (Megagametophytes), E (Embryogenic cells).

## Gene Structures and Comparative Analysis with Angiosperms

The gene structure (exon and introns regions) of *P. glauca* genes was determined by mapping the complete cDNA onto the genomic sequence (BACs or shotgun contigs) for 35 genes. Homologs were retrieved from three well-characterized angiosperm genomes, *Arabidopsis thaliana* [19], *Populus trichocarpa*[18] and *Zea mays*[20]. The comparative analyses considered all of the genes together and also as two separate groups, i.e. genes highly expressed and genes related to secondary cell-wall formation and nitrogen metabolism. On average, the protein coding sequence similarity between *P. glauca* and *A. thaliana* was 76%, 78% with *P. trichocarpa* and 75% with *Z. mays*.

The number of exons and introns was well conserved between homologous genes among the different species (Table 1). The average length of exons was also well conserved between homologs among species (average of 240 bp, median of 155 bp) and varied only slightly between the two sub-groups genes (Table 1 and Additional file 2: Figure S2). Pairwise comparisons of matching exons also indicated conservation of length among the species considered (not shown). These observations indicate that exon structure is generally well conserved.

**Table 1:** Average number and length of exons in genes used for comparative analyses

| | Highly expressed genes[1] | | | Secondary cell-wall formation and nitrogen metabolism genes[2] | | |
|---|---|---|---|---|---|---|
| | Exon number | Exon length | Standard deviation | Exon number | Exon length | Standard deviation |
| *Arabidopsis thaliana* | 5.9 | 220.8 | 215.0 | 9.1 | 228.9 | 189.8 |
| *Populus trichocarpa* | 6.2 | 241.5 | 253.3 | 9.4 | 261.1 | 263.5 |
| *Zea mays* | 6.1 | 244.5 | 236.7 | 9.0 | 257.6 | 274.8 |
| *Picea glauca* | 6.2 | 236.5 | 226.0 | 9.5 | 223.9 | 217.8 |

[1]Data were obtained from 18 different genes and an average total of 109 exons per species.

[2]Data were obtained from 17 different genes and an average total of 157 exons per species.

In contrast, introns revealed much more variation between species. Our analyses included comparisons of individual introns and of total intronic sequences in each gene. The average length of individual introns (in bp) was 144, 295, 454, and 532 for *A. thaliana, P. trichocarpa, Z. mays and P. glauca*, respectively (Figure 2 and Additional file 2: Figure S2). The average intron length varied significantly among *P. glauca* and the three species; pairwise contrasts were significant with *A. thaliana* and *Z. mays,* and nearly significant with *P. trichocarpa* (Figure 2). In *P. glauca, P. trichocarpa* and *Z. mays*, we also observed that intron lengths were more heterogeneous as shown by differences between low and upper quartiles, minimum and maximum lengths and outliers of large size (Figure 2). The average length of the longest intron per gene was 382 bp in *A. thaliana*, 806 bp in *P. trichocarpa* ,1652 bp in *Z. mays* and 2022 bp in *P. glauca*.

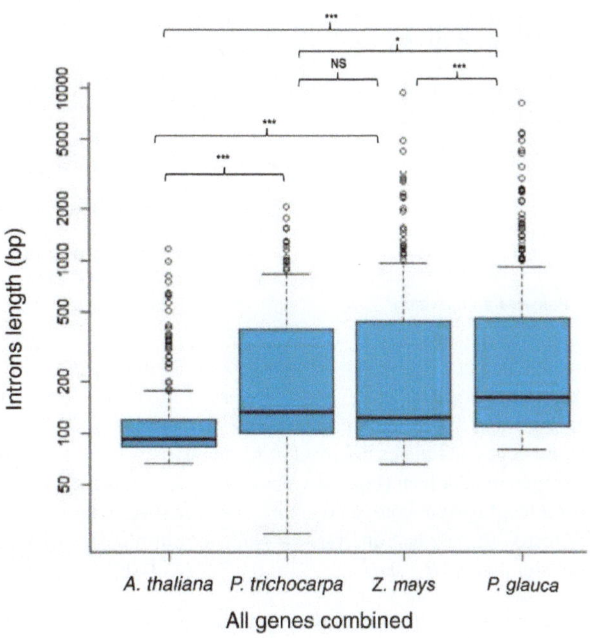


**Figure 2:** Comparative analysis of individual intron length in *P. glauca* , *A. thaliana P. trichocarpa* and *Z. mays* . Box plots represent intron length data for all of the introns of the 35 genes used in comparative analyses. Intron lengths were compared among the four spe-

cies by Kruskal-Wallis test with post-test analysis by Dunn's multiple comparisons: NS, not significant ($P \geq 0.06$); *P=0.06; **$P<0.01$; ***$P<0.001$.

Comparison of the total length of intronic sequences on a gene-by-gene basis showed that on average, *P. glauca* genes had 4.1 times more intronic sequences than *A. thaliana,* 2.2 times more than *P. trichocarpa* and 1.8 times more than *Z. mays* (Figure 3A). The total length of intron sequences and length ratio was calculated for each gene in pairwise comparisons between all of the species. Comparisons between *P. glauca* and *A. thaliana* gene sets were statistically significant (Figure 3); the ratios were close to five on average in highly expressed genes and three in genes associated with secondary cell-wall formation and nitrogen metabolism (Figure 3B). In contrast, the ratio of total intron lengths between *P. glauca* compared to *P. trichocarpa* and *Z. mays* was constant at around two-fold and the total length of intronic sequence per gene was not statistically different. Results also indicated that *A. thaliana* has significantly less intronic sequence than *P. trichocarpa* and*Z. mays* and that their ratios were most different for the highly expressed genes and more similar for the genes involved in secondary cell-wall formation and nitrogen metabolism (Figure 3B). A significant difference of intron lengths was also observed between the two expression groups within *P. glauca* (p<0.05).

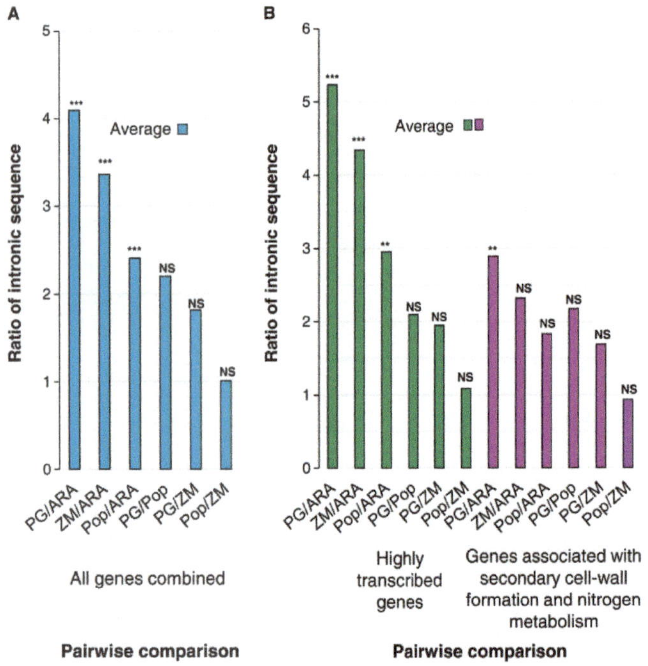


**Figure 3:** Comparative analysis of total intron length in *P. glauca* , *A. thaliana, P. trichocarpa* and *Z. mays* . Average ratio of total length of intron sequences in pair-wise comparisons in: **A-** all genes; **B-** highly expressed genes and genes involved in secondary cell-wall formation and nitrogen metabolism (For individual ratios, see Figure 4). The total intron lengths were compared among the four species by Kruskal-Wallis test with post-test analysis by Dunn's multiple comparisons: NS, not significant ($P \geq 0.05$); **$P < 0.01$; ***$P < 0.001$.

The variation in the ratios of total intron sequence per genes was quite striking, for both of the gene expression groups (Figure 4). For instance, depending on the gene, the ratios ranged from 0.2 to 10. This high level of heterogeneity in pairwise comparisons is likely to account for the lack of statistically significant differences. In addition, the intron length ratios were not consistent across species (Figure 4A and B).

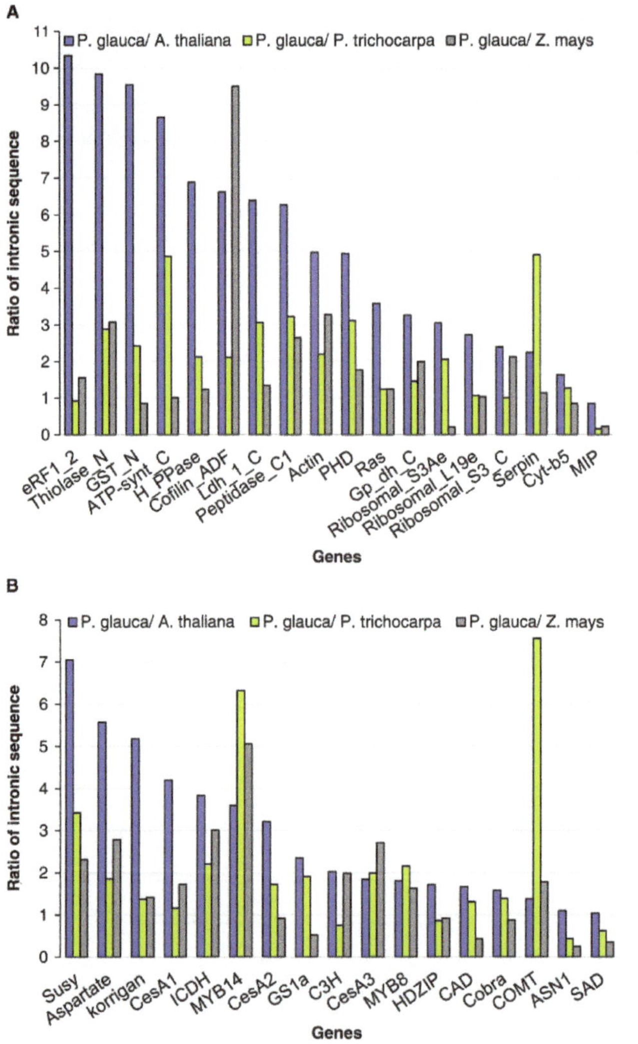


**Figure 4:** Gene by gene pair-wise comparisons of total length of intronic sequences in *P. glauca* , *A. thaliana, Populus trichocarpa* and *Z. mays.* **(A)** highly expressed genes and **(B)** genes associated with secondary cell-wall formation and nitrogen metabolism.

In this study, we show that much of the divergence in the total length of intron sequences per gene was related to a few long introns. Very long introns were observed in a few *P. glauca* genes such as PHD, Peptidase_C1 and Thiolase. Structure plots showed that introns in *A. thaliana* generally had uniform lengths whereas the other species had introns that were highly heterogeneous in lengths (Figure 5 and Additional file3: Figure S3). While most of the *P. glauca* genes only had a few (1–3) very long introns (>1000 bp), gene sequences such as those for sucrose synthase (Susy) had many introns of moderate size (Figure 5). The longest introns in *P. glauca* were most often in a different position than long *Z. mays* and *P. trichocarpa*introns. In addition, we did not observe a trend of increased length in first introns in 5′ UTRs as reported for several eukaryotes [31], as the long introns in *P. glauca* appeared to be randomly distributed.

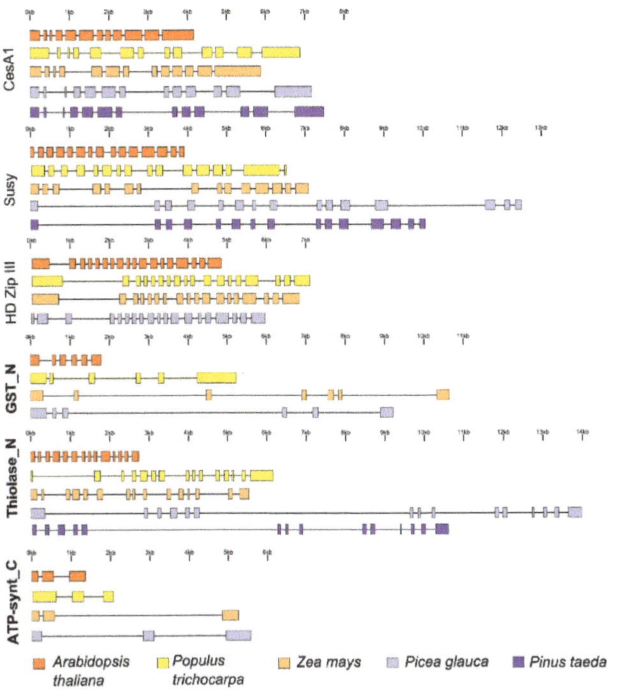


**Figure 5: Gene structure of six genes from different angiosperm and gymnosperm species.** The first three genes are associated with secondary cell-wall formation and nitrogen metabolism; and highly expressed genes are bolded.

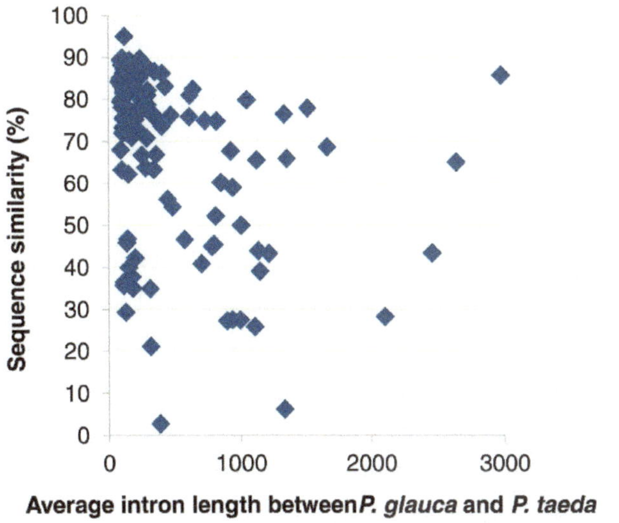


**Figure 6: Relationship between intron size and sequence similarity of introns from *P. glauca* and *P. taeda*.** A total of 138 introns were obtained from 22 genes and sequence alignments were produced with the Needle software (see Methods).

## Comparative analysis of gene structures between *Picea glauca* and *Pinus taeda*

A total of 23 different genes were submitted to pairwise comparisons between *Picea glauca and Pinus taeda,*which are both of the Pinaceae (for details, see Methods). A high level of similarity was observed for coding sequences (91% on average) indicating that they were likely orthologous genes (Additional file 1: Table S4), and gene structure was conserved between the two conifers, with almost identical numbers of exons. The total intronic sequences per gene did not vary significantly at 3.13 and 3.17 Kbp for *P. glauca* and *P. taeda*, respectively (Additional file 1: Table S1). Pairwise comparison of introns indicated that the majority of individual introns were similar in length in the two species, despite the fact that the two genera diverged ca. 140 million years ago [32, 33] (Figure 5). Although these observations are based on a set of only 23 genes, they provide an indication that intron length is mostly conserved between these two conifer genera.

The 138 intron sequences of the 22 genes (PAL gene do not have introns) were aligned between spruce and pine; sequence similarity ranged quite broadly among homologous introns (Figure 6).We observed that highly conserved introns generally were short, and that longer introns had highly variables levels of sequence similarity, except for two introns that were both long and highly conserved.

## Repeat Elements in *Picea Glauca*genes

The possible origin of long introns as observed in conifer genomes was investigated by searching for the presence of repeated sequences including transposable elements.

First, the repetitive element content of the BACs was estimated based on a repetitive library constructed with*P. glauca* data (see Methods) as a baseline. It was 55% on average, but it varied considerably among the BAC clones, ranging from 18% to 83%. Additional file 4: Figure S1 shows that around half of repetitive sequences were classified as LTR-RT elements and the other half as unknown elements (without significant hits in Repbase and nr genbank).

We then analyzed the sequences of the 35 *P. glauca* genes described above including those identified in BACs, representing a total of 238 introns. The gene structures of these genes were screened for repeat elements using a *P. glauca* repeat library (see Methods). We found repetitive elements in 10 of the genes for a total of 24 unclassified fragments with no significant hits in RepBase; 22 of the fragments produced no hits in genbank and were 179 bp on average and only two had significant hits in nr genbank (Additional file 1: Table S8).

We also extended our analysis to include an additional set of genomic sequences obtained by targeted gene space sequencing based on sequence capture (see Methods, for details). Complete genomics sequences spanning the entire known mRNA sequence were recovered for 5970 complete genes, 1836 of which contained one or more introns. The different repetitive elements identified in introns and exons were then estimated. The proportion of genes harbouring repetitive elements in their introns was 32.4% and was only 3.2% in exons. The repetitive elements represented 2.94% and 0.74% of the intronic and exonic sequences, respectively (Table 2). The repetitive sequences that were identified ranged from 31 to 1142 bp (median 117 bp) in exons and from 17 to 1189 bp (median 114bp) in introns. The unclassified elements were the most numerous, representing on average 80% of the hits in both introns and exons (Table 2). Class I LTR transposons were the most abundant group of classified repetitive elements and were only represented by incomplete elements. The LTRs were accounted for the higher repetitive element sequence representation in introns; however, on average, the sequences identified as Copia and Gypsy elements were longer in exons than in introns.

**Table 2:** Abundance of repetitive elements in *P. glauca* genes obtained from sequence capture

| Class | Exons (%) | Introns (%) |
|---|---|---|
| Copia | 0.09 | 0.24 |
| Gypsy | 0.09 | 0.19 |
| LINE | 0.03 | 0.15 |
| UNK[1] | 0.03 | 0.07 |
| NHF[2] | 0.49 | 2.29 |

A total of 5970 genes were analyzed, 1836 contained one or more introns.

[1]No significant hit in RepBase but significant hits in nr genbank.

[2]No significant hit in RepBase and nr genbank.

## DISCUSSION

This study reports on the detailed gene structure analysis of 35 genes from the conifer *Picea glauca* obtained from BAC sequencing and genome assembly. Recent analyses of the *Picea abies* and *Pinus taeda* genomes have analyzed individual introns and reported among the highest average intron lengths, the longest introns and highest average among long introns [9, 12]. We aimed to develop an understanding of the gene structure in conifers through a detailed analysis of entire genes taking into account gene expression profiles, with a particular emphasis on the potential impact of longer introns on gene structure trough comparative analyses. Our findings were also derived from the analysis of nearly 6000 gene sequences obtained from sequence capture sequencing. We present an interpretation of our findings in regard to the evolution of gene structure.

### Evolution of gene Structure in Plants

Analyses over a broad phylogenetic spectrum in eukaryotes showed that increases in genome size correlate with increases in the average intron length [2, 13]. A strong relationship between intron length and genome size was observed from studies in humans and pufferfish [14], species of *Drosophilla*[15], and from studies of plants with small genomes [2, 13].

Our study compared the gene structure (introns and exons) of 35 homologous genes between four seed plant species with very different genome sizes. The conifer *P. glauca* has the largest genome with 19.8 Gbp [34]; among angiosperms, the monocot *Z. mays* has a genome of 2.3 Gbp [24], and dicots represent smaller plant genomes in this set, i.e. *P. trichocarpa* with genome of 484 Mbp [18] and *A. thaliana* with the smallest genome of 125 Mbp [19]. In the present study, the average exon length was similar between the four species, but the overall length of genes varied owing to longer introns in *P. glauca, P. trichocarpa* and *Z. mays*. For the set of sequences analyzed, *P. glauca* had 4.1 times more intron sequence per gene than *Arabidopsis*, 2.2 times more than poplar and 1.8 times more than maize (Figures 3 and 4); however, the statistical significance of these differences was variable.

The landscape of intron sizes in plants appears rather complex. A significant number of V*itis vinifera* introns were shown to be uncommonly large for its genome size of 416 Mbp, compared to other plants [17]. In *Gossypium,* after multiple inferred rounds of genome expansion and contraction, intron size remained unchanged [16]. Such a pattern may be expected, given that genome size increase by polyploidy is sudden and fundamentally different than other types of genome size variation such as the gradual accumulation or loss of repeat elements over time. Taken together, observations from different plants indicated that events resulting in the expansion or contraction of intergenic regions are not clearly reflected by shifts in introns length. It thus appears that the evolution of intron length and genome size may be uncoupled in plants or alternatively, that the evolution of intron length is lineage specific (Figure 7).

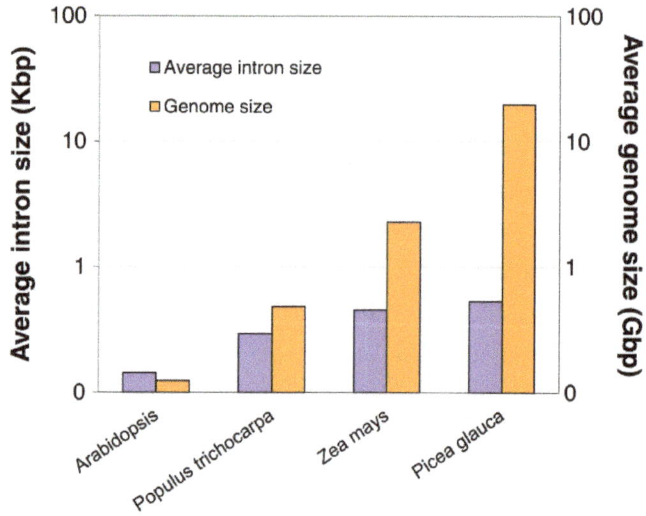


**Figure 7:** Variation in introns length and genome size in 35 target genes. Average intron size for the *Arabidopis*, *P. trichocarpa*, *Z. mays* and *P. glauca* determined from the analysis of 35 homologous genes. Note that Y- axes are in log 10 scale.

Even though our study was based on 35 genes, our results are consistent with variations of intron size reported for *A. thaliana*, *P. trichocarpa* and *Z. mays* genomes [9, 12, 18, 19, 20]. We concluded that the increased intron length in *P. glauca, P. trichocarpa* and *Z. mays* was heterogeneous compared to *A. thaliana*. Even in genes with many introns, only a few introns were very long, whereas in *Arabidopsis*, genes exhibited a more uniform intron length, suggesting that introns expansion or contraction within a gene may be independent across species.

Comparisons between the *A. thaliana* (125 Mbp) and *A. lyrata* (~200 Mbp) genomes, which diverged about 10 million years ago, showed that most of the difference in genome size was due to hundreds of thousands of small deletions, mostly in noncoding DNA [35]. The authors concluded that evolution toward genome compaction is occurring in *Arabidopsis*. Conifers such as species of *Picea* and *Pinus* have large amounts of repetitive elements in intergenic regions and apparently more intronic sequence per gene in comparison to many angiosperms. Our results do not reveal whether the *P. glauca* genome and introns are expanding, or alternatively evolving at slower pace, than other plant genomes which are contracting. Some evidence like the presence of very ancient retrotransposon elements [9, 36] and the lack of gene rearrangements since before their split from extant angiosperms [8] lend credence to the paradigm that conifer genomes are slowly evolving.

## Repetitive Sequences in Gene Evolution

Transposable elements play a role in plant genes as was shown by the abundance of TE- gene chimeras in*Arabidopsis* which was reported as 7.8% of expressed genes [37]. The abundance of TEs may be especially high in long introns as recently shown in *Picea abies* where most of the introns were longer than 5 Kbp, representing 5% of the total intron count [9]. This trend was also observed in other repeat rich genomes as *V. vinifera* and *Z. mays*[20, 21, 38].

We isolated *P. glauca* BAC clones each containing a different complete transcription unit for 21 target genes. In each the BACs (average 131 Kb), only one intact gene sequence was identified, which is indicative of large intergenic regions as reported for other conifers [39, 40, 41]. Previous studies on conifer trees have considered only two targeted genes (from terpenoid biosynthesis) isolated from *P. glauca* BAC clones [40] and only a few other intact genes with complete coding sequence isolated from BACs in pines [7, 39, 41].

Complete sequencing of the *P. glauca* BACs showed that the repetitive element content is not distributed uniformly in proximal intergenic regions, as indicated by the variable proportion of repetitive elements among the different BACs. A study in 10 *P. taeda* BACs, sequences similar to eukaryote repeat elements (according to Repbase) represented 23% of the sequence on average, and ranged from 19% to 33% [7]. In *P. glauca*, 26% of BAC sequences were classified as LTR-RT repetitive elements on average and ranged from 8% to 47%, while *P. taeda* had an average of 18.8% of LTR-RT [7]. Furthermore, an average 26% of the *P. glauca* BAC sequences were unknown repeat elements. Results in spruce and pine indicate a relatively low abundance of TEs in gene proximal sequences compared to whole genomes at 70% in the *Picea abies* genome [9] and around 80% in *Pinus taeda*[12].

*Picea* and *Pinus* genomes are reported to have among the highest average for the longest intron per gene, when compared to angiosperms of diverse genome sizes [9]. We verified whether insertions of repetitive elements could be responsible for the length of introns in *P. glauca* in a set of more than 1800 genes sequences, and found that more genes harboring repetitive elements in introns were 10 times more frequent than genes harboring repetitive elements in exons, i.e. 29.8% vs 3.2%. The vast majority of the repetitive elements were short fragments, suggesting that they were remnants or fragments of TE insertions that have not persisted and could represent ancient insertion events. Importantly, interpretation of our findings in *P. glauca* must take into account the fact that the sequences were derived from a sequence capture study and that nearly all of the introns in the data set were <1 Kbp. Thus we show that TE sequences are ubiquitous even in genes that do not harbor long introns, suggesting that their presence has been very widespread during the evolution of conifer genes. An analysis of intact LTR TE in *Picea* genomic sequences showed that most insertions date back to 10 MYA or more, with a maximum around 20–25 MYA [9]. The TE remnants that we detected in *P. glauca* indicate that many genes introns contained TE in a more or less distant past. In this report and in recent analyses of conifer genomes, an emphasis has been place on long introns; however the median intron length in conifers is very similar to other plant species, most of which have a median between 100 bp and 200 bp. Therefore our findings on intron are relevant for a large majority of introns rather than a small fraction represented by large or very large introns.

## Slow Evolution of Conifer Genes

Analyses of the gene structure of 23 orthologous genes between *P. glauca* and *P. taeda* clearly showed the conservation of gene structure and the distribution of intron sizes in spite a divergence time of 100 to 140 MYA [32, 33]. The conservation of long introns was also observed across gymnosperm taxa, where a group of long introns in *P. abies* was identified as orthologous to long introns in *P. sylvestris* and *Gnetum gnemon*[9]. We suggest that the long introns observed in *P. glauca* likely date back to a period predating the divergence of major conifer groups. As more conifers genomes become available [9, 10, 11] and assembly contiguities are improved it will be possible to extend this analyses of orthologous gene structures among conifers.

We also observed that the sequence of many introns was highly similar between spruce and pine, and that shorter introns were more conserved on average. Between humans and chimpanzee, a strong positive correlation was found between intron length and divergence [42]. The pattern found in conifers as well as observations in primates lead to the hypothesis that shorter introns could be under stronger selection pressure than longer introns, which could be explained by factors such as the maintenance of functional regulatory elements in shorter introns or impacts on RNA transcript processing and stability. In our analysis of sequence similarity between *Picea* and *Pinus*, 20 of the introns were longer than 1 Kbp and only two of them had high sequence similarity. Future studies with more long introns are required to confirm the hypothesis that shorter introns are more conserved in conifers. Despite the fact that introns are assumed to be non-coding, conserved introns may play a functional role related to gene expression.

## Costs and Benefits Associated With Intron Size

There is also considerable debate about other factors that may impact the evolution of introns, aside from transposable elements. Lynch [43] stated that the reduced efficiency of selection in regions of low recombination may lead to an increase in intron size if small introns provide a slightly improved transcription efficiency or splicing accuracy. On the other hand, Comeron and Kreitman [3] proposed that there might be situations in which a longer intron is selectively advantageous as an explanation for intron persistence and increased lengths. If so, there would be indirect selection for large introns in regions of low recombination because they can reduce the load caused by deleterious mutations by increasing the recombination rate. It was proposed that conifers have low recombination rates at both the genome and within-gene scales [44]. Their low recombination rates may explain at least in part, the accumulation of longer introns.

The high degree of sequence conservation that we observed in short introns between spruce and pine may also depend on the recombination rate within genes, where small introns would be under stronger selection because of efficiency in transcription and splicing, and long introns in regions of low recombination diverge because of reduced selection pressure. Another factor underlying the evolution of intron size is that intron length would be constrained by energy use during transcription, given that large introns represent a higher cost of transcription, the so-called "economy" or low-cost transcription hypothesis [4]. In the present study, the 35 *P. glauca* genes analyzed were divided in two groups based on their expression profiles, i.e. 17 genes associated with secondary cell-wall formation or nitrogen metabolism, many of which had tissue preferential expression, and 18 genes that were highly transcribed in a large range of tissues (based on Raherison et al. [30]). The highly expressed genes had more intronic sequences per gene on average than the more specialized subset of genes (4,182 bp versus 3,013 bp). We also observed a large variation among genes in each group, i.e. from 446 to 12,009 bp in highly transcribed genes and 440 to 9,847 bp in the set of more specialized genes. These observations do not support the "economy" hypothesis in *P. glauca* as there appears to be no clear rule governing the relationship between intron size and expression levels or profiles. In humans, genes contained total intronic sequences are ~5,500 bp per gene on average [45], which more than any plant described so far. It was observed that intron length declines steadily as the expression level increases in humans, in agreement with the low-cost transcription hypothesis [4]. Considering the smaller amount of intron sequences in plant genes including conifers compared to humans, it may be that the economy rule does not impact their introns as strongly as in vertebrates and that other evolutionary forces are main drivers of intron size evolution. This interpretation is consistent with the findings reported for the *P. abies* genome [9]. Future studies with more genes are needed to confirm this hypothesis.

## CONCLUSIONS

Our results indicate that *P. glauca* has longer introns than *Arabidopsis*, *P. trichocarpa* and *Z. mays* on average due the presence of few long introns. Intron size and the position of long introns within genes were variable between plant lineages but well conserved in conifers. Our findings are consistent with recent reports indicating that conifers accumulate very long introns but we point out that long introns represent a relatively small fraction of the overall intronic content, which is reflected by the median length of similar size to other plants. We show that RE sequences are detected at a high frequency (32%) even in introns <1 Kbp, indicating their ubiquitous presence in conifer genes over the course of evolution.

Taken together, our observations and the recent literature suggest that the evolution of plant gene structure is determined by more interacting forces than classically expected. The pattern is reminiscent of the heterogeneity of rates of evolution at the genetic, genomic and morphological levels seen among seed plants including angiosperms, conifers, annual and perennial taxa. It stands to reason that the distinctive features of the conifer genome, such as its large size and relatively small occupancy of the gene space, its conserved macro-structure, the large numbers of repetitive elements, and long introns, represent the product of the intricate evolutionary history of conifers.

## METHODS

### *Picea glauca* BAC isolation and validation

A BAC library developed from the single *Picea glauca* (Moench) Voss individual PG29 from BC Ministry of Forests was utilized. The non-arrayed library consisted of approximately 1.1 million BAC clones with an average insert size of 140 kbp, representing approximately 3× coverage of *P. glauca* genome [40]. The library was screened by quantitative PCR (qPCR) through successive steps using BAC super-pools and pools, serial dilutions and clone identify verification by amplicon sequencing. The BAC isolation and sequencing are reported here for the first time.

We isolated 21 BAC clones each containing a different single copy gene from *P. glauca* (see list of genes and accession numbers in Additional file 1: Table S2). Each of the genes screened was represented by a unique FL-cDNA clone in *P. glauca* as described in Rigault et al. [29]. The selected genes encoded enzymes and transcriptional regulators involved in secondary cell-wall formation and nitrogen metabolism and were subject to manual curation. They were chosen as to facilitate comparison with BAC isolation studies conducted in other conifers species (e.g. [21]). Two sets of gene specific primers were designed for each gene based on the cDNA sequence available in *P. glauca* gene catalogue [29]. The genomic sequence obtained was used to design two additional primers such that two small amplicons of 120–200 bp could be amplified by quantitative PCR (qPCR) (Additional file 1: Table S3). All of the primer sets were verified by PCR and qPCR using the genomic DNA from *P. glauca,* genotype PG-653, and then they were used to screen the BAC library in three steps. See PCR conditions in Additional file 5.

The BAC library was subdivided into pools with a titer 1000 BACs on average, which were arrayed into ten 96 deep-well plates. Each plate was inoculated in 96-well culture plates with 1 ml of terrific broth (TB) and 20 μg/mL of chloramphenicol and grown in a 37°C shaker at 300 rpm overnight. The same TB medium and growing conditions were utilized to culture bacteria throughout the screening steps. Bacterial cultures from each of the columns and rows within a plate were combined in a total of 200 super-pools for DNA isolation as described in Osoegawa et al. [46].

The first step followed Jeukens et al. [47]. Briefly, the super-pool DNA was amplified by the two small amplicons for each target gene by qPCR using QuantiTect SYBR Green master mix as described in Boyle et al. [48]. The intersection of a positive row and a positive column was indicative of positive wells on the original plate. The presence of target genes in the positive super-pools was verified by qPCR in 30 μL reactions using QuantiTect SYBR Green master [48]. We performed PCR of the long amplicon and its purification for gene sequence validation by Sanger sequencing (Additional file 1: Table S3). The second step of the screening relied on serial dilutions of the super-pools to inoculate 50 bacteria from the positive per well in a 96 deep-well plate. DNA super-pools were extracted and screened by qPCR using the same conditions as in the first step. Then, we extracted DNA from 1 μL of bacterial culture from each well of a positive column to test it by qPCR and determine the positive well in the column. From the positive well of the same bacterial culture plate, we proceeded with serial dilutions and we inoculated a 96 deep-well plate with one isolated colony per well. The third step of the screening consisted to pool columns and rows of bacterial cultures. We identified positive wells by qPCR and plated the culture of each positive well on a different Petri dish and one colony per dish was inoculated in 5 mL TB with chloramphenicol. DNA was extracted from 2 mL of each culture. Positive isolated clones were validated by PCR, qPCR and resequencing of the long amplicon. The validation steps to confirm gene identity and integrity proved essential as conifers such as conifers contain many pseudogenes that reduce the efficiency of targeted BAC isolation [39].

The 21 isolated BAC clones, each identified by screening for a different gene (for accessions numbers, see Additional file 1: Table S2) were sequenced by Roche 454 FLX pyrosequencing at McGill University and Genome Québec Innovation Centre, Montreal, Canada. Sequences were assembled *de novo* into contigs using the GS *De novo* Assembler module of Newbler version 2.3 (Roche) [49]. In this analysis, the BAC vector and *E. coli* genome were trimmed and the assembly parameters were a minimum overlap of 200 bp length, minimum overlap identity of 98% and minimum contig length of 500 bp. In general, more than one contig per BAC was obtained; therefore, the order of the contigs within each BAC was tested by PCR.

To determine gene structure (introns and exons), cDNAs were mapped onto the BAC contigs containing the respective gene using est2genome incorporated in the annotation software MAKER [50]. Four of the genes were eliminated from the comparative gene structure analyses because they were either incomplete, lacked introns or identifiable homologs in the species targeted for comparative analyses.

## *Pinus taeda* Orthologous Sequences

Seven BAC clones of *Pinus taeda* containing orthologs of *P. glauca* genes were identified by BLAST [51] using an e-value threshold of 1e-20 and sequence identity>90% (Additional file 1: Table S4). An additional 16 sequences were identified by BLAST [51] using an e-value threshold of 1e-20 in the whole genome shotgun assembly of *Pinus taeda*[11]. Their gene structures were defined using est2genome [50] and *P. taeda* cDNA or*P. glauca* cDNA when *P. taeda* complete cDNA was not available. Accessions numbers available in Additional file 1: Table S4. A pairwise alignment of all corresponding intron and exon sequences of orthologous genes between *P. glauca* and *P. taeda* was conducted, followed by the estimation of their similarity with the software Needle, part of the analysis package EMBOSS [52]. The BAC clones containing the LIM gene in *P. glauca* and the Korrigan, Peptidase_C, Thiolase, Gp_dh_C and eRF1_2 genes in *P. taeda* lacked an intron and exon; these missing exons and introns were excluded from the comparison between *P. glauca* and *P. taeda*.

## Screening for Highly Expressed Genes in Whole Genome Shotgun Assembly

Based on transcript profiles (PiceaGenExpress database [30]) a set of 500 gene sequences each representing a unique FL-cDNA clone that was highly expressed in all tissues was identified from the *P. glauca* gene catalog [29]. A preliminary assembly of the *P. glauca* genome assembly described by Birol et al. [10] was screened with each of these sequences. The screening was performed by exonerate and est2genome model [53], which considers intron/exon boundaries. The cDNA/genome alignments were further filtered based on the identity and length coverage to retain only complete alignments with entire cDNAs; i.e., genes with complete genomic sequence. We randomly selected 18 the genomic sequences thus identified as containing complete structures of highly expressed genes. As for genes contained into the BACs, genes annotations were generate automatically and curated manually individual reciprocal BLASTs and sequence alignments.

## Identification of Closest Homologs in Angiosperms

Homologous sequences to *P. glauca* genes were identified in *Arabidopsis thaliana and P. trichocarpa* by BLASTX [51] with threshold e-value of 1e-10. Reciprocal analysis (BLASTX) between the *A. thaliana, P. trichocarpa* sequence and the *P. glauca* gene catalogue was used to verify that the genes were the closest homologs among known sequences. In *Zea mays*, the closest homolog was identified based on *A. thaliana*sequences by BLASTX, with a threshold e-value of 1e-10, and orthology was verified in the Maize Genome Project Sequencing database (Additional file 1: Table S5). We also performed a BLASTX of *P. glauca* against *Z. mays* sequences and we verified that the closest homologs identified between *Z.mays* and *A. thaliana* were among the best hits. Gene structures were recovered from the databases of TAIR 10 [54], Phytozome (JGI v3.0 gene annotation of assembly v3 of *P. trichocarpa*) [18, 55] and the Maize Genome Sequencing Project [56]. Accession numbers available in Additional file 1: Table S5.

From the 21 genes contained into the *P. glauca* BAC clones, four genes were eliminated from the gene structure comparative analyzes between *P. glauca, A. thaliana, P. trichocarpa* and *Z. mays*: (1) Dof5 because a clear A. *thaliana* homolog could not be identified, (2) asparaginase because a clear *Z. mays* homolog could not be identified, (3) PAL because it lacked introns and (4) LIM because it presented an incomplete sequence (cDNA). A total of 35 genes had their gene structure compared with closest homologs in angiosperms: 17 genes related to secondary cell wall formation and nitrogen metabolism and 18 highly transcribed genes with little tissue-specific expression.

## Statistical Analyses of Introns

Intron lengths were compared between *P. glauca*, *A. thaliana*, *P. trichocarpa* and *Z. mays* by nonparametric Kruskal-Wallis tests with post-test analysis by Dunn's multiple comparisons, because intron length did not follow a Gaussian distribution. Comparisons of two groups of genes used by Wilcoxon rank sum test with continuity correction; this test was used to compare total intron sequences in *P. glauca* genes belonging to the two expression groups and to compare *P. glauca* with *P. taeda*. Data analyses were performed using the R packages coin and multcomp [57, 58, 59].

## Gene Space Obtained From Sequence Capture Technology

Sequences were obtained by using genomic DNA hybridizations

on a custom *P. glauca* chip containing oligonucleotide baits for 23,864 genes. The method development, the DNA sequence isolation and analysis procedure and the resulting sequence data are reported in this manuscript for the first time.

DNA was extracted from needles of the *P. glauca* individual 77111 from the Canadian Forest Service as described in Pelgas et al. [60] using the DNeasy Plant mini kit according to the manufacturer's instructions (QIAGEN). One microgram of high quality DNA was used to prepare a GS-FLX rapid library according to the manufacturer instructions (Roche). The library was amplified by ligation-mediated PCR using 454 A and B primers as described in the NimbleGen SeqCap EZ Library LR User's guide.

Custom probes were designed by Nimblegen based on the cDNA sequences and ESTs from the *P. glauca* gene catalogue [29]. We used a Newbler (gsAssembler module v2.5.3) assembly of sequences from random genomic sequencing (0.15× of coverage) from *P. glauca*[29] to identify highly repetitive elements and to filter out probes representing such elements as they were expected to reduce the efficiency of the sequence capture. Next, a comparative genomic hybridization (Array CGH) experiment was conducted in collaboration with Nimblegen (Madison, WI, USA) to eliminate probes with abnormally high level of hybridization that could not be identified with *in silico* approaches. Throughout the process, probes within genes harboring abnormally high capture levels were eliminated. The final design covered 23,864 genes. The target enrichment procedures including quantitative PCR assessments are described in Additional file 5.

Emulsion PCR and GS-FLX Titanium sequencing was performed according to manufacturer's instructions at the Plateforme d'Analyses Génomiques of the Institut de Biologie Intégrative et des Systèmes (Université Laval, Quebec, Canada). Raw sequencing reads were *de novo* assembled using the gsAssembler module of Newbler v2.5.3. Contigs were screened for complete gene structures based on the *P. glauca* gene catalogue [29]. Technical details are available in Additional file 5.

## *Picea glauca* Repetitive Library and Identification of Repeat Elements

For repeat identification, a random sample of 100,000 *P. glauca* 454 reads from randomly sheared DNA was searched *de novo* for repeats using the software RepeatScout [61]. The results were filtered by removing low complexity sequences and sequences shorter than 100 nt, and retaining only repeats having at least 10 matches when mapped onto the original 454 set using RepeatMasker [62]. Since RepeatScout is tailored to analyze complete genomes or at least large scaffolds, its output is usually fragmented when the program is run on random sheared reads. In order to reduce fragmentation, we merged the repeats belonging to the same element running cap3 [63] under relaxed settings (-o 30, -p 80, -s 500) on the RepeatScout output. Finally, the entire set of repeated sequences was clustered using the software cd-hit-est [64] by collapsing all the repeats sharing at least 80% similarity in order to remove redundancies.

Repeat characterization proceeded by similarity searches were used to associate candidate repeats to known TE families and to remove repeats showing similarity to gene sequences and being possibly part of gene families. In particular, the repeat candidates from each species were searched against RepBase [65] using TBLASTX [57] and setting as significance threshold an e-value of 1e-5. Repeats that did not provide significant hits were used as queries in BLASTX searches against the non-redundant (nr) division of Genbank. Those having significant hits with genes were removed from the library while those having significant hits with TEs were labeled accordingly and the remaining repeats were considered as unclassified.

The search for repeat elements in all BAC contigs and in the gene space obtained from sequence capture was conducted using RepeatMasker [62] using the *Picea glauca* repetitive library and default parameters.

# DECLARATIONS

## Acknowledgements

The authors thank D. Peterson (Mississippi State Univ., USA) and G. Claros and F. Canovas (Univ. de Málaga, Spain) for sharing information on gene targets and strategies for BAC isolation in pines. Technical assistance of S. Caron, É. Fortin, G. Tessier (Univ. Laval, Canada) is acknowledged for BAC screening. F. Belzile, R. Lévesque, L. Bernatchez (Univ. Laval, Canada) are acknowledged for valuable discussions and suggestions at the project planning stage. Funding for the project was received from Génome Québec for a Genome exploration grant (JM, JBou, PR, KR), from Genome Canada, Génome Québec and Genome British Columbia for the SmarTForests project (JM, JBoh, JBou, IB, KR, SJ) and NSERC of Canada (JM). JS received partial funding from Univ. Laval.

## Competing interests

The authors declare that they have no competing interests.

# REFERENCES

1. Lynch M, Conery JS: The origins of genome complexity. Science. 2003, 302: 1401-1404. 10.1126/science.1089370.
2. Deutsch M, Long M: Intron-exon structures of eukaryotic model organisms. Nucleic Acids Res. 1999, 27: 3219-3228. 10.1093/nar/27.15.3219.
3. Comeron JM, Kreitman M: The correlation between intron length and recombination in *Drosophila*: dynamic equilibrium between mutational and selective forces. Genetics. 2000, 156: 1175-1190.
4. Castillo-Davis CI, Mekhedov SL, Hartl DL, Koonin EV, Kondrashov FA: Selection for short introns in highly expressed genes. Nat Genet. 2002, 31: 415-418.
5. Murray BG, Leitch IJ, Bennett MD: Gymnosperm DNA C-values Database. 2004, http://www.kew.org/cvalues/,
6. Morse AM, Peterson DG, Islam-Faridi MN, Smith KE, Magbanua Z, Garcia SA, Kubisiak TL, Amerson HV, Carlson JE, Nelson CD, Davis JM: Evolution of genome size and complexity in *Pinus*. PLoS One. 2009, 4: e4332-10.1371/journal.pone.0004332.
7. Magbanua ZV, Ozkan S, Bartlett BD, Chouvarine P, Saski CA, Liston A, Cronn RC, Nelson CD, Peterson DG:

Adventures in the enormous: A 1.8 million clone BAC library for the 21.7 Gb genome of loblolly pine. PLoS One. 2011, 6: e16214-10.1371/journal.pone.0016214.

8. Pavy N, Pelgas B, Laroche J, Rigault P, Isabel N, Bousquet J: A spruce gene map infers ancient plant genome reshuffling and subsequent slow evolution in the gymnosperm lineage leading to extant conifers. BMC Biol. 2012, 10: 84-10.1186/1741-7007-10-84.

9. Nystedt B, Street NR, Wetterbom A, Zuccolo A, Lin Y-C, Scofield DG, Vezzi F, Delhomme N, Giacomello S, Alexeyenko A, Vicedomini R, Sahlin K, Sherwood E, Elfstrand M, Gramzow L, Holmberg K, Hällman J, Keech O, Klasson L, Koriabine M, Kucukoglu M, Käller M, Luthman J, Lysholm F, Niittylä T, Olson Å, Rilakovic N, Ritland C, Rosselló JA, Sena J, et al: The Norway spruce genome sequence and conifer genome evolution. Nature. 2013, 497: 579-584. 10.1038/nature12211.

10. Birol I, Raymond A, Jackman SD, Pleasance S, Coope R, Taylor GA, Yuen MMS, Keeling CI, Brand D, Vandervalk BP, Kirk H, Pandoh P, Moore RA, Zhao Y, Mungall AJ, Jaquish B, Yanchuk A, Ritland C, Boyle B, Bousquet J, Ritland K, Mackay J, Bohlmann J, Jones SJM: Assembling the 20 Gb white spruce (*Picea glauca*) genome from whole-genome shotgun sequencing data. Bioinformatics. 2013, 29: 1492-1497. 10.1093/bioinformatics/btt178.

11. Neale DB, Wegrzyn JL, Stevens KA, Zimin AV, Puiu D, Crepeau MW, Cardeno C, Koriabine M, Holtz-Morris AE, Liechty JD, Martínez-García PJ, Vasquez-Gross HA, Lin BY, Zieve JJ, Dougherty WM, Fuentes-Soriano S, Wu L-S, Gilbert D, Marçais G, Roberts M, Holt C, Yandell M, Davis JM, Smith KE, Dean JF, Lorenz WW, Whetten RW, Sederoff R, Wheeler N, McGuire PE, et al: Decoding the massive genome of loblolly pine using haploid DNA and novel assembly strategies. Genome Biol. 2014, 15: R59-10.1186/gb-2014-15-3-r59.

12. Wegrzyn JL, Liechty JD, Stevens KA, Wu L-S, Loopstra CA, Vasquez-Gross HA, Dougherty WM, Lin BY, Zieve JJ, Martínez-García PJ, Holt C, Yandell M, Zimin AV, Yorke JA, Crepeau MW, Puiu D, Salzberg SL, Dejong PJ, Mockaitis K, Main D, Langley CH, Neale DB: Unique features of the Loblolly Pine (Pinus taeda L.) Megagenome revealed through sequence annotation. Genetics. 2014, 196: 891-909. 10.1534/genetics.113.159996.

13. Vinogradov AE: Intron–genome size relationship on a large evolutionary scale. J Mol Evol. 1999, 49: 376-384. 10.1007/PL00006561.

14. McLysaght A, Enright AJ, Skrabanek L, Wolfe KH: Estimation of synteny conservation and genome compaction between pufferfish (*Fugu*) and human. Yeast. 2000, 17: 22-36. 10.1002/(SICI)1097-0061(200004)17:1<22::AID-YEA5>3.3.CO;2-J.

15. Moriyama EN, Petrov DA, Hartl DL: Genome size and intron size in *Drosophila*. Mol Biol Evol. 1998, 15: 770-773. 10.1093/oxfordjournals.molbev.a025980.

16. Wendel JF, Cronn RC, Alvarez I, Liu B, Small RL, Senchina DS: Intron size and genome size in plants. Mol Biol Evol. 2002, 19: 2346-2352. 10.1093/oxfordjournals.molbev.a004062.

17. Jiang K, Goertzen LR: Spliceosomal intron size expansion in domesticated grapevine (Vitis vinifera). BMC Res Notes. 2011, 4: 52-10.1186/1756-0500-4-52.

18. Tuskan GA, DiFazio S, Jansson S, Bohlmann J, Grigoriev I, Hellsten U, Putnam N, Ralph S, Rombauts S, Salamov A, Schein J, Sterck L, Aerts A, Bhalerao RR, Bhalerao RP, Blaudez D, Boerjan W, Brun A, Brunner A, Busov V, Campbell M, Carlson J, Chalot M, Chapman J, Chen G-L, Cooper D, Coutinho PM, Couturier J, Covert S, Cronk Q, et al: The genome of black cottonwood, Populus trichocarpa (Torr. & Gray). Science. 2006, 313: 1596-1604. 10.1126/science.1128691.

19. Arabidopsis Genome Initiative: Analysis of the genome sequence of the flowering plant *Arabidopsis thaliana*. Nature. 2000, 408: 796-815. 10.1038/35048692.

20. Haberer G, Young S, Bharti AK, Gundlach H, Raymond C, Fuks G, Butler E, Wing RA, Rounsley S, Birren B, Nusbaum C, Mayer KFX, Messing J: Structure and architecture of the maize genome. Plant Physiol. 2005, 139: 1612-1624. 10.1104/pp.105.068718.

21. Ren X-Y, Vorst O, Fiers MWEJ, Stiekema WJ, Nap J-P: In plants, highly expressed genes are the least compact. Trends Genet. 2006, 22: 528-532. 10.1016/j.tig.2006.08.008.

22. Kumar A, Bennetzen JL: Plant retrotransposons. Annu Rev Genet. 1999, 33: 479-532. 10.1146/annurev.genet.33.1.479.

23. Feschotte C, Jiang N, Wessler SR: Plant transposable elements: where genetics meets genomics. Nat Rev Genet. 2002, 3: 329-341.

24. Schnable PS, Ware D, Fulton RS, Stein JC, Wei F, Pasternak S, Liang C, Zhang J, Fulton L, Graves TA, Minx P, Reily AD, Courtney L, Kruchowski SS, Tomlinson C, Strong C, Delehaunty K, Fronick C, Courtney B, Rock SM, Belter E, Du F, Kim K, Abbott RM, Cotton M, Levy A, Marchetto P, Ochoa K, Jackson SM, Gillam B, et al: The B73 maize genome: complexity, diversity, and dynamics. Science. 2009, 326: 1112-1115. 10.1126/science.1178534.

25. Ralph SG, Chun HJ, Kolosova N, Cooper D, Oddy C, Ritland CE, Kirkpatrick R, Moore R, Barber S, Holt RA, Jones SJ, Marra MA, Douglas CJ, Ritland K, Bohlmann J: A conifer genomics resource of 200,000 spruce (Picea spp.) ESTs and 6,464 high-quality, sequence-finished full-length cDNAs for Sitka spruce (Picea sitchensis). BMC Genomics. 2008, 9: 484-10.1186/1471-2164-9-484.

26. Bedon F, Grima-Pettenati J, Mackay J: Conifer R2R3-MYB transcription factors: sequence analyses and gene expression in wood-forming tissues of white spruce (*Picea glauca*). BMC Plant Biol. 2007, 7: 17-10.1186/1471-2229-7-17.

27. Cañas RA, de la Torre F, Cánovas FM, Cantón FR: High levels of asparagine synthetase in hypocotyls of pine seedlings suggest a role of the enzyme in re-allocation of seed-stored nitrogen. Planta. 2006, 224: 83-95. 10.1007/s00425-005-0196-6.

28. Nairn CJ, Lennon DM, Wood-Jones A, Nairn AV,

Dean JFD: Carbohydrate-related genes and cell wall biosynthesis in vascular tissues of loblolly pine (*Pinus taeda*). Tree Physiol. 2008, 28: 1099-1110. 10.1093/treephys/28.7.1099.

29. Rigault P, Boyle B, Lepage P, Cooke JEK, Bousquet J, MacKay JJ: A white spruce gene catalog for conifer genome analyses. Plant Physiol. 2011, 157: 14-28. 10.1104/pp.111.179663.
30. Raherison E, Rigault P, Caron S, Poulin P-L, Boyle B, Verta J-P, Giguère I, Bomal C, Bohlmann J, MacKay J: Transcriptome profiling in conifers and the PiceaGenExpress database show patterns of diversification within gene families and interspecific conservation in vascular gene expression. BMC Genomics. 2012, 13: 434-10.1186/1471-2164-13-434.
31. Bradnam KR, Korf I: Longer first introns are a general property of eukaryotic gene structure. PLoS One. 2008, 3: e3093-10.1371/journal.pone.0003093.
32. Savard L, Li P, Strauss SH, Chase MW, Michaud M, Bousquet J: Chloroplast and nuclear gene sequences indicate late Pennsylvanian time for the last common ancestor of extant seed plants. Proc Natl Acad Sci U S A. 1994, 91: 5163-5167. 10.1073/pnas.91.11.5163.
33. Wang X-Q, Tank DC, Sang T: Phylogeny and divergence times in Pinaceae: evidence from three genomes. Mol Biol Evol. 2000, 17: 773-781. 10.1093/oxfordjournals.molbev.a026356.
34. Ohri D, Khoshoo TN: Genome size in gymnosperms. Plandt Syst Evol. 1986, 153: 119-132. 10.1007/BF00989421.
35. Hu TT, Pattyn P, Bakker EG, Cao J, Cheng J-F, Clark RM, Fahlgren N, Fawcett JA, Grimwood J, Gundlach H, Haberer G, Hollister JD, Ossowski S, Ottilar RP, Salamov AA, Schneeberger K, Spannagl M, Wang X, Yang L, Nasrallah ME, Bergelson J, Carrington JC, Gaut BS, Schmutz J, Mayer KFX, Van de Peer Y, Grigoriev IV, Nordborg M, Weigel D, Guo Y-L: The Arabidopsis lyrata genome sequence and the basis of rapid genome size change. Nat Genet. 2011, 43: 476-481. 10.1038/ng.807.
36. Morgante M, De Poali E: Toward the conifer genome sequence. Genetics, Genomics and Breeding of Conifers Trees. Edited by: Plomion C, Bousquet J, Kole C. 2011, Enfield: Science Publishers, 389-403.
37. Lockton S, Gaut BS: The contribution of transposable elements to expressed coding sequence in *Arabidopsis thaliana*. J Mol Evol. 2009, 68: 80-89. 10.1007/s00239-008-9190-5.
38. Jaillon O, Aury J-M, Noel B, Policriti A, Clepet C, Casagrande A, Choisne N, Aubourg S, Vitulo N, Jubin C, Vezzi A, Legeai F, Hugueney P, Dasilva C, Horner D, Mica E, Jublot D, Poulain J, Bruyère C, Billault A, Segurens B, Gouyvenoux M, Ugarte E, Cattonaro F, Anthouard V, Vico V, Fabbro CD, Alaux M, Gaspero GD, Dumas V, et al: The grapevine genome sequence suggests ancestral hexaploidization in major angiosperm phyla. Nature. 2007, 449: 463-467. 10.1038/nature06148.
39. Kovach A, Wegrzyn JL, Parra G, Holt C, Bruening GE, Loopstra CA, Hartigan J, Yandell M, Langley CH, Korf I, Neale DB: The*Pinus taeda* genome is characterized by diverse and highly diverged repetitive sequences. BMC Genomics. 2010, 11: 420-10.1186/1471-2164-11-420.
40. Hamberger B, Hall D, Yuen M, Oddy C, Hamberger B, Keeling CI, Ritland C, Ritland K, Bohlmann J: Targeted isolation, sequence assembly and characterization of two white spruce (*Picea glauca*) BAC clones for terpenoid synthase and cytochrome P450 genes involved in conifer defence reveal insights into a conifer genome. BMC Plant Biol. 2009, 9: 106-10.1186/1471-2229-9-106.
41. Bautista R, Villalobos DP, Díaz-Moreno S, Cantón FR, Cánovas FM, Claros MG: Toward a *Pinus pinaster* bacterial artificial chromosome library. Ann For Sci. 2007, 64: 855-864. 10.1051/forest:2007060.
42. Gazave E, Marqués-Bonet T, Fernando O, Charlesworth B, Navarro A: Patterns and rates of intron divergence between humans and chimpanzees. Genome Biol. 2007, 8: R21-10.1186/gb-2007-8-2-r21.
43. Lynch M: Intron evolution as a population-genetic process. Proc Natl Acad Sci U S A. 2002, 99: 6118-6123. 10.1073/pnas.092595699.
44. Jaramillo-Correa JP, Verdú M, González-Martínez SC: The contribution of recombination to heterozygosity differs among plant evolutionary lineages and life-forms. BMC Evol Biol. 2010, 10: 22-10.1186/1471-2148-10-22.
45. Sakharkar MK, Chow VTK, Kangueane P: Distributions of exons and introns in the human genome. In Silico Biol (Gedrukt). 2004, 4: 387-393.
46. Osoegawa K, de Jong PJ, Frengen E, Ioannou PA: Construction of bacterial artificial chromosome (BAC/PAC) libraries. Current Protocols in Molecular Biology. Edited by: Ausubel FM, Brent R, Kingston RE, Moore DD, Seidman JG, Smith JA, Struhl K. 2001, Hoboken, NJ, USA: John Wiley & Sons Inc
47. Jeukens J, Boyle B, Kukavica-Ibrulj I, St-Cyr J, Lévesque RC, Bernatchez L: BAC library construction, screening and clone sequencing of lake whitefish (*Coregonus clupeaformis,* Salmonidae) towards the elucidation of adaptive species divergence. Mol Ecol Resour. 2011, 11: 541-549. 10.1111/j.1755-0998.2011.02982.x.
48. Boyle B, Dallaire N, MacKay J: Evaluation of the impact of single nucleotide polymorphisms and primer mismatches on quantitative PCR. BMC Biotech. 2009, 9: 75-10.1186/1472-6750-9-75.
49. Margulies M, Egholm M, Altman WE, Attiya S, Bader JS, Bemben LA, Berka J, Braverman MS, Chen Y-J, Chen Z, Dewell SB, Du L, Fierro JM, Gomes XV, Goodwin BC, He W, Helgesen S, Ho CH, Irzyk GP, Jando SC, Alenquer MLI, Jarvie TP, Jirage KB, Kim J-B, Knight JR, Lanza JR, Leamon JH, Lefkowitz SM, Lei M, Li J, et al: Genome sequencing in open microfabricated high density picoliter reactors. Nature. 2005, 437: 376-380.
50. Cantarel BL, Korf I, Robb SMC, Parra G, Ross E, Moore B, Holt C, Alvarado AS, Yandell M: MAKER: An easy-to-use annotation pipeline designed for emerging model organism genomes. Genome Res. 2008, 18: 188-196.
51. Altschul SF, Gish W, Miller W, Myers EW, Lipman DJ: Basic local alignment search tool. J Mol Biol. 1990, 215:

403-410. 10.1016/S0022-2836(05)80360-2.

52. Rice P, Longden I, Bleasby A: EMBOSS: the European Molecular Biology Open Software Suite. Trends Genet. 2000, 16: 276-277. 10.1016/S0168-9525(00)02024-2.

53. Slater GS, Birney E: Automated generation of heuristics for biological sequence comparison. BMC Bioinforma. 2005, 6: 31-10.1186/1471-2105-6-31.

54. The Arabidopsis Information Resource. http://arabidopsis.org.

55. Goodstein DM, Shu S, Howson R, Neupane R, Hayes RD, Fazo J, Mitros T, Dirks W, Hellsten U, Putnam N, Rokhsar DS: Phytozome: a comparative platform for green plant genomics. Nucleic Acids Res. 2012, 40 (D1): D1178-D1186. 10.1093/nar/gkr944.

56. Maize Genome Sequencing Project. http://www.maizesequence.org,

57. Hothorn T, Bretz F, Westfall P: Simultaneous inference in general parametric models. Biom J. 2008, 50: 346-363. 10.1002/bimj.200810425.

58. Hothorn T, Hornik K, van de Wiel MA, Zeileis A: Implementing a class of permutation tests: the coin package. J Stat Softw. 28 (8): 1-23. URL http://www.jstatsoft.org/v28/i08/

59. R project. http://www.r-project.org,

60. Pelgas B, Bousquet J, Meirmans PG, Ritland K, Isabel N: QTL mapping in white spruce: gene maps and genomic regions underlying adaptive traits across pedigrees, years and environments. BMC Genomics. 2011, 12: 145-10.1186/1471-2164-12-145.

61. Price AL, Jones NC, Pevzner PA: De novo identification of repeat families in large genomes. Bioinformatics. 2005, 21 (Suppl 1): i351-i358. 10.1093/bioinformatics/bti1018.

62. Smit AFA, Hubley R, Green P: RepeatMasker Open-3.0. 1996–2010 http://www.repeatmasker.org

63. Huang X, Madan A: CAP3: a DNA sequence assembly program. Genome Res. 1999, 9: 868-877. 10.1101/gr.9.9.868.

64. Huang Y, Niu B, Gao Y, Fu L, Li W: CD-HIT Suite: a web server for clustering and comparing biological sequences. Bioinformatics. 2010, 26: 680-682. 10.1093/bioinformatics/btq003.

65. Jurka J, Kapitonov VV, Pavlicek A, Klonowski P, Kohany O, Walichiewicz J: Repbase update, a database of eukaryotic repetitive elements. Cytogenet Genome Res. 2005, 110: 462-467. 10.1159/000084979.

CHAPTER 7

# A SPRUCE GENE MAP INFERS ANCIENT PLANT GENOME RESHUFFLING AND SUBSEQUENT SLOW EVOLUTION IN THE GYMNOSPERM LINEAGE LEADING TO EXTANT CONIFERS

Nathalie Pavy, Betty Pelgas, Jérôme Laroche, Philippe Rigault, Nathalie Isabel and Jean Bousquet

**Keywords:** Angiosperm, duplication, evolution, gene families, genetic map, gymnosperm, phylogenomics, Picea, spruce, structural genomics

## BACKGROUND

Gene duplication plays an important role in providing raw material to evolution [1]. In plants, gene duplicates arise through diverse molecular mechanisms, ranging from whole-genome duplication to more restricted duplications of smaller chromosomal regions [2]. The evolution of the flowering plant genomes has been intensively studied since the completion of the genome sequence for several angiosperm species. Lineage-specific whole-genome duplications greatly contributed to the expansion of plant genomes and gene families (for examples, see [3, 4, 5, 6, 7, 8, 9]), with whole-genome duplications found in basal angiosperms, monocots and eudicots [9, 10, 11, 12].

Little is known about the large-scale evolutionary history of gene duplications for other seed plants, as well as before the origin of angiosperms. Spermatophytes encompass the angiosperms and the gymnosperms, whose seeds are not enclosed in an ovary. The two groups diverged around 300 million years ago (Mya) in the Late Carboniferous [13, 14]. Contrary to angiosperms, which underwent massive adaptive radiation to supplant the gymnosperms as the dominant vascular plant group [15, 16], extant gymnosperms are divided into a relatively small number of groups including the Pinophyta (conifers), Cycadophyta (cycads), Gnetophyta (gnetophytes) and Ginkgophyta (*Ginkgo*), and they contain about 1,000 species [17]. Polyploidy is rare in gymnosperms. Only 5% of them, and 1.5% of the subgroup conifers, have been reported as polyploid species [18, 19], as indicated by cytological analysis [18], distributions of synonymous substitution rates [19, 20] or phylogenetic analysis [20]. Nevertheless, the genomes of some gymnosperms, such as in the conifer family Pinaceae, are among the largest of all known organisms [21], with haploid genome sizes up to 37 Gb for *Pinus gerardiana* [22, 23].

Several issues need to be addressed regarding the evolution of the seed plant genome, and that of the plant genome predating the gymnosperm-angiosperm (GA) divergence. How many gene duplications are shared between angiosperms and gymnosperms, which would predate their divergence and make them ancient? How frequently have gene duplications occurred solely in gymnosperms since their split from angiosperms? Are ancient duplicates, those preceding the GA split, relatively more abundant and more translocated through the gymnosperm genome than most recent duplicates specific to the gymnosperms?

These questions could be addressed through a phylogenomic approach, where the members of different gene families are mapped in a gymnosperm taxon with these families further sampled in completely sequenced angiosperm taxa to reconstruct their multiple phylogenies. Given that the gene complement of model angiosperms has been entirely determined by complete genome sequencing, but not that of

**Citation:** Nathalie Pavy, Betty Pelgas, Jérôme Laroche, Philippe Rigault, Nathalie Isabel and Jean Bousquet, A spruce gene map infers ancient plant genome reshuffling and subsequent slow evolution in the gymnosperm lineage leading to extant conifers, DOI: 10.1186/1741-7007-10-84.

a gymnosperm taxon, such gene phylogenies would give rise to mixed angiosperm-gymnosperm nodes and gymnosperm-specific nodes. With respect to the divergence time between pro-angiosperms and pro-gymnosperms (approximately 300 Mya), different grouping of gene duplicates could help determine the relative age of duplications, such that mixed angiosperm-gymnosperm nodes predating the split between angiosperms and gymnosperms would indicate ancient duplications, while gymnosperm-specific nodes postdating this split would indicate more recent duplications. The various proportions of these nodes over a large number of gene phylogenies would provide a glance at the relative frequency of ancient to recent gene duplications in the gymnosperm lineage, and the mapping of these duplicates on a gymnosperm genome would allow assessment of their possible translocation. Because of the incomplete nature of gene inventories in gymnosperms, such an analysis from the perspective of the angiosperm lineage is still not possible, given that nodes containing angiosperm duplicates only might not be truly angiosperm-specific. On a smaller scale, similar approaches have been applied to investigate the deep phylogenies of a few seed plant gene families completely sequenced in the conifers. They have indicated that, while some gene duplications deemed ancient predated the split between gymnosperms and angiosperms, some duplications have occurred more recently that are specific to the gymnosperm lineage (for example, [24]).

Based on such a phylogenetic approach together with gene mapping, one could also ask if the spread of gene families over the gymnosperm genome is more likely for ancient duplicates predating the GA split than for more recent duplicates postdating this split. Theoretical and empirical approaches have shown that duplicated regions should be translocated with time [9]. As such, one would expect the more recent gymnosperm-specific duplicates to be physically less spread across the genome than more ancient duplicates predating the GA split. Altogether, the relative age of gymnosperm-specific gene duplicates and their degree of translocation would allow an assessment of whether the conservation of genome macro-structure parallels the recognised archaic nature of gymnosperms in terms of morphology, the reproductive system and other phenotypic attributes [25]. Testing these hypotheses requires large catalogues of gene sequences, which have recently become available in conifers [26], and mapping of a large number of genes in a gymnosperm.

In this study, we assembled a map involving 1,801 spruce-expressed genes and examined the distribution of gene families onto the spruce genome and its level of conservation across Pinaceae and angiosperm genomes. We asked whether ancient gene duplicates shared with angiosperms are more numerous and more reshuffled than more recent duplicates occurring in the gymnosperm lineage leading to extant conifers. We also investigated how stable the genome macro-structure has been between the conifers *Picea* and *Pinus*since their divergence 120 to 140 Mya [13, 14, 27], a period of time corresponding to tremendous reshuffling of the angiosperm genome.

## RESULTS

### Spruce Gene Map

We generated a spruce consensus linkage map for the white spruce (*Picea glauca* (Moench) Voss) and black spruce (*Picea mariana* (Mill.) B.S.P.) genomes (Figure 1). This map encompassed 2,270 loci including 1,801 genes spread over the 12 linkage groups of spruce and corresponding to the haploid number of 12 chromosomes prevalent in the Pinaceae, including *Picea* (Figure 1). These genes represented a large array of molecular functions and biological processes (Figure 2). Map length was 2,083 centiMorgan (cM). The number of mapped genes is more than twice that of the most complete spruce gene map available to date [28] and is in the same range as the map available for the loblolly pine genome, which includes 1,816 genes mapped over 1,898 cM [29]. Map length and the number of gene loci per chromosome thus appeared similar in spruce and loblolly pine.

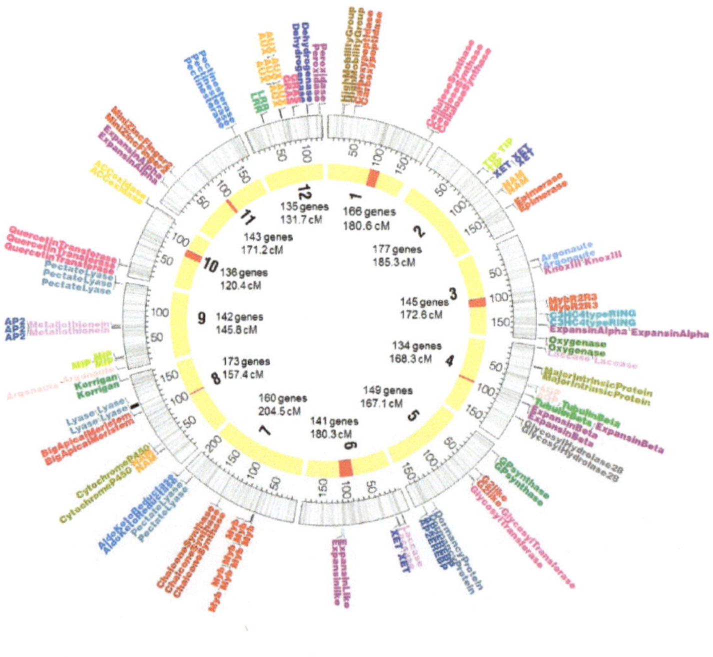


**Figure 1:** Map of the spruce genome and tandemly arrayed genes. The 12 spruce chromosomes were plotted with Circos [100]. From inside

to outside: gene-rich regions in red; the 12 chromosomes with ticks representing the genes mapped along the spruce linkage groups, and with genetic distances in cM (Kosambi); distribution of the tandemly arrayed genes. The chromosome nomenclature and numbers of genes mapped are inside the circle. For the complete names of tandemly arrayed genes.

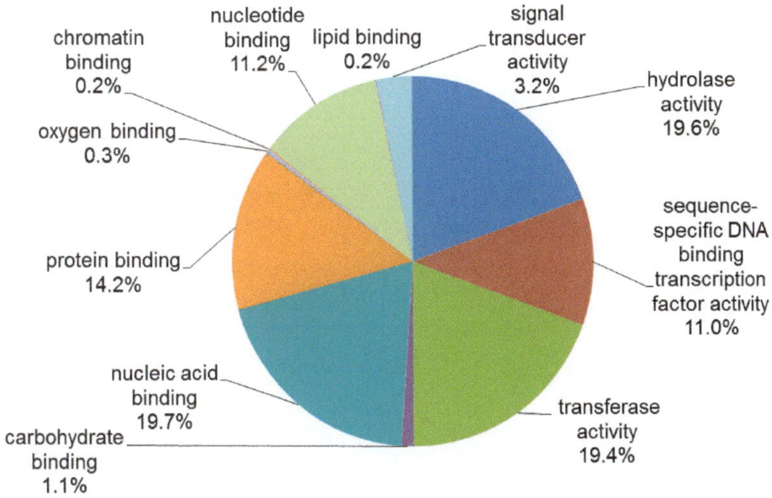


**Figure 2:** Molecular functions of the genes incorporated in the phylogenetic analyses. The pie chart includes the molecular functions assigned at level three of the Gene Ontology classification for the 527 sequences from the 157 gene families represented by two or more mapped genes in spruce and used in phylogenetic analyses.

## Gene Density

Our analyses revealed instances of gene clustering. Using Kolmogorov-Smirnov tests, the gene distribution deviated significantly from a uniform distribution for nine ($P \leq 0.01$) or ten ($P \leq 0.05$) of the 12 spruce chromosomes (Table 1). To localise gene-rich regions (GRRs), we conducted analyses of gene distribution relying on various bandwidths using kernel density estimation. The effect of the bandwidth upon the spread of the GRRs was weak (data not shown). At $P \leq 0.01$, only two GRRs were found on chromosomes 6 and 10; they included 1.3% of the genes (24) over 0.6% of the map length (14.7 cM). At $P \leq 0.05$, seven GRRs, including 9.2% of the mapped genes (166 out of 1,801), were found on seven chromosomes and represented 4.0% of the map length (Figures 1 and 3). In GRRs, gene density was about twice (1.78 gene/cM) that in the rest of the map (0.78 gene/cM). Tandemly arrayed genes (TAGs, see below) were not responsible for the higher gene density of the GRRs. There was no significant difference ($P > 0.05$) in the molecular functions represented by genes lying in the GRRs compared with the remainder of the map. However, regarding biological processes, the GRRs were enriched in Gene Ontology (GO) terms corresponding to metabolism (carbohydrate metabolic processes), reproduction, growth and regulation of anatomical structure ($P \leq 0.05$).

**Table 1:** Testing for gene clustering within spruce chromosomes using the Kolmogorov-Smirnov statistics ($D_n$ and $D^*_n$).

| Chromosome | Number of mapped genes | Chromosome length (cM) | $D_n$ | $D^*_n$ | $P$ |
|---|---|---|---|---|---|
| 1 | 166 | 180.6 | 0.0990 | 1.2812 | ≤ 0.01 |
| 2 | 177 | 185.3 | 0.0444 | 0.5948 | > 0.15 |
| 3 | 145 | 172.6 | 0.1483 | 1.7946 | ≤ 0.01 |
| 4 | 134 | 168.3 | 0.1286 | 1.4967 | ≤ 0.01 |
| 5 | 149 | 167.1 | 0.0490 | 0.6013 | > 0.15 |
| 6 | 141 | 180.3 | 0.1373 | 1.6451 | ≤ 0.01 |
| 7 | 160 | 204.5 | 0.0961 | 1.2207 | ≤ 0.01 |
| 8 | 173 | 157.4 | 0.0774 | 1.0226 | ≤ 0.02 |
| 9 | 142 | 145.8 | 0.0914 | 1.0953 | ≤ 0.01 |
| 10 | 136 | 120.4 | 0.1208 | 1.4219 | ≤ 0.01 |
| 11 | 143 | 171.2 | 0.0933 | 1.1219 | < 0.01 |
| 12 | 135 | 131.7 | 0.1032 | 1.2051 | ≤ 0.01 |

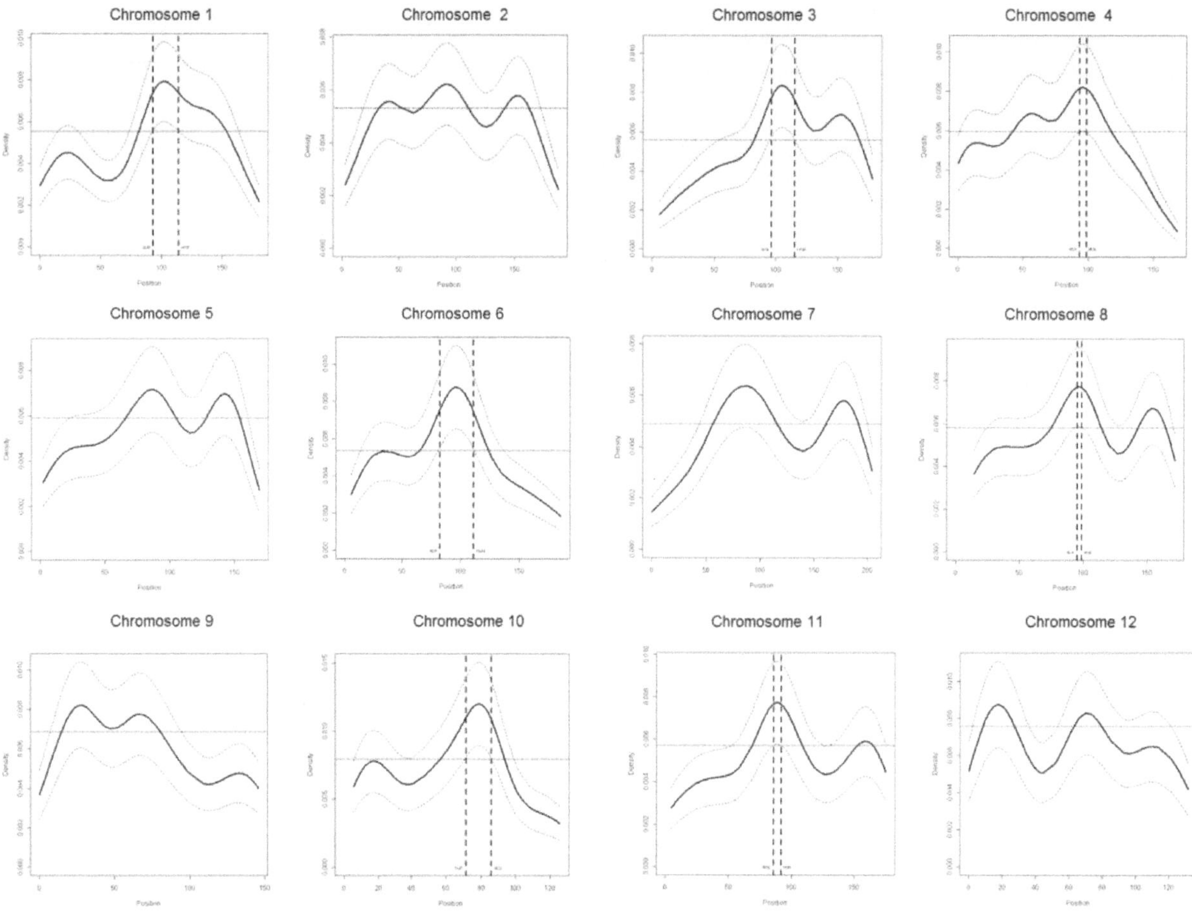


**Figure 3:** Kernel density estimation for the spruce chromosomes. On each plot, the curve in bold is the kernel density function and the dotted curves represent the limits of the confidence interval. The horizontal line represents the expected density of the uniform gene distribution. The vertical dotted lines represent the boundaries of the gene-rich regions: chromosomal regions for which the lower limit of the confidence interval of the density function is above the uniform function.

## Tandemly Arrayed Genes

A total of 125 family members were organised into 51 TAGs (31 arrays within 1 cM and 20 arrays within 5 cM; Figure 1). Most of the arrays included two genes, but arrays were identified including up to eight genes, such as the *myb-r2r3* array on chromosome 7 (Figure 1). Based on the GO classification, genes coding for extracellular proteins and cell wall proteins, and genes involved in DNA-binding functions and in secondary metabolism were over-represented among spruce gene arrays. To test whether this distribution could be observed by chance alone, we randomly redistributed the 664 gene family members and counted the number of chromosomes represented for each family. This simulation was replicated 1,000 times. The observed and the simulated distributions were found to be significantly different ($\chi^2 = 35.7$, degrees of freedom = 11, $P = 0.00018$). The main contribution to the $\chi^2$ value was from the families with members mapping to a single chromosome. Seventeen gene families were found to be associated with a unique chromosome more often than would be expected by chance alone. The TAGs were the major contributors to this distribution.

## Co-Localizing Genes

Within 32 gene groups representing 71 genes (3.9% of the mapped genes), no recombinants were observed out of 500 white spruce progeny. These groups encompassed a variety of molecular functions with no significant deviation from the composition of the overall dataset. In 20 groups, genes were related neither in sequence nor in function. By contrast, 12 groups were made of functionally related genes, including five tandem arrays and seven groups of genes from different families. These twelve groups involved three main functions: metabolism (six groups), regulation of transcription (three groups) and transport (three groups).

We obtained assessments of gene expression for co-localizing genes from 10 groups [30]. In three groups, the co-localizing genes were co-regulated across eight tissues (mature xylem, juvenile xylem, phelloderm (including phloem), young needles, vegetative buds, megagametophytes, adventitious roots and embryogenic cells). The first group included one citrate synthase involved in carbohydrate metabolism and a calcineurin B-like protein involved in transduction through calcium binding. The second group included one reductase involved in histidine catabolic process, which was co-expressed with a ribosomal 30S protein. The third group consisted of two chalcone synthases.

## Intergeneric Map Comparisons

We compared spruce and pine gene sequences and their respective localizations on linkage maps, using that of *Pinus taeda* L. (loblolly pine) with 1,816 gene loci [31] and that of *Pinus pinaster* Ait. (maritime pine) with 292 gene loci [32]. In total, 212 gene loci were shared between spruce and pines. Out of them, 12 gene loci were syntenic between the three genomes, 51 were found between spruce and maritime pine, and 149 others were found between spruce and loblolly pine. Remarkably, the vast majority of the conserved pairs of gene sequences found among pairs of species could be mapped on homoeologous chromosomes. Out of 165 genes mapped on both maps from spruce and loblolly pine, 161 (97.5%) were syntenic, of which 88.8% were collinear (Figure 4).

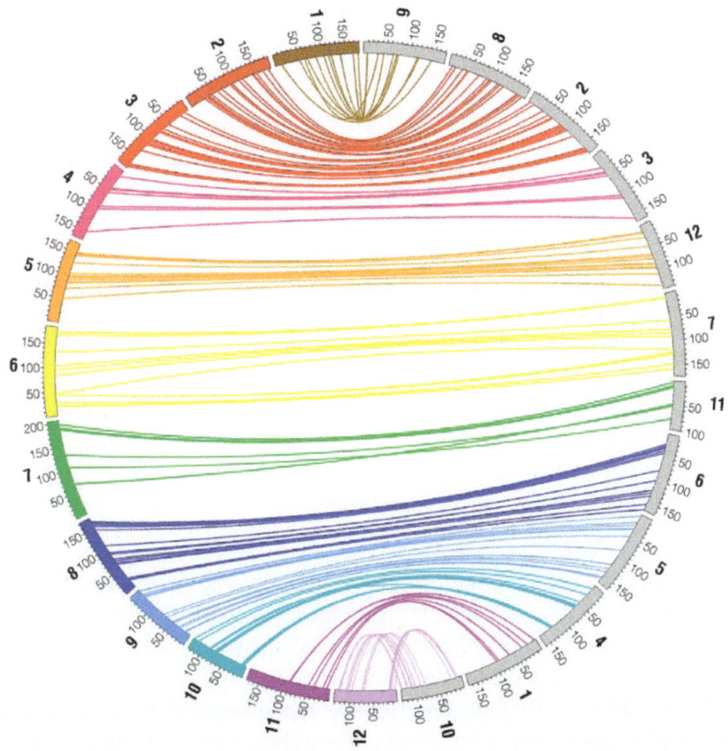


**Figure 4:** A spruce/loblolly pine comparative map. The syntenic positions of the 161 homologous genes mapped on both spruce and loblolly pine genomes were plotted with Circos [100] and are indicated by colour-coded lines connecting the spruce (in colour) and the loblolly pine chromosomes (in grey). The chromosome numbers are indicated outside the circle.

Macro-synteny was spread all along the genomes with large conserved segments (Figure 4). The conserved positions of homologous genes allowed us to delineate the respective positions of homoeologous chromosomal regions in spruce, loblolly pine and maritime pine. The conserved regions represented 82.0% and 86.5% of the lengths of the spruce and loblolly pine maps, respectively (Figure 4). The portion of 82.0% of the spruce map conserved with the loblolly pine map could be extended to 87.6% when conservation with the maritime pine map was also considered. Thus, map comparison with maritime pine provided a significant enrichment in shared genes and homoeologous regions among maps. This high level of conservation enabled us to draw the first comprehensive map for a sizeable part of the gene space of the Pinaceae.

## Phylogenetic Analyses of 157 Gene Families

In total, 527 spruce genes were considered in the phylogenetic analyses. They were distributed in 157 families each containing at least two genes mapped on the spruce genome. These families were distributed across diverse molecular functions, representative of the distribution of expressed genes found in white spruce (Figure 2, see Methods).

The phylogenetic trees for all analysed gene families. Figure 5 shows the unrooted tree representative of the strict consensus between majority-rule bootstrap parsimony (MP) and majority-rule bootstrap neighbour-joining (NJ) trees obtained for the quercetin 3-O-methyltransferase family. In this example, two pairs of genes (*Pg6-29*/*Pg2-68* and *Pg10-23*/*Pg10-26*) resulted from recent duplications after the GA split (Figure 5). One pair clustered on chromosome 10, while the two other genes were translocated on chromosomes 2 and 6 of spruce (Figure 5). Another more ancient duplication giving rise to the two gene lineages leading to *Pg2-68*/*Pg6-29* and *Pg10-23*/*Pg10-26* occurred before the GA split, with the two groups located on different spruce chromosomes, implying at least one translocation (Figure 5).

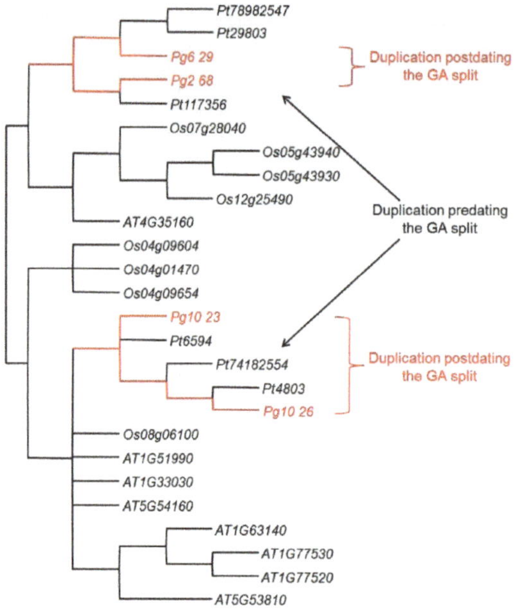


**Figure 5:** Quercetin 3-O-methyltransferase gene family tree. Unrooted phylogenetic tree obtained from the strict consensus of 50%-bootstrap consensus neighbour-joining and parsimony trees and indicating two spruce gene duplications post-dating the gymnosperm-angiosperm split (no intervening *Arabidopsis* or rice sequence between spruce sequences) and one spruce gene duplication predating the gymnosperm-angiosperm split (with intervening*Arabidopsis* or rice sequences between spruce sequences). Sequences are from spruce (Pg), pine (Pt), *Arabidopsis*(AT) and rice (Os). GA: gymnosperm-angiosperm split, estimated at around 300 Mya [13].

Using the strict consensus of majority-rule bootstrap NJ and MP phylogenetic trees for each of the 157 gene families, we evaluated, in a similar fashion, the relative age of duplications for a total of 992 gene pairs (nodes) relative to the GA split. Topological differences between NJ and MP trees affected 115 gene pairs (11.6%) whereas 877 gene pairs (88.4%) were positioned identically by the two analytical approaches, relative to the GA split. Out of these 877 congruent results, 688 pairs (78.4%) diverged before the GA split, 87 pairs (9.9%) diverged after the GA split and the divergence of 102 pairs (11.6%) could not be determined because of lack of support (polytomies). In other words, there were about eight ancient duplications for each recent one (Figure6).

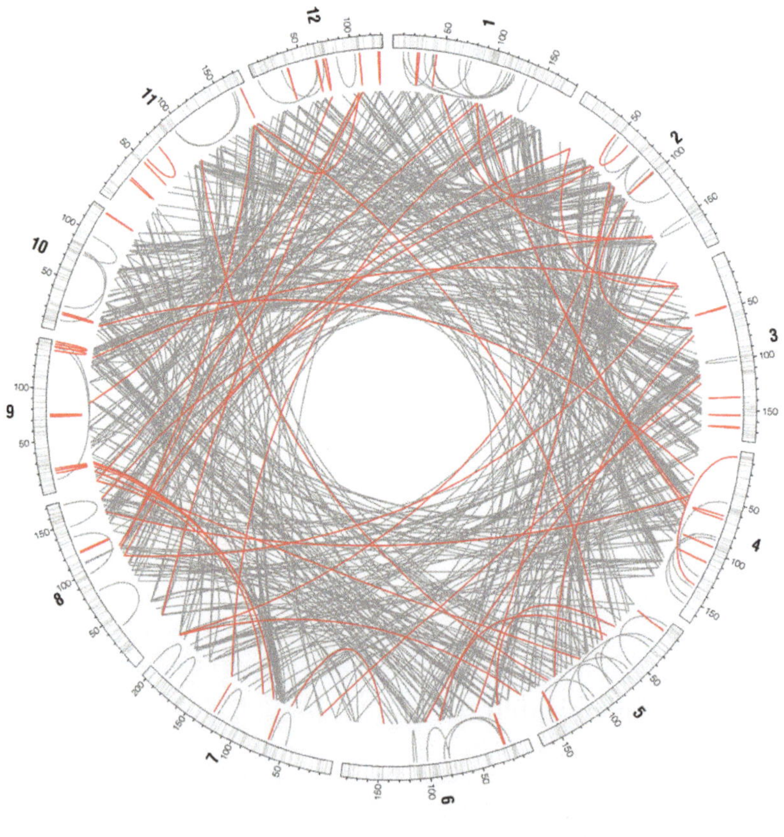

**Figure 6:** Organization of the spruce gene space and duplications. Genome representation with spruce chromosomes (1 to 12) showing

from outside to inside: the 12 chromosomes with ticks representing the genes mapped along the spruce linkage groups, and with genetic distances in cM (Kosambi); links between genes representing duplications within chromosomes and duplications followed by interchromosomal translocations. Links in grey illustrate ancient and links in red illustrate recent, referring to before or after the gymnosperm-angiosperm split, around 300 Mya [13].

## Distribution and Relative Age of Gene Pairs

We analysed the distribution patterns found among the gene pairs on the spruce genome. Most spruce gene pairs were translocated (86.3%) and most of these translocations occurred before the GA split (94.5%). We counted the number of duplicates found on each of the 12 chromosomes, and compared the observed distribution to a theoretical distribution that would be expected by chance alone. Out of 688 gene pairs (or nodes) representing 'ancient' duplications, 56 pairs (8.1%) were located on the same chromosome and 632 pairs (91.9%) were duplicates involving a translocation to another chromosome. This difference was highly significant ($\chi^2 = 482.2$; $P < 2.2e^{-16}$), indicating that ancient gene pairs have been highly dispersed. Out of 87 pairs of genes representing 'recent' duplications, only 37 pairs (42.5%) were translocated and 50 pairs (57.5%) were located on the same chromosome. This difference was not significant ($\chi^2 = 1.9$; $P = 0.16$). For each pair of genes, we computed the distance between the duplicates found on a same chromosome. The mean distance between duplicates arising from a recent duplication event was 4.3 cM; whereas this distance was 47.0 cM between duplicates derived from ancient duplication events. This 10-fold difference was highly significant (Welch t-test $t = -7.8$; $P = 1.1e^{-11}$). Many gene copies found on the same chromosome were forming arrays of genes tandemly duplicated within 5 cM. Within the 51 tamdemly gene arrays that incorporated 6.9% of the mapped genes, 125 gene pairs (duplications) could be classified relative to the GA split: 44 were classified as recent, only 5 were ancient, and 76 were undetermined. Overall, only four gene arrays could be accounted for by ancient duplications predating the GA split (*BAM*, *expansin-like*, *pectinesterase*, *tonoplast intrinsic protein*), whereas 29 other arrays were generated by duplications after the GA split ($\chi^2 = 18.9$; $P = 1.3e^{-05}$; Figure 6). Thus, the more recent origin of these closely-spaced duplicates has apparently resulted in less time and opportunity for them to be dispersed or translocated.

## DISCUSSION

The completion of several genome sequencing projects in angiosperms has resulted in improved knowledge of the content and organisation of the flowering plant genomes. In gymnosperms, in the absence of a completely sequenced and ordered genome, recent efforts have been put toward improving knowledge of the gene space through several EST sequencing projects [33]; but the structural organisation of this gene space on the genome remains largely undetermined [34]. The spruce genetic map and analyses presented herein allow better comprehension of the genome macro-structure for a gymnosperm. These results combined with phylogenies reveal the relative proportion of gene duplications shared between angiosperms and gymnosperms or unique to gymnosperms, and how the seed plant genome has been reshuffled over time from a conifer perspective.

## Gene Distribution and Density

To localise the GRRs, we implemented a statistical approach based on the kernel density function. This represents a technical improvement compared with existing methodologies given that we used an adaptive kernel approach to avoid the use of an arbitrarily fixed bandwidth. This approach allowed us to take into account the density observed locally to compute the bandwidth size. Because the number of genes currently positioned on the spruce genome represents around 6% of the estimated total number of genes [26], we applied stringent parameters in these analyses to reduce the rate of false positives. Thus, we may have underestimated the extent of GRRs. Besides these significant peaks, a few other peaks of kernel density that do not currently reach significance (Figure 3) may do so with an increased number of mapped genes. Indeed, Kolmogorov-Smirnov tests of homogeneity of gene distribution indicated that nine chromosomes had a significantly non-uniform distribution. Even so, there does not seem to be a widespread occurrence of GRRs on the spruce genome. In addition, the seven significant GRRs were distributed among seven chromosomes. This peculiar distribution suggests that GRRs may correspond to centromeric regions where, on genetic maps, markers tend to cluster due to more limited recombination.

In angiosperms, species with small genomes tend to be made of GRRs alternating with gene-poor regions. For example, the genic space of *Arabidopsis thaliana* represents 45% of the genome while the remaining 55% is 'gene-empty' and interspersed among genes as blocks ranging in size from a few hundred base pairs to 50 kb [35]. By contrast, plant species with larger genomes do not show such a contrasted gene distribution, in line with the pattern found here for the large spruce genome. Rather, they harbour a gradient of gene density along chromosomes, such as in maize [36], soybean [37] and wheat [38, 39]. In the soybean genome, a majority of the predicted genes (78%) are found in chromosome ends, whereas repeat-rich sequences are found in centromeric regions [40]. In conifers, retroelements have been reported as a large component of the genome, with some families well dispersed while others occur in centromeric or peri-centromeric regions (for example, see [41, 42, 43, 44, 45]). Thus, they might have participated in shaping the distribution of genes along chromosomes by reducing the occurrence of GRRs.

The type of gene distribution along the genome bears consequences for the planning of genome sequencing strategies. For instance, a gene distribution of 'island' type implies that a deeper sequencing effort is necessary to reach a majority of the genes [38]. Though genetic distance does not equate physical distance, the pattern seen here in spruce indicates that genetic maps alone that would include most of the gene complement will be insufficient to anchor a significant portion of physical scaffolds, especially if these are small. In conifers, little is known about physical gene density in genomic sequences. In spruce, two partially sequenced BAC clones had a single gene per 172 kbp and 94 kbp, respectively, which represents a density at least 10-fold lower than the average gene density of the sequenced genomes of *Arabidopsis*, rice, poplar or grapevine [46]. In addition, the sequencing of four other randomly selected BAC clones in spruce failed to report any gene [45].

## Tandemly Arrayed Genes and Functional Clusters

In the present analysis, we identified two types of gene clusters: arrays of gene duplicated in tandem and arrays of unrelated sequences sharing functional annotations. There were 51 arrays (TAGs) encompassing genes from the same family that were duplicated within 5 cM. They incorporated 6.9% (125) of the mapped genes and they could be indicative of small segmental duplications. Such TAGs were also reported in genomic sequences of model angiosperms: they involve 11.7% of the *Arabidopsis* genes and 6.7% of the rice genes [47]. Most of the spruce arrays (78.0%) included only two genes. Similar proportions were found in genome sequences of model angiosperms [47]. The largest spruce array found consisted of eight *myb-r2r3* genes on chromosome 7 (Figure 1). Interestingly, seven of these *myb-r2r3* belong to the same subgroup Sg4C [48]. The other genes belonging to Sg4C were not positioned on this linkage map. Spruce TAGs were significantly enriched in functions related to DNA binding, secondary metabolism and structural proteins (Figure 1). In*Arabidopsis* and rice, TAGs are under-represented among transcription factors and over-represented in enzymes [47]. GO analyses and expression data showed a strong correlation between tandem duplicates and biotic stress genes in *Arabidopsis* [49], leading the authors to suggest that 'tandem duplicates are likely important for adaptive evolution to rapidly changing environments'. In the *myb-r2r3* gene array, the three genes named *PgMyb5*, *PgMyb10* and *PgMyb13* exhibited very different expression patterns [50]. The lack of co-expression of the genes mapped in arrays did not support a gene arrangement oriented by co-regulation. In these arrays, a majority of the genes were derived from duplications occurring after the GA split but were shared between spruce and pine, indicating from the perspective of geological time that expression divergence may occur quite rapidly after gene duplication [51]. Such a pattern is in accordance with the observation that if a new function is not acquired rapidly through neo-functionalisation, one duplicate tends to evolve towards a pseudogene and disappear [7]. A high frequency of pseudogenes has indeed been reported in conifer genomes [26, 42, 52]. However, there are exceptions to this neo-functionalisation trend among surviving duplicates, such as in the conifer *Knox-I* family. In this family, the closely located *kn1* and *kn2* arose from a duplication postdating the divergence between gymnosperms and angiosperms; nevertheless, neo-functionalisation has not happened yet between these duplicates in spite of the duplication occurring before the divergence between the spruce and pine lineages, more than 100 Mya [24]. Sub-functionalisation of these duplicates has been noted [24], conferring partial functional redundancy that might enhance survival and adaptation in these long-lived perennials. Several other cases of conifer-specific duplications might exist that imply partial redundancy of function instead of neo-functionalisation.

We found three clusters made of co-expressed gene sequences that were similar to operon-like structures: two cases made of non-homologous sequences and a third one made of tandemly duplicated chalcone synthases. In angiosperms, only five such structures have been described and were associated with secondary metabolism and defence mechanisms [53]. Such metabolic clusters have emerged as a new and growing theme in plant biology [54]. The three clusters found in our study were similar to these functional clusters, except that their roles were not restricted to secondary metabolism. Among our data, two other cases for which we could not obtain expression data were also good candidates for functional clusters. On chromosome 11, there were also two co-localising pectin methylesterases. Moreover, a single group of two non-homologous sequences was clearly involved in the secondary metabolism. This group encompassed one flavonol synthase and one glutathione synthase on chromosome 6 of *Picea*. Glutathione plays several important roles in the defence of plants against environmental threats. It is a substrate for glutathione *S*-transferases, enabling neutralisation of potentially toxic xenobiotics [55]. Thus, these flavonol and glutathione synthases may belong to a cluster of functionally related but non-homologous genes. Similarly, 80 co-expression clusters sharing the same GO term were described along the 3B wheat chromosome, suggesting the existence of adaptive complexes of co-functional genes [39]. In spruce, exhaustive transcriptomic resources have recently been developed [30]. Their analysis combined with the positioning on the genome of additional genes should allow us to pinpoint whether adaptation at the metabolic level has contributed to shaping the organisation of the gene space.

## Highly Conserved Organization of the Gene Space between Spruce and Pine

Before conifer gene catalogues were available, the number of available orthologous markers to enable comparative studies of genome macro-structure between conifer species was highly limited [34]. A substantial conservation of the Pinaceae genome macro-structure was nevertheless suspected [34, 56, 57]. Estimating the extent of conservation in genome macro-structure was more exhaustive in our study, because we identified a much enlarged set of orthologous mapped gene loci (over 200) between spruce and pine. Synteny and collinearity between the spruce and loblolly pine genomes were very high. Lower collinearity was noted with the maritime pine genome, which resulted from likely lower accuracy of the gene order based on the use of a smaller mapping population for this species [32]. Therefore, it is safe to assume that the organisation of the conifer gene space has been largely maintained over a period dating back 120 to 140 Mya, since the early diversification of Pinaceae in its main lineages in the Early Cretaceous [13, 14, 27]. Such a high level of conservation of the genome macro-structure has also been reported among angiosperm genomes from the Rosids and Asterids clades [58, 59], which diverged about 115 Mya [58, 59, 60], a time period similar to that of the pine-spruce divergence. By contrast, since the monocot-eudicot split 140 to 150 Mya [61], which slightly preceded the pine-spruce divergence, synteny has been largely disrupted between model monocots and dicots [62]. Such large discrepancies in apparent rate of evolution of genome macro-structure are largely conspicuous among angiosperm lineages, where it has been shown that the genomes of perennial species such as grape and poplar evolved slower than those of annual species such as *Arabidopsis* and rice [63]. These differences in evolutionary rates are also reminiscent of those in substitution rates between annual and perennial or woody seed plants, where various hypotheses related to mutation rate, generation time, population size and fixation rate have been proposed [64, 65, 66, 67, 68].

## Age and Organization of Gene Duplicates

The phylogenetic analysis of 157 gene families indicated a large imbalance in favour of ancient duplications predating the GA split versus more recent duplications postdating the GA split. Since the genes sampled in the present study were identified after sequencing ESTs, one could argue that the sample might be biased towards expression patterns that are possibly related to high sequence conservation (for example, [69]), hence artificially increasing the ratio of ancient versus recent gene duplications detected in the present study. First, the ratio was highly asymmetric (eight to one) and we showed that the genes and families involved in our study were representative of a large array of molecular functions and biological processes seen in the most recent spruce gene catalogue, and implicating conserved and less conserved gene families [26]. Gene annotations in conifers [26, 70, 71] also do not favour this hypothesis. Indeed, the most complete catalogue of expressed genes for a conifer, which was based on a large effort involving the sequencing of 23,589 full-length cDNA inserts, has recently enabled the reporting of the most exhaustive comparison of homologous genes sequenced both in angiosperms and gymnosperms [26]. The results indicated that the spruce protein families were largely overlapping with those of angiosperm model plants completely sequenced [26]. Comparing the occurrence of the Pfam domains in the spruce gene catalogue with genomes completely sequenced from model plant species showed that only 28 protein domains were statistically over-represented in spruce and most of them were involved in metabolism, stress response and retrotransposition. Moreover, the gene coding portion of the spruce genome was evaluated at around thirty thousand transcribed genes, a number in the same range as that observed for model angiosperm genomes [26]. The in-depth study of a few transcription factor families also showed that conifers lack some members in specific subfamilies while containing more genes in closely related subfamilies that were derived from duplication events postdating the GA split [24, 48, 72]. These various observations suggest that the conifer genes that are highly divergent from their angiosperm's homologues are rare in the sequence resources developed so far, in spite of the fact that these resources relied on the investigation of a diversity of tissues and conditions [26]. In the future, the availability of the genome sequence may allow the discovery of more conifer-specific genes that could be highly duplicated; but we would (do?) not expect to find them in abundance, as suggested by the present phylogenetic analysis.

A large majority of spruce gene pairs were translocated and most of these translocations occurred before the GA split, affecting a large majority of the 157 gene families analysed. By contrast, genes duplicated after the GA split were located overwhelmingly in close proximity on the same chromosome and often organised in tandem. These trends were consistent with the observation that the physical distances between duplicates on the *Caenorhabditis elegans* genome increase with time, due to chromosomal rearrangements and other mutational events [73]. Nevertheless, this pattern is not always clear, for instance in *Arabidopsis* gene families where the occurrence of tandem duplications and segmental duplications are negatively correlated [47, 74]. In this model plant, unequal cross-over and gene loss were proposed as possible mechanisms leading to the counter-selection of tandem duplications [74].

The observed large excess of ancient duplications predating the GA split over more recent duplications postdating the GA split is consistent with the hypothesis of relative stasis in the gymnosperm lineage leading to conifers and the little evidence for a recent large expansion of the gene space. While a single whole genome duplication has been hypothesised to have affected the common ancestor of seed plants around 350 Mya [20], evidence for more recent widespread polyploidy in the gymnosperm lineage after its divergence from the angiosperm lineage was not found in the present study, in agreement with results from cytological studies reporting a rare occurrence of this phenomenon in gymnosperms [18, 75]. Variation in basic chromosome number in the diploid gymnosperms would rather be the result of chromosome fusion or fission [75]. For instance, such fission would have led to the additional chromosome seen in Douglas fir, relative to other Pinaceae [56]. If so, some of the translocations hypothesised in the present study after the GA split could also be the result of ancient chromosomal fissions increasing the basic chromosome number in the lineage leading to spruce and pine.

While more recent duplications specific to the gymnosperm lineage leading to extant conifers were detected, the stasis of genome macro-structure noted in this lineage is in concordance with that observed between the spruce and pine genomes, and corresponding to a period exceeding 100 My since the last common ancestor of Pinaceae [13, 14, 27]. Such slow rates of genome evolution parallel the slow rates of speciation and patterns of reticulate evolution noted in Pinaceae taxa [76, 77] and their archaic morphological features and life history [14, 78]. Such multiple coincidences reinforce the idea that perennial nature and large historical population sizes are key factors to the slow evolution of conifers [65]. The large excess of ancient duplications detected in the present study also indicates that much gene expansion has occurred in the plant lineage before the divergence between gymnosperms and angiosperms in the Late Carboniferous. Part of this expansion in the primitive land plants might be related to the major burst of duplications noted in transcription factors and coinciding with the water-to-land transition of plants [79]. Further studies implicating other divisions of green plants are needed to better comprehend the temporal dynamic of gene family expansions and reshuffling of the plant genome before the emergence of modern seed plants.

## Perspectives

The high levels of synteny, collinearity and similar ranges of linkage distances noted between the spruce and pine genomes will provide opportunities to transfer genomic information between these genera, especially if such conservation is maintained at a finer scale. The reported genomic features are likely to extend to other genera of the Pinaceae and therefore save time and energy in the deployment of genomic resources necessary to identify orthologous regions in these ecological and economic important species. Such mutual enrichment of genetic maps across species is also significant, with respect to the complexity underlying the analysis of the large conifer genomes. Increasing the number of homologous genes mapped on both the spruce and pine genomes will increase the resolution of comparative mapping efforts, which in turn should highlight the extent of micro-arrangements since the divergence of Pinaceae taxa. Such dense gene maps and high structural correspondence between them will also help identify homoeologous quantitative trait loci responsible for adaptation [28] and other complex characters, between species and genera. While successfully applied to dissect nitrogen use efficiency in cereals [80], these efforts in comparative structural genomics should provide insight into the evolutionary trajectories of the conifer genome at the functional level.

## METHODS

### Genotyping of Gene SNPs

A collection of 27,720 white spruce cDNA clusters [26] was used to develop gene SNPs. *In silico* SNP discovery in expressed white spruce (*Picea glauca* (Moench) Voss) genes was conducted after Sanger resequencing and according to parameters previously reported [81]. A GoldenGate (Illumina, San Diego, CA, USA) SNP array (PgLM2) specifically designed to map additional white spruce genes was constructed with 1,536 attempted SNPs dispersed through 1,509 expressed genes. The assay was conducted following procedures previously described [82]. Genotyping was done at the Genome Québec Innovation Centre (McGill University, Montreal, QC, Canada, team of A. Montpetit) and by using 250 ng of template DNA per sample (at a rate of 50 ng/μL). The results were analysed with the BeadStudio software (Illumina). We retained the 1,292 SNPs exhibiting a GenTrain quality score of 0.25 or more. Of these, 1,121 (73.0%) SNPs representative of 1,098 genes segregated among the 500 progeny of the white spruce cross D (C94-1-2516, ♀77111 × ♂2388). The average call rate per valid SNP was 99.48%. In total, 773,395 new genotypes were obtained from valid SNPs. Using replicated positive controls, the rate of reproducibility was estimated at 99.994%.

### Annotation of the Mapped Genes

The mapped genes were representative of a large array of molecular functions and biological processes including wood formation, growth, vascular tissues development and differentiation, responses to abiotic and biotic stresses and adaptation [83, 84]. They were mostly derived from expressional and functional studies [83,84] as well as outlier detection [85].

GO annotation was performed with the Blast2GO software [86]. GO terms were assigned based on the top 10*blastx* hits found against the non redundant (nr) protein database with an E-value below $1e^{-10}$. GO annotation was run based on the terms from the PlantGO-Slim classification for molecular functions, biological processes and cellular components. Annotations were used to assess GO term enrichment using the Fisher exact test function implemented in Blast2GO.

In total, 6,923 GO terms including molecular functions, biological processes and cellular components were assigned to the 1,801 mapped genes, for an average of 3.9 GO terms per gene. There were 1,285 sequences associated with a molecular function. At level two of the molecular function classification, most of the terms fell in the 'binding' category (49.0%) or in the 'catalytic activity' category (40.0%). At the level three of the classification, five categories each represented about 15% to 20% of the genes; therefore, most of the terms were related either to hydrolase activity, transferase activity, nucleic acid binding, protein binding or lipid binding. Also, 1,027 sequences were associated with biological processes with a large diversity of 84 terms involved in the annotations.

### Spruce Gene Linkage Maps

Two parental linkage maps for white spruce were estimated *de novo* from the 500 progeny above for cross D. Parental maps were assembled as previously described [56], using the genotyping data of the PgLM2 array, as well as data from a previous GoldenGate SNP array on the same 500 progeny [28]. Anonymous markers and gene markers from previous linkage mapping projects involving a different white spruce cross (cross P: C96-1-2516, ♀80112 × ♂80109; 260 progeny) were also considered [56, 82]. The white spruce maps derived from both crosses D and P [28] were merged with the JoinMap 4.0 function 'Combine Groups for Map Integration'. Before marker ordering within each integrated linkage group, we compared the recombination frequencies estimated for homologous markers from both data sets. For this purpose, a 'heterogeneity test' was conducted with JoinMap4.0. As a result, pairs of loci showing significantly different recombination frequencies ($P < 0.01$) were eliminated to avoid erroneous marker ordering.

A black spruce (*Picea mariana* (Mill.) B.S.P.) map [82] was also used to position 58 additional genes (3% of the total mapped genes) that were not mapped in white spruce. Because the genomes of the two species are highly syntenic and collinear [82], white spruce and black spruce genetic maps were aligned based on 258 gene markers mapped in common. The white spruce composite map was taken as the reference map; then, black spruce gene loci not positioned in white spruce were transferred onto the white spruce composite map. The position of a transferred gene on the linkage group was estimated as the middle of the interval between the two closest white spruce anchor loci. Hence, the spruce composite map included 1,743 genes from white spruce and 58 genes from black spruce for a total of 1,801 positioned genes.

### Gene Distribution along the Chromosomes

We tested whether the genes were uniformly distributed along the chromosomes or whether they were clustered. For each chromosome, we compared the observed gene distribution with a uniform distribution by using the Kolmogorov-Smirnov test. We computed the maximum difference (noted $D_n$) between the observed distribution $F(x)$ and the empirical distribution $F_n(x)$. With large sample sizes such as those considered here (in terms of genes), $D_n$ follows a complex distribution and the critical value depends on the sample size. D'Agostino and Stephens proposed a modified statistic named $D_n$* with critical values independent of the sample size [87]. The critical values of statistical significance used for $D_n$* were 0.895 ($\alpha = 0.05$) and 1.035 ($\alpha = 0.01$).

### Gene Density Analysis

To identify the location and extent of GRRs, we used the kernel density estimation, a non-parametric technique based on the kernel density function [88]. The choice of the bandwidth is arbitrary and affects the smoothness of the distribution. The function *asciker* available in the software Stata® 10 (College Station, TX, USA) was used to compute the non-parametric density estimator as well as the confidence intervals at two significance levels ($\alpha = 0.05$ and 0.01) [89]. We used the adaptive kernel method [90], which is based on a varying bandwidth instead of an arbitrarily fixed one. This approach enabled us to take into account the density observed locally to compute the bandwidth size. The density and the 95% confidence intervals were calculated with the *akdensity* function in Stata®10. Then, we compared the position of the confidence interval with the uniform function in the R package. If the lower bound of the confidence interval was greater than the uniform distribution, the region was declared as a GRR ($P < 0.05$).

## Synteny with Pine Genomes

The gene-based linkage map for the loblolly pine genome [29] was downloaded from the Dendrome database [31]. Out of 1,816 genes mapped onto the loblolly pine genome, a dataset of 1,666 genes was retrieved from the Dendrome database (accession TG091). These sequences were compared with 27,720 spruce unigenes [91]) and the 1,801 spruce mapped loci by using the *blastn* program. Also, 426 mapped loci were collected from maritime pine [32]. For each pine sequence, the best hit among the spruce genes with various thresholds from 70% to 95% of identity was retrieved. All the spruce sequences matching a loblolly pine sequence with at least 95% of identity were located on a homoeologous chromosome. However, to increase the number of matches analysed, a minimum identity level of 80% was retained. Under these circumstances, 7.2% of the matches were not found on homoeologous pairs of chromosome. Moreover, we performed the reciprocal comparison involving mapped spruce genes against the pine sequence database. If one link was found between the pine and spruce chromosomes in both reciprocal analyses, we declared this link as homologous. In a few cases, one gene from one species could match several genes from the other species on the homoeologous chromosome. In such cases, a single link was retained between the two homoeologous chromosomes.

## Gene Families Used in Phylogenetic Analyses

Phylogenetic analyses were conducted by including angiosperm and gymnosperm (conifer) sequences. These involved 527 mapped spruce genes from 157 families. To assess the representativeness of this sample (Figure2), we compared the distribution of the molecular function GO terms found across these genes and across the white spruce GCAT gene catalogue [26] with a two-tailed Fisher exact test. Out of the seven terms assigned at the level two of the molecular function's classification, only two terms were over-represented among the mapped families compared with the gene catalogue (False Discovery Rate (FDR) $\leq$ 0.01). These terms were 'binding' (GO:0005488) and 'transcription regulator activity' (GO:0030528). Indeed, 31 families encompassing 180 sequences were related to transcription activity or regulation. The four largest groups of mapped transcriptional regulators included 21 genes from the *myb-r2r3* family, 17 genes from the *b-hlh* family, 14 *nam*genes and 11 *aux-iaa* genes. At the level three of this classification, only two terms were differentially represented among the mapped families and the spruce gene catalogue (FDR $\leq$ 0.01). Genes related to nucleic acid binding (GO:0003676) were over-represented while the ones related to transferase activity transferring glycosyl groups (GO:0016757) were under-represented among the mapped genes. Thus, there was no large molecular functional class missing in the sample of 157 families used for the phylogenomic analysis, which was quite representative of the overall relative diversity and abundance of molecular functions seen in the spruce GCAT gene catalogue [26].

## Phylogenetic Analyses

The phylogenetic analyses aimed to classify duplications involving mapped spruce genes as recent or ancient by considering the GA split as a reference (see below). We translated the spruce sequences with the *getorf*program available in the EMBOSS package [92]. Among all possible open reading frames, we selected the sequence with a match with a protein sequence from another species. The *Arabidopsis* protein dataset (TAIR7 release) from TAIR [93]) and the rice protein dataset from the rice annotation database [94]) were retrieved. We also built a protein sequence dataset from pine. Although not a strict requirement for the present purpose of dating spruce gene duplications, we also retrieved two datasets of EST contigs derived from *Pinus pinaster*and *Pinus taeda* [95]. We concatenated the 12,901 sequences from *Pinus pinaster* and the 72,928 sequences from *Pinus taeda* into a single file. These sequences were translated with the *getorf* program to obtain a set of all possible protein sequences. For each spruce sequence, we conducted three *blastp* searches against the protein datasets from *Arabidopsis*, rice and pine. In the *blastp* outputs, we screened the five best hits. We aligned the spruce, pine, *Arabidopsis* and rice homologous sequences with the *kalign* program [96] and we selected manually conserved domains for further phylogenetic analysis. Because the considered angiosperm sequences were from complete genomic sequences, it was not necessary to filter out sequences based on absolute sequence similarity. However, the observed e-values in the *blastp* searches of the spruce sequences against the *Arabidopsis* proteome were in the range of $e^{-50}$ to $e^{-100}$.

Phylogenetic analyses were conducted for 157 gene families. To be considered, a family had to contain at least two mapped members on the spruce genome and had to be represented in *Pinus* as well as in both models and completely sequenced angiosperms *Arabidopsis* and rice, respectively an eudicot and a monocot. Both the NJ method [97], an approach based on matrices of substitution rates, and MP analysis [98] were employed using the package Phylip 3.6 [99]. We used MP and NJ methods instead of other more computer-intensive approaches such as Bayesian or maximum-likelihood algorithms given that the metric that we wanted to estimate, the proportion of ancient versus recent spruce gene duplications relative to the GA split, was based on a large sample of gene families (157), and given that our interpretation of the topologies was conservative because it was based on the strict consensus from MP and NJ trees for each gene family. Results obtained for a few gene families using other more computer-intensive phylogenetic methods resulted in essentially the same consensus trees (results not shown). To estimate NJ trees, distance matrices were calculated using the *protdist* program with the JTT amino acid substitution matrix and submitted to the program *neighbor*. Parsimony trees were estimated using *protpars*. For each family and each method, the robustness of the topologies obtained was assessed by means of 500 bootstraps using the program *seqboot*. For each method, the consensus tree derived from the bootstrap analysis was the majority-rule consensus generated with the program *consense*. Then, for each gene family phylogeny, only nodes supported minimally by 50% of bootstraps and in concordance between the two phylogenetic methods were retained, that is, the strict consensus of two majority-rule bootstrap MP and NJ trees.

For each gene family phylogeny, we used the unrooted strict consensus of the MP and NJ trees and estimated the relative age of spruce gene duplications by determining if they occurred before or after the GA split. Nodes involving spruce gene sequences but with no intervening angiosperm sequences from *Arabidopsis* or rice indicated duplications post-dating the GA split and were referred to as recent duplications. On the other hand, when spruce gene sequences were separated by intervening nodes involving angiosperm sequences from rice and/or *Arabidopsis*, these duplications predated the GA split and were referred to as ancient. In some cases, ancient gene duplicates produced before the GA split may have been lost both in the *Arabidopsis* and rice genomes, which would bias upward the number of duplications that would be declared specific to the gymnosperm lineage leading to spruce. Based on the literature, one may argue that this bias should be negligible, given that loss of gene duplicates mostly occurs quickly after duplication if processes such as sub-functionalisation or neo-functionalisation do not occur [7]. At the same time, sampling two divergent and completely sequenced angiosperm genomes (representative of the eudicot and the monocot lineages) should keep such a bias low because gene loss would have to occur independently in both angiosperm lineages after their split, or in their common ancestor during the relatively short period between the GA split (around 300 Mya) [13] and the divergence of monocots from other angiosperms (around 150 Mya) [61]. For instance, the sister lineage of the large *knox-1* gene family in conifers has been lost in rice and other monocots but conserved in *Arabidopsis* and other eudicots [24]. Finally, because this bias would result in the scoring of a number of false-positive gymnosperm-specific duplications, it would tend to reduce the true proportion between ancient duplications predating the GA split and more recent duplications post-dating this split. If real, the bias is likely negligible, given that the observed value of this proportion was already highly skewed toward ancient duplications (ratio of eight to one, see Results).

All circular genetic maps were drawn with the Circos software [100].

## DECLARATIONS

### Acknowledgements

This work was supported by Genome Canada and Génome Québec for the Arborea project. We thank the two anonymous reviewers for their useful comments, S. Beauseigle and S. Blais (Université Laval) for their help in the laboratory, G. Daigle (Université Laval) for support with statistical analyses, P. Belleau (Centre de Recherche du Centre Hospitalier Universitaire de Québec) for helpful discussions, A. Deschênes (Université Laval) for help with GO annotation and gene density analyses, A. Montpetit and his team (McGill University) for performing the genotyping assay and J. Mackay (Université Laval) for sharing gene expression data ahead of publication.

### Competing interests

The authors declare that they have no competing interests.

## REFERENCES

1. Lynch M: The origins of genome architecture. 2007, Sunderland, MA: Sinauer Associates
2. Freeling M: Bias in plant gene content following different sorts of duplication: tandem, whole-genome, segmental, or by transposition. Annu Rev Plant Biol. 2009, 60: 433-453. 10.1146/annurev.arplant.043008.092122.
3. Li WH: Evolution of duplicated genes. Evolution of Genes and Proteins. Edited by: Nei M, Koehin RK. 1983, Sunderland, MA: Sinauer Associates, 14-37.
4. Lynch M, Conery JS: The evolutionary fate and consequences of duplicate genes. Science. 2000, 290: 1151-1155. 10.1126/science.290.5494.1151.
5. Lynch M, O'Hely M, Walsh B, Force A: The probability of preservation of a newly arisen gene duplicate. Genetics. 2001, 159: 1789-1804.
6. Bowers JE, Chapman BA, Rong J, Paterson AH: Unravelling angiosperm genome evolution by phylogenetic analysis of chromosomal duplication events. Nature. 2003, 422: 433-438. 10.1038/nature01521.
7. Lynch M, Conery JS: The evolutionary demography of duplicate genes. J Struct Funct Genomics. 2003, 3: 35-44. 10.1023/A:1022696612931.
8. Nei M, Rooney AP: Concerted and birth-and-death evolution of multigene families. Annu Rev Genet. 2005, 39: 121-152. 10.1146/annurev.genet.39.073003.112240.
9. Proost S, Pattyn P, Gerats T, van de Peer Y: Journey through the past: 150 million years of plant genome evolution. Plant J. 2011, 66: 58-65. 10.1111/j.1365-313X.2011.04521.x.
10. Jiao Y, Wickett NJ, Ayyampalayam S, Chanderbali AS, Landherr L, Ralph PE, Tomsho LP, Hu Y, Liang H, Soltis PS, Soltis DE, Clifton SW, Schlarbaum SE, Schuster SC, Ma H, Leebens-Mack J, dePamphilis CW: Ancestral polyploidy in seed plants and angiosperms. Nature. 2011, 473: 97-100. 10.1038/nature09916.
11. Fawcett JA, Maere S, van de Peer Y: Plants with double genomes might have had a better chance to survive the Cretaceous-Tertiary extinction event. Proc Natl Acad Sci USA. 2009, 106: 5737-5742. 10.1073/pnas.0900906106.
12. Wood TE, Takebayashi N, Barker MS, Mayrose I, Greenspoon PB, Rieseberg LH: The frequency of polyploid speciation in vascular plants. Proc Natl Acad Sci USA. 2009, 106: 13875-13879. 10.1073/pnas.0811575106.
13. Savard L, Li P, Strauss SH, Chase MW, Michaud N, Bousquet J: Chloroplast and nuclear gene sequences indicate Late Pennsylvanian time for the last common ancestor of extant seed plants. Proc Natl Acad Sci USA. 1984, 91: 5163-5167.
14. Gernandt DS, Willyard A, Syring JV, Liston A: The conifers (Pinophyta). Genetics, Genomics and Breeding of Conifer Trees. Edited by: Plomion C, Bousquet J,

Kole C. 2011, New York: Edenbridge Science Publishers and CRC Press, 1-39.

15. Sanderson MJ, Donoghue MJ: Shifts in diversification rate with the origin of angiosperms. Science. 1994, 264: 1590-1593. 10.1126/science.264.5165.1590.

16. Magallon SA, Sanderson MJ: Absolute diversification rates in angiosperm clades. Evolution. 2001, 55: 1762-1780.

17. Beck CB: Origin and Evolution of Gymnosperms. 1988, New York: Columbia University Press

18. Khoshoo TN: Chromosome numbers in gymnosperms. Silvae Genetics. 1961, 10: 1-9.

19. Cui L, Wall PK, Leebens-Mack JH, Lindsay BG, Soltis DE, Doyle JJ, Soltis S, Carlson JE, Arumuganathan K, Barakat A, Albert VA, Ma H, dePamphilis CW: Widespread genome duplications throughout the history of flowering plants. Genome Res. 2006, 16: 738-749. 10.1101/gr.4825606.

20. Jiao Y, Wickett NJ, Ayyampalayam S, Chanderbali AS, Landherr L, Ralph PE, Tomsho LP, Hu Y, Liang H, Soltis PS, Soltis DE, Clifton SW, Schlarbaum SE, Schuster SC, Ma H, Leebens-Mack J, dePamphilis CW: Ancestral polyploidy in seed plants and angiosperms. Nature. 2011, 473: 97-100. 10.1038/nature09916.

21. Murray BG: Nuclear DNA amounts in gymnosperms. Ann Bot. 1998, 82: 3-15.

22. Ahuja MR: Polyploidy in gymnosperms: revisited. Silvae Genetica. 2005, 54: 59-69.

23. Ahuja MR, Neale DB: Evolution of genome size in conifers. Silvae Genetica. 2005, 54: 126-137.

24. Guillet-Claude C, Isabel N, Pelgas B, Bousquet J: The evolutionary implications of knox-I gene duplications in conifers: correlated evidence from phylogeny, gene mapping, and analysis of functional divergence. Mol Biol Evol. 2004, 21: 2232-2245. 10.1093/molbev/msh235.

25. Judd WS, Campbell CS, Kellog EA, Stevens PF, Donoghue MJ: Plant Systematics, a Phylogenetic Approach. 2008, Sunderland, MA: Sinauer Associates

26. Rigault P, Boyle B, Lepage P, Cooke JE, Bousquet J, MacKay JJ: A white spruce gene catalogue for conifer genome analyses. Plant Phys. 2011, 157: 14-28. 10.1104/pp.111.179663.

27. Wang XQ, Tank DC, Sang T: Phylogeny and divergence times in Pinaceae: evidence from three genomes. Mol Biol Evol. 2000, 17: 773-781. 10.1093/oxfordjournals.molbev.a026356.

28. Pelgas B, Bousquet J, Meirmans PG, Ritland K, Isabel N: QTL mapping in white spruce: gene maps and genomic regions underlying adaptive traits across pedigrees and environments. BMC Genomics. 2011, 12: 145-10.1186/1471-2164-12-145.

29. Eckert AJ, van Heerwaarden J, Wegrzyn JL, Nelson CD, Ross-Ibarra J, Gonzalez-Martinez SC, Neale DB: Patterns of population structure and environmental associations to aridity across the range of loblolly pine (*Pinus taeda* L., Pinaceae). Genetics. 2010, 185: 969-982. 10.1534/genetics.110.115543.

30. Raherison E, Rigault P, Caron S, Poulin P-L, Boyle B, Verta J-P, Giguère I, Bomal C, Bohlmann J, Cooke J, Mackay J: Transcriptome profiling in conifer trees reveals interspecific conservation in the vascular gene expression and patterns of gene family diversification. BMC Genomics. 2012, 13 (1): 434-10.1186/1471-2164-13-434.

31. [http://dendrome.ucdavis.edu/cmap/]

32. Chancerel E, Lepoittevin C, Le Provost G, Lin Y-C, Jaramillo-Correa JP, Eckert A, Wegrzyn JL, Zelenika D, Boland A, Frigerio J-M, Chaumeil P, Garnier-Géré P, Boury C, Grivet D, Gonzalez-Martinez SC, Rouzé P, van de Peer Y, Neale DB, Cervera MT, Kremer A, Plomion C: Development and implementation of a highly multiplexed SNP array for genetic mapping in maritime pine and comparative mapping with loblolly pine. BMC Genomics. 2011, 12: 368-10.1186/1471-2164-12-368.

33. MacKay JJ, Dean JFD: Transcriptomics. Genetics, Genomics and Breeding of Conifers Trees. Edited by: Plomion C, Bousquet J, Kole C. 2011, New York: Edenbridge Science Publishers and CRC Press, 323-357.

34. Ritland K, Krutovsky K, Tsumura Y, Pelgas B, Isabel N, Bousquet J: Genetic mapping in conifers. Genetics, Genomics and Breeding of Conifer Trees. Edited by: Plomion C, Bousquet J, Kole C. 2011, New York: Edenbridge Science Publishers and CRC Press, 196-238.

35. The Arabidopsis Genome Initiative: Analysis of the genome sequence of the flowering plant *Arabidopsis thaliana*. Nature. 2000, 408: 796-815.36

36. Schnable PS, Ware D, Fulton RS, Stein JC, Wei F, Pasternak S, Liang C, Zhang J, Fulton L, Graves TA, Minx P, Reily AD, Courtney L, Kruchowski SS, Tomlinson C, Strong C, Delehaunty K, Fronik C, Courtney B, Rock SM, Belter E, Du F, Kim K, Abbott RM, Cotton M, Levy A, Marchetto P, Ochoa K, Jackson SM, Gillam B, et al: The B73 maize genome: complexity, diversity, and dynamics. Science. 2009, 326: 1112-1115. 10.1126/science.1178534.

37. Choi IY, Hyten DL, Matukumalli LK, Song Q, Chaky JM, Quigley CV, Chase K, Lark KG, Reiter RS, Yoon M-S, Hwang E-Y, Yi S-I, Young ND, Shoemaker RC, van Tassel CP, Specht JE, Cregan PB: A soybean transcript map: gene distribution, haplotype and single-nucleotide polymorphism analysis. Genetics. 2007, 176: 685-696. 10.1534/genetics.107.070821.

38. Choulet F, Wicker T, Rustenholz C, Paux E, Salse J, Leroy P, Schlub S, Le Paslier M-C, Magdelenat G, Gonthier C, Couloux A, Budak H, Breen J, Pumphrey M, Liu S, Kong X, Jia J, Gut M, Brunel D, Anderson JA, Gill BS, Appels R, Keller B, Feuillet C: Megabase level sequencing reveals contrasted organization and evolution patterns of the wheat gene and transposable element spaces. Plant Cell. 2010, 22: 1686-1701. 10.1105/tpc.110.074187.

39. Rustenholz C, Choulet F, Laugier C, Safar J, Simkova H, Dolezel J, Magni F, Scalabrin S, Cattonaro F, Vautrin S, Bellec A, Bergès H, Feuillet C, Paux E: A 3000-loci transcription map of chromosome 3B unravels the structural and functional features of gene islands in

hexaploid wheat. Plant Phys. 2011, 157: 596-1608.

40. Schmutz J, Cannon SB, Schlueter J, Ma J, Mitros T, Nelson W, Hyten DL, Song Q, Thelen JJ, Cheng J, Xu D, Hellsten U, May GD, Yu Y, Sakurai T, Umezawa T, Bhattacharyya MK, Sandhu D, Valliyodan B, Lindquist E, Peto M, Grant D, Shu S, Goodstein D, Barry K, Futrell-Griggs M, Abernathy B, Du J, Tian Z, Zhu L, et al: Genome sequence of the palaeopolyploid soybean. Nature. 2010, 463: 178-183. 10.1038/nature08670.

41. Kamm A, Doudrick RL, Heslop-Harrison JS, Schmidt T: The genomic and physical organization of Ty1-copia-like sequences as a component of large genomes in *Pinus elliottii* var *elliottii* and other gymnosperms. Proc Natl Acad Sci USA. 1996, 93: 2708-2713. 10.1073/pnas.93.7.2708.

42. Morse AM, Peterson DG, Islam-Faridi MN, Smith KE, Magbanua Z, Garcia SA, Kubisiak TL, Amerson HV, Carlson JE, Nelson CD, Davis JM: Evolution of genome size and complexity in *Pinus*. PloS One. 2009, 4: 1-11. 10.1371/journal.pone.0005361.

43. Kovach A, Wegrzyn JL, Parra G, Holt C, Bruening GE, Loopstra CA, Hartigan J, Yandell M, Langley CH, Korf I, Neale DB: The*Pinus taeda* genome is characterized by diverse and highly diverged repetitive sequences. BMC Genomics. 2010, 11: 420-10.1186/1471-2164-11-420.

44. Magbanua ZV, Ozkan S, Bartlett BD, Chouvarine P, Saski CA, Liston A, Cronn RC, Nelson CD, Peterson DG: Adventures in the enormous: a 1.8 million clone BAC library for the 21.7 Gb genome of loblolly pine. PloS One. 2011, 6: e16214-10.1371/journal.pone.0016214.

45. Morgante M, De Paoli E: Toward the conifer genome sequence. Genetics, Genomics and Breeding of Conifers Trees. Edited by: Plomion C, Bousquet J, Kole C. 2011, New York: Edenbridge Science Publishers and CRC Press, 389-403.

46. Hamberger B, Hall D, Yuen M, Oddy C, Hamberger B, Keeling CI, Ritland C, Ritland K, Bohlmann J: Targeted isolation, sequence assembly and characterization of two white spruce (*Picea glauca*) BAC clones for terpenoid synthase and cytochrome P450 genes involved in conifer defence reveal insights into a conifer genome. BMC Plant Biol. 2009, 9: 106-10.1186/1471-2229-9-106.

47. Rizzon C, Ponger L, Gaut BS: Striking similarities in the genomic distribution of tandemly arrayed genes in *Arabidopsis*and rice. PLoS Comput Biol. 2006, 2: e115-10.1371/journal.pcbi.0020115.

48. Bedon F, Bomal C, Caron S, Levasseur C, Boyle B, Mansfield SD, Schmidt A, Gershenzon J, Grima-Pettenati J, Séguin A, MacKay J: Subgroup 4 R2R3-MYBs in conifer trees: gene family expansion and contribution to the isoprenoid- and flavonoid-oriented responses. J Exp Bot. 2010, 61: 3847-3864. 10.1093/jxb/erq196.

49. Hanada K, Zou C, Lhti-Shiu MD, Shinozaki K, Shiu S-H: Importance of lineage-specific expansion of plant tandem duplicates in the adaptive response to environmental stimuli. Plant Phys. 2008, 148: 993-1003. 10.1104/pp.108.122457.

50. Bedon F, Grima-Pettenati J, Mackay J: Conifer R2R3-MYB transcription factors: sequence analyses and gene expression in wood-forming tissues of white spruce (*Picea glauca*). BMC Plant Biol. 2007, 30: 17-

51. Moore RC, Purugganan MD: The early stages of duplicate gene evolution. Proc Natl Acad Sci USA. 2003, 100: 15682-15687. 10.1073/pnas.2535513100.

52. Kinlaw CS, Neale DB: Complex gene families in pine genomes. Trends Plant Sci. 1997, 2: 356-359. 10.1016/S1360-1385(97)84624-9.

53. Osbourn A: Gene clusters for secondary metabolic pathways: an emerging theme in plant biology. Plant Phys. 2010, 154: 531-535. 10.1104/pp.110.161315.

54. Chu HY, Wegel E, Osbourn A: From hormones to secondary metabolism: the emergence of metabolic gene clusters in plants. Plant J. 2011, 66: 66-79. 10.1111/j.1365-313X.2011.04503.x.

55. Marrs KA: The functions and regulation of glutathione S-transferases in plants. Annu Rev Plant Physiol Plant Mol Biol. 1996, 47: 127-158. 10.1146/annurev.arplant.47.1.127.

56. Pelgas B, Beauseigle S, Achere V, Jeandroz S, Bousquet J, Isabel N: Comparative genome mapping among *Picea glauca*, *P. mariana x P. rubens* and *P. abies*, and correspondence with other Pinaceae. Theor Appl Genet. 2006, 113: 1371-1393. 10.1007/s00122-006-0354-7.

57. Krutovsky KV, Troggio M, Brown GR, Jermstad KD, Neale DB: Comparative mapping in the Pinaceae. Genetics. 2004, 168: 447-461. 10.1534/genetics.104.028381.

58. Xu X, Pan S, Cheng S, Zhang B, Mu D, Ni P, Zhang G, Yang S, Li R, Wang J, Orjeda G, Guzman F, Torres M, Lozano R, Ponce O, Martinez D, De la Cruz G, Chakrabarti SK, Patil VU, Skryabin KG, Kuznetsov BB, Ravin NV, Kolanova TV, Beletsky AV, Mardanov AV, Di Genova A, Bolser DM, Martin DMA, Li G, Yang Y, et al: Genome sequence and analysis of the tuber crop potato. Nature. 2011, 475: 189-195. 10.1038/nature10158.

59. Cenci A, Combes MC, Lashermes P: Comparative sequence analyses indicate that *Coffea* (Asterids) and *Vitis* (Rosids) derive from the same paleo-hexaploid ancestral genome. Mol Genet Genomics. 2010, 283: 493-501. 10.1007/s00438-010-0534-7.

60. Wang H, Moore MJ, Soltis PS, Bell CD, Brockington SF, Alexandre R, Davis CC, Latvis M, Manchester SR, Soltis DE: Rosid radiation and the rapid rise of angiosperm-dominated forests. Proc Natl Acad Sci USA. 2009, 106: 3853-3858. 10.1073/pnas.0813376106.

61. Chaw SM, Chang CC, Chen HL, Li WH: Dating the monocot-dicot divergence and the origin of core eudicots using whole chloroplast genomes. J Mol Evol. 2004, 58: 424-441. 10.1007/s00239-003-2564-9.

62. Liu H, Sachidanandam R, Stein L: Comparative genomics between rice and *Arabidopsis* shows scant collinearity in gene order. Genome Res. 2001, 11: 2020-2026. 10.1101/gr.194501.

63. Zuccolo A, Bowers JE, Estill JC, Xiong Z, Luo M, Sebastian A, Goicoechea JL, Collura K, Yu Y, Jiao Y, Duarte J, Tang H, Ayyampalayam S, Rounsley S, Kudma D, Paterson AH, Pires JC, Chanderbali A, Soltis DE,

Chamala S, Barbazuk B, Soltis PS, Albert VA, Ma H, Mandoli D, Banks J, Carlson JE, Tomkins J, Depamphilis CW, Wing RA, et al: A physical map for the*Amborella trichopoda* genome sheds light on the evolution of angiosperm genome structure. Genome Biol. 2011, 12: R48-10.1186/gb-2011-12-5-r48.

64. Bousquet J, Strauss SH, Doerksen AH, Price RA: Extensive variation in evolutionary rate of *rbcL* sequences among seed plants. Proc Natl Acad Sci USA. 1992, 89: 7844-7848. 10.1073/pnas.89.16.7844.

65. Bouillé M, Bousquet J: Trans-species shared polymorphisms at orthologous nuclear gene loci among distant species in the conifer *Picea* (Pinaceae): implications for the long-term maintenance of genetic diversity in trees. Am J Bot. 2005, 92: 63-73. 10.3732/ajb.92.1.63.

66. Gaut B, Yang L, Takuno S, Eguiarte LE: The patterns and causes of variation in plant nucleotide substitution rates. Ann Rev Ecol Evol Syst. 2011, 42: 245-266. 10.1146/annurev-ecolsys-102710-145119.

67. Petit RJ, Hampe A: Some evolutionary consequences of being a tree. Ann Rev Ecol Evol Syst. 2006, 37: 187-214. 10.1146/annurev.ecolsys.37.091305.110215.

68. Smith SA, Donoghue MJ: Rates of molecular evolution are linked to life history in flowering plants. Science. 2008, 322: 86-89. 10.1126/science.1163197.

69. Drummond DA, Bloom JD, Adami C, Wilke CO, Arnold FH: Why highly expressed proteins evolve slowly. Proc Natl Acad Sci USA. 2005, 102: 14338-14343. 10.1073/pnas.0504070102.

70. Pavy N, Laroche J, Bousquet J, Mackay J: Large-scale statistical analysis of secondary xylem ESTs in pine. Plant Mol Biol. 2005, 57: 203-224. 10.1007/s11103-004-6969-7.

71. Pavy N, Paule C, Parsons L, Crow JA, Morency M-J, Johnson JE, Noumen E, Guillet-Claude C, Butterfield Y, Barber S, Yang G, Liu J, Stott J, Kirkpatrick R, Siddiqui A, Holt R, Marra M, Seguin A, Retzel E, Bousquet J, Mackay J: Generation, annotation, analysis and database integration of 16,500 white spruce EST clusters. BMC Genomics. 2005, 6: 144-10.1186/1471-2164-6-144.

72. Côté CL, Boileau F, Roy V, Ouellet M, Levasseur C, Morency M-J, Cooke JEK, Séguin A, Mackay JJ: Gene family structure, expression and functional analysis of *HD-Zip III* genes in angiosperm and gymnosperm forest trees. BMC Plant Biol. 2010, 10: 273-10.1186/1471-2229-10-273.

73. Katju V, Lynch M: The structure and early evolution of recently arisen gene duplicates in the *Caenorhabditis elegans*genome. Genetics. 2003, 165: 1793-1803.

74. Cannon SB, Mitra A, Baumgarten A, Young ND, May G: The roles of segmental and tandem gene duplication in the evolution of large gene families in *Arabidopsis thaliana*. BMC Plant Biol. 2004, 4: 10-10.1186/1471-2229-4-10.

75. Fuchs J, Brandes A, Schubert I: Systematics and evolution telomere sequence localization and karyotype evolution in higher plants. Plant Syst Evol. 1995, 196: 227-241. 10.1007/BF00982962.

76. Perron M, Perry DJ, Andalo C, Bousquet J: Evidence from sequence-tagged-site markers of a recent progenitor-derivative species pair in conifers. Proc Natl Acad Sci USA. 2000, 97: 11331-11336. 10.1073/pnas.200417097.

77. Bouillé M, Senneville S, Bousquet J: Discordant mtDNA and cpDNA phylogenies indicate geographic speciation and reticulation as driving factors for the diversification of the genus *Picea*. Tree Genet Genomes. 2011, 7: 469-484. 10.1007/s11295-010-0349-z.

78. Miller CN: The origin of modern conifer families. Origin and Evolution of Gymnosperms. Edited by: Beck CB. 1988, New York: Columbia University Press, 448-486.

79. Lang D, Weiche B, Timmerhaus G, Richardt S, Riano-Pachon DM, Corrêa R, Mueller-Roeber B, Rensing SA: Genome-wide phylogenetic comparative analysis of plant transcriptional regulation: a timeline of loss, gain, expansion, and correlation with complexity. Genome Biol Evol. 2010, 2: 488-503. 10.1093/gbe/evq032.

80. Quraishi UM, Abrouk M, Murat F, Pont C, Foucrier S, Desmaizieres G, Confolent C, Rivière N, Charmet G, Paux E, Murigneux A, Guerreiro L, Lafarge S, Le Gouis J, Feuillet C, Salse J: Cross-genome map based dissection of a nitrogen use efficiency ortho-metaQTL in bread wheat unravels concerted cereal genome evolution. Plant J. 2011, 65: 745-756. 10.1111/j.1365-313X.2010.04461.x.

81. Beaulieu J, Doerksen T, Boyle B, Clément S, Deslauriers M, Beauseigle S, Blais S, Poulin P-L, Lenz P, Caron S, Rigault P, Bicho P, Bousquet J, MacKay J: Association genetics of wood physical traits in the conifer white spruce and relationships with gene expression. Genetics. 2011, 188: 197-214. 10.1534/genetics.110.125781.

82. Pavy N, Pelgas B, Beauseigle S, Blais S, Gagnon F, Gosselin I, Lamothe M, Isabel N, Bousquet J: Enhancing genetic mapping of complex genomes through the design of highly-multiplexed SNP arrays: application to the large and unsequenced genomes of white spruce and black spruce. BMC Genomics. 2008, 9: 21-10.1186/1471-2164-9-21.

83. Bomal C, Bedon F, Caron S, Mansfield SD, Levasseur C, Cooke JEK, Blais S, Tremblay L, Morency M-J, Pavy N, Grima-Pettenati J, Séguin A, MacKay J: Involvement of *Pinus taeda* MYB1 and MYB8 in phenylpropanoid metabolism and secondary cell wall biogenesis: a comparative in planta analysis. J Exp Bot. 2008, 59: 3925-3939. 10.1093/jxb/ern234.

84. Pavy N, Boyle B, Nelson C, Paule C, Giguère I, Caron S, Parsons LS, Dallaire N, Bedon F, Bérubé H, Cooke J, MacKay J: Identification of conserved core xylem gene sets: conifer cDNA microarray development, transcript profiling and computational analyses. New Phytol. 2008, 180: 766-786. 10.1111/j.1469-8137.2008.02615.x.

85. Namroud MC, Beaulieu J, Juge N, Laroche J, Bousquet J: Scanning the genome for gene single nucleotide polymorphisms involved in adaptive population differentiation in white spruce. Mol Ecol. 2008, 17: 3599-3613. 10.1111/j.1365-294X.2008.03840.x.

86. Conesa A, Götz S: Blast2GO: a comprehensive suite for functional analysis in plant genomics. Int J Plant

Genomics. 2008, 2008: 619832-

87. D'Agostino RB, Stephens MA: Goodness-of-fit Techniques. 1986, New York: Marcel-Dekker Inc.
88. Parzen E: On estimation of a probability density function and mode. Ann Math Stat. 1962, 33: 1065-1076. 10.1214/aoms/1177704472.
89. Fiorio CV: Confidence intervals for kernel density estimation. Stata Journal. 2004, 4: 168-179.
90. Van Kerm P: Adaptative kernel density estimation. Stata Journal. 2003, 3: 148-156.
91. [http://www.arborea.ulaval.ca/research/gene_discovery/gene_catalogue]
92. Rice LI, Bleasby A: EMBOSS: The European Molecular Biology Open Software Suite. Trends Genet. 2000, 16: 276-277. 10.1016/S0168-9525(00)02024-2.
93. [http://www.arabidopsis.org]
94. [http://www.tigr.org/tdb/e2k1/osa1/]
95. Dong Q, Schlueter SD, Brendel V: PlantGDB, plant genome database and analysis tools. Nucleic Acids Res. 2004, 32: D354-359. 10.1093/nar/gkh046.
96. Lassmann T, Sonnhammer EL: Kalign--an accurate and fast multiple sequence alignment algorithm. BMC Bioinformatics. 2005, 6: 298-10.1186/1471-2105-6-298.
97. Saitou N, Nei M: The neighbor-joining method: a new method for reconstructing phylogenetic trees. Mol Biol Evol. 1987, 4: 406-425.
98. Swofford DL, Olsen GJ, Waddell PJ, Hillis DM: Phylogenetic inference. Molecular Systematics. Edited by: Hillis DM, Moritz C, Mable BK. 1996, Sunderland: Sinauer, 407-514. 2
99. Felsenstein J: Phylip 3.6. 2005, [http://evolution.genetics.washington.edu/phylip.html]
100. Krzywinski M, Schein J, Birol I, Connors J, Gascoyne R, Horsman D, Jones SJ, Marra MA: Circos: an information aesthetic for comparative genomics. Genome Res. 2009, 19: 1639-1645. 10.1101/gr.092759.109.

CHAPTER 8

# MODULAR ORGANIZATION OF THE WHITE SPRUCE (PICEA GLAUCA) TRANSCRIPTOME REVEALS FUNCTIONAL ORGANIZATION AND EVOLUTIONARY SIGNATURES

Elie S. M. Raherison, Isabelle Giguère, Sébastien Caron, Mebarek Lamara, John J. MacKay

**Keywords:** Conifers, modular organization, tissue differentiation, transcriptional network, transcriptome, white spruce (Picea glauca).

## INTRODUCTION

Transcriptome analysis is among the most widely used large-scale experimental approaches to investigate genome function and to help annotate genome sequences (Pickrell *et al.*, **2010**; Prasad*et al.*, **2013**; Ellison *et al.*, **2014**). The transcriptome as a whole is highly dynamic as a consequence of quantitative variations over spatial and temporal scales. The observed changes in transcript levels result from the combinatorial action of genetically encoded regulatory pathways that orchestrate development and responses to environmental stimuli. The accumulation of individual gene transcripts may be viewed as a bridge between the genotype and the phenotype and, as such, may be acted upon through natural selection (Romero *et al.*, **2012**; Prasad *et al.*, **2013**).

The fundamental interest in and application of transcriptome analysis are reflected in its broad application in nonmodel organisms, including many taxa such as conifer trees where large genome size or complexity represents a technical barrier to genomic investigation. Conifer tree genomes are highly repetitive and, with sizes of 18–30 Gb, they are several fold larger than most angiosperm plant genomes sequenced to date, including *Arabidopsis thaliana* (Arabidopsis Genome Initiative, **2000**), black cottonwood (*Populus trichocarpa*; Tuskan *et al.*, **2006**), maize (*Zea mays*; Schnable*et al.*, **2009**), and eucalyptus (*Eucalyptus grandis*; Myburg *et al.*, **2014**) as well as animal genomes such as human (IHGSC, **2004**). Three conifer genomes were recently sequenced and include white spruce (*Picea glauca*; Birol *et al.*, **2013**), Norway spruce (*Picea abies*; Nystedt *et al.*, **2013**), and loblolly pine (*Pinus taeda*; Neale *et al.*, **2014**). By contrast, cDNA and transcriptome sequencing in conifers has been ongoing for over 15 yr (Allona *et al.*, **1998**), has led to many functional investigations, and produced outcomes that enable genetic mapping research, association studies and investigations of evolutionary adaptation (MacKay *et al.*, **2012**). Despite this extensive body of work, a general framework describing transcriptome organization in conifers is lacking and would help to develop a more unified understanding of genome function.

Conifer transcriptome studies have been conducted to gain insight into the molecular underpinnings of growth, development and stress responses. Conifers are long-lived and need to cope with diverse biotic and abiotic stress factors over time, which has led to a number of studies on the theme of resistance and acclimation. For example, Sun *et al.* (**2011**) identified genes responsive to fungal elicitation in Scot's pine (*Pinus sylvestris*) while Verne *et al.* (**2011**) identified up- and down-regulated genes in white pine weevil-resistant individuals in interior spruce (*P. glauca* × *Picea engelmannii*). In the context of global environmental change, Yeaman *et al.*

**Citation:** Elie S. M. Raherison, Isabelle Giguère, Sébastien Caron, Mebarek Lamara, John J. MacKay, Modular organization of the white spruce (Picea glauca) transcriptome reveals functional organization and evolutionary signatures, DOI: 10.1111/nph.13343.

(**2014**) investigated patterns of gene expression related to acclimation across environmental conditions in lodgepole pine (*Pinus contorta*) and interior spruce; and Baldi *et al.* (**2011**) identified candidate genes for cold tolerance in cypress (*Cupressus sempervirens*). Considering the ecological and economic importance of conifer woods, another area of interest is transcriptome variation and reorganization during secondary xylem growth. Genes with significant changes in expression between earlywood and latewood formation were identified in loblolly pine trees from two seed sources (Yang & Loopstra, **2005**), in maritime pine (*Pinus pinaster*) during a growing season (Paiva*et al.*, **2008**), and in radiata pine (*Pinus radiata*) at different stages of developmental aging (Li*et al.*, **2010**). A few studies have addressed related evolutionary aspects by identifying genes that are conserved in expression patterns and sequence between species such as white spruce and *A. thaliana* (Pavy *et al.*, **2008**), among different spruce species (Raherison *et al.*, **2012**) and between pine and spruce (Yeaman *et al.*, **2014**).

A majority of transcriptome profiling experiments in conifers and plants in general have aimed to identify and functionally describe expression variations in subsets of genes associated with biological processes or genotypic differentiation (Lorenz *et al.*, **2011**; Palle *et al.*, **2011**). Genes whose transcript levels do not vary across samples have been less investigated, despite their potentially key housekeeping roles (Aceituno *et al.*, **2008**; Chang *et al.*, **2011**; Eisenberg & Levanon, **2013**). Furthermore, many studies have focused on a single or a few tissues or organs; however, recent studies in flowering plants show that tissue differentiation is the major driver of transcriptome reorganization in species such as *A. thaliana* (Ma *et al.*, **2005**), maize (Downs *et al.*, **2013**), barley (*Hordeum vulgare*; Downs *et al.*, **2013**), and tobacco (*Nicotiana tabacum*; Druka*et al.*, **2006**). The conservation of tissue-specific gene expression patterns is also well established in vertebrates including mammals (Prasad *et al.*, **2013**) and nonmammals (Chan *et al.*, **2009**). Taken together, these observations indicate that studies considering several tissues are a powerful approach to delineate transcriptome organization across a whole organism. Furthermore, the rate of gene expression evolution can differ depending on tissue or organ sources in angiosperms (Yang & Wang, **2013**) and in mammals (Brawand *et al.*, **2011**).

We investigated transcriptome structure by interrogating 23 853 genes on a custom-designed microarray profiling platform (Raherison *et al.*, **2012**) in two different controlled experiments with biological replications. We have previously shown a high level of concordance between results obtained with this microarray and RNA-Seq data (Raherison *et al.*, **2012**). Other recent studies used RNA-Seq profiling (Yeaman *et al.*, **2014**) and were able to interrogate nearly identical numbers of genes (23 889 in spruce and 23 519 in pine). Our objectives were threefold. (1) We first aimed to characterize the organization of the white spruce transcriptome based on multiple tissue data generated from a single experiment enabling statistical model analyses and the identification of high-confidence coexpression groups. This led us to define and functionally analyse expressional groups including invariant and variable genes across vegetative tissues of white spruce. (2) Our second objective was to investigate expressional and functional diversity of vascular tissue preferential genes which were conserved among *Picea* species. (3) Our third objective was to delineate gene coexpression networks and identify genes involved in secondary cell wall development during wood formation, and study network reorganization over the course of a growth season.

## MATERIALS AND METHODS

We used microarray data sets from two independent experiments (Expts 1 and 2) using the same microarray platform as described in Raherison *et al.* (**2012**). These experiments are described in the Plant material section. We also used gene expression results in shoot secondary xylem and phelloderm tissues of white spruce (*Picea glauca* (Moench) Voss), black spruce (*Picea mariana* (M) Britton) and Sitka spruce (*Picea sitchensis* (Bong.) Carrière) previously described in Raherison *et al.* (**2012**). Unless specified otherwise, all statistical analyses were performed using the base R packages; the Benjamini & Hochberg method was employed where appropriate to correct sets of*P*-values for multiple comparisons (Benjamini & Hochberg, **1995**) using the multtest R package, and the hypergeometric test for under- or overrepresentation analysis.

### Plant Material

We used young spruce trees from open-pollinated seed lots that were grown for 6–16 wk in a glasshouse under natural light conditions. Tissue samples were collected either at one (Expt 1) or at four different time-points (Expt 2) in a growing season. In Expt 1, in July we collected from 3-yr-old seedlings of white spruce seven vegetative tissues: shoot apex, needles of young branches (young foliage), shoot or root secondary xylems, shoot or root phelloderms, and root tips (Fig. 1a). The shoot apex sample was *c.* 1 cm in length and contained the shoot tip with the apical bud as well as primary needles. Secondary xylem and phelloderm tissues lay inside and outside the vascular cambium, respectively (Raherison *et al.*, **2012**). Four biological replicates were used for each tissue sample, and each replicate was a pool of samples from five independent trees. In Expt 2, shoot secondary xylem tissues of 4-yr-old seedlings of Norway and white spruces were sampled on 22 June, 13 July, 17 August and 21 September. Samples were collected from six individual trees for each time-point. All of the samples were immediately frozen in liquid nitrogen and stored at −80°C until RNA isolation.

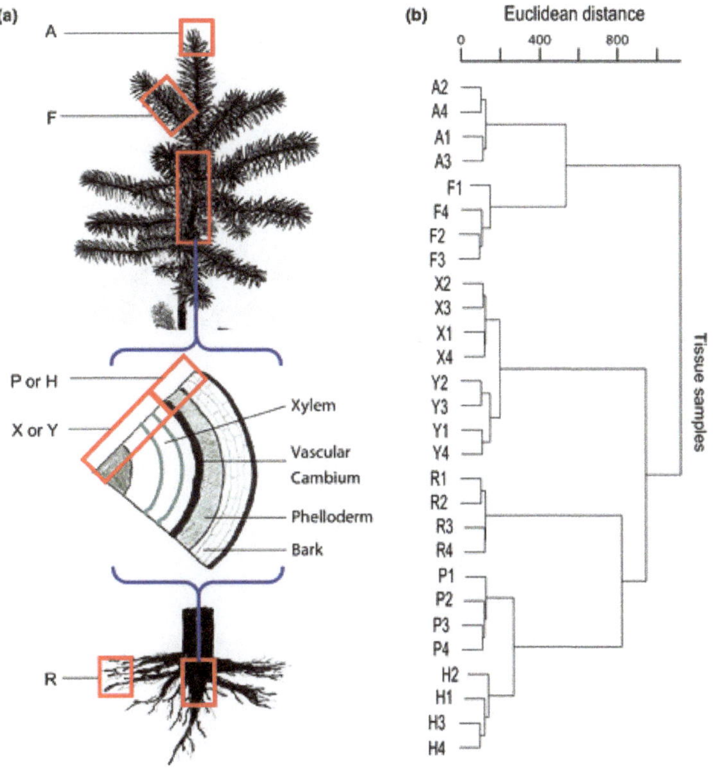


**Figure 1.** Classification of collected tissues based on gene expression. (a) Tissue sampling (image provided by Dr A. Séguin). (b) Hierarchical clustering of white spruce (*Picea glauca*) tissues based on microarray data ($log_2$ scale). The analysis used the *hclust* function of R with Ward's method (Ward, **1963**) and based on expression data of detected genes. A, shoot apex; F, young foliage; X, shoot secondary xylem; Y, root secondary xylem; R, root tips; P, shoot phelloderm; H, root phelloderm. Each tissue was replicated four times (1–4).

## RNA isolation, Labelling and Hybridization

Total RNA was isolated from tissue samples using the cetyltrimethyl ammonium bromide (CTAB) extraction method of Chang *et al.* (**1993**) as described in Pavy *et al.* (**2008**). Total RNA of each sample replicate was amplified using the Amino Allyl MessageAmp II aRNA amplification kit (Ambiom by Life Technologies, Carlsbad, CA, USA) following the manufacturer's instructions. In Expt 1, dye-swap technical replicates for each biological replicate were performed. Each sample of amplified RNA (aRNA) was paired and labelled with Alexa Fluor 555 and 647 dyes (Invitrogen, Carlsbad, CA, USA). In Expt 2, aRNAs were labelled with Alexa Fluor 555 dye. RNA amplification, aRNA labelling, hybridization, array scanning and image processing were performed as described in Raherison *et al.* (**2012**).

## Microarray Preprocessing and Statistical Analyses

In Expt 1, intensity data for single dyes of the two-dye experiment were analysed separately. Replacement of abnormal or flagged spots, data normalization, and identification of positive probes and detected genes were performed as described in Raherison *et al.* (**2012**). A total of 24 850 probes corresponding to 23 623 unique white spruce genes were detected at least in one tissue (Table 1). Using the nlme R package, we applied to detected genes a mixed analysis of variance (ANOVA) model with tissue, dye and the interaction tissue × dye as fixed effects, and the interaction tree × tissue as a nested random effect. A set of 846 genes were not included in the ANOVA because their expression levels were nonnormally distributed (Anderson–Darling's test; nortest R package), had nonhomogenous variances (Levene's test; car R package), or were affected by the factor dye or the interaction tissue × dye. We identified genes that had changes (variable genes; adjusted $P$-value $< 0.05$) or were unchanged (invariant genes; adjusted $P$-value $> 0.05$) across expression tissues from the remaining detected genes. These remaining detected genes were also used for further analyses. In Expt 2, flagged spots of expression data were removed from the analysis. Data normalization and identification of positive probes were carried out as described in Raherison *et al.* (**2012**). A two-way ANOVA with species and time as factors was performed to identify the 1366 temporally variable genes or genes differentially expressed in secondary xylem across time-points in a growing season. Temporally variable genes were those with one or more statistically significant probes (adjusted $P$-value $< 0.05$). The average expression values in two dye technical replicates for each tissue sample (Expt 1) or six replicates in two species for each time-point (Expt 2) were used for weighted correlation network analysis (wgcna R package; Langfelder & Horvath, **2008**), template matching (Harris *et al.*, **2009**), and correlational and hierarchical clustering analyses. In wgcna, we used soft-thresholding power values of 14 and 18 for Expts 1 and 2, respectively, to identify expression modules. An expression module is a cluster of highly (positively and negatively) correlated genes. It reveals two negatively correlated expression profiles. Each expression profile is associated with a group of genes that we referred to as a coexpression group in Expt 1 and a temporally variable cluster in Expt 2. The wgcna script used is included in Methods S1. The raw and preprocessed data have been deposited in the National Center for Biotechnology Information – Gene Expression Omnibus (NCBI-GEO) with accession numbers GSE60277 and GSE51884 for Expts 1 and 2, respectively.

**Table 1.** Numbers of genes detected and classification as variable and invariant based on transcript levels

| Category | Total genes | Annotated genes (%)[1] |
|---|---|---|
| Interrogated genes | 23 853 | 15 724 (66) |
| Detected genes[2] | 23 623 | 15 586 (66) |
| Statistically tested genes (ANOVA)[3] | 22 857 | 15 096 (66) |
| Invariant genes | 4805 | 2410 (50) |
| Variable genes | 18 052 | 12 726 (70) |
| High-confidence genes (adjusted *P*-value < 0.0001) | 10 548 | 7651 (73) |
| Low-confidence genes (0.05 < adjusted P-value > 0.0001) | 7504 | 5075 (68) |

[1]Genes that matched an *Arabidopsis thaliana* sequence (BLASTX; *E*-value < $10^{-10}$).
[2]Transcripts were detected above background levels in at least one tissue of white spruce (*Picea glauca*).
[3]Genes that met assumptions for ANOVA testing (for details, see the Materials and Methods section).

## Functional Annotation Analysis

The cDNA sequences from which the oligonucleotide probes were designed (Raherison *et al.*, **2012**) were annotated against the *A. thaliana* protein data set (The Arabidopsis Information Resource 10) using BLASTX (*E*-value < $10^{-10}$). Functional classes presented in Table S2 were developed as follows: first, we identified and clustered enriched biological process gene ontology (GO) terms using the Database for Annotation, Visualization and Integrated Discovery (DAVID; enrichment at*P*-value < 0.05; Huang *et al.*, **2009**); second, enriched GO terms were classified into functional categories of plant GO slim using cateGOrizer (Hu *et al.*, **2008**); and finally, based on the GO term itself, its cluster and its functional category, each GO term was manually assigned to a consensus functional class. Six sequences that we analysed represented *P. glauca* MYB proteins. These sequences were named in this report: BT112255, PgMYB29; BT117714, PgMYB30; BT119291, PgMYB31; DR571012, PgMYB32; BT108182, PgMYB33; and BT106711, PgMYB34.

## Reverse Transcription Quantitative PCR (RT-qPCR)

We performed RT-qPCR on 28 genes to verify results from microarray analyses (Table S3). Gene-specific primers were designed using Primer3Plus (Untergasser *et al.*, **2007**), evaluated withOligoCalc (Kibbe, **2007**) and checked for uniqueness in the white spruce gene catalogue (Rigault*et al.*, **2011**). The same RNA samples that were used for microarray experiments were used for reverse transcription and RT-qPCR as described in Boyle *et al.* (**2009**) using the QuantiFast SYBR Green PCR Kit (Qiagen, Carlsbad, CA, USA) with 5 ng of cDNA in a final volume of 15 μl. The number of molecules was determined following the method of Rutledge & Stewart (**2008**) with modifications described in Boyle *et al.* (**2009**). A one-way ANOVA (for the factor tissue) and the template matching method were used to determine whether results from microarray and RT-qPCR analyses were statistically matched with regard to gene category and expression profile, respectively.

# RESULTS

## Most White Spruce Genes Vary Significantly Across Vegetative Tissues

We analysed the transcript levels of 23 853 unique genes in seven vegetative tissues during normal development of white spruce (Expt 1). These tissues represent aboveground (apex, young foliage or needles, shoot secondary xylem, and shoot phelloderm) and belowground (root secondary xylem, root phelloderm and root tips) organs. Microarray analysis of each tissue sample comprised four biological replicates that were analysed separately. Dye-swap technical replicates were also used. We identified 18 052 genes whose expression level varied significantly across tissues (variable genes) and 4805 whose expression level did not vary across tissues (invariant genes) (Tables 1, S1, S4). A total of 10 548 variable genes were identified as high confidence (adjusted *P*-value < 0.0001) and 7504 genes (0.0001 < adjusted *P*-value < 0.05) as low confidence (Tables 1, S1, S4).

## The White Spruce Transcriptome Is Organized Into Coexpression Groups Classifying Tissues Based On Their Types or Physiological Functions

As a first step to explore expression data and transcriptome organization, we used hierarchical clustering. Biological samples of each tissue were clustered within a single unique group, thus indicating high repeatability of transcript level determinations as well as uniformity among samples of each type (Fig. 1b). Identical results were obtained when analysing either the whole set of detected genes (Fig. 1b) or only the set of variable genes (Fig. S1). The clustering clearly formed groups according to tissue type. Root xylem clustered closely with shoot xylem, and root phelloderm with shoot phelloderm, thus indicating a strong similarity between tissues from above- and belowground organs. The cluster mostly similar to foliage (i.e. needles) was the shoot apex, an organ that contains primordial leaves.

To further characterize the organization of the white spruce transcriptome, we clustered variable genes and identified 22 coexpression groups or sets of genes sharing a similar expression profile, ranging in size from 48 to 1929 genes (Figs 2a, S2; Tables S1, S4). Gene clustering to delineate the coexpression groups proceeded in two steps. First, we clustered high-confidence genes into 11 modules each containing two coexpression groups with the wgcna package of R and defined the expression profile associated with each of the groups. Second, we used the resulting average expression level within each of the 22 groups as a template to determine membership of low-confidence genes within the coexpression groups (adjusted $P$-value $< 0.05$). In this study, expression modules (M1–M11) are comprised of genes whose expression was positively or negatively correlated, whereas members of a coexpression group had only positively correlated profiles (M1a–M11b; 22 groups in total), as shown in heatmaps (Figs 2a, S2a–j). A principal component analysis (eigengene modules) was used to show the differentiation of tissues within each expression module (Figs 2, S2). Eigengene groups were used to study correlation between groups and tissues (Figs 2b, S2k) and between tissues (Fig. 2c).

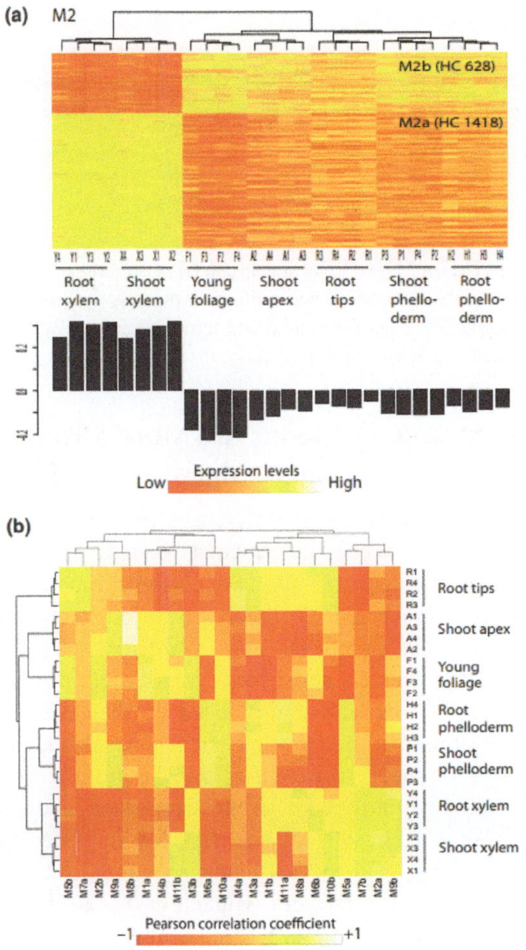


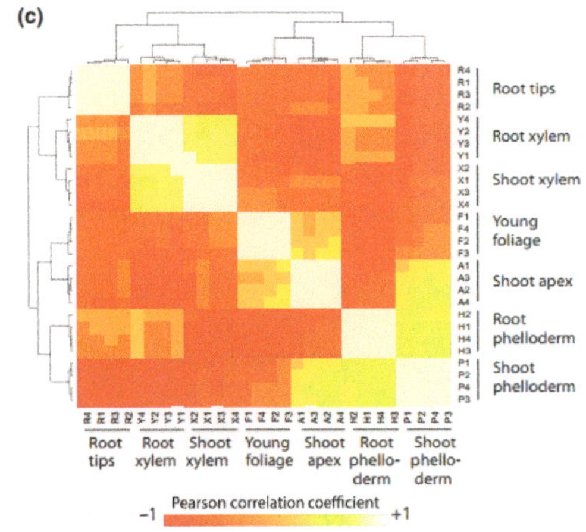


**Figure 2.** Expression module and coexpression groups across tissues in white spruce (*Picea glauca*). (a) Heatmaps show transcript abundance ($\log_2$ scale) of high-confidence variable genes in the M2 expression module and M2a,b coexpression groups determined with the wgcna package of R. Number in parentheses, number of high-confidence (HC) variable genes in the coexpression group. Total and low-confidence variable gene numbers are listed in Supporting Information Table S4. Rows ($y$-axis); columns ($x$-axis), each of the four replicates of tissues (P1–4, shoot secondary phelloderm; Y1–4, root secondary xylem; R1–4, root tip; X1–4, shoot secondary xylem; F1–4, young foliage; A1–4, shoot apex; H1–4, shoot secondary phelloderm). Bar plots display the eigengene modules (or the first principal components). See Fig. S2 for a graphical representation of other expression modules and coexpression groups. (b) Correlation between coexpression groups (M1a–M11b) and tissues based on eigengene groups. (c) Correlation between tissues based on eigengene groups.

Our results indicate a modular organization of the transcriptome that is linked to differentiation between tissue types or physiological functions (Figs 2a,b, S2; Table S4). For example, four coexpression groups containing 42% of the variable genes were positively correlated with a single tissue (average Pearson correlation coefficient $R > 0.7$; shoot apex in M4a, M8b, M9a and M11b; and needles in M5b) or tissue type ($R > 0.6$; secondary xylem from shoots and roots in M2a, M7b and M9b; and phelloderm from shoots and roots in M6a and M10a). The profile of the M1a group comprising 11% of the variable genes was positively correlated with photosynthetic tissues ($R = 0.8$).

The identification of coexpression groups also supports a strong similarity between tissues from above- and belowground organs. For example, 14 profiles comprising 70% of variable genes closely clustered xylem samples from the shoot and the root, and similar phelloderm samples from the shoot and the root (M2–M7 and M10). By contrast, only 21% of variable genes were separately clustered for vascular tissues from the root and the shoot (M1 and M11). Analysis based on eigengene groups revealed highly positive correlation between vascular tissues from the shoot and the root (Fig. 2c).

To confirm microarray analysis results, we determined and statistically analysed transcript levels of selected genes using RT-qPCR, including 24 variable genes and four invariant genes (Table S3). Results were strongly consistent between the two methods, with 92% of tested genes having similar expression profiles and similar categories (invariant and variable genes) (Table S3).

## GO Term Enrichments Show Functional Diversity among Coexpression Groups

Functional analysis was performed for the 22 coexpression groups and invariant genes based on annotations and enrichment for biological process GO terms for genes. Most of the detected genes (66%) were similar to an *A. thaliana* gene sequence (Table 1) and were annotated accordingly. The proportion of white spruce sequences that matched an *A. thaliana* sequence was somewhat higher among variable genes (70% for the high-confidence group) than for the invariant genes (50%; Table 1). We found 4324 genes associated with enriched GO terms, representing 50% of the genes with assigned GO term annotations (Figs 3a, S3; Table S2).

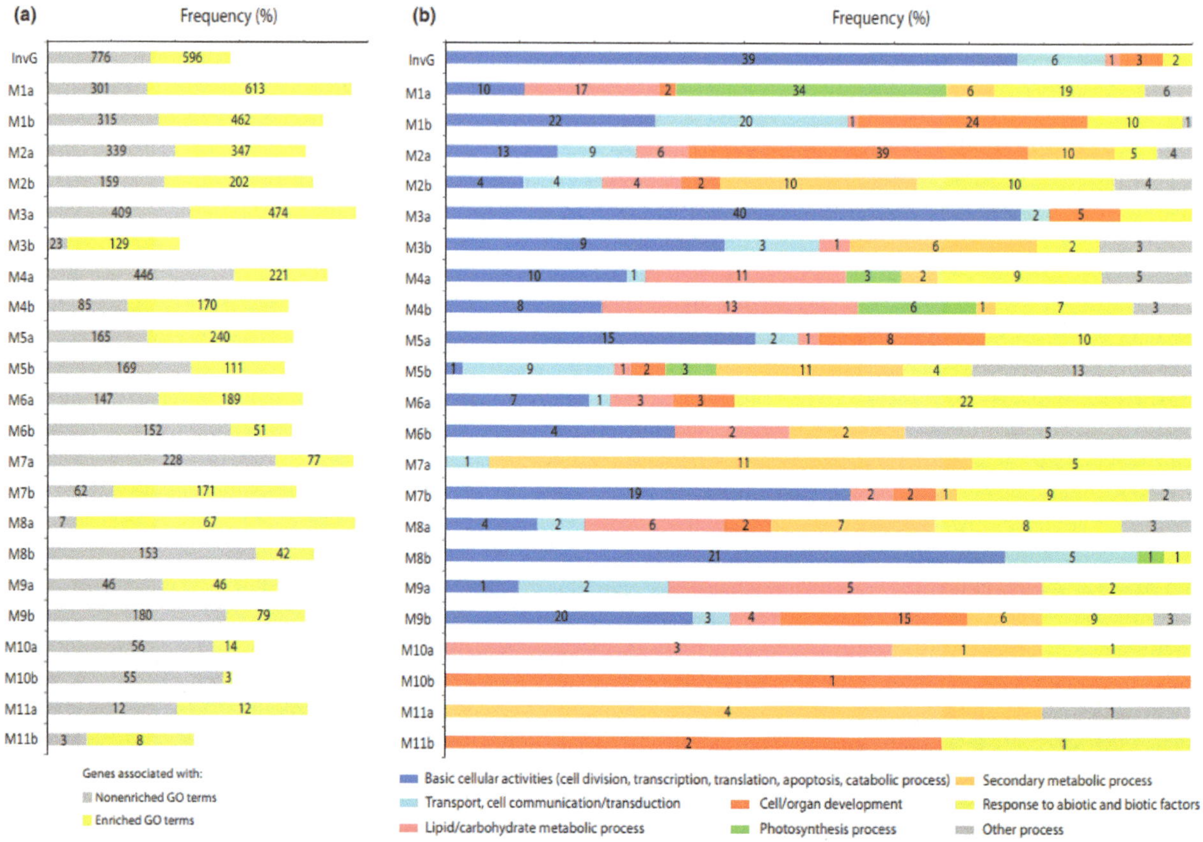


**Figure 3.** Functional analysis of invariant genes (InvG) and coexpression groups (M1a–M11b) in white spruce (*Picea glauca*). (a) *x*-axis, frequency (%) of genes associated with each of the two annotation categories: enriched and nonenriched biological process gene ontology (GO) terms. Numbers inside bars are the number of genes per category. (b) *x*-axis, frequency (%) of enriched GO terms associated with consensus functional classes. Numbers inside bars are the number of enriched GO terms per consensus functional class.

The GO term enrichment results showed that the dominant enriched biological processes varied across the coexpression groups (Fig. 3b). In addition, strong or preferential expression in a tissue type may be associated with several coexpression groups from different modules, and thus with enrichment of different processes. For example, group M6a with genes strongly expressed in phelloderm tissues compared with all other tissues was enriched for stress response. Group M10a was also highly expressed in phelloderm tissues but contrasted with M6a through stronger expression in needles; it was enriched for terms related to cell/organ development. Observations such as these show how assignment to a coexpression group is a useful first step for developing hypotheses for different groups of phelloderm preferential genes. Other similar examples were observed for groups of genes with strong expression in apex tissue (M4a, M8b, M9a and M11b) or in xylem tissues (M2a, M7a and M9b). These findings revealed the diversity of functions among genes preferentially expressed in a tissue type.

## Conifer-Specific Genes Are More Strongly Associated With Secondary Meristematic Tissues

Functional and structural differences have been extensively documented in secondary vascular tissues when comparing conifer and angiosperms trees. For example, functions of water transport and mechanical support are attributed to a single cell type in conifers (tracheids) and to two distinct cell types in angiosperms (vessels and fibres); in addition, different types of hemicelluloses and lignins make up their secondary cell walls. We thus explored to what extent genes that are identified as unique or specific to conifers may be more represented in coexpression groups with strong expression in these tissue types. We used results of searches for sequence similarity to *A. thaliana* sequences (this work) and to Sitka spruce and loblolly pine sequences (Rigault *et al.*, **2011**) to identify putative conifer-specific genes and analysed their distribution across coexpression groups. We found that conifer-specific genes were overrepresented in groups with a preference for expression in secondary vascular tissues (M2a and M6a), and underrepresented in groups without a clear preference in any tissue (M1b and M3a) or with high expression in tissues more closely derived from the shoot meristem (M1a and M8b) (Table S5).

## Highly Conserved Invariant Genes Are Associated With High Expression and Basic Cell Functions

We observed that invariant genes spanned a broad range of expression levels (Table 2) and that fewer of them had functional annotations compared with the variable genes (Figs 3a, S3). Overall, the 596 invariant genes within enriched GO terms were predominantly associated with basic cellular activities and to a lesser extent with transport, cell communication/transduction, cell/organ development, and stress response (Fig. 3b), suggesting an overrepresentation of housekeeping genes. We analysed the group further with the aim of uncovering relationships between expression levels and functional annotations or sequence similarities. Results showed that genes that were conserved between conifers and *A. thaliana* were more strongly expressed as they were overrepresented among the high-expression classes. Conversely, the putative conifer-specific genes were overrepresented in low-expression classes (Table 2). Genes associated with the major enriched functions were distributed in similar proportions in all expression level classes except for the stress response category, which was statistically overrepresented in the lowest expression class. This latter group probably represents genes whose expression may vary in response to a stress factor and had a low basal level in tissues grown in permissive conditions.

**Table 2.** Function and cross-species conservation of invariant genes according to their expression levels

| | | Cross-species sequence conservation[2] | | | Functional classes[3] | | | |
|---|---|---|---|---|---|---|---|---|
| Expression classes[1] | Total | Conifers | Conifers–*Arabidopsis thaliana* | White spruce–*Arabidopsis thaliana* | A (%) | B (%) | C (%) | D (%) |
| Low (< Q1) | 1202 | **142*** | *225** | *110** | 79 | 7 | **16*** | 11 |
| Low to moderate (Q1, Q2) | 1201 | **147*** | *227** | *89** | 94 | 6 | 8 | 10 |
| Moderate to high (Q2, Q3) | 1201 | 109 | **419*** | *134** | 94 | 10 | 7 | 13 |
| High (> Q3) | 1201 | *82** | **686*** | **397*** | 90 | 9 | 10 | 20 |
| Total in number | 4805 | 480 | 1557 | 730 | | | | |

[1]Based on their mean expression from all tissue samples of white spruce (*Picea glauca*), invariant genes were divided into four equal classes delimited by the first, second and third quartiles (Q1, Q2 and Q3, respectively).
[2]Conifers: number of genes with conserved sequences between white spruce (*Picea glauca*), sitka spruce (*P. sitchensis*) and loblolly pine (*Pinus taeda*); Conifers–*Arabidopsis thaliana*: number of genes with conserved sequences between one of the three conifer species and *A. thaliana*; and White spruce–*A. thaliana*: number of genes with conserved sequences between white spruce and *A. thaliana* based on the sequence similarity approach (Rigault *et al.*, 2011; this work; BLASTX; *E*-value $< 10^{-10}$).
[3]Frequency (%) of genes associated with each of the functional classes: A, basic cellular activities; B, cell/organ development; C, response to abiotic and biotic factors; D, transport, cell communication/transduction. The sum of percentages per row exceeds 100% because a gene may be involved in one or more functional classes.
[2,3]Hypergeometric tests performed for expressional classes by column: *, statistically significant (adjusted *P*-value < 0.05), for overrepresentation (bold numbers) or underrepresentation (italic numbers).

## Expressional and Functional Diversity among Conserved Secondary Xylem and Phelloderm Preferential Genes

We used the defined coexpression groups to characterize sets of genes that we previously identified as preferentially expressed in secondary xylem (2537 genes) and phelloderm (2800 genes) in shoots and were conserved in white, black and Sitka spruces (Raherison *et al.*, **2012**). We found that 95% of the conserved xylem and phelloderm preferential genes were distributed among all coexpression groups, indicating that their modular organization is represented by diverse expression profiles (Fig. 4). Their respective distributions among coexpression groups were highly contrasted and matched closely with the profiles of the coexpression groups with regard to expression in xylem and phelloderm tissues, except for group M1a (with similar expression levels in both xylem and phelloderm tissues). The two tissue types are derived from different but neighbouring secondary stem meristems; the xylem is derived from the cambial zone (located more internally) and the phelloderm is derived from the phellogen (located more externally). Our results show the extent of transcriptome reorganization resulting from differentiation pathways from these two meristems.

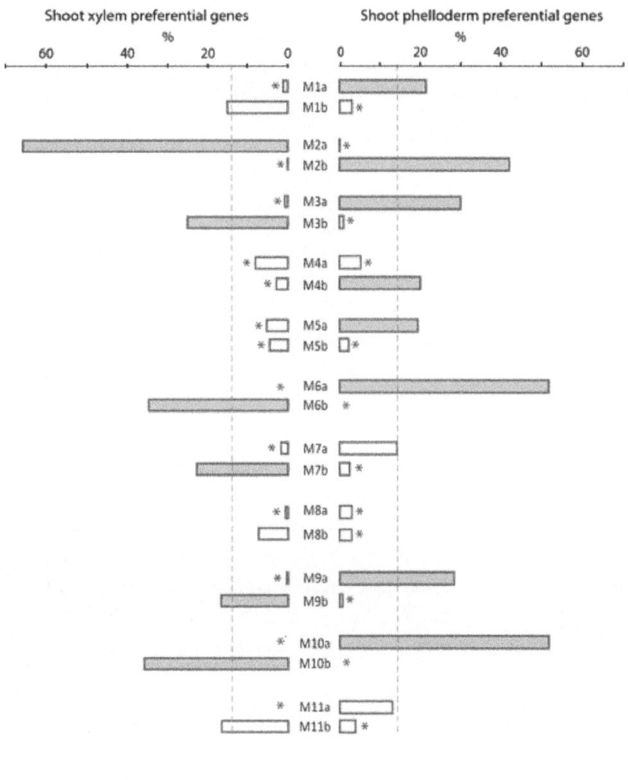


**Figure 4.** Distribution of secondary xylem or phelloderm preferential genes conserved amongPicea species across coexpression groups.

Vascular tissue preferential genes in white (*P. glauca*), Norway (*P. abies*) and black (*P. mariana*) spruces were identified in Raherison *et al.* (**2012**). Dashed grey vertical lines represent the expected frequency based on chance alone (%) of preferential genes across coexpression groups. Overrepresentation of xylem or phelloderm preferential genes compared with chance alone (hypergeometric test; adjusted *P*-value < 0.05) is shown by closed bars and underrepresentation of xylem or phelloderm preferential genes is shown by open bars with an asterisk (hypergeometric test; adjusted *P*-value < 0.05). Each of the over- and underrepresentations of xylem and phelloderm preferential genes closely matched with the expression profiles of the corresponding coexpression groups as shown in Fig. 2(a) and Supporting Information Fig. S2(a–j).

## Distinct Expression Patterns Characterize Wood Formation at Different Stages during a Growth Season

Secondary vascular development and growth in woody plants follow an annual cycle that is characterized by an initial rapid growth phase (earlywood formation), followed by a slower phase that transitions to dormancy (latewood formation). We carried out a time-course microarray profiling experiment (Expt 2) and identified 1366 genes with significant temporal variation in white and Norway spruces (adjusted *P*-value < 0.05). The genes were clustered with wgcna into six temporal coexpression groups of 41–507 genes (Fig. 5a). Overall, transcript levels did not change between June and July, remained stable (Fig. 5a; cluster T1; 726 genes) or varied progressively (Fig. 5a; cluster T2; 547 genes) in August and changed dramatically from August to September. Transcripts accumulating preferentially during June–July and September represented 51 and 43%, respectively, of the genes (Fig. 5a). By contrast, only 3% of the genes were preferentially expressed in August.

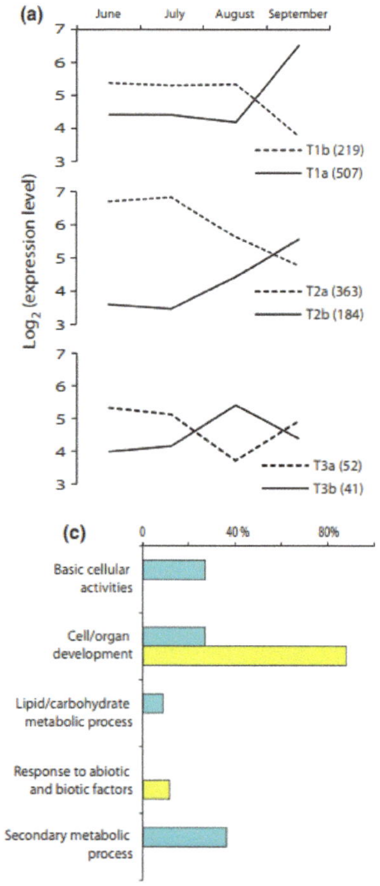


(b)

| T1a | T1b | T2a | T2b | T3a | T3b | |
|---|---|---|---|---|---|---|
| 33 | 9 | 6 | 21 | 1 | 5 | M1a |
| 20 | 19 | 39 | 3 | 9 | 2 | M1b |
| 13 | 69 | 138 | 5 | 7 | 1 | M2a |
| 79 | 1 | 1 | 34 | 2 | 1 | M2b |
| 56 | 11 | 25 | 14 | 5 | 1 | M3a |
| 4 | 12 | 14 | 5 | 1 | 6 | M3b |
| 24 | 7 | 22 | 6 | 6 | 0 | M4a |
| 13 | 4 | 4 | 1 | 0 | 2 | M4b |
| 18 | 15 | 23 | 2 | 2 | 0 | M5a |
| 13 | 6 | 10 | 2 | 3 | 2 | M5b |
| 86 | 4 | 3 | 30 | 1 | 1 | M6a |
| 7 | 4 | 9 | 4 | 0 | 0 | M6b |
| 17 | 3 | 3 | 5 | 3 | 1 | M7a |
| 1 | 8 | 10 | 4 | 2 | 0 | M7b |
| 0 | 5 | 3 | 0 | 1 | 1 | M8a |
| 8 | 0 | 4 | 4 | 4 | 0 | M8b |
| 2 | 0 | 4 | 3 | 0 | 0 | M9a |
| 8 | 1 | 3 | 0 | 3 | 0 | M9b |
| 35 | 0 | 1 | 10 | 0 | 0 | M10a |
| 0 | 4 | 12 | 2 | 0 | 0 | M10b |
| 2 | 0 | 0 | 0 | 0 | 0 | M11a |
| 1 | 2 | 2 | 0 | 0 | 0 | M11b |

**Figure 5.** Distribution of temporally variable xylem expressed genes across coexpression groups. (a) Expression profiles of temporally variable genes in secondary xylem of white (*Picea glauca*) and Norway (*P. abies*) spruces. Numbers in brackets are the number of genes within temporally variable clusters (T1a–T3b). (b) Number of temporally variable genes across tissue-based coexpression groups (M1a–M11b). Grey and blue/yellow areas, under- and overrepresentation (hypergeometric test; adjusted *P*-value < 0.05), respectively; blue and yellow areas, earlywood and latewood genes, respectively, that is, June–July (T1b and T2a) and September (T1a and T2b) preferential genes that were overrepresented in M2a and M10b, and in M2b, M6a and M10a, respectively. (c) Functional annotation analysis based on gene ontology (GO) term enrichment of earlywood (blue bars) and latewood (yellow bars) genes.

Next, we analysed the temporally variable gene clusters for membership of the 22 coexpression groups derived from our comparison of vegetative tissues in white spruce (Fig. 5b). We found that June–July preferential genes (average Pearson correlation coefficient $R = 0.6$) were overrepresented in groups exhibiting high expression in xylem tissues (M2a and M10b); these genes are called earlywood genes. This observation is consistent with the fact that the secondary xylem tissues used to define the tissue-based coexpression groups were sampled in July during earlywood formation (Fig. 5b). Conversely, September preferential genes ($R$ = 0.8) were overrepresented in groups with a strong preference in phelloderm tissues or in other tissues than xylem (M2b, M6a, and M10a); these genes are referred to as latewood genes (Fig. 5b). These observations indicate that many of the earlywood genes were specific to xylem tissues at the start of the growing season, while latewood genes were more active in other parts of the trees during early phases of the growth season. Analysis of functional annotations indicated that basic cellular activities, lipid/carbohydrate reserves and secondary metabolic process were specific to earlywood genes and that stress responses were specific to latewood genes. Cell/organ development is a common function with a higher representation in latewood genes (Fig. 5c; Table S2).

## Variation in Secondary Cell Wall Regulatory Network Hub Genes Is Associated With Xylem Transcriptome Reorganization during the Growth Season

We developed three unweighted and signed networks targeting two xylem preferential modules (M2 and M7), June–July and September preferential genes. In these analyses, two genes whose expression profiles had an absolute Pearson correlation coefficient above 0.9 were connected. We computed the number of genes connected to each gene (i.e. degree) to identify a hub(s), which represent(s) the most connected gene(s).

We identified the enriched pathways for the M2–7 network, which are shown in Table S6. Among the enriched pathways, we investigated the branch of the phenylpropanoid biosynthesis pathway leading to lignin biosynthesis (Fig. 6). Based on gene expression and sequence similarity, 12 of the 23 represented genes were preferentially expressed in xylem tissues and candidate genes for lignin biosynthesis. The remaining genes may be involved in the biosynthesis of phenolic compounds for purposes other than secondary cell wall lignification or in other branches of the phenylpropanoid metabolic pathway.

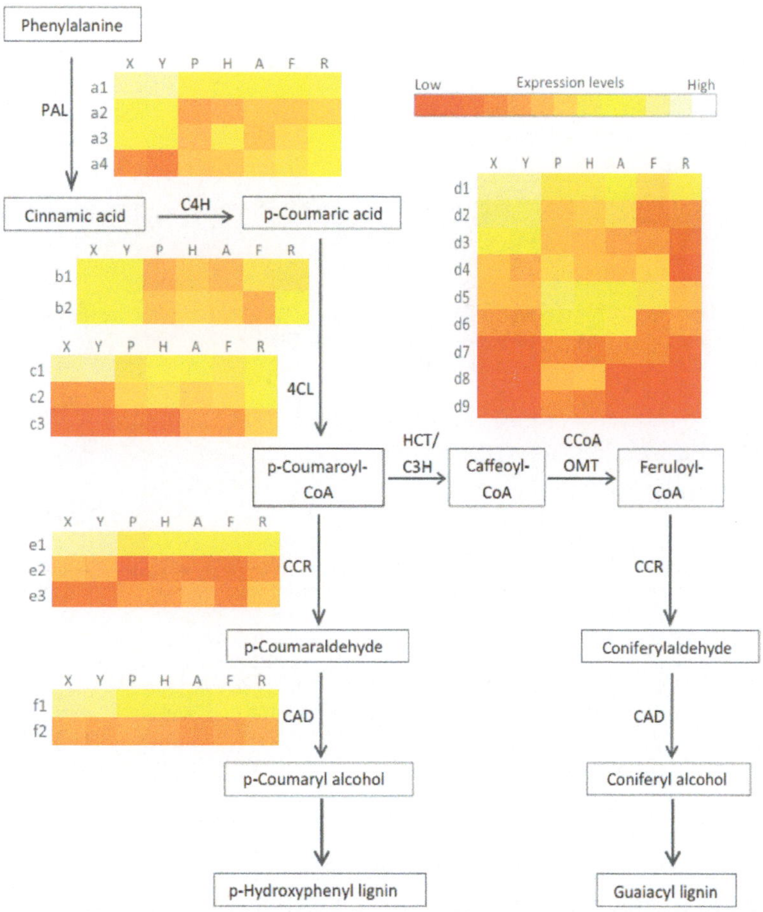


**Figure 6.** Phenylpropanoid pathway leading to lignin production. Boxes, metabolite; arrows, enzyme reactions; enzymes are reported using a three-letter code: PAL, phenylalanine ammonia lyase; C4H, cinnamate 4-hydroxylase; 4CL, 4-coumarate-CoA ligase; HCT/C3H, shikimate/quinate hydroxycinnamoyltransferase/p-coumarate 3-hydroxylase; CCoAOMT, caffeoyl-CoA O-methyltransferase; CCR, cinnamoyl-CoA reductase; CAD, cinnamoyl alcohol dehydrogenease. Heatmaps next to enzyme name indicate expression value of genes (a1-4, b1-2, c1-3, d1-9, e1-3, f1-2) across tissues (X, Y, P, H, A, F, R); Each expression value is the average of four the biological replicates per tissue; a1-4, BT112211 (GenBank ID), BT106538, BT119163, BT114680; b1-2, BT116118, BT117086; c1-3, BT106671, BT116171, DR551141; d1-9, BT106698, BT106398, BT101243, BT117977, BT110631, BT102905, BT112333, BT104804, BT114515; e1-3, DR573886, BT111802, BT112289; f1-2, BT112280, BT116920; X, shoot secondary xylem; Y, root secondary xylem; P, shoot phelloderm; H, root phelloderm; A, shoot apex; F, young foliage; R, root tips.

The xylem preferential (M2–7) network was scale free (Fig. S4) and had *PgNAC-7* (no apical meristerm (NAM), *Arabidopsis* transcription activation factor (ATAF) and cup-shaped cotyledon (CUC) transcription factor 007 in *Picea glauca*) as the most connected hub gene (Table 3a). A subnetwork containing common interactions (connections) and genes between xylem (M2–7) and June–July preferential networks also had *PgNAC-7* as the most connected hub gene (Tables 3a, S1). PgNAC-7 was recently identified as a key regulator of secondary cell wall development in white spruce, which is functionally analogous to VND6/7 (vascular-related NAC-domain 6/7; Duval *et al.*, **2014**). Functional annotation analysis was carried out among genes positively and negatively connected to *PgNAC-7.* Genes positively connected to *PgNAC-7* were specifically enriched for basic cellular, development and reserve metabolisms, whereas stress response and secondary metabolism terms were enriched in both negatively and positively connected genes (Fig. 7a,b).

**Table 3.** Functional annotation of the top ten most connected genes (hubs) of the networks; (a) subnetwork (xylem (M2–7) and June–July preferential network) which contains 185 genes; (b) September preferential network which contains 691 genes

| GenBank accession no. | Cluster ID[1] | Degree[2] | Rank[3] | PFAM description[4] | Functional annotation[5] |
|---|---|---|---|---|---|
| (a) | | | | | |
| BT102049 | GQ0165_B14 | 53 | 1 | No apical meristem (NAM) protein | PgNAC-7, NAC 007 |
| CO235762 | WS0021_I12 | 43 | 2 | | |
| BT117400 | GQ03818_L16 | 40 | 3 | Subtilase family | Subtilase family protein |
| DR591085 | WS00830_F12 | 40 | 3 | | |
| BT108414 | GQ03121_H22 | 39 | 5 | Myb domain protein | PgMYB2* |
| BT106274 | GQ02906_F17 | 36 | 6 | | |
| BT107150 | GQ03103_E24 | 35 | 7 | Xyloglucan endo-transglycosylase (XET) C-terminus; glycosyl hydrolase family | Xyloglucan endotransglucosylase/hydrolase |
| BT107836 | GQ03113_F02 | 34 | 8 | Aldehyde dehydrogenase family | Aldehyde dehydrogenase 3F1 |
| DR594044 | WS00839_J22 | 33 | 9 | Subtilase family; protease-associated domain; peptidase inhibitor I9 | Subtilisin-like serine endopeptidase family protein |
| BT106749 | GQ03009_M04 | 32 | 10 | | Ring/U-box superfamily protein |
| (b) | | | | | |
| BT105806 | GQ02828_L20 | 496 | 1 | Tubby C 2 (TUB) | Protein of unknown function (DUF567) |
| BT112527 | GQ03316_N22 | 496 | 1 | Leucine-rich repeat | ADR1, disease resistance protein (CC-NBS-LRR class) family |
| BT117952 | GQ03918_F10 | 496 | 1 | Dehydrin (DHR) | |
| BT103744 | GQ02801_H14 | 495 | 4 | ThiF family | Ubiquitin-activating enzyme |
| BT109195 | GQ03205_L14 | 494 | 5 | Glycosyl hydrolase family, N-terminal domain | Glucuronidase |
| BT113137 | GQ03326_D07 | 494 | 5 | Dehydrin | |
| BT114680 | GQ03519_E06 | 494 | 5 | Aromatic amino acid lyase | Phenylalanine ammonia-lyase 4 |
| BT104565 | GQ02811_O09 | 493 | 8 | Embryo-specific protein 3 (ATS3) | Embryo-specific protein 3 (ATS3) |
| BT115870 | GQ03707_L13 | 493 | 8 | Glycosyl hydrolase family | β-glucosidase |
| BT101047 | GQ0065_L03 | 492 | 10 | | |
| BT104876 | GQ02816_C10 | 492 | 10 | Hsp20/alpha crystallin family | Heat shock protein 17.4 |
| BT104881 | GQ02816_D12 | 492 | 10 | | Sucrose synthase 3 |
| EX329848 | GQ02830_C21 | 492 | 10 | Dehydrin (DHR) | |
| BT114833 | GQ03603_F10 | 492 | 10 | Dehydrin (DHR) | |
| BT116286 | GQ03715_F15 | 492 | 10 | Pyruvate phosphate dikinase, PEP/pyruvate binding domain | Pyruvate phosphate dikinase, PEP/pyruvate binding domain |
| DR574449 | WS00737_M23 | 492 | 10 | | Cox19 family protein (CHCH motif) |

[1]Rigault *et al.* (2011).
[2]The number of genes connected to each gene within the network.
[3]The rank of genes based on their degree within the network; genes with similar degrees have the same rank.
[4]PFAM, the protein families database; ATS3, *Arabidopsis thaliana* seed gene 3; PEP, phosphoenolpyruvate.
[5]Functional annotation based on The Arabidopsis Information Resource version 10 (TAIR10); DUF, domain of unknown function; NAC 007, no apical meristerm (NAM), Arabidopsis transcription activation factor (ATAF) and cup-shaped cotyledon (CUC) transcription factor 007; PgNAC-7, name of the NAC 007 gene in white spruce (*Picea glauca*) according to Duval *et al.* (2014); ADR1, activated diesease resistance 1; CC-NBS-LRR, coiled coil–nucleotide-binding-site–leucine-rich-repeat.

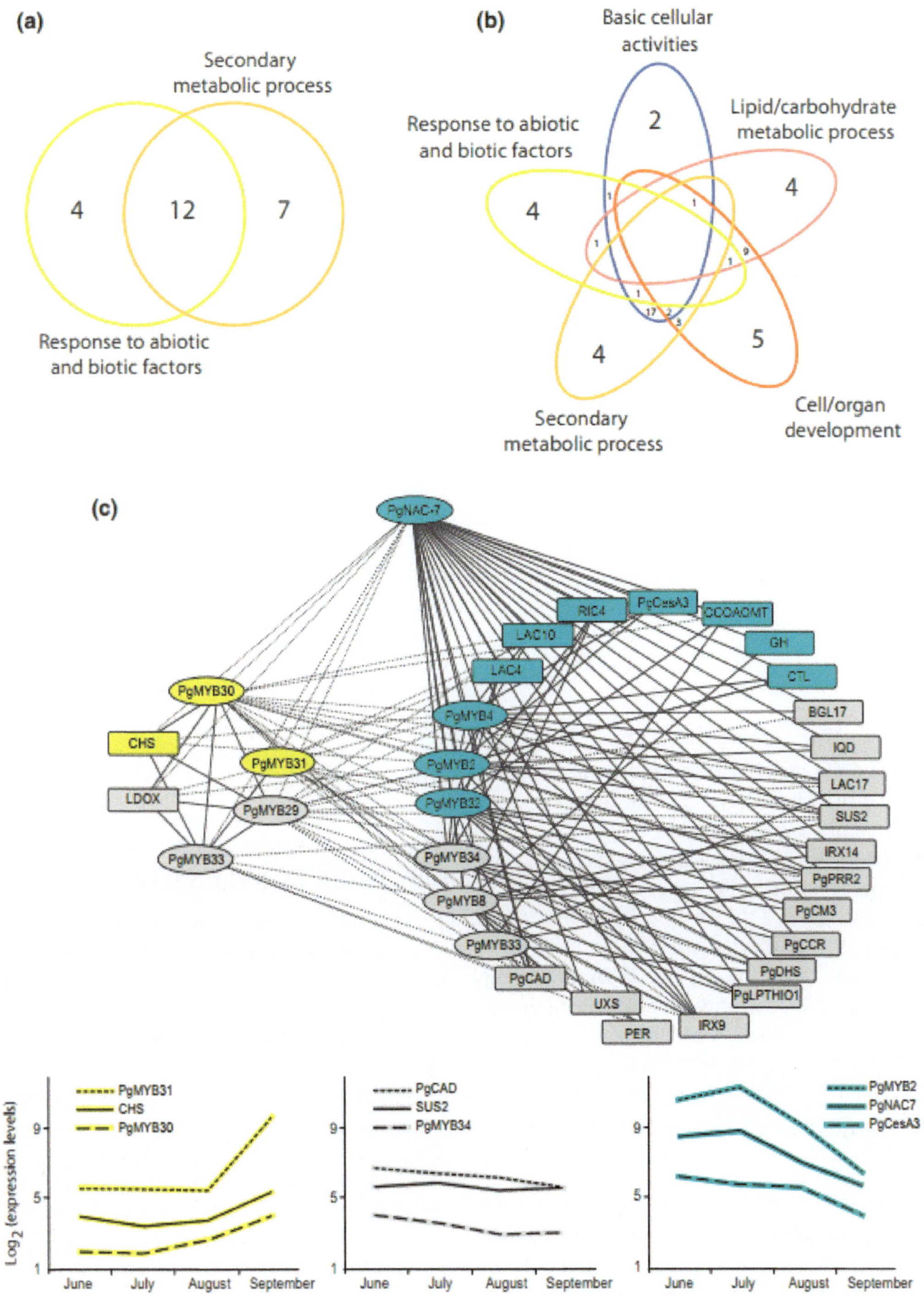


**Figure 7.** Functional annotations, network analysis and temporal expression patterns of *PgNAC-7*(no apical meristerm (NAM), *Arabidopsis* transcription activation factor (ATAF) and cup-shaped cotyledon (CUC) transcription factor 007 in *Picea glauca*) connected genes. (a, b) Venn diagrams show the distribution of *PgNAC-7* (a) negatively and (b) positively connected genes among functional classes. (c) Upper panel, an unweighted and signed network shows connections of *PgNAC-7* with MYB transcription factors (oval-shaped nodes) and putative genes (rectangle-shaped nodes) associated with secondary cell wall formation (Mizrachi *et al.*, **2012**; Duval *et al.*, **2014**). *PgNAC-7*/MYB transcription factors and genes. LAC4, 10, 17, laccase4, 10 and 17; RIC4, rop-interactive cdc42/rac-interactive binding (CRIB) motif-containing protein 4; CCOAMT, putative caffeoyl-CoA O-methyltransferase; GH, glycosyl hydrolase; CTL, chitinase-like protein; BGL17, β-glucosidase 17; IQD, protein IQ-domain; SUS2, sucrose synthase 2; IRX9, 14, irregular xylem (probable β-1,4-xylosyltransferase) 9 and 4; PER, peroxidase; UXS, UDP-xylose synthase; CHS, chalcone synthase; LDOX, leucoanthocyanidin dioxygenase; PgMYB2, 8, 29, 31, 32, 33, 34, MYB transcription factor 2, 8, 29, 31, 32, 33 and 34 in *P. glauca*; PgPRR2, pinoresinol reductase 2 in *P. glauca*; PgCesA3, cellulose synthase A 3 in *P. glauca*; PgCM3, chorismate mutase 3 in P. *glauca*; PgCCR, cinnamoyl CoA reductase in *P. glauca*; PgDHS, 3-deoxy-d-arabino-heptulosonate 7-phosphate synthase in *P. glauca*; PgLPTHIO1, esterase/lipase/thioesterase family protein (lysophospholipase 1); PgCAD, cinnamyl alcohol dehydrogenase in *P. glauca*. Edges or connections are coexpression between genes with a Pearson correlation coefficient $< -0.9$ (dashed line) or $\geq +0.9$ (solid line). Lower panel, temporal variation of genes; node and graph colours, temporal expression of genes in secondary xylem tissue of white (*Picea glauca*) and Norway (*P. abies*) spruces during a growing season: blue and yellow, highly expressed genes in June, July or August (earlywood genes) and in September (latewood genes), respectively; grey, temporally invariable genes.

We present a network showing *PgNAC-7* and genes involved in secondary cell wall formation as reviewed in Mizrachi *et al.* (**2012**) or shown to be transactivated in conifers by regulators of secondary cell wall formation including MYB transcription factors (Bomal *et al.*, **2008**; Duval *et al.*, **2014**; Fig. 7c; Table S7). The network includes temporally variable and nonvariable genes (Fig. 7c). Temporally variable genes included *PgNAC-7*, five of the 10 MYB genes and seven of the 22 genes encoding enzymes, indicating that the network may vary over the course of a growth season. We observed that all of the earlywood genes were positively connected to *PgNAC-7* and the latewood genes were negatively connected to *PgNAC-7*.

A network was constructed for September preferential genes. The most connected hub genes were three stress-responsive genes, that is, a nucleotide binding site–leucine-rich repeat (NBS-LRR)-encoding gene (activated disease resistance 1 (*ADR1*)), a dehydrin-encoding gene (*DHR*), and a tubby protein-encoding gene (*TUB*) (Tables 3b, S1). Several other dehydrins were also coregulated in this network. These findings are consistent with changes in environmental factors, including water deficit and temperature decrease, that occur at the end (September) of a growth season and influence latewood formation in conifers.

## DISCUSSION

Our findings describe the modular organization of the white spruce transcriptome, which we divided into invariant genes and coexpression groups of variable genes across tissues. We assigned 18 052 genes to 22 coexpression groups that are statistically robust and identified 4805 genes as being invariant across tissues. We also showed the complexity, conservation and reorganization of the transcriptome associated with secondary vascular tissue formation in spruces. These findings relating to transcriptome organization are discussed in the context of plant evolution, gene functions and expression networks.

### Evolutionary Signatures of Tissue Differentiation

We measured transcript levels in seven vegetative tissues of above- and belowground organs of white spruce using a microarray platform and found that a large proportion of genes were differentially expressed across tissues with strong (10 548 high-confidence genes) or moderate statistical support (7504 low-confidence genes). Tissue differentiation has generally been linked to deeper transcriptome reorganization than developmental stages or environmental conditions in plants and animals. Our results from the comparison of white spruce tissues identified a much larger proportion of variable genes than comparisons of biological conditions or developmental stages in conifer trees, including interior hybrid spruce (Verne *et al.*, **2011**), maritime pine (Paiva*et al.*, **2008**), radiata pine (Li *et al.*, **2010**) and white spruce (this work). Similar findings have been reported in other plants including *A. thaliana* (Ma *et al.*, **2005**), tobacco (Edwards *et al.*, **2010**) and maize (Downs *et al.*, **2013**). In *A. thaliana*, Aceituno *et al.* (**2008**) indicated that tissue type was a more influential variable for transcriptome variation than any of the tested experimental conditions. In mammals, tissue affiliation explained generally at least 10 times more expression variance than perturbations or diseases (Prasad *et al.*, **2013**). In the mouse, Su *et al.* (**2002**) identified 78% of genes with differential expression across tissues vs 9–10% across time-points (Kwon *et al.*, **2012**) or 1–7% between influenza-infected lungs and control (Pommerenke *et al.*, **2012**). Tissue differentiation is evolutionarily ancient and has been linked to transcriptome signatures that are conserved across species (Prasad *et al.*, **2013**). In our previous work, we found that only 60 of the 5407 vascular tissue differential genes had different expression patterns when comparing three spruce species (Raherison *et al.*, **2012**). These findings demonstrate the way in which transcriptome signatures highlight the ancient evolutionary origins of tissue differentiation (*c.* 380 million yr ago; Kenrick & Crane, **1997**).

In light of these observations and the high level of genomic conservation among spruces (Ritland*et al.*, **2011**), expression profiles may be regarded as a fundamental property of genes and their orthologues across the genus. It is also to be expected that profiles will extend to other genera within the Pinaceae by extrapolation from comparison of white spruce and loblolly pine showing a high level of conservation in gene coding sequences (Rigault *et al.*, **2011**) and in genome structure (Pavy *et al.*, **2013**). The response to abiotic stress factors is also well conserved between pine and spruce, as shown by Yeaman *et al.* (**2014**), who reported that the expression pattern was conserved for 74% of genes identified as sequence orthologues.

### Modular Transcriptome Organization Underpins Tissue Differentiation

The features of sequences that make up the different tissue transcriptomes largely overlap but their quantitative variations in transcript amounts reveal the underlying transcriptome reorganization. These quantitative changes are underpinned by a complex modular organization that was represented here by coexpression groups (Figs 2, S2; Tables S3, S4). The modular organization may be illustrated by considering secondary xylem and phelloderm from the stem. Transcripts were detected for 22 970 and 23 296 genes in xylem and phelloderm, respectively, and were distributed among invariant genes and all of the 22 coexpression groups (Table S8). This modular organization showed the expressional diversity of gene profiles in a given tissue.

Tissue preferential expression has been used to classify genes, and to carry out functional studies and evolutionary investigations of tissue differentiation (Ma *et al.*, **2005**; Pavy *et al.*, **2008**; Yang & Wang, **2013**; Duval *et al.*, **2014**). We analysed xylem and phelloderm preferential genes conserved in three spruce species and found that they were distributed among all coexpression groups as well as being differentially represented in many of them (Fig. 4). These observations show the diversity of expression profiles among genes identified as preferentially expressed in a given tissue.

### Coexpression is Associated with Functional Similarity

We report a high rate of GO term enrichment among coexpression groups. Almost 50% of the genes with assigned GO terms were associated with GO term enrichment in the different groups (Figs 3a, S3). Coexpression groups were for the most part very distinct in regard to predominant enriched biological functions. Taken together, these observations indicated that several members within coexpression groups shared functions in addition to expression profiles, which supports the notion that members of coexpression groups participate in similar or common processes. For example, enriched terms for photosynthesis, stress response and lipid metabolic processes were largely observed in the group of genes highly expressed in foliage and shoot apex tissues (M1a in Fig. 3b; Table S2). The shoot apex requires lipids to support intense cell division activity and a high growth rate (Jordy, **2004**). Both shoot apex and needles are photosynthetic tissues and are regularly exposed to stress conditions. In turn, cell division, cell/organ development and cell transduction were the predominant enriched functions in profiles with higher expression in root vascular tissues (M1b in Fig. 3b; Table S2). Root tips contain a meristematic zone in an active state of cell division (Reddy & Chary, **2003**; Verbelen *et al.*, **2006**); root growth and development are controlled by cell signal transduction pathways (Perrin *et al.*, **2005**; Grieneisen *et al.*, **2007**; Laskowski *et al.*, **2008**).

Coexpression groups with opposite profiles within an expression module may have contrasting GO term enrichments (Figs 2a, S2). For example, the group with strong expression in secondary xylem tissues (M2a) had higher representation of terms related to cell/organ development than the group with low expression in secondary xylem (M2b). The opposite was observed for the terms related to carbohydrate/lipid reserves and defence mechanisms. Coexpression groups with high (M7a) and low (M7b) expression in root tips and xylem were enriched for secondary metabolic processes and for basic cellular mechanisms, respectively; and both were enriched for stress responsive genes.

Considering these observations, enriched GO terms may represent indirect information on the biological function of coexpressed genes which lack GO terms or functional annotations (i.e. guilt by association). Genes highly coexpressed have a greater chance of sharing a common regulatory mechanism and therefore have similar functions (Allocco *et al.*, **2004**). In this regard, a few of the coexpression profiles had a relatively low proportion of genes with associated GO terms (e.g. M3b and M11b) where < 25% of genes with assigned GO terms were associated with enriched GO terms. Genes without predicted functions are numerous in conifers (Rigault *et al.*, **2011**) and represent significant numbers even in model organisms. Coexpression groups may help not only to functionally annotate these genes but also to improve inference of gene function across species. According to Prasad *et al.* (**2013**), an analysis combining tissue expression profiles with sequence similarity provides a better prediction of functions of homologous genes across species than using sequence information alone.

We identified 4805 genes that did not vary across tissues (invariant genes) and thus could be designated as housekeeping genes according to Chang *et al.* (**2011**) and Eisenberg & Levanon (**2013**). The number of putative housekeeping genes decreases to 1201 when using expression level criteria (moderate to high expression genes; Table 2) defined by Aceituno *et al.* (**2008**) in *A. thaliana*. We found about the same number of housekeeping genes in white spruce as reported for humans (Chang *et al.*, **2011**) and three-fold more than reported for *A. thaliana* (Aceituno *et al.*, **2008**), both of which studies considered many more tissue types and biological conditions. As relevant for housekeeping functions, the majority of invariant genes had a biological function related to basic cellular activities, in agreement with previous reports (Chang *et al.*, **2011**; Eisenberg & Levanon, **2013**). Based on sequence similarity analysis, we also found that most of the highly expressed invariant genes were conserved between *A. thaliana* and white spruce (Table 2). This finding is consistent with previous reports indicating high expression of the majority of genes conserved between closely (mouse and human; Liao & Zhang, **2006**) or distantly related species (mouse and fruit fly; Tamura *et al.*, **2004**). Highly expressed genes were shown to have lower evolutionary rates of change in protein sequences (Tamura *et al.*, **2004**; Liao & Zhang, **2006**) and to be under stronger selection for mRNA folding (Park *et al.*, **2013**).

## Temporal Reorganization of Gene Expression Networks in the Vascular Transcriptome

Network analysis identified *PgNAC-7* as a network hub that is positively connected to several genes involved in secondary cell wall formation. These findings are consistent with the role of *PgNAC-7*and downstream MYB genes (*PgMYB1*, *-2*, *-4* and *-8*) which have been shown in white spruce to act as secondary cell wall regulators in transient assays (Duval *et al.*, **2014**) and in transgenic trees (Bomal *et al.*, **2008**). This network is functionally similar to that regulated by class IIB NAC proteins VND4, VND5 and VND6 of *A. thaliana* (Duval *et al.*, **2014**) and several downstream genes of MYBs (Zhong & Ye, **2009**; Zhong *et al.*, **2010**), which is also conserved in woody angiosperms including poplar (*Populus trichocarpa*) and eucalyptus (*Eucalyptus gunnii*) (Zhong & Ye, **2009**; Zhong *et al.*, **2010**; Hussey *et al.*, **2013**).

Temporal analysis in white spruce indicated that *PgNAC-7* and three of the spruce MYBs had expression profiles strongly related to earlywood, while no temporal variation was observed for other MYBs and most of the downstream genes encoding enzymes involved in secondary cell wall assembly. These observations indicate that the regulatory network may be reconfigured over the course of a growth season. A scan of the latewood preferential cluster identified one other NAC sequence and four other MYBs but these sequences have not been tested functionally. In turn, they represent candidate genes to be investigated for involvement in the regulation of members of this network.

Remodelling of the regulatory network in latewood was supported by the finding that the most connected hubs are stress responsive genes, that is, *ADR1*, a tubby gene and a dehydrin gene.*ADR* genes encode NBS-LRR proteins involved in receptor-mediated signalling (Bonardi *et al.*, **2011**). The network also contained many other coregulated dehydrin genes. ADR1, tubby protein and dehydrin proteins enhance drought tolerance in plants (Chini *et al.*, **2004**; Puhakainen *et al.*, **2004**; Wardhan *et al.*, **2012**). Our findings are consistent with the transition to dormancy that is initiated during latewood formation and involves several cellular changes including shifts in osmotic processes (Paiva *et al.*, **2008**). Differences between earlywood and latewood formation in terms of biological function and transcriptional regulatory network indicate transcriptional changes associated with seasonal network reorganization during a growing season in conifers.

## CONCLUDING REMARKS

We classified expressed white spruce genes into coexpression groups that are indicative of the modular organization of the transcriptome. Our results showed that deeper transcriptome reorganization is associated with tissue differentiation than with developmental stages or environmental conditions, and that patterns are conserved between spruce species, as might be expected given the ancient evolutionary origins of tissue differentiation. Time-course analyses of wood formation indicated network reorganization and a transition from cell wall formation towards defence-related genes. This report extends previous work on the PiceaGeneExpress database (Raherison *et al.*, **2012**) by statistical modelling of transcriptome organization leading to network analysis. We have developed a framework of information for spruce genes and an approach that could be readily applied to any number of tree and other plant genomes. Combined with other types of information, coexpression groups may enable basic discoveries to be made relating to genome function, regulatory networks, biomarkers and candidate genes in addition to supporting the development of applications for breeding and the conservation of genetic diversity.

## Acknowledgements

We thank Dr M. Mazerolle (Centre for Forest Research) for his advice on the mixed ANOVA statistical test, Dr A. Séguin (Centre de Foresterie des Laurentides) for the tissue illustration in Fig. 1(a), and Dr K. Plourde and G. Germanos (Centre for Forest Research) for sample preparations. Funding was received by J.J.M. from Genome Canada and Genome Québec as a part of the SMarTForests project and from an NSERC Discovery grant.

## SUPPORTING INFORMATION

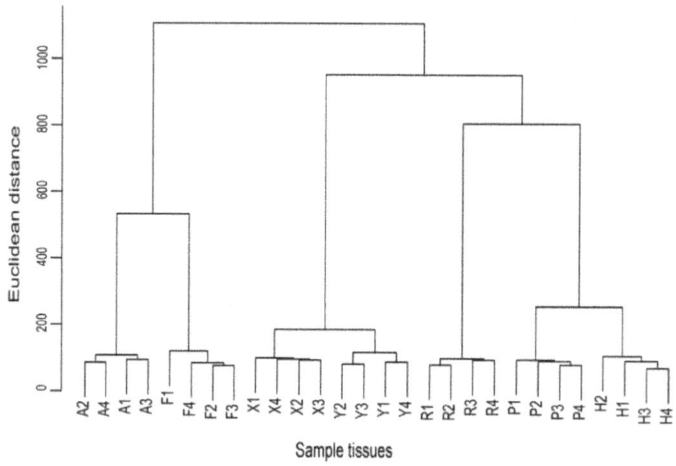


**Figure S1:** Hierarchical clustering of white spruce (Picea glauca) tissues based on variable genes. Analysis was performed using the hclust function of R with Ward's method (Ward, 1963) and microarray data ($log_2$ scale) of variable genes. A, shoot apex; F, young foliage; X, shoot secondary xylem; Y, root secondary xylem; R, root tips; P, shoot phelloderm; H, root phelloderm. Each tissue was replicated four times (1–4).

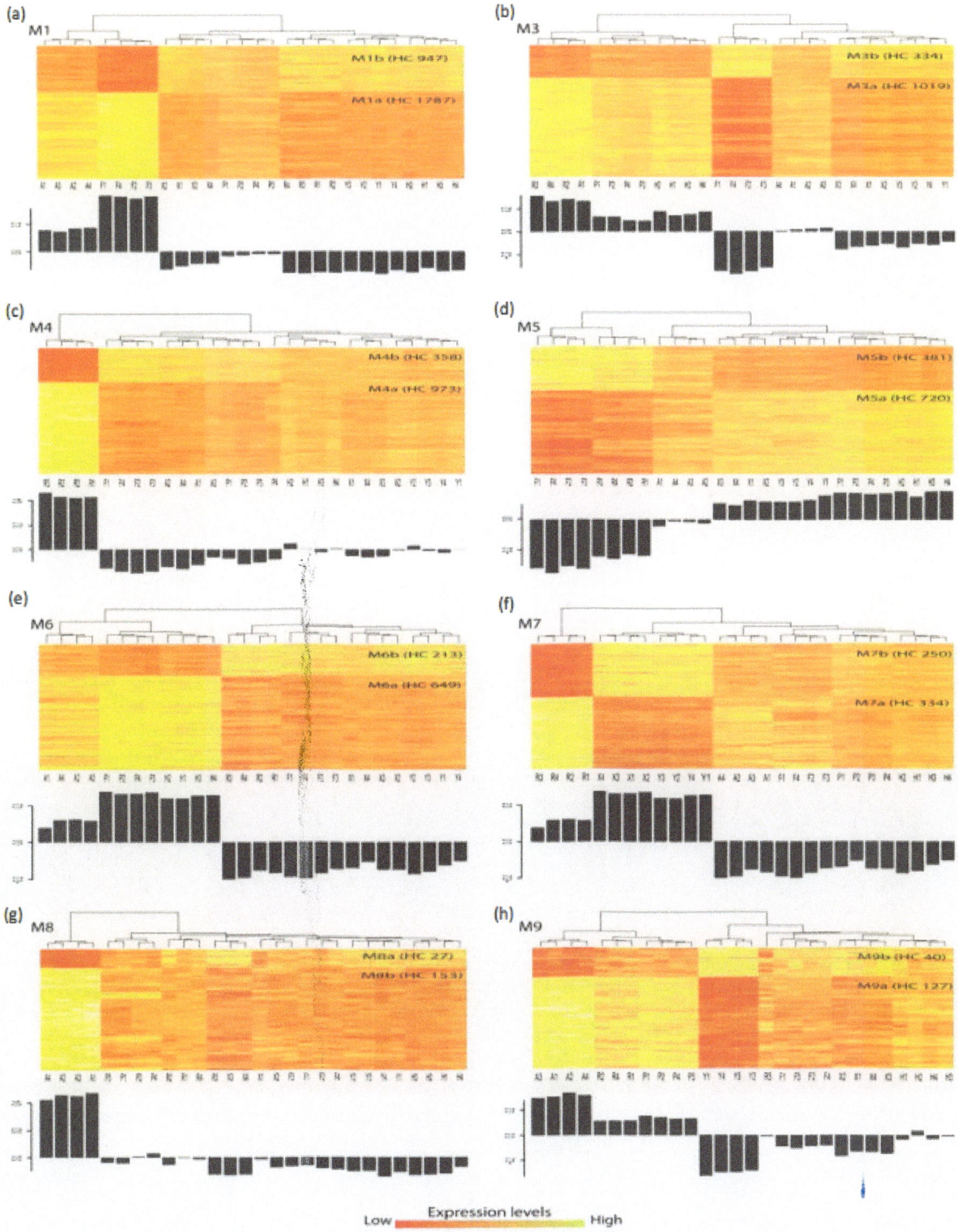


**Figure S2:** Representation of coexpression groups and their correlation to white spruce (Picea glauca) tissues. M1 and M3–M9, expression modules; M1a,b and M3a–M9b, coexpression groups. (a–h) Heatmaps show transcript abundance (log2 scale) of highconfidence (HC) variable genes in expression modules and coexpression groups determined with the WGCNA package of R. Number in parentheses, number of high-confidence variable genes in the coexpression group. Total and low-confidence variable (a) (b) (c) (d) (e) (f) (g) (h) gene numbers are listed in Table S2. Rows (y-axis), columns (x-axis), tissues (A, shoot apex; P, shoot phelloderm; H, root phelloderm; R, root tips; F, young foliage; X, shoot secondary xylem; Y, root secondary xylem). Each tissue was represented by four biological replicates (1–4). Bar plots display the module eigengenes (or the first principal components).

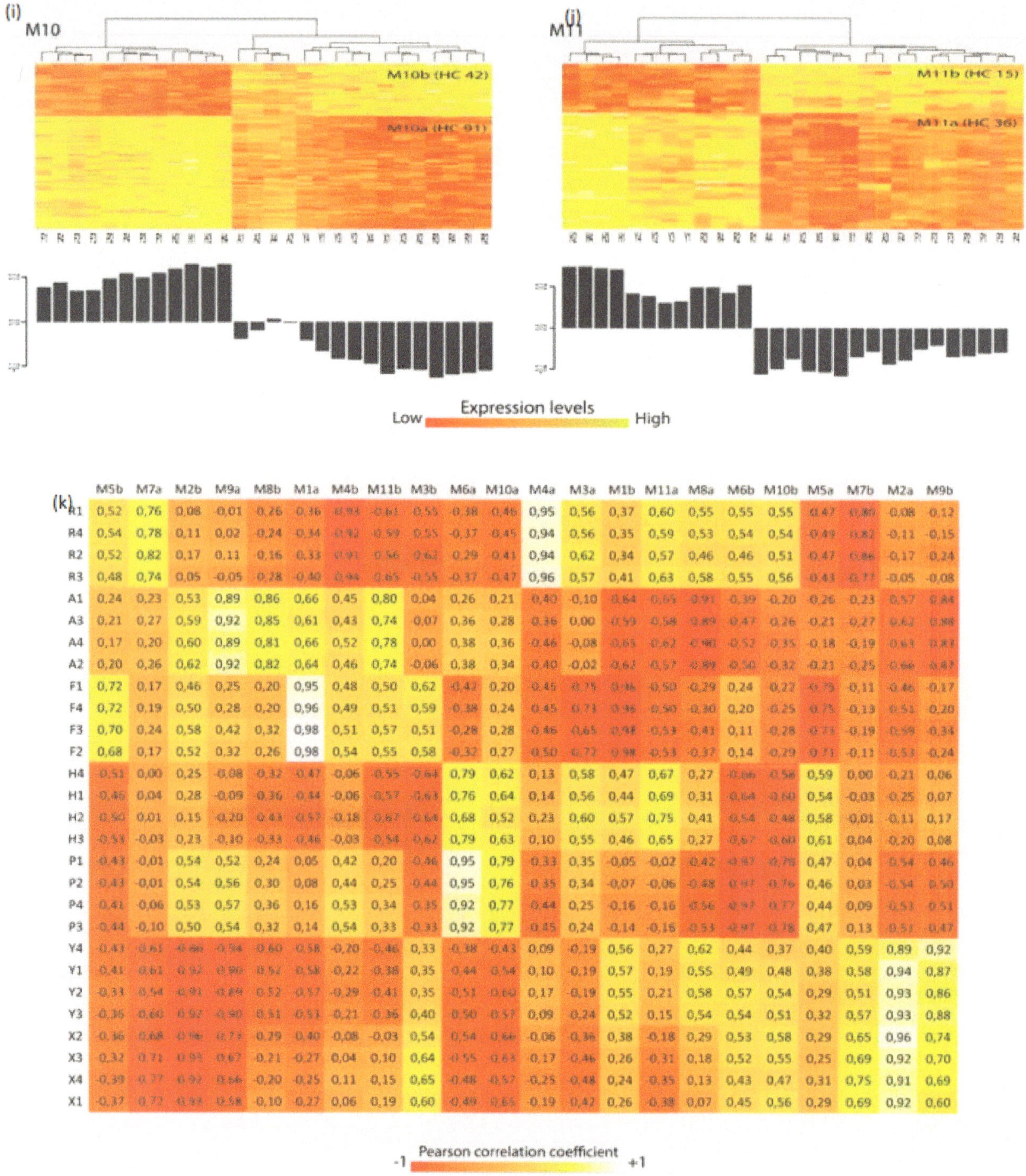


**Figure S2:** (continued) Representation of coexpression groups and their correlation to white spruce (Picea glauca) tissues. M10 and M11, expression modules; M10a–M11b, coexpression groups. (i, j) Heatmaps show transcript abundance (log2 scale) of highconfidence (HC) variable genes in expression modules and coexpression groups determined with the WGCNA package of R. Number in parentheses, number of highconfidence variable genes in the coexpression group. Total and low-confidence variable gene numbers are listed in Table S2. rows (y-axis), columns (x-axis), tissues (F, young foliage; P, shoot phelloderm; H, root phelloderm; A, shoot apex; Y, root secondary xylem; X, shoot secondary xylem; R, root tips). Each tissue was represented by four biological replicates (1–4). Bar plots display the module eigengenes (or the first principal components). (k) Heatmap shows the Pearson correlation coefficient between coexpression groups and tissue samples.

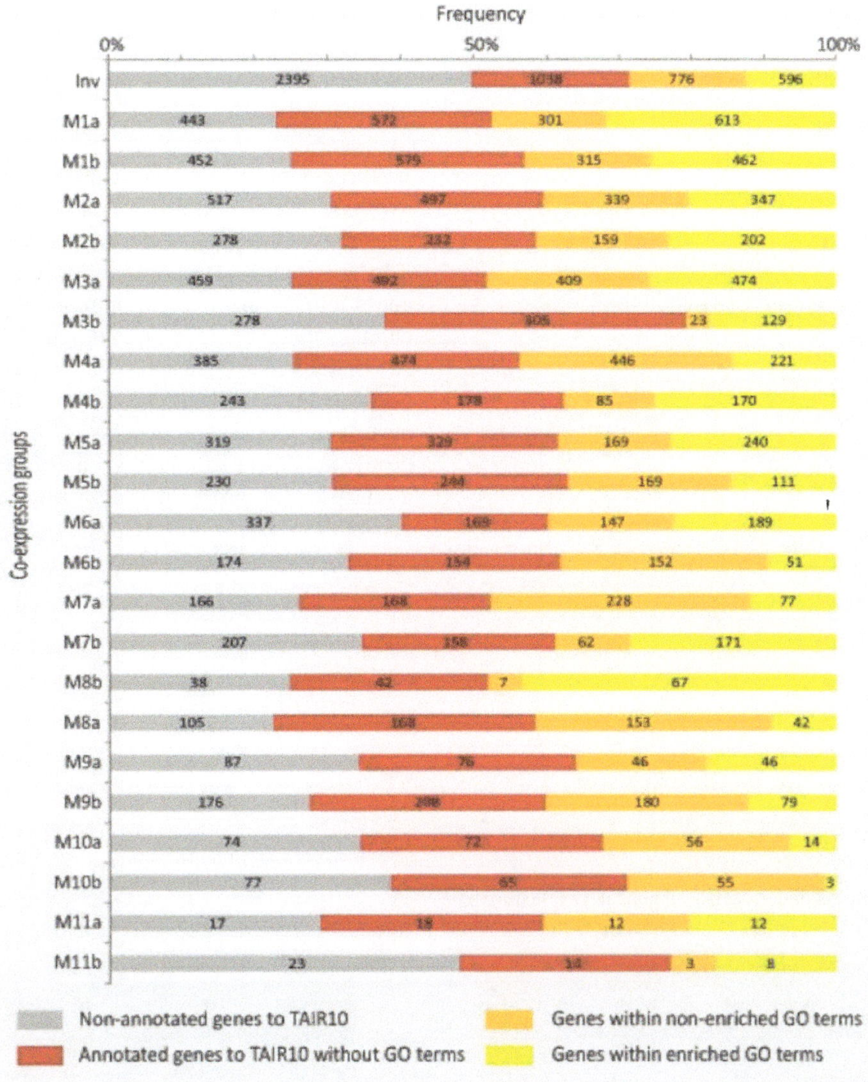


**Figure S3:** Functional annotation of invariant genes and coexpression groups. Inv, invariant genes; M1a–M11b, coexpression groups. x-axis, frequency (%) of genes that were associated to each of the four annotation categories: non-annotated genes to The Arabidopsis Information Resource version 10 (TAIR10, www.arabidopsis.org); annotated genes to TAIR10 but without gene biological process gene ontology (GO) terms; genes within enriched and non-enriched GO terms. Numbers inside each bar are the numbers of genes in each category. Enrichment test was performed using Database for Annotation, Visualization and Integrated Discovery (DAVID) tool (http://david.abcc.ncifcrf.gov, Huang et al., 2009).

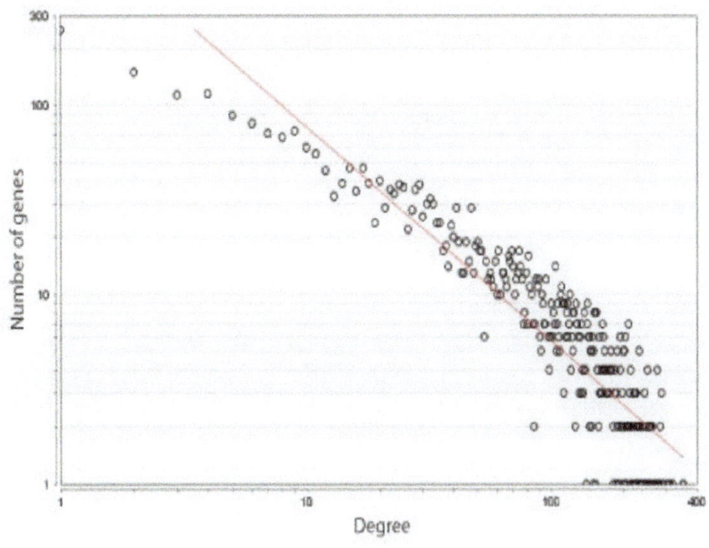


**Figure S4:** Gene degree distribution in a double log scale of the xylem (M2–7) preferential network. The M2–7 network had a scale-free topology since its gene degree distribution fitted to a power-law distribution (i.e. few hubs hold together numerous small-degree nodes; Barabási et al., 2001). Figure as output directly from the Cytoscape software (Assenov et al., 2008). June–July and September preferential networks were not scale free.

**Table S4:** Number of genes per coexpression group in white spruce (Picea glauca)

| Coexpression group | High-confidence gene | Low-confidence gene | Total |
|---|---|---|---|
| M1a | 1787 | 142 | 1929 |
| M1b | 947 | 861 | 1808 |
| M2a | 1418 | 282 | 1700 |
| M2b | 628 | 243 | 871 |
| M3a | 1019 | 815 | 1834 |
| M3b | 334 | 401 | 735 |
| M4a | 973 | 553 | 1526 |
| M4b | 358 | 318 | 676 |
| M5a | 720 | 333 | 1053 |
| M5b | 381 | 373 | 754 |
| M6a | 649 | 193 | 842 |
| M6b | 213 | 318 | 531 |
| M7a | 334 | 305 | 639 |
| M7b | 250 | 348 | 598 |
| M8a | 153 | 1 | 154 |
| M8b | 27 | 441 | 468 |
| M9a | 127 | 128 | 255 |
| M9b | 40 | 603 | 643 |
| M10a | 91 | 125 | 216 |
| M10b | 42 | 158 | 200 |
| M11a | 36 | 23 | 59 |
| M11b | 15 | 33 | 48 |

**Table S7:** Members of the secondary cell wall gene network shown to be transactivated by PgNAC-7 or MYB transcription factors in white spruce (Picea glauca)

| GenBank Accession No. | Cluster ID[1] | Gene name[2] | Function | Trans activation[3] | PgNAC-7 (R>0.9)[4] | PgMyb8 (R>0.9)[4] | Coexpression group | Temporal cluster[5] | Preferential expression in spruces[6] |
|---|---|---|---|---|---|---|---|---|---|
| BT102049 | GQ0165_B14 | PgNAC-7 | NAC-domain transcription factor | | | x | M2a | T2a | Xylem |
| BT108136 | GQ03117_E18 | PgMYB8 | myb domain protein | *a, b* | x | | M2a | None | Xylem |
| BT108631 | GQ03124_J09 | PgMYB1 | myb domain protein | *a, b* | | | M3b | T2a | Xylem |
| BT108414 | GQ03121_H22 | PgMYB2 | myb domain protein | *b* | x | | M2a | T2a | Xylem |
| BT107883 | GQ03113_N22 | PgMYB4 | myb domain protein | *b* | x | | M2a | T2a | Xylem |
| BT116706 | GQ03805_C07 | PgDHS-2 | 3-deoxy-d-arabino-heptulosonate 7-phosphate synthase | *a, b* | x | | M2a | None | Xylem |
| BT115945 | GQ03709_D11 | PgCM-3 | Chorismate mutase-1 | *b* | x | | M2a | None | Xylem |
| BT112280 | GQ03312_O11 | PgCAD | cinnamyl alcohol dehydrogenase | *a, b* | x | x | M2a | None | Xylem |
| BT106671 | GQ03007_K12 | Pg4CL | 4-coumarate:CoA ligase | *a, b* | | | M2a | None | Xylem |
| BT111802 | GQ03303_N02 | PgCCR-1 | cinnamoyl coa reductase | *b* | x | | M2a | None | Xylem |
| BT111350 | GQ03236_G10 | PgPRR-2 | pinoresinol reductase | *b* | x | x | M2a | None | Xylem |
| BT116976 | GQ03810_K09 | PgCesA-3 | Cellulose synthase family protein | *a, b* | x | | M2a | T1b | Xylem |
| BT106155 | GQ02904_A20 | PgTUA-1 | tubulin alpha-4 chain | *a, b* | | | M6a | None | None |
| BT116258 | GQ03714_M12 | PgXTH8-1 | xyloglucan endotransglucosylase / hydrolase | *a, b* | | | M4a | T1b | Phelloderm |

**Table S8:** Distribution of detected genes across invariant gene and coexpression groups of white spruce (Picea glauca)

WGCNA is an R package that can be briefly described as follow. A pair-wise Pearson correlation matrix was computed and then transformed into an adjacency matrix using a power function: adjacency value of two genes = $|\text{correlation between two genes}|^{\beta}$, where $\beta$ is the weight. The adjacency matrix was transformed into a distance matrix using the topological overlap measure (Li & Horvath, 2007). A hierarchical clustering analysis was applied to the dissimilarity matrix (1 - distance matrix). Dendrogram leaves were cut and separated genes into distinct expression modules using the dynamic tree cut algorithm (Langfelder et al., 2008). An expression module can be defined as a cluster of highly (positively and negatively) correlated genes (Langfelder & Horvath, 2008). It reveals two negatively correlated expression profiles. Each expression profile is associated with a group of genes that we referred to a coexpression group in experiment 1 or temporally variable cluster in experiment 2. Below is the WGCNA script we used:

| | Foliage | Shoot apex | Shoot phelloderm | Shoot xylem | Root phelloderm | Root xylem | Root tips |
|---|---|---|---|---|---|---|---|
| Inv | 4554 | 4414 | 4638 | 4511 | 4606 | 4375 | 4492 |
| M1a | 1928 | 1925 | 1908 | 1882 | 1892 | 1818 | 1840 |
| M1b | 1797 | 1791 | 1807 | 1808 | 1807 | 1806 | 1806 |
| M2a | 1675 | 1671 | 1686 | 1700 | 1685 | 1696 | 1661 |
| M2b | 871 | 869 | 868 | 849 | 868 | 838 | 860 |
| M3a | 1807 | 1824 | 1834 | 1816 | 1833 | 1810 | 1831 |
| M3b | 734 | 733 | 730 | 732 | 726 | 728 | 715 |
| M4a | 1469 | 1472 | 1492 | 1472 | 1512 | 1471 | 1524 |
| M4b | 675 | 675 | 672 | 669 | 672 | 661 | 654 |
| M5a | 1042 | 1049 | 1052 | 1051 | 1053 | 1048 | 1032 |
| M5b | 754 | 748 | 743 | 733 | 738 | 725 | 749 |
| M6a | 831 | 839 | 842 | 830 | 842 | 819 | 822 |
| M6b | 524 | 519 | 526 | 521 | 522 | 523 | 530 |
| M7a | 623 | 632 | 631 | 609 | 633 | 594 | 636 |
| M7b | 591 | 591 | 596 | 598 | 597 | 597 | 584 |
| M8a | 139 | 154 | 145 | 136 | 134 | 105 | 124 |
| M8b | 467 | 465 | 467 | 467 | 467 | 467 | 468 |
| M9a | 249 | 255 | 254 | 247 | 254 | 237 | 252 |
| M9b | 637 | 633 | 641 | 641 | 640 | 642 | 634 |
| M10a | 215 | 212 | 216 | 210 | 216 | 210 | 210 |
| M10b | 196 | 196 | 196 | 198 | 196 | 191 | 199 |
| M11a | 57 | 54 | 58 | 53 | 59 | 57 | 59 |
| M11b | 48 | 48 | 48 | 48 | 47 | 46 | 46 |
| Unclassified | 500 | 502 | 506 | 496 | 497 | 487 | 492 |
| Tested_genes | 22383 | 22271 | 22556 | 22277 | 22496 | 21951 | 22220 |
| Detected_genes | 23104 | 22951 | 23296 | 22970 | 23225 | 22608 | 22913 |

Inv, invariant genes; M1a–M11b, coexpression groups; unclassified, variable genes that could not be classified to any coexpression groups using WGCNA or template matching methods; tested_genes, detected gene that met assumptions for ANOVA testing (for details, see the Materials and Methods section); detected_genes, genes whose transcript levels were detected above background levels in the tissue.

**Methods S1:** Weighted correlation network analysis (WGCNA) script.

```
# Loading expression data
# Expression data is a matrix with genes in rows and tissues in columns
getwd();
workingDirectory="C:/Users/...";
setwd(workingDirectory);
library(WGCNA);
options(stringsAsFactors = FALSE);
datExp=read.table("ExpressionData.txt")

# Transposing data and removing unnecessary data
datExpr0 = as.data.frame(t(datExp[,]));
names(datExpr0)=datExpr0[1,]
datExpr0=datExpr0[-1,]

# Detecting outlier samples with hierarchical clustering analysis
# no sample of experiments 1 and 3 was removed after the outlier verification
sampleTree = flashClust(dist(datExpr0), method = "average");
plot(sampleTree)

# Choosing the soft-thresholding power: analysis of network topology
powers = c(c(1:10), seq(from = 12, to=20, by=2));
sft = pickSoftThreshold(datExpr0, powerVector = powers, verbose = 5);
plot(sft$fitIndices[,1], -sign(sft$fitIndices[,3])*sft$fitIndices[,2],
xlab="Soft Threshold (power)", ylab="Scale Free Topology Model Fit,
signed R^2",type="n", main = paste("Scale independence"));
text(sft$fitIndices[,1], -sign(sft$fitIndices[,3])*sft$fitIndices[,2],
labels=powers,cex=cex1,col="red");

abline(h=0.86,col="red")
plot(sft$fitIndices[,1], sft$fitIndices[,5],
xlab="Soft Threshold (power)",ylab="Mean Connectivity", type="n",
main = paste("Mean connectivity"))
text(sft$fitIndices[,1], sft$fitIndices[,5], labels=powers, cex=cex1,col="red")


# Identifying expression modules
# We used power = 14 and 18 to identify modules across tissues (Experiment 1)
# and times (Experiment 3), respectively
net = blockwiseModules(datExpr0, power = 14, minModuleSize = 30,
maxBlockSize=12000, reassignThreshold = 0, mergeCutHeight = 0.25,
numericLabels = TRUE, pamRespectsDendro = FALSE, saveTOMs = TRUE,
saveTOMFileBase = "TOM1", verbose = 3)

# Calculating eigengene of modules
MEs = moduleEigengenes(datExpr0, moduleColors)$eigengenes
MET =orderMEs(MEs)
```

## REFERENCES


1. Aceituno FF, Moseyko N, Rhee SY, Gutiérrez RA. 2008. The rules of gene expression in plants: organ identity and gene body methylation are key factors for regulation of gene expression in *Arabidopsis thaliana*. *BMC Genomics* **9**: 438.
2. Allocco DJ, Kohane IS, Butte AJ. 2004. Quantifying the relationship between co-expression, co-regulation and gene function. *BMC Bioinformatics* **5**: 18.
3. Allona I, Quinn M, Shoop E, Swope K, Cyr SS, Carlis J, Riedl J, Retzel E, Campbell MM, Sederoff R. 1998. Analysis of xylem formation in pine by cDNA sequencing. *Proceedings of the National Academy of Sciences, USA* **95**:9693–9698.
4. Arabidopsis Genome Initiative. 2000. Analysis of the genome sequence of the flowering plant *Arabidopsis thaliana*. *Nature* **408**: 796.
5. Baldi P, Pedron L, Hietala AM, La Porta N. 2011. Cold tolerance in cypress (*Cupressus sempervirens* L.): a physiological and molecular study. *Tree Genetics & Genomes* **7**: 79–90.
6. Benjamini Y, Hochberg Y. 1995. Controlling the false discovery rate: a practical and powerful approach to multiple testing. *Journal of the Royal Statistical Society Series B (Methodological)* **57**: 289–300.
7. Birol I, Raymond A, Jackman SD, Pleasance S, Coope R, Taylor GA, Saint Yuen MM, Keeling CI, Brand D,Vandervalk BP. 2013. Assembling the 20 Gb white spruce (*Picea glauca*) genome from whole-genome shotgun sequencing data. *Bioinformatics* **29**: 1492–1497.
8. Bomal C, Bedon F, Caron S, Mansfield SD, Levasseur C, Cooke JE, Blais S, Tremblay L, Morency M-J, Pavy N.2008. Involvement of *Pinus taeda* MYB1 and MYB8 in phenylpropanoid metabolism and secondary cell wall biogenesis: a comparative *in planta* analysis. *Journal of Experimental Botany* **59**: 3925–3939.
9. Bonardi V, Tang S, Stallmann A, Roberts M, Cherkis K, Dangl JL. 2011. Expanded functions for a family of plant intracellular immune receptors beyond specific recognition of pathogen effectors. *Proceedings of the National Academy of Sciences, USA* **108**: 16463–16468.
10. Boyle B, Dallaire N, MacKay J. 2009. Evaluation of the impact of single nucleotide polymorphisms and primer mismatches on quantitative PCR. *BMC Biotechnology* **9**: 75.
11. Brawand D, Soumillon M, Necsulea A, Julien P, Csárdi G, Harrigan P, Weier M, Liechti A, Aximu-Petri A, Kircher M. 2011. The evolution of gene expression levels in mammalian organs. *Nature* **478**: 343–348.
12. Chan ET, Quon GT, Chua G, Babak T, Trochesset M, Zirngibl RA, Aubin J, Ratcliffe M, Wilde A, Brudno M. 2009.Conservation of core gene expression in vertebrate tissues. *Journal of Biology* **8**: 33.
13. Chang C-W, Cheng W-C, Chen C-R, Shu W-Y, Tsai M-L, Huang C-L, Hsu IC. 2011. Identification of human housekeeping genes and tissue-selective genes by microarray meta-analysis. *PLoS ONE* **6**: e22859.

14. Chang S, Puryear J, Cairney J. 1993. A simple and efficient method for isolating RNA from pine trees. *Plant Molecular Biology Reporter* **11**: 113–116.

15. Chini A, Grant JJ, Seki M, Shinozaki K, Loake GJ. 2004. Drought tolerance established by enhanced expression of the *CC–NBS–LRR* gene, *ADR1*, requires salicylic acid, EDS1 and ABI1. *Plant Journal* **38**: 810–822.

16. Downs GS, Bi Y-M, Colasanti J, Wu W, Chen X, Zhu T, Rothstein SJ, Lukens L. 2013. A developmental transcriptional network for *Zea mays* defines coexpression modules. *Plant Physiology* **161**: 1830–1843.

17. Druka A, Muehlbauer G, Druka I, Caldo R, Baumann U, Rostoks N, Schreiber A, Wise R, Close T, Kleinhofs A.2006. An atlas of gene expression from seed to seed through barley development. *Functional & Integrative Genomics* **6**: 202–211.

18. Duval I, Lachance D, Giguère I, Bomal C, Morency M-J, Pelletier G, Boyle B, MacKay JJ, Séguin A. 2014. Large-scale screening of transcription factor–promoter interactions in spruce reveals a transcriptional network involved in vascular development. *Journal of Experimental Botany* **65**: 2319–2333.

19. Edwards KD, Bombarely A, Story GW, Allen F, Mueller LA, Coates SA, Jones L. 2010. TobEA: an atlas of tobacco gene expression from seed to senescence. *BMC Genomics* **11**: 142.

20. Eisenberg E, Levanon EY. 2013. Human housekeeping genes, revisited. *Trends in Genetics* **29**: 569–574.

21. Ellison CE, Kowbel D, Glass NL, Taylor JW, Brem RB. 2014. Discovering functions of unannotated genes from a transcriptome survey of wild fungal isolates. *mBio* **5**: e01046–01013.

22. Grieneisen VA, Xu J, Marée AF, Hogeweg P, Scheres B. 2007. Auxin transport is sufficient to generate a maximum and gradient guiding root growth. *Nature* **449**: 1008–1013.

23. Harris LW, Lockstone HE, Khaitovich P, Weickert CS, Webster MJ, Bahn S. 2009. Gene expression in the prefrontal cortex during adolescence: implications for the onset of schizophrenia. *BMC Medical Genomics* **2**: 28.

24. Hu Z-L, Bao J, Reecy J. 2008. CateGOrizer: a web-based program to batch analyze gene ontology classification categories. *Online Journal of Bioinformatics* **9**: 108–112.

25. Huang DW, Sherman BT, Lempicki RA. 2009. Bioinformatics enrichment tools: paths toward the comprehensive functional analysis of large gene lists. *Nucleic Acids Research* **37**: 1–13.

26. Hussey SG, Mizrachi E, Creux NM, Myburg AA. 2013. Navigating the transcriptional roadmap regulating plant secondary cell wall deposition. *Frontiers in Plant Science* **4**: 325.

27. IHGSC. 2004. Finishing the euchromatic sequence of the human genome. *Nature* **431**: 931–945.

28. Jordy MN. 2004. Seasonal variation of organogenetic activity and reserves allocation in the shoot apex of *Pinus pinaster* Ait. *Annals of Botany* **93**: 25–37.

29. Kenrick P, Crane PR. 1997. The origin and early diversification of land plants. A cladistic study. Washington DC, USA: Smithsonian Institute Press.

30. Kibbe WA. 2007. OligoCalc: an online oligonucleotide properties calculator. *Nucleic Acids Research* **35**:W43–W46.

31. Kwon E-Y, Shin S-K, Cho Y-Y, Jung UJ, Kim E, Park T, Park JHY, Yun JW, McGregor RA, Park YB. 2012. Time-course microarrays reveal early activation of the immune transcriptome and adipokine dysregulation leads to fibrosis in visceral adipose depots during diet-induced obesity. *BMC Genomics* **13**: 450.

32. Langfelder P, Horvath S. 2008. WGCNA: an R package for weighted correlation network analysis. *BMC Bioinformatics* **9**: 559.

33. Laskowski M, Grieneisen VA, Hofhuis H, Colette A, Hogeweg P, Marée AF, Scheres B. 2008. Root system architecture from coupling cell shape to auxin transport. *PLoS Biology* **6**: e307.

34. Li X, Wu HX, Southerton SG. 2010. Seasonal reorganization of the xylem transcriptome at different tree ages reveals novel insights into wood formation in *Pinus radiata*. *New Phytologist* **187**: 764–776.

35. Liao B-Y, Zhang J. 2006. Evolutionary conservation of expression profiles between human and mouse orthologous genes. *Molecular Biology And Evolution* **23**: 530–540.

36. Lorenz WW, Alba R, Yu Y-S, Bordeaux JM, Simões M, Dean JF. 2011. Microarray analysis and scale-free gene networks identify candidate regulators in drought-stressed roots of loblolly pine (*P. taeda* L.). *BMC Genomics***12**: 264.

37. Ma L, Sun N, Liu X, Jiao Y, Zhao H, Deng XW. 2005. Organ-specific expression of *Arabidopsis* genome during development. *Plant Physiology* **138**: 80–91.

38. MacKay J, Dean JF, Plomion C, Peterson DG, Cánovas FM, Pavy N, Ingvarsson PK, Savolainen O, Guevara MA,Fluch S. 2012. Towards decoding the conifer giga-genome. *Plant Molecular Biology* **80**: 555–569.

39. Mizrachi E, Mansfield SD, Myburg AA. 2012. Cellulose factories: advancing bioenergy production from forest trees. *New Phytologist* **194**: 54–62.

40. Myburg AA, Grattapaglia D, Tuskan GA, Hellsten U, Hayes RD, Grimwood J, Jenkins J, Lindquist E, Tice H, Bauer D. 2014. The genome of *Eucalyptus grandis*. *Nature* **510**: 356–362.

41. Neale DB, Wegrzyn JL, Stevens KA, Zimin AV, Puiu D, Crepeau MW, Cardeno C, Koriabine M, Holtz-Morris AE,Liechty JD. 2014. Decoding the massive genome of loblolly pine using haploid DNA and novel assembly strategies. *Genome Biology* **15**: R59.

42. Nystedt B, Street NR, Wetterbom A, Zuccolo A, Lin Y-C, Scofield DG, Vezzi F, Delhomme N, Giacomello S,Alexeyenko A. 2013. The Norway spruce genome sequence and conifer genome evolution. *Nature* **497**:579–584.

43. Paiva J, Garnier-Géré P, Rodrigues J, Alves A, Santos

S, Graça J, Le Provost G, Chaumeil P, Silva-Perez D, Bosc A.2008. Plasticity of maritime pine (Pinus pinaster) wood-forming tissues during a growing season. *New Phytologist* **179**: 1180–1194.

44. Palle SR, Seeve CM, Eckert AJ, Cumbie WP, Goldfarb B, Loopstra CA. 2011. Natural variation in expression of genes involved in xylem development in loblolly pine (*Pinus taeda* L.). *Tree Genetics & Genomes* 7: 193–206.

45. Park C, Chen X, Yang J-R, Zhang J. 2013. Differential requirements for mRNA folding partially explain why highly expressed proteins evolve slowly. *Proceedings of the National Academy of Sciences, USA* **110**: E678–E686.

46. Pavy N, Boyle B, Nelson C, Paule C, Giguère I, Caron S, Parsons LS, Dallaire N, Bedon F, Bérubé H. 2008. Identification of conserved core xylem gene sets: conifer cDNA microarray development, transcript profiling and computational analyses. *New Phytologist* **180**: 766–786.

47. Pavy N, Gagnon F, Rigault P, Blais S, Deschênes A, Boyle B, Pelgas B, Deslauriers M, Clément S, Lavigne P. 2013. Development of high-density SNP genotyping arrays for white spruce (Picea glauca) and transferability to subtropical and nordic congeners. *Molecular Ecology Resources* **13**: 324–336.

48. Perrin RM, Young L-S, Murthy UN, Harrison BR, Wang Y, Will JL, Masson PH. 2005. Gravity signal transduction in primary roots. *Annals of Botany* **96**: 737–743.

49. Pickrell JK, Marioni JC, Pai AA, Degner JF, Engelhardt BE, Nkadori E, Veyrieras J-B, Stephens M, Gilad Y, Pritchard JK. 2010. Understanding mechanisms underlying human gene expression variation with RNA sequencing.*Nature* **464**: 768–772.

50. Pommerenke C, Wilk E, Srivastava B, Schulze A, Novoselova N, Geffers R, Schughart K. 2012. Global transcriptome analysis in influenza-infected mouse lungs reveals the kinetics of innate and adaptive host immune responses. *PLoS ONE* 7: e41169.

51. Prasad A, Kumar SS, Dessimoz C, Jaquet V, Bleuler S, Laule O, Hruz T, Gruissem W, Zimmermann P. 2013. Global regulatory architecture of human, mouse and rat tissue transcriptomes. *BMC Genomics* **14**: 716.

52. Puhakainen T, Hess MW, Mäkelä P, Svensson J, Heino P, Palva ET. 2004. Overexpression of multiple dehydrin genes enhances tolerance to freezing stress in *Arabidopsis*. *Plant Molecular Biology* **54**: 743–753.

53. Raherison E, Rigault P, Caron S, Poulin P-L, Boyle B, Verta J-P, Giguère I, Bomal C, Bohlmann J, MacKay J. 2012.Transcriptome profiling in conifers and the PiceaGenExpress database show patterns of diversification within gene families and interspecific conservation in vascular gene expression. *BMC Genomics* **13**: 434.

54. Reddy S, Chary S. 2003. University botany II:(gymnosperms, plant anatomy, genetics, ecology). New Delhi, India:New Age International (P) Ltd.

55. Rigault P, Boyle B, Lepage P, Cooke JE, Bousquet J, MacKay JJ. 2011. A white spruce gene catalog for conifer genome analyses. *Plant Physiology* **157**: 14–28.

56. Ritland K, Krutovsky KV, Tsumura Y, Pelgas B, Isabel N, Bousquet J. 2011. Genetic mapping in conifers. In:Plomion C, Bousquet J, Kole C, eds. Genetics, genomics and breeding of conifers. New York, NY, USA: Edenbridge Science Publishers & CRC Press, 196–238.

57. Romero IG, Ruvinsky I, Gilad Y. 2012. Comparative studies of gene expression and the evolution of gene regulation. *Nature Reviews Genetics* **13**: 505–516.

58. Rutledge RG, Stewart D. 2008. A kinetic-based sigmoidal model for the polymerase chain reaction and its application to high-capacity absolute quantitative real-time PCR. *BMC Biotechnology* **8**: 47.

59. Schnable PS, Ware D, Fulton RS, Stein JC, Wei F, Pasternak S, Liang C, Zhang J, Fulton L, Graves TA. 2009. The B73 maize genome: complexity, diversity, and dynamics. *Science* **326**: 1112–1115.

60. Su AI, Cooke MP, Ching KA, Hakak Y, Walker JR, Wiltshire T, Orth AP, Vega RG, Sapinoso LM, Moqrich A. 2002.Large-scale analysis of the human and mouse transcriptomes. *Proceedings of the National Academy of Sciences, USA* **99**: 4465–4470.

61. Sun H, Paulin L, Alatalo E, Asiegbu FO. 2011. Response of living tissues of *Pinus sylvestris* to the saprotrophic biocontrol fungus *Phlebiopsis gigantea*. *Tree physiology* **31**: 438–451.

62. Tamura K, Subramanian S, Kumar S. 2004. Temporal patterns of fruit fly (*Drosophila*) evolution revealed by mutation clocks. *Molecular Biology And Evolution* **21**: 36–44.

63. Tuskan GA, Difazio S, Jansson S, Bohlmann J, Grigoriev I, Hellsten U, Putnam N, Ralph S, Rombauts S, Salamov A. 2006. The genome of black cottonwood, *Populus trichocarpa* (Torr. & Gray). *Science* **313**: 1596–1604.

64. Untergasser A, Nijveen H, Rao X, Bisseling T, Geurts R, Leunissen JA. 2007. Primer3Plus, an enhanced web interface to Primer3. *Nucleic Acids Research* **35**: W71–W74.

65. Verbelen J-P, De Cnodder T, Le J, Vissenberg K, Baluška F. 2006. The root apex of *Arabidopsis thaliana* consists of four distinct zones of growth activities: meristematic zone, transition zone, fast elongation zone and growth terminating zone. *Plant Signaling & Behavior* **1**: 296.

66. Verne S, Jaquish B, White R, Ritland C, Ritland K. 2011. Global transcriptome analysis of constitutive resistance to the white pine weevil in spruce. *Genome Biology and Evolution* **3**: 851–867.

67. Ward JH. 1963. Hierarchical grouping to optimize an objective function. *Journal of the American Statistical Association* **58**: 236–244.

68. Wardhan V, Jahan K, Gupta S, Chennareddy S, Datta A, Chakraborty S, Chakraborty N. 2012. Overexpression of CaTLP1, a putative transcription factor in chickpea (*Cicer arietinum* L.), promotes stress tolerance. *Plant Molecular Biology* **79**: 479–493.

69. Yang R, Wang X. 2013. Organ evolution in angiosperms driven by correlated divergences of gene sequences and expression patterns. *The Plant Cell* **25**: 71–82.

70. Yang S-H, Loopstra CA. 2005. Seasonal variation in gene

expression for loblolly pines (*Pinus taeda*) from different geographical regions. *Tree Physiology* **25**: 1063–1073.

71. Yeaman S, Hodgins KA, Suren H, Nurkowski KA, Rieseberg LH, Holliday JA, Aitken SN. 2014. Conservation and divergence of gene expression plasticity following *c*. 140 million years of evolution in lodgepole pine (*Pinus contorta*) and interior spruce (*Picea glauca* × *Picea engelmannii*). *New Phytologist* **203**: 578–591.

72. Zhong R, Lee C, Ye ZH. 2010. Evolutionary conservation of the transcriptional network regulating secondary cell wall biosynthesis. *Trends in Plant Science* **15**: 625–632.

73. Zhong R, Ye ZH. 2009. Transcriptional regulation of lignin biosynthesis. *Plant Signaling & Behavior* **4**: 1028–1034.

CHAPTER 9

# FOSSIL GINKGOPHYTE SEEDLINGS FROM THE TRIASSIC OF FRANCE RESEMBLE MODERN GINKGO BILOBA

Kathleen Bauer, Lea Grauvogel-Stamm, Evelyn Kustatscher and Michael Krings

**Keywords:** Ontogeny, Germination, Cotyledon, Ginkgophyta, Gymnosperms, Middle Triassic

## BACKGROUND

Fossil evidence of gymnosperm ontogeny is exceedingly rare and mostly limited to isolated specimens of embryos and/or seedlings [1]. For example, structurally preserved araucariaceous embryos *in situ* have been described from the Jurassic of Patagonia [2, 3], and impression fossils of araucariaceous seedlings are known from the Jurassic of Brazil [4]. Compressed seedlings attributed to the herbaceous conifer *Aethophyllum stipulare* have also been reported from the Middle Triassic Voltzia Sandstone of France [5], and a conifer embryo with cotyledons has been described from the Upper Pennsylvanian or Lower Permian of North America [6]. Finally, Rydin et al. [7] and Dilcher et al. [8] report on Welwitschiaceae seedlings from the Cretaceous of Brazil.

The ginkgophytes (Ginkgophyta) are a distinctive lineage of gymnosperms that is hypothesized as originating in the late Palaeozoic [9]. The group is considered poorly diversified during the Permian, with a major radiation commencing in the Middle or Late Triassic [10, 11]. Today Ginkgophytes are represented by a single species, *Ginkgo biloba*, the maidenhair tree. The early ontogenetic development of *G. biloba* has been studied and documented in great detail [12, 13, 14, 15, 16]. Sprecher [13] defines four distinct ontogenetic stages in the early development of *G. biloba*, beginning with germination and ending with the development of the first fully differentiated leaves. To date, the early ontogenetic development of fossil representatives of the ginkgophytes has remained unknown due to a lack of fossils that could be used to reconstruct this process.

The occurrence of well-preserved fossil seedlings in the early Middle Triassic Voltzia Sandstone of the Vosges (northeastern France) has been noted by Grauvogel [17] and Grauvogel-Stamm & Grauvogel [18]. However, none of these fossils have been described in detail and compared to modern equivalents. In this paper, we describe an assemblage of 20 compression specimens of seedlings from a single bedding plane in the Voltzia Sandstone. The seedlings are assignable to the ginkgophytes based on leaf morphological features, and thus represent the first record of seedling development of a fossil ginkgophyte. What makes these fossils even more significant is that the specimens span a number of different ontogenetic stages thus making it possible to compare several stages of seedling development directly with those of the modern taxon.

## RESULTS

The assemblage consists of 20 specimens, 18 of which are seedlings in different stages of development (Table 1).

**Citation:** Kathleen Bauer, Lea Grauvogel-Stamm, Evelyn Kustatscher and Michael Krings, Fossil Ginkgophyte Seedlings from the Triassic of France Resemble Modern Ginkgo Biloba, DOI: 10.1186/1471-2148-13-177.

**Table 1:** Measurements of seeds and seedlings

| Specimen | Seed | Cotyledons | Primary leaves | Number of bifurcations | Taproot | "True" leaves |
|---|---|---|---|---|---|---|
| Ba25I | 8x8 mm | 2 | not yet | not yet | not yet | not yet |
| Ba25II | 6x5,5 mm | 2 | not yet | not yet | not yet | not yet |
| Ba03 | not preserved | ?2 bases | fragmentary | 2 | no | not yet |
| Ba15 | not preserved | 2 bases | fragmentary | 2 | not preserved | not yet |
| Ba17 | not preserved | 1 base preserved | fragmentary | ? | not preserved | not yet |
| Ba18 | not preserved | not preserved | ? | ? | not preserved | not yet |
| Ba23 | 3,4x3,6 mm | 2 bases | 2 with linear segments | 2 | no secondary roots | not yet |
| Ba22 | not preserved | 2 bases | 2 with linear segments | 2 | with secondary roots | not yet |
| Ba30 | not preserved | not preserved | fragmentary | 1(?2) | not preserved | not yet |
| Ba31 | not preserved | not preserved | 3 | 2 | no secondary roots | not yet |
| Ba21I | not preserved | none | 2 with linear segments | 2 | with secondary roots | 1 |
| Ba21II | not preserved | none | 2 with linear segments | 2 | fragmentary | fragmentary |
| Ba21III | not preserved | none | 2 with linear segments | 2 | fragmentary | 1 |
| Ba24 | not preserved | 2 bases | 2 with linear segments | 2 | with secondary roots | not yet |
| Ba26 | not preserved | none | 3 | 2 | with secondary roots | 1 |
| Ba27 | not preserved | 2 bases | ? | 3 | not preserved | 4 |
| Ba28 | not preserved | 2 bases | 2 with linear segments | 2 | with secondary roots | not yet |
| Ba29 | not preserved | fragmentary | 2 with linear segments | 3 | with secondary roots | not yet |
| Ba32 | not preserved | fragmentary | 2 with linear segments | 3 | with secondary roots | not yet |
| Ba33 | not preserved | fragmentary | fragmentary | 2 | with secondary roots | not yet |

Five specimens (Ba03, 15; 17; 18; 23, Figure 1B) show the earliest stage of development. The seedlings range from 12.8–29 mm long, and are characterized by a small axis, 2.5–17 mm long and 1–4 mm wide. Each axis bears two primary leaves that are twice-bifurcate; leaf segments are linear, 20–22 mm long, and slightly tapered toward the tip. The remnant of the seed, if still present, is located near the axil of the cotyledon and primary leaf (Ba23, Figure 1B).

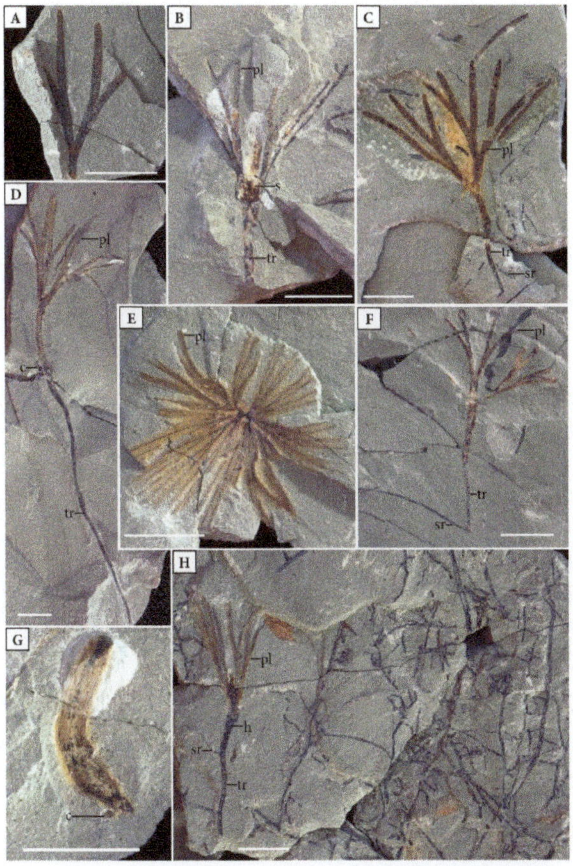


**Figure 1: Scale bar – 1 cm. A** Ginkgophyte leaf, Hangviller (Ba19). **B** Ginkgophyte seedling, stage II,, Adamswiller (Ba23). **C**Ginkgophyte seedling in transition from stage III to IV, taproot broken, Adamswiller (Ba32). **D** Ginkgophyte seedling, transition from stage III to IV, Adamswiller (Ba22). **E** Ginkgophyte seedling, stage IV, top of Figure 1G, Adamswiller (Ba27). **F** Ginkgophyte seedling, transition from stage II to IV, Adamswiller (Ba26). **G** Ginkgophyte seedling, stage IV, Adamswiller (Ba27, back side). **H** Ginkgophyte seedling in transition from stage III to IV, Adamswiller (Ba 24). (c) attachment of cotyledon; (tr) taproot; (sr) secondary root; (s) seed; (h) hypocotyl); (pl) primary leaves.

Other specimens demonstrate a slightly more advanced stage of development (Ba22, 30, 31; Figure 1D). Two or three primary leaves are attached to the axis (17–19.8 mm long) that are once- or twice-bifurcate and 17–22 mm long (Ba22, Figure 1D). The linear leaf segments are between 1.35–1.6 mm wide and approximately 10–12 mm long; the tips are acute to slightly rounded (Ba22, Ba30, Figure 1D). What is interpreted as the attachment area of a cotyledon or the seed (Figure 1H) occurs near the transition from the hypocotyl to the epicotyl in the form of a small structure (2 × 3 mm). In some specimens it appears that the hypocotyl region is slightly thickened (Figure 1C, D, H). Whether or not this thickening corresponds to the basal lignotuber formation in *Ginkgo biloba* (Figure 2B, D, F; [14, 16]) cannot be determined at this time. The taproot is at least 56 mm long and 1.1 mm wide proximally, but decreases in width towards the tip.

Most of the seedlings (i.e. Ba21I–III; Ba24, 26–29, 32, 33, Figure 1E-H) are 30–50 mm long and possess 2–4 primary leaves, which are bifurcated twice or three times. The taproots are 7.5–39 mm long and approximately 1 mm wide at the hypocotyl. Secondary roots are narrower (Ba24, Figure 1H). The hypocotyl is well differentiated. Tangled masses of ramifying delicate rootlets occur in close association with some of these seedlings, but none appear to be physically connected (Ba24, Figure 1H).

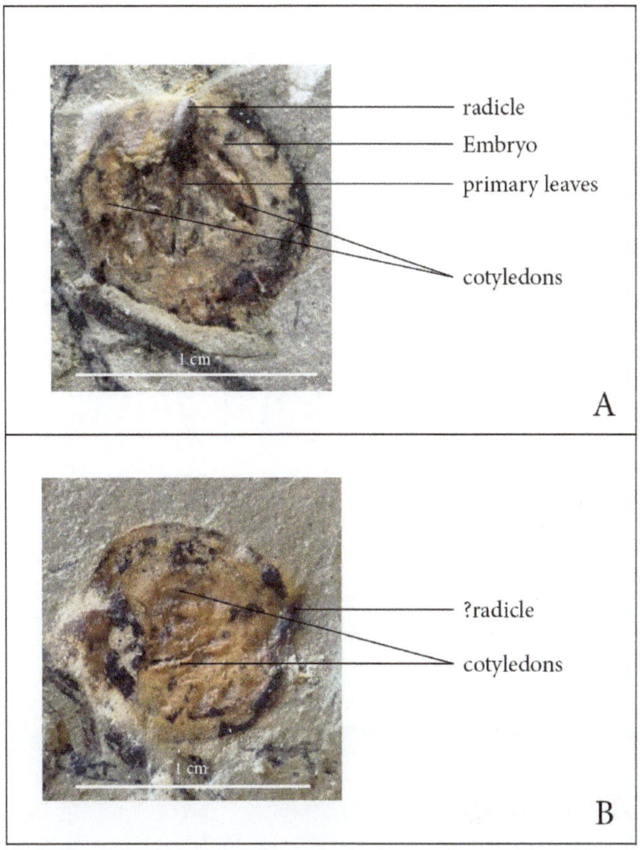


**Figure 2: Scale bar – 1 cm. A** *Ginkgo biloba* seed, stage 0–I. **B** *Ginkgo biloba* seedling, stage IV showing taproot, first leaves and differentiated leaves. **C** *Ginkgo biloba* seedling, stage IV with oppositely standing leaves. **D** *Ginkgo biloba*seedling, stage II with not ramified taproot and first leaves. **E** *Ginkgo biloba* seedling, stage IV (primary leaves of (B) enlarged). **F** *Ginkgo biloba*, stage IV (enlarged taproot of (B) showing attachment points of secondary roots).

One specimen (Ba27; Figure 1E, G) probably possesses 4 leaves, each of which bifurcates three times. These leaves are linear with the individual segments slightly tapering toward the tips. Part of the shoot axis, as well as the transition from the hypocotyl to the epicotyl, is recognizable within the sediment. In this specimen, the root is not preserved. Adhering to the lower portion of the axis is a small structure that might be the attachment of a cotyledon.

Two slabs contain isolated seeds of uncertain systematic affinities. The seeds are elongate-elliptical and up to 0.9 cm long and 0.8 cm wide (Figure 3A, B). In one of the seeds, outlines of interior structures probably resembling the embryo with cotyledons and vegetative primordial of the primary leaves are faintly visible.

**Figure 3: Scale bar – 1 cm.** Putative ginkgophyte seed (**A**, **B** part and counterpart), showing embryo and early germination (Stage I).

## DISCUSSION

The assemblage of Middle Triassic seedlings from the Voltzia Sandstone of France provides a remarkable sequence of stages in seedling morphology of a Mesozoic gymnosperm. Of particular interest is the fact that the specimens are not all preserved at the same stage of development, but rather represent a variety of ontogenetic stages which can be used to trace the sequence of morphological patterns for a single taxon.

### Affinities

To date six different plant groups have been reported from the Voltzia Sandstone, i.e. lycophytes, sphenophytes, ferns, conifers, cycadophytes, and ginkgophytes. Lycophytes, sphenophytes, and ferns can be ruled out as the producers of the seedlings based on their life history biology. Cycads can also be eliminated because the cotyledons of cycad seedlings remain within the seed coat until well after the root system is fully established in the soil [19]. Moreover, upon emergence, the cotyledons of cycads often remain intimately fused for some time. In addition, the leaves of older cycadophyte seedlings are pinnate [20]. On the other hand, the primary leaves of conifers occur singly and are not bifurcated [5]. Finally, the epigeal seedlings of conifers typically possess more than two cotyledons [4, 21], while the fossil seedlings reported here (where preserved) have two cotyledons. Based on this analysis the seedlings from the Voltzia Sandstone were most likely produced by a ginkgophyte. Their precise affinity within the ginkgophytes remains unresolved, however, we hypothesise that the morphology of the differentiated leaves in some of the fossils (e.g., Ba22, 23, 32, Figure 1B-D), together with the linear shape of the leaf segments, strongly suggests affinities with the genus *Baiera*. Interestingly *Baiera* leaves have been described previously from the Voltzia Sandstone (e.g., [5], pl.5, 3; [22]). Moreover, the overall similarity of the primary leaves, together with the close spatial co-occurrence of the fossils, suggests that all seedlings belong to the same species, with the possible exception of Ba27, which is characterized by four leaves that are three-times bifurcated, whereas all other seedlings have two or three leaves, each of which twice bifurcate.

The systematic affinities of the seeds (Figure 3) co-occurring with the seedlings in the Voltzia Sandstone cannot be determined. However, it is likely that they also belong to the ginkgophytes based on similarities in size and overall morphology to seeds assigned to this group of plants from elsewhere (e.g., [23, 24, 25, 26]).

### Taphonomy

Sediment structure and fossil content of the Voltzia Sandstone are suggestive of a deltaic sedimentary environment and semi-arid, seasonal climate [27, 28]. Thick lenses of fine-grained sandstone correspond to point bars deposited in sinuous channels, whereas silt-clay lenses represent deposits of finer material in brackish ponds scattered between the channels. The fact that most of the seedlings are preserved intact suggests that they were not submerged in water over extended periods of time and were probably not transported over long distances after having been washed out from their place(s) of growth. Undoubtedly, a fortunate (and thus very rare) combination of depositional environment, germination time, and ecological conditions has resulted in the quality of preservation displayed by these fossils. The seedlings are embedded in fine silt-clay sediment, which probably represents a deposit formed in the small ponds and pools. We are uncertain whether this indicates that the seedlings grew along the margins of these bodies of water, or grew at some distance and became washed into the ponds during a flooding event. However, the fidelity of preservation of delicate structures such as seedlings would appear to argue in favour of the former.

### Ontogenetic Development of the Fossil Ginkgophytes

The seedlings from the Voltzia Sandstone reflect three subsequent stages in the early ontogenetic development of a Mesozoic ginkgophyte. While pre-germination conditions and germination (stage I) cannot be unequivocally documented for this ginkgophyte based on the fossils at hand (see below), the second stage (stage II) is characterized by the development of a short shoot axis, a typical gymnosperm taproot, and the appearance of the first 2–4 leaves (the primary leaves). These leaves already possess the typical ginkgophyte bifurcation and are once to three-times subdivided into linear segments. Remnants of the seed are still attached to the seedling. During the subsequent stage III, the taproot thickens and produces delicate lateral (secondary) rootlets (Figure 1H). In stage IV, the shoot axis (or epicotyl, see Taylor et al. [9]: 1037) continues to elongate. Moreover, toward the end of the early ontogenetic development, the first "true" leaves emerge (Figure 1E). Like the primary leaves, these leaves are fully differentiated and characterized by multiple linear segments (Figure 1E). While the attachment points of the cotyledons are still recognizable on the shoot axis in the oldest seedlings (stage IV) from the assemblage, remnants of the seed coat are absent (Figure 1G). At the end of the sequence, the juvenile plant is established.

If the isolated seeds found in close association with the seedlings (Figure 3A, B) in fact belong to the same plant species as the seedlings, then they would indicate that the earliest phase of the ontogenetic sequence (stage I) in this plant is characterized by the seed containing the embryo with two cotyledons and vegetative primordial of the primary foliage. Moreover, germination (i.e. emergence of the radicle and shoot axis from the seed) marks the end of stage I. The sequence of early developmental stages depicted above based on fossil seedlings from the Voltzia sandstone closely resemble the ontogenetic stages seen in *Ginkgo biloba* (Figure 2A-F: stage I: 2A; stage II: 2D, stage IV: 2B-C, 2E-F), with the exception that the primary leaves in *G. biloba* are smaller and have a more simple morphology. Moreover, the primary leaves in *G. biloba* lack the typical bifurcation evident in the primary leaves of Mesozoic ginkgophytes. All other aspects of the early developmental stages of the fossil are very similar, if not identical, to that observed in *G. biloba*. Sprecher [13] defines the developmental stages in *G. biloba* as follows: During stage I (germination), the radicle and axis emerge from the seed. Subsequently, the shoot axis elongates, the first leaves (primary foliage) appear, and a taproot is developed (stage II). Secondary roots begin to appear during stage III. Remnants of the seed remain attached to the seedling during almost the entire early development, and first disappear following the maturation of the first fully differentiated ("true") leaves (stage IV).

## CONCLUSIONS

The Triassic ginkgophyte seedlings reported here provide a new complement of morphological and developmental data to our understanding of the paleobiology of the ginkgophytes. The sequence of events representing the early ontogenetic development of plants is governed by a complex series of biochemical and physiological interactions that are in turn directed by the interplay between the genetics of the organism and multiple factors operating within the ecosystem. The discovery of these seedlings provides a rare view of developmental stages of a fossil organism that can be directly compared with those of a well understood living plant. Not only do these fossils indicate that morphological information about seedling developmental stages are preserved, but this discovery also offers another perspective that can be used to indicate similarity between the fossil and extant relative. Although the precise systematic affinities of the ginkgophyte seedlings from the Voltzia sandstone remain unresolved, we hope that additional specimens, including more advanced juvenile plants, will be discovered. Information about seedling morphology provides a new data set that will help contribute to a more robust discussion of not only seed plant phylogeny, but also to our understanding about developmental processes that occurred in fossil plants after embryo formation and during seedling development.

## METHODS

### Stratigraphical Setting and Age

The Voltzia Sandstone (Grès à Voltzia) belongs to the Upper Buntsandstein (Bithynian, early Anisian) [29, 30]. The sandstone was deposited in a deltaic environment on the western margin of the German Basin. The succession is ~20 m thick and consists of the Grès à meules and the Grès argileux, which constitutes the transition to the marine Muschelkalk [31]. The assemblage of ginkgophyte seedlings has been collected from a single bedding plane within the Grès à meules (lower Voltzia Sandstone), in which three different facies have been recognized [27, 28, 32]: (a) thick lenses of fine-grained sandstone, grey or pink but more often multi-colored, containing plant debris and stegocephalian bone fragments; (b) greenish or reddish silt-clay lenses, generally consisting of a succession of thin laminae (each a few mm thick) with well-preserved fossils of aquatic and terrestrial organisms, as well as plant remains; and (c): beds of calcareous sandstone with a sparse marine fauna. The sediments crop out in several quarries in northeastern France that have been actively visited by Louis Grauvogel between 1936 and 1980. Ginkgophyte fossils, including isolated leaves, thus far have been discovered from the silt-clay lenses of facies b in the quarries of Adamswiller, Bust, Gottenhouse, Soultz-les-Bains, Hangviller and Arzviller. The seedlings described in this report come from a single, 60 cm thick, silt-clay lens in Adamswiller (Bas-Rhin, France); all have been collected from the same thin bedding planes. For detailed information on the Adamswiller silt-clay lens, including biota, refer to Gall ([27]; *Baiera* in Figure twenty eight) and Selden and Nudds ([33]; Figure one hundred twenty) in which this taxon is merged in "plant debris". For comparison with the fossils, nine seeds of *Ginkgo biloba* were potted and cultivated in a greenhouse over a period of 3 months. Germination and early seedling development were closely monitored and documented photographically (Figure 2A-F).

### Storage and Documentation

The material is deposited in the collection of Louis Grauvogel in Ringendorf (France); specimens are preceded by "Ba". Digital images were captured with a Canon Eos D550 digital camera.

## DECLARATIONS

### Acknowledgements

The research presented here has been conducted within the framework of the project "The Permian-Triassic ecological crisis in the Dolomites: extinction and recovery dynamics in terrestrial ecosystems" funded by the Promotion of Educational Policies, University and Research Department of the Autonomous Province of Bolzano – South Tyrol. E.K. acknowledges Alexander von Humboldt-Stiftung (3.3-ITA/1141759STP) for financial support. We thank Rainer Butzmann (Munich) for assistance with literature and Thomas N. Taylor (Lawrence, KS) for proofreading of the manuscript. The paper greatly benefited from the critical remarks and comments of two anonymous referees.

## Competing interests

The authors declare that they have no competing interests.

## REFERENCES


1. Baskin CC, Baskin JM: Seeds – Ecology, Biogeography, and Evolution of Dormancy and Germination. 2001, San Diego: Academic Press
2. Stockey RA: Seeds and embryos of *Araucaria mirabilis*. Am J Bot. 1975, 62: 856-868. 10.2307/2441898.
3. Stockey RA: Reproductive biology of the Cerro Cuadrado (Jurassic) fossil conifers: *Paraaraucaria patagonica*. Am J Bot. 1977, 64: 733-744. 10.2307/2441726.
4. Stockey RA, Taylor TN: On the structure and evolutionary relationships of the Cerro Cuadrado fossil conifer seedlings. Bot J Linn Soc. 1978, 76: 161-176. 10.1111/j.1095-8339.1978.tb01504.x.
5. Grauvogel-Stamm L: La flore du Grès à Voltzia (Buntsandstein supérieur) des Vosges du Nord (France). Anatomie, interprétations phylogénique et paléogéographique. Sci Géol Mém. 1978, 50: 1-225.
6. Mapes G, Rothwell GW, Haworth MT: Evolution of seed dormancy. Nature. 1989, 337: 645-646. 10.1038/337645a0.
7. Rydin C, Mohr B, Friis EM: *Cratonia cotyledon* gen. et. sp. nov.: a unique Cretaceous seedling related to Welwitschia. Biol Lett. 2003, 270: 1-4.
8. Dilcher DL, Bernardes-De-Oliveira ME, Pons D, Lott TA: Welwitschiaceae from the Lower Cretaceous of Northeatern Brazil. Am J Bot. 2005, 92 (8): 1294-1310. 10.3732/ajb.92.8.1294.
9. Taylor TN, Taylor EL, Krings M: Paleobotany: The Biology and Evolution of fossil plants. 2009, London: Elsevier
10. Zhou Z, Wu X: The rise of ginkgoalean plants in the early Mesozoic: a data analysis. Geol J. 2006, 41: 363-375. 10.1002/gj.1049.
11. Zhou Z: An overview of fossil Ginkgoales. Palaeoworld. 2009, 18: 1-22. 10.1016/j.palwor.2009.01.001.
12. Seward AC, Gowan J: The maidenhair tree (*Ginkgo biloba* L.). Ann Bot. 1900, 14: 109-154.
13. Sprecher A: Le Ginkgo biloba. 1907, Genova: Imprimerie Atar
14. Del Tredici P: Lignotuber formation in *Ginkgo biloba*. Ginkgo biloba – A Global Treasure. Edited by: Hori T, Ridge RW, Tulecke W, Del Tredici P, Tremouillaux-Guiller J, Tobe H. 1997, Tokyo: Springer, 119-126.
15. Del Tredici P: The evolution, ecology, and cultivation of *Ginkgo biloba*. Ginkgo biloba. Edited by: Van Bek TA. 2000, Amsterdam: Harwood Academic Publishers, 7-23.
16. Del Tredici P: The phenology of sexual reproduction in *ginkgo biloba*: ecological and evolutionary implications. Bot Rev. 2007, 73: 267-278. 10.1663/0006-8101(2007)73[267:TPOSRI]2.0.CO;2.
17. Grauvogel L: Note préliminaire sur la flora du Grès à Voltzia. CR Geosci. 1947, 3: 62-63.
18. Grauvogel-Stamm L, Grauvogel L: Présence de plantules de *Baiera* (Gingkophytes) dans le Grès à Voltzia (Anisien) des Vosges du Nord (France). Conséquences paléoécologiques [abstract]. Eur Palaeobotanical Conf, Montpelier abstr book. 1983, 1: 18-
19. Norstog KJ, Nicholls TJ: The biology of the cycads. 1997, New York: Cornell University Press
20. Hill TG, De Fraine E: On the seedling structure of gymnosperms. II. Ann Bot. 1909, 23: 189-227.
21. Bresinsky A, Körner C, Kadereit JW, Neuhaus G, Sonnewald U: Strasburger Lehrbuch der Botanik. 2008, Heidelberg: Spektrum
22. Firtion F: Note sur quelques gisements fossilifères du Grès à Voltzia d'Alsace. Bull Serv Carte géol Als Lorr. 1936, 3: 13-20.
23. Harris TM, Millington W: Ginkgoales. The Yorkshire Jurassic Flora IV. 1974, London: Trustees of the British Museum (Natural History)
24. Wu X, Yang X, Zhou Z: Ginkgoalean ovulate organs and seeds associated with *Baiera furcata*-type leaves from the Middle Jurassic of Qinghai Province, China. Rev Palaeobot Palynol. 2006, 138: 209-225. 10.1016/j.revpalbo.2006.01.009.
25. Yang X, Friis EM, Zhou Z: Ovule-bearing organs of *Ginkgo ginkgoidea* (Tralau) comb. Nov., and associated leaves from the Middle Jurassic of Scania, South Sweden. Rev Palaeobot Palynol. 2008, 149: 1-17. 10.1016/j.revpalbo.2007.09.005.
26. Fischer TC, Meller B, Kustatscher E, Butzmann R: Permian Ginkgophyte fossils from the Dolomites resemble extant *O-ha-tsuki*, aberrant leaf-like fructifications of *Ginkgo biloba* L. BMC Evol Biol. 2010, 10: 337-10.1186/1471-2148-10-337.
27. Gall JC: Faunes et paysages du Grès à Voltzia du Nord des Vosges. Essai paléoécologique sur le Buntsandstein supérieur. Mém Carte géol d'Alsace et Lorraine. 1971, 34: 1-318.
28. Gall JC: Ancient sedimentary environments and the habitats of living organisms. 1983, Berlin: Springer
29. Kozur HW, Mahler H, Sell J: Stratigraphic and paleobiogeographic importance of the latest Olenekian and early Anisian conchostracans of Middle Europe. New Mex Sci Bull. 1993, 3: 255-259.
30. Bachmann GH, Kozur HW: The German Triassic: correlations with international chronostratigraphic scale, numerical ages and Milankovitch cyclicity. Hallesches Jahrb für Geowissenschaften. 2004, B26: 17-62.
31. Gall JC, Grauvogel-Stamm L: The early Middle Triassic "Grès à Voltzia" formation of eastern France: a model of environmental refugium. C R Palevol. 2005, 4: 637-652. 10.1016/j.crpv.2005.04.007.
32. Gall JC: Fluvial depositional environment evolving into deltaic setting with marine influences in the Buntsandstein of Northern Vosges (France). Lect Notes Earth Sci. 1985, 4: 449-477. 10.1007/BFb0010532.
33. Selden PA, Nudds JR: Evolution of fossil ecosystems. 2004, London: Manson Publishing

CHAPTER 10

# PERMIAN GINKGOPHYTE FOSSILS FROM THE DOLOMITES RESEMBLE EXTANT O-HA-TSU-KI ABERRANT LEAF-LIKE FRUCTIFICATIONS OF GINKGO BILOBA L

Thilo C Fischer, Barbara Meller , Evelyn Kustatscher , Rainer Butzmann

## BACKGROUND

In 1869 Van Tieghem [1] provided a first interpretation of the female organs of *Ginkgo biloba* (figure 1) in his work on comparative anatomy of fructifications. In his view the whole stalk, cupule and ovule are homologous to a sporophyll. Later (1896), Fujii [2] provided a scientific description of the long-known *Ginkgo biloba O-ha-tsuki* leaves with seeds (figure 2-1, 2-2: female, 2-4: male counterpart). Based on the rare presence of an axillary bud in multi-ovulate forms of the fructification (figure 2-3), the basal part of the stalk was interpreted by him as being shoot-derived. The stalklet, collar and ovule would correspond to petiole, reduced lamina and ovule of a sporophyll. Other opinions, mainly the interpretation as a two-flowered inflorescence (Strasburger), are also summarized and cited in Fujii [2]. Wettstein [3], Sakisaka [4] and others (cited therein) essentially agreed with Fujii's view and extended the argumentation by considering also the vascularisation of the organs. Vegetative leaves of *Ginkgo biloba* have two vascular bundles at the petiole base, whereas the stalk of the common fructification with two ovules has four (or correspondingly more in the rare forms with multiple ovules). *O-ha-tsuki* leaves possess two bundles, like vegetative leaves. This provides further evidence to homologise them only with the stalklet/collar/ovule - distal part of the common *Ginkgo biloba* fructification. Van Tieghem's view, in contrast, that the whole female organs are homologous to a sporophyll could be supported by interpreting the rare axillary bud as an ectopic development of a meristem, but the vascular system of the stalk with four bundles could hardly be explained. However, the *O-ha-tsuki* form was also interpreted only as a case of retroconvergent morphology, arguing with the absence of fossil occurrences of "leafy ovuliphores" in the Mesozoic [5].

**Figure 1:** *Fertile shoot of female* Ginkgo biloba. The common female fructification of *Ginkgo biloba* is a brachyblast (spur shoot) with sterile leaves (trophophylls) and ovules on stalks, the latter either being interpreted as structures derived from sporophylls, or as shoots, or as organs composed of both.

**Citation:** Thilo C Fischer, Barbara Meller , Evelyn Kustatscher, Rainer Butzmann, Permian ginkgophyte fossils from the Dolomites resemble extant O-ha-tsuki aberrant leaf-like fructifications of Ginkgo biloba L, DOI: 10.1186/1471-2148-10-337.

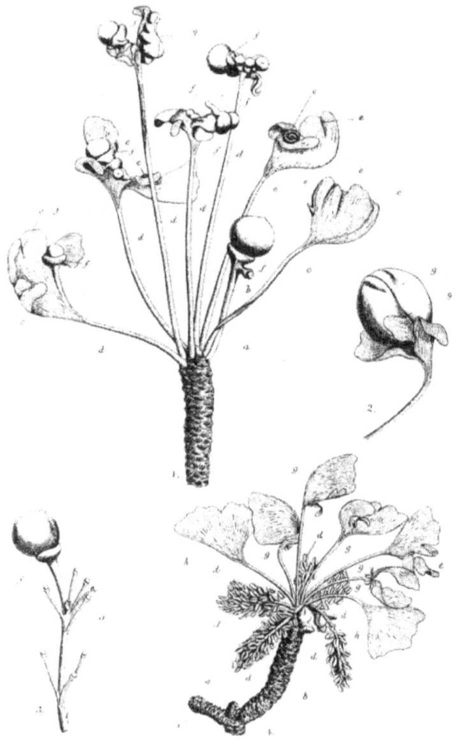

**Figure 2:** *Historical figure by Fujii (1896) on the* O-ha-tsuki *phenomenon of* Ginkgo biloba. Plate from Fujii (1896) demonstrating rare female *O-ha-tsuki* ("seed at leaf") type aberration of *Ginkgo biloba* fructifications (1-1: whole fertile brachyblast, 1-2: single ovule at leaf, its male analogue (1-4), and a rare multi-ovulate fructification with a bud (1-3)).

Fossil ginkgophyte fructifications are rare in the fossil record [6]. Most closely related to *Ginkgo biloba* are ovulate structures attached to brachyblasts (spur shoots) carrying also *Ginkgo*-like vegetative leaves (or associations of both in the fossiliferous sediments) from the Mesozoic and Cenozoic: *Ginkgo yimaensis*(Jurassic) [7], *Ginkgo apodes* (Late Jurassic to Early Cretaceous) [8], and *Ginkgo adiantoides* leaves/*Ginkgo geissertii* ovule (Cenozoic) [9, 10]. *Nehvizdyella bipartita* (Late Cretaceous) [11] has similar fructifications, but non-divided leaves. The fossil taxa affiliated to *Ginkgo* can be ordered in an evolutionary series with successive reduction of individual stalks, number of ovules and accompanying increase in the size of ovules and in the width of the leaf segments, reflecting also ontogenetic sequences in *Ginkgo biloba* [12]. Palaeozoic fructifications of ginkgophytes are also known, but much more difficult to interpret. *Trichopitys heteromorpha* from the Lower Permian [13] is the oldest generally accepted ginkgophyte [[14], and other authors]. It consists of axes carrying dichotomous, non-laminar leaves with zones of dividing stalks with terminal ovules which are possibly recurved. *Karkenia* is a genus with multi-ovulate axes or multi-ovulate globular fructifications; the first species (associated with *Ginkgoites tigrensis* leaves) was described by Archangelsky from the Lower Cretaceous of Patagonia [15]. Another well known member is the Liassic *Karkenia hauptmannii*, represented by brachyblasts with globular fructifications and *Sphenobaiera* leaves [16]. *Avatia bifurcata* and *Hamshawvia longipedunculata* from the Triassic of South Africa [17] also possess globular multi-ovulate fructifications (on dichotomous stalks). *Hamshawvia* fructifications are found attached to brachyblasts with *Sphenobaiera* vegetative leaves.

Recently, Naugolnykh [18] has reviewed "foliar" seed-bearing organs of Paleozoic ginkgophytes. Some of the taxa he included in the ginkgophytes, e.g., *Arberia*, have been interpreted differently i.e. as glossopterid [14]. Taxa that can be accepted as ginkgophyte seed-bearing organs clearly have a dichotomous structure like*Trichopitys*, *Karkenia*, *Toretzia*, and *Grenana*. These genera show stalks carrying ovules. If accepted as being homologous to leaves (sporophylls), these either would represent non-laminar leaves, or, alternatively, the laminar part would have been already reduced. The latter possibility led Naugolnykh to suppose "pre-karkeniaceous" ancestors of ginkgophytes.

Here we report on Permian laminar ginkgophyte leaves with putatively attached seeds.

## RESULTS

### Description of the Fossils from the Upper Permian of the Dolomites and Comparison with Ginkgophyte Characters

#### *Morphology*

Numerous specimens of vegetative *Sphenobaiera* leaves (wedge-shaped, without petiole) were discovered from a lens of fossiliferous argillaceous siltstone to fine-grained sandstone at the Upper Permian Bletterbach locality [19]. Among these, some specimens were identified where also seeds are preserved (the described specimens are called "seeds", even if "ovules" is often used for ginkgophyte seeds). Figure 3 and 4 illustrate a specimen of a wedge-shaped, dichotomously divided leaf with all characteristics of the genus *Sphenobaiera*, but with two seemingly attached seeds. Especially in the case of the larger seed, which is in a lateral position at the leaf margin, the organic connection to the leaf segment is assumed. There is no basal collar-like structure in the fossil like at extant *Ginkgo biloba* ovules or like the irregular basal swellings in the *O-ha-tsuki* form. This type of fossil seeds occurred only on some of the approximately one hundred blocks of sediments from the Bletterbach collection with *Sphenobaiera* leaves, in close association in each case to such leaves, and not randomly on other blocks of sediment with various plant fossils. None of this type of seed showed a stalk or remains of such.

**Figure 3:** Ginkgophyte leaf with putatively attached seeds from the Upper Permian of the Bletterbach (Dolomites). Morphology of the specimen (inventory number PAL-1368), drawing of it performed with binocular control. Evident structures drawn as lines, supposed structural outline as dotted lines. Scale bar 10 mm.

**Figure 4:** Ginkgophyte leaf with putatively attached seeds from the Upper Permian of the Bletterbach (Dolomites), detail. Detail of PAL-1368 at the assumed attachment site of the seed. Scale bar 10 mm.

## *Cuticular Structures*

Cuticles were prepared from the leaf lamina region of the counterpart of the first specimen (figure 3).

The abaxial cuticle of the leaf (figure 5) is formed by cells of different shapes and sizes: polygonal, rectangular, isodiametric and elongated ((15-)25-75(-100 μm)) which are partly arranged in irregular rows. One longitudinal band with two rows of elongated cells is visible in the middle of the figure, probably representing the costal area of a vein; the cells bear small papillae. Two other bands at the right and left margin are less distinct but may represent additional costal areas. Anticlinal walls (this is the cell outline on the prepared cuticles) are straight, sometimes slightly curved. Stomatal complexes are not all well preserved. They are arranged in bands (putative intercostal regions) and are surrounded by more polygonal-isodiametric cells. The stomatal complexes are about 50-60 μm in diameter, rarely up to 90 μm. The 4-6 subsidiary cells (figure 5, 6a) have small marginal papillae almost covering the stomatal pit. Guard cells are rarely recognizable. Similar cuticle characteristics have been found for other fossil ginkgophytes [7, 9, 20] - [23]. Under the scanning electron microscope (SEM), the outside of the abaxial cuticle appears to be nearly smooth (figure 6c). The papillae are visible as flat bulges. The inner side not always distinctly displays the anticlinal walls, but stomata are clearly recognizable (figure 6b, d). The adaxial cuticle of the leaf shows only few preserved rectangular anticlinal walls with the light microscope (LM) (not figured) and indistinct anticlinal walls with the SEM. Also one stomatal complex is visible with the SEM (figure 7a, b).

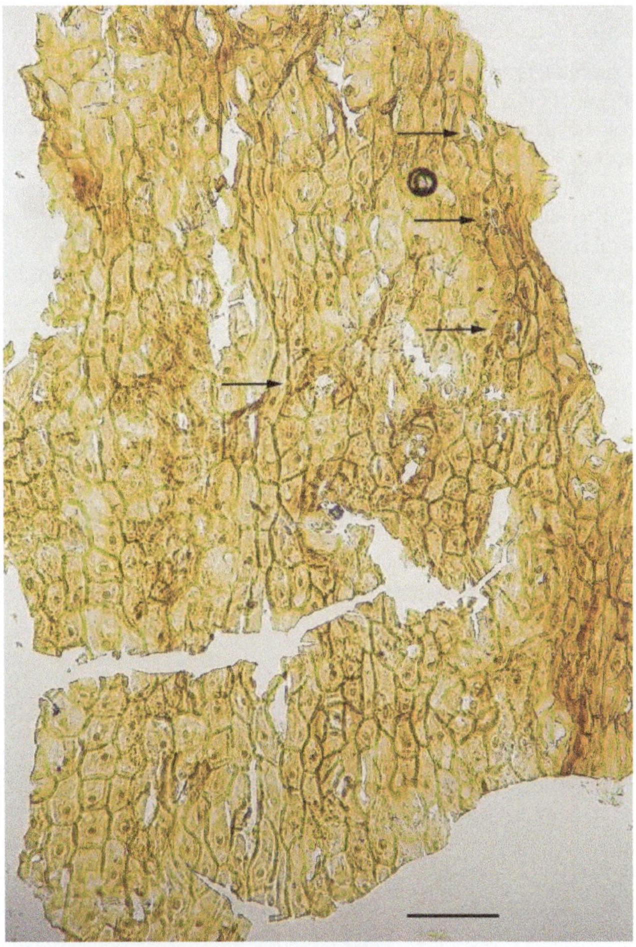

**Figure 5:** Abaxial cuticle prepared from the leaf lamina region of ginkgophyte leaf with putatively attached seeds. Abaxial cuticle prepared from the leaf lamina region of the specimen PAL-1368. The arrows point to stomatal complexes. Scale bar 100 μm.

**Figure 6:** Details of abaxial cuticles prepared from the leaf lamina region of ginkgophyte leaf with putatively attached seeds. a: Stomatal complexes of figure 5in detail. Scale bar 100 µm. b: Inner side of the abaxial cuticle of the lamina; at the right margin the surface of the outside is also visible (detail in figure 6 c) (SEM). Scale bar 100 µm. c: The nearly smooth outside of the abaxial leaf cuticle with papillae, indicating a stoma slit. The whitish spots on the surface probably derive from the wax layer (SEM). Scale bar 50 µm. d: Inner side of the abaxial leaf cuticle with two stomata (SEM). Scale bar 50 µm.

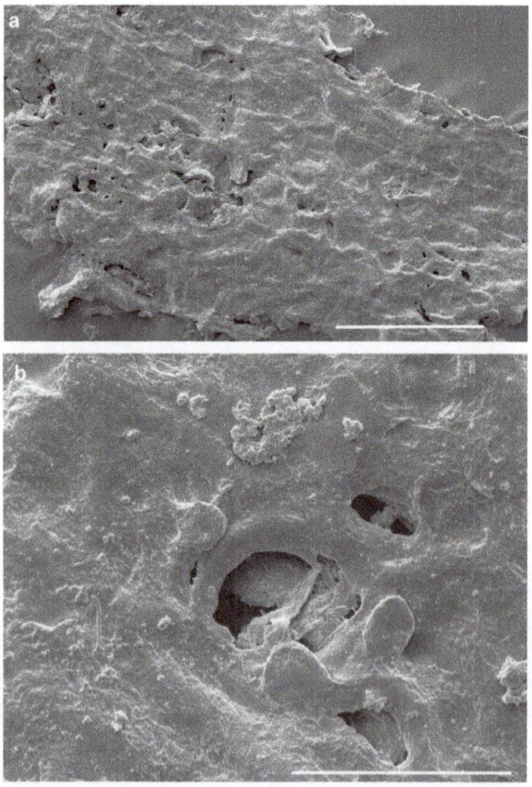

**Figure 7:** Details of adaxial cuticles prepared from the leaf lamina region of ginkgophyte leaf with putatively attached seeds. a: Inner side of adaxial cuticle of the lamina (SEM). Scale bar 50 µm. b: Outside of the adaxial cuticle of the lamina with one stoma, surrounded by papillae (SEM). Scale bar 20 µm.

The cuticle of the seed (figure 8a-f) shows small cells with mainly straight, rarely slightly curved anticlinal walls. They are of polygonal shape, irregularly arranged and without papillae. The cell diameter varies from (14-)25 to 55 µm, rarely up to 70 µm. The stomatal complexes with the subsidiary cells are 60 to 95 µm in diameter. The 5-6 subsidiary cells of mainly trapezoid shape (diameters about 14-39 µm) are distinctly thickened at the inner margin reducing, thus, the stomatal pit (figure 8a, b). Details of the guard-cells are rarely visible (figure8b). Other circular to elliptical pits, about 30-35 µm in diameter, which are also enclosed by six cells but are not thickened, are idioblasts (figure 8d). The shown outside surface of the seed cuticle (figure 8f) is almost smooth with at least one stomatal complex with weakly developed papillae and two tubes, which might derive from the wax layer.

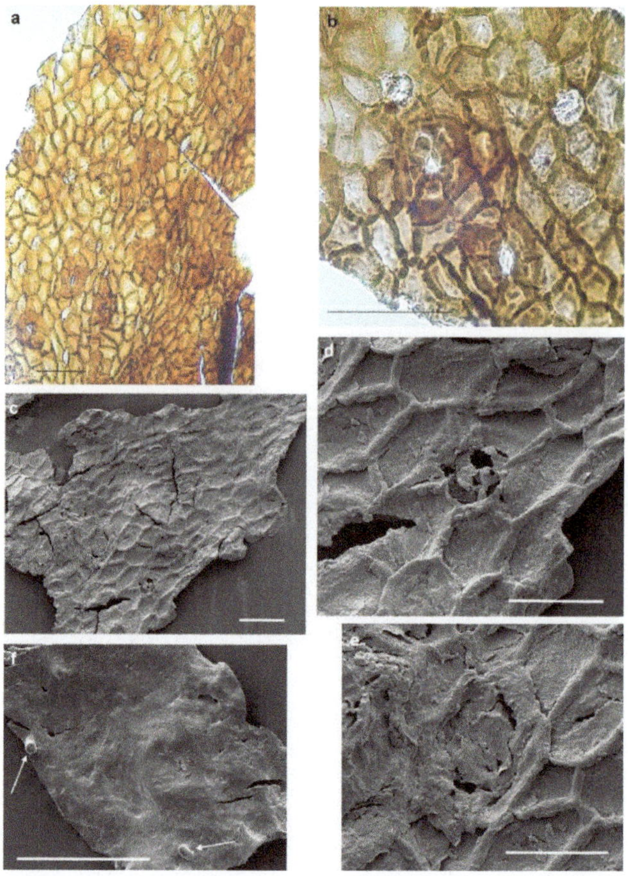

**Figure 8:** Details of cuticle prepared from the seed of ginkgophyte leaf with putatively attached seeds. a: Cuticle prepared from the seed of the specimen with several stomatal complexes and idioblasts. Scale bar 100 µm. b: Seed cuticle with stomatal complexes in detail; the lower one shows also the guard-cells. In the upper part of the picture two idioblasts are shown in detail. Scale bar 100 µm. c: Inner side of the seed cuticle (SEM). Scale bar 100 µm. d: Detail of figure 8 c showing an idioblast (SEM). Scale bar 50 µm. e: Detail of figure 8 c showing one stomatal complex. Scale bar 50 µm. f: Outside of the cuticle with two tubes (arrows) and at least one stomatal complex with very weakly developed papillae. Scale bar 100 µm.

## Comparison to Cuticles of Other Ginkgophytes

### *Cuticles of extant Ginkgo biloba (O-ha-tsuki form)*

The abaxial cuticle of the extant *O-ha-tsuki* leaf of *Ginkgo biloba* (figure 9a-d) shows the distinct pattern typical of *Ginkgo* leaves: narrow costal rows without stomata and broader areas in between (intercostal fields) with numerous stomata. The narrow costal fields are 100-125 μm wide and consist of three to five rows of longitudinal, mostly rectangular cells. The anticlinal walls are indistinct but seem to be straight to very slightly undulating. Papillae occur not on every cell, but often. The intercostal areas are much broader than the costal areas, at least 200 μm or more. Numerous stomata are randomly oriented and are irregularly distributed as well as the papillae. The anticlinal walls between the stomata are indistinct, but mainly straight or curved. The stomata are enclosed by 5-6 (-7) subsidiary cells with prominent marginal papillae; sometimes the papillae are less strong or not observable. The stomata are elongated, their pores do not exceed 40 μm in length, and the inner walls of the guard-cells are slightly thickened in the middle part. The aperture is about 20-25 μm. The outside surface of the whole abaxial cuticle shows a fine granulated structure.

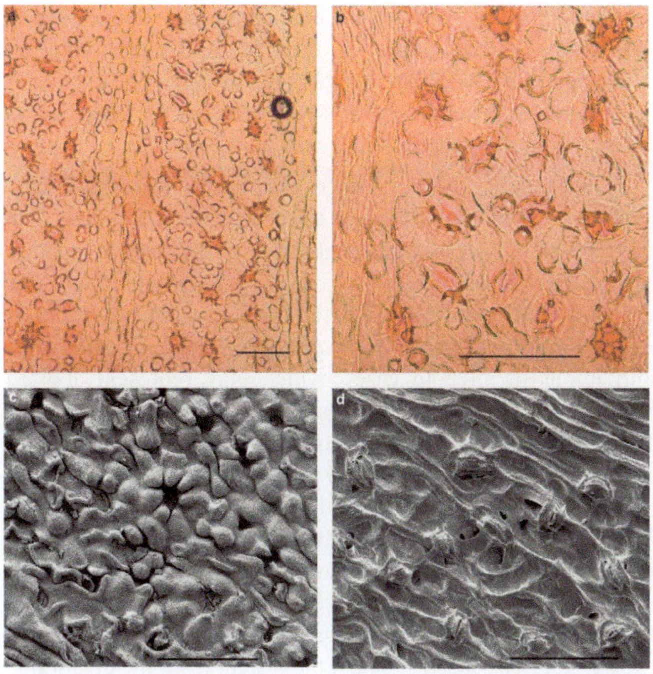

**Figure 9:** ***Abaxial cuticles of the* O-ha-tsuki *leaf of* Ginkgo biloba**. a: Abaxial cuticle of the lamina with two costal rows without stomata and two intercostal areas with numerous stomatal complexes (inventory number Herbarium BOZ PVASC15174). Scale bar 100 μm. b: Detail of the abaxial cuticle of the lamina. Note the variable thickening and cutinization of the papillae of the subsidiary cells. Scale bar 100 μm. c: Outside of the abaxial cuticle of the lamina with numerous stomatal complexes, which are variable in the numbers and shape of the papillae (SEM). Scale bar 100 μm. d: Inner side of the abaxial cuticle of the lamina with several stomatal complexes (SEM). Scale bar 100 μm.

The adaxial cuticle of the leaf (figure 10a, b) is very thin and without or almost devoid of stomata and papillae. The cuticle consists of narrow rows of mainly rectangular to narrow elongated cells (about 90 μm long and 35 mm wide) (costal areas) and broader areas (intercostal areas) with wider cells of rectangle elongated shape (about 70 μm long and 35 μm wide). The anticlinal walls are mainly undulating.

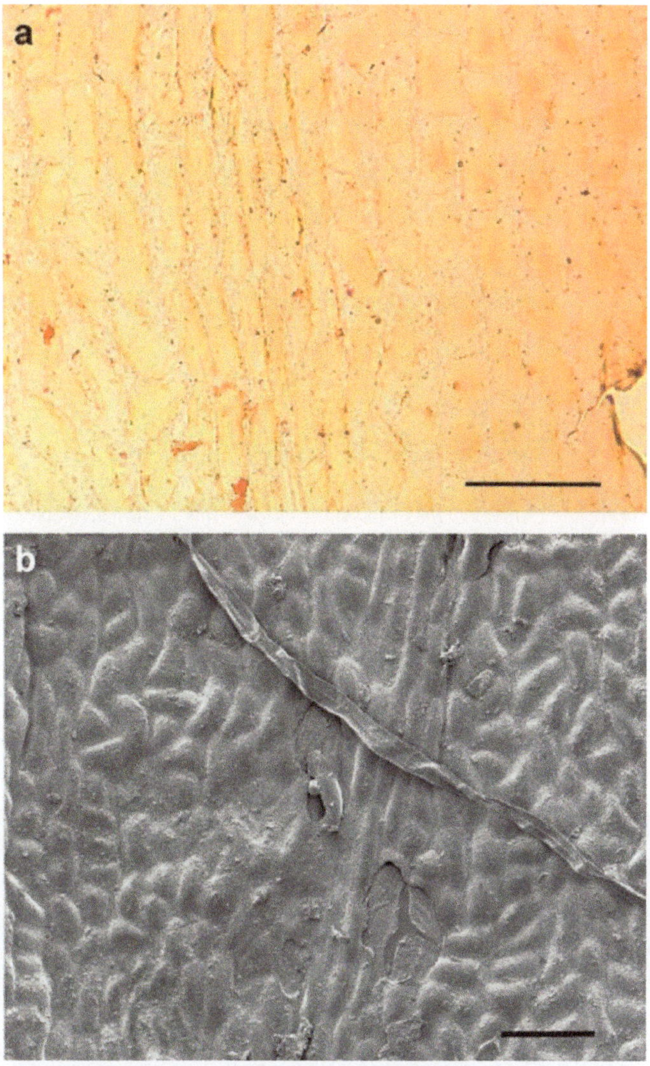

**Figure 10:** *Adaxial cuticles of the* O-ha-tsuki *leaf of* Ginkgo biloba. a: The adaxial thin cuticle of the lamina. Scale bar 100 μm. b: Outside of the adaxial cuticle of the lamina with one costal area and two intercostal areas. The pit in the centre of the picture might indicate a stoma (SEM). Scale bar 100 μm.

The cuticle of the *O-ha-tsuki* ovule epidermis (figure 11a-f) consists of isodiametric and polygonal to rectangular cells, which are irregularly arranged. The anticlinal walls are straight to curved. Papillae have not been observed with the LM. The cell diameter varies from 25 to 45 μm. The stomatal complex is built by 5-6 subsidiary cells and two small guard cells. The subsidiary cells are partly irregular in shape and up to 60 μm in diameter; their periclinal walls sometimes seem striate (LM), the inner anticlinal walls are distinctly thickened. The stomata are about 20-25 μm long. Idioblasts are also regularly observable. The outside surface of the ovule cuticle is mainly smooth, but some papillae surrounding a putative stomatal pore are observable (figure11 f: SEM).

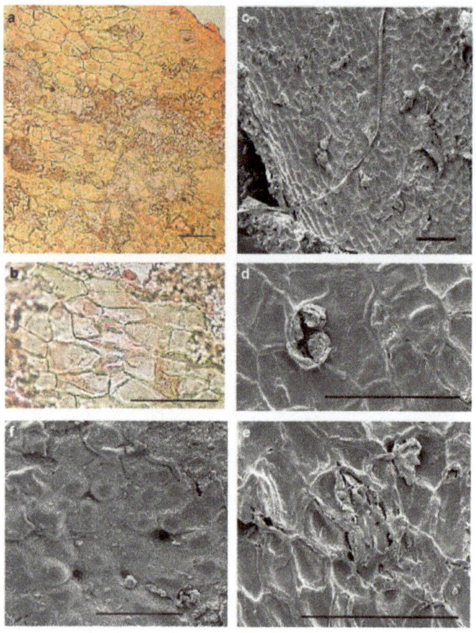

**Figure 11:** *Cuticles of the* O-ha-tsuki *ovule of* Ginkgo biloba. a: Overview of the ovule cuticle with some stomatal complexes. Scale bar 100 µm. b: Two stomatal complexes next to each other in detail and one star-like idioblast above the scale bar. Scale bar 100 µm. c: Inner side of the ovule cuticle with some idioblasts and stomatal complexes (SEM). Scale bar 100 µm. d: An idioblast in detail (SEM). Scale bar 50 µm. e: One stomatal complex (SEM). Scale bar 100 µm. f: Outside of the ovule cuticle with papillae. Some are arranged around pits, indicating stomata (SEM). Scale bar 100 µm.

## Further Specimens of Sphenobaiera Leaves Associated with Seeds from the Bletterbach

A second specimen of a *Sphenobaiera* leaf which shows putatively attached seeds is shown in figure 12. This leaf has multiple segments and three seeds are preserved in positions apical to the respective leaf segments. In total, seven specimens of *Sphenobaiera* leaves with putatively attached or associated seeds have been found at the Bletterbach locality (others not shown) where at least the position of the seeds suggests the attachment to the leaves, but the individual preservation does not allow proving organic connection in these cases.

**Figure 12:** Second specimen of a ginkgophyte leaf with putatively attached ovules from the Upper Permian of the Bletterbach. Morphology of the specimen and drawing (inventory number PAL-1369). The specimen is preserved within a sediment layer with several plant remains. Scale bar 10 mm.

## DISCUSSION

The systematic position of the leaves with putatively attached seeds from the Bletterbach could principally be supposed among the cycads, the pteridosperms, or the ginkgophytes. Cycadophytes, however, are characterized by pinnately (or rarely bipinnately) compound leaves, most unlike the Bletterbach fossils. Among pteridosperms the character of interest "leaf-borne seeds" is found, but leaflets are mostly not dichotomous. There are Y-shaped fronds which, in case of forms with entire margin (non-pinnate) like the Triassic *Dicoidium dutoitii*, show similarity in gross morphology ([14], figure 15.20) to once-dichotomous leaves like PAL-1368, but not to the multiple-dichotomous forms of *Sphenobaiera* from the Bletterbach (figure 12). Most importantly, these leaves possess central veins (the rachis). In contrast, in well preserved *Sphenobaiera* specimens from the Bletterbach dichotomous/parallel venation can be observed; and a prominent midvein should be observable in PAL-1368 but is absent. The Permian pteridosperm *Dichophyllum flabellifera* ([14], figure 15.57) possesses divided pinnules similar to *Sphenobaiera* leaves, but pinnules are tongue-shaped and have midveins. Based on the described morphology an affiliation of PAL-1368 to the pteridosperms can be excluded. The leaves with putatively attached seeds can be identified as belonging to the ginkgophyte genus *Sphenobaiera*.

The results of the cuticle studies on PAL-1368 also support the ginkgophyte origin of the fossil leaf and seed. The differences of leaf and seed cuticles are comparable to these differences in modern *Ginkgo biloba* or its *O-ha-tsuki* form. The cuticle structures of the fossil seed and those of extant *O-ha-tsuki* ovules show similar isodiametric to rectangular cell shapes and idioblasts in both cases, which may represent aborted stomata, trichome bases or glands. For *Ginkgo biloba* leaves Florin [24] mentions the rare occurrence of trichomes. With the SEM it is visible, that the cuticle is covered with wax. The indistinct parts of the LM pictures are caused by that. Two tubular structures (SEM) assumed to consist of wax were observed, this has also been described for *Ginkgo biloba* and other gymnosperms [25]. In contrast to the impression from the LM picture, the stomata are also surrounded by papillae on the surface. The fossil seed cuticle differs from the extant one only by the distinctly thicker anticlinal walls and the thickened subsidiary cells.

Cuticle structures especially of Permian *Sphenobaiera* leaves have only been described by Florin [24] from the Lower Permian of France (*Sphenobaiera raymondi*) and by Schweitzer [26] from the Upper Permian of Germany (*Sphenobaiera digitata*). The cuticle of the *Sphenobaiera raymondi* specimen is not well preserved. It seems to be composed mainly of isodiametric cells instead of longitudinal ones, possesses straight cell walls, and stomatal complexes are specified as haplocheilic with narrowing of the stomatal pore. *Sphenobaiera digitata* cuticles are described as amphistomatic, with a cuticle showing longitudinally oriented cells, rows of stomata towards the center of the lamina and less regular cells and stomata distribution towards the margin. Papillae are present only on subsidiary cells. Stomatal complexes are described as dicyclic, but seem to be mainly monocyclic as is shown in the drawing given in [26].

Fossil ginkgophyte leaves are often amphistomatic with rare stomata in the adaxial cuticle, but hypostomatic leaves have also been described [24] and have been found also among common *Ginkgo biloba* leaves as well as varying stomata abundances in apical and basal leaf parts. Furthermore, stomata index for *Ginkgo biloba*was reported to be inversely correlated with $CO_2$ concentration [27]. Varying stomata abundance is also described for a fossil ginkgophyte [28]. Hypostomatic ginkgophyte leaves are represented by a *Sphenobaiera*species from the Triassic Molteno Formation in South Africa [[23], p. 133] and by *Baiera* cf. *furcata* from the Middle Jurassic of China [29]. From the Jurassic Yorkshire Flora [21] *Sphenobaiera ophioglossum* shows similar cuticle structures (cell types of costal and intercostal fields, stomatal complexes, presence of papillae) [[21], figure 17 D], but is designated as the adaxial cuticle. The abaxial cuticle of *Sphenobaiera schenckii* from the Triassic of South Africa [23] is seemingly identical, but its adaxial cuticle shows abundant stomatal complexes (not present in the adaxial cuticle of the Bletterbach specimen). Several *Sphenobaiera* leaves from the Bletterbach have to be studied in future for detailed taxonomic comparisons and specific identification of these leaves.

With respect to the cuticle of the fossil seed depicted here, the cells of the outer cuticle of the integument of*Ginkgo yimaensis* ovules [7, plate 3, figure four] are very similar and the description also concurs. One distinct difference is the size of the stomatal complex, which is much larger (150-175 × 7.5-17.5 (-35) μm) and less circular than in the specimen from Bletterbach. The cuticle of the outer integument of the ovules, described for *Yimaia recurva* (associated with leaves of *Baiera hallei*) [30] differs by the rounded corners of the cells, the larger stomatal complexes and the unspecialized subsidiary cells. The cuticle of the Jurassic ovule *Yimaia qinghaiensis* [31] closely resembles that of the seed from the Bletterbach with respect to its general reticulate structure formed by irregular-polygonal and irregularly arranged cells, the absence of papillae on these cells, and its scattered stomatal complexes; even if any of these characters is not uncommon among gymnosperms. The Cretaceous ovuliferous organ *Nehvizdyella bipartita* [11] possesses the same cuticular characteristics as the Jurassic ovule *Yimaia qinghaiensis* and as the cuticle of the *Sphenobaiera* seed from Bletterbach. *Ginkgo ginkgoidea* (Tralau) Yang, Friis et Zhou [31] from the Jurassic of Sweden shows also alike cuticle structures.

Ontogenetic developmental aberrations can resemble primitive phylogenetic states of organs, commonly called "atavisms". Ideally, the aberrant character closely resembles the primitive character, which can either be known by the fossil record of the group, or by identification of the primitive state of the character by comparison with living representatives of sister groups and outgroups for the clade. For the aberrant *O-ha-tsuki* leaves of *Ginkgo biloba* the underlying genetics is completely unknown. An alternative interpretation of the phenomenon to atavism would be its description as a case of ectopic organ development, which offers no mechanistic explanation. Rothwell expected fossil occurrence of ginkgophyte brachyblasts (spur shoots) with sporophylls like the *O-ha-tsuki* leaves [[32], p.101]. Anderson and Anderson [17] also led attention to such "anomalous strobili with leafy expansions" (= *O-ha-tsuki* form) of *Ginkgo biloba* comparing them with their Triassic *Avatia bifurcata* fructification, even without a distinct laminar structure of the considered fossil. However, it can not be excluded that these Triassic fructifications represent rare aberrant forms.

*Sphenobaiera* represents a heterogeneous group of leaf fossils with wide stratigraphical distribution. The leaf morphogenus comprises dichotomous and wedge-shaped leaves, which do not possess a petiole like extant*Ginkgo* and the fossil leaf morphogenera *Ginkgo*, *Baiera* and *Ginkgoites*. Most species of *Sphenobaiera* are thought to represent ginkgophytes [14].

The *Sphenobaiera* leaves with putatively attached seeds described here from the Bletterbach locality could possibly represent cases of ectopic organ development, present for an unknown reason. However, this seems unlikely since rare aberrations would not be expected in smaller collections of fossils. The specimen PAL-1368 reveals striking similarity especially with *O-ha-tsuki* leaves with few or only one well developed ovule (figure13). *O-ha-tsuki* leaves with multiple ovule formation often only show malformed ovules.

**Figure 13:** O-ha-tsuki *leaf of* Ginkgo biloba *with well developed ovule.* Courtesy H Okada (inventory number Herbarium BOZ PVASC15174).

Zhou and Zheng [12] suggested an evolutionary scheme starting with Jurassic *Ginkgo yimaensis* with multiple small ovules and multiple divided vegetative leaves, subsequent reduction of ovule number and number of leaf segments in Cretaceous *Ginkgo apodes* and Cainozoic *Ginkgo adiantoides*, ending with *Ginkgo biloba* with predominantly only one ovule per stalk and bilobate to undivided vegetative leaves ("reduction hypothesis", "peramorphosis"). Given that the interpretation of the presented Permian fossils from

Bletterbach as sporophylls is correct; these would represent an ancestral state of female ginkgophyte sporophylls before reduction of the sporophyll lamina had taken place, predating the reconstructed evolutionary series.

## CONCLUSIONS

The *Sphenobaiera* fossils with seemingly attached seeds from the Upper Permian of the Bletterbach are hypothetically interpreted as ancestral ginkgophyte sporophylls with laminar structure, as would be expected from aberrant *O-ha-tsuki* leaves with seeds of extant *Ginkgo biloba*. A formal description or affiliation of the fossils at the species level will be provided together with the one for the numerous vegetative *Sphenobaiera*leaves and with the other floral elements. Ongoing excavations at the Bletterbach locality, at other Permian localities, but also screening of museum collections could provide more and possibly better preserved ginkgophyte leaves with ovules/seeds. Especially specimens with completely preserved vascular systems would be highly desirable for comparison with vegetative and fertile organs of extant *Ginkgo biloba*.

## METHODS

### Excavation of the Fossils and Deposition

The fossils were recovered during excavations in the years 2003-2009. The figured specimens are kept in the palaeontological collection of the Museum of Nature South Tyrol (Bozen/Bolzano, figure three: no. PAL-1368, figure twelve: no. PAL-1369).

### Cuticular Analysis

Small pieces of organic material were removed from the fossil with a scalpel, rinsed in water and treated with a drop of 10% HF until all sediment particles were dissolved. The material was rinsed in water and incubated in conc. $HNO_3$ with $KClO_3$ (Schulze's Reagent) until the organic material became transparent. The material was rinsed with water, briefly treated with 5% KOH, rinsed with water, and transferred into glycerol for study and storage. The samples of the *O-ha-tsuki* leaf and seed were taken from one specimen, from the middle part of the lower and upper half of the leaf and seed and directly put into Schulze's Reagent. One piece was studied with the LM, the other with the SEM. The *O-ha-tsuki* leaf sample for the LM has been stained with safranin. The cuticles were studied with a Nikon eclipse 80i light microscope (LM), the pictures had been taken with a Samsung digimax V70. For the scanning electron microscope FEI (SEM), the macerated wet cuticles were transferred to the SEM-stub with a carbon adhesive tape, where they dried. One piece of the fossil leaf cuticle (slide one) was removed from the glycerol, washed with water and ethanol and then picked onto carbon tape on the SEM-stub. During the drying process a rather distinct shrinkage of the cuticles has been observed, which results in partly smaller cell sizes of the SEM cuticles. Both, fossil and extant leaf cuticles are very thin and fragile; the cuticles of the seeds are thick. These thick cuticles tended to roll in and a strong mechanical pressure was necessary to get a nearly flat cuticle. The pictures were slightly adjusted with Adobe Photoshop 7.0 in brightness, contrast and frame.

### Drawing of Fossils

Drawings of fossils were performed using a binocular to discriminate biological structures from those produced by local destruction.

### *O-ha-tsuki* Specimens

The specimens have been conserved in aqueous formaldehyde solution and are kept in the Herbarium of the Museum of Nature South Tyrol BOZ (PVASC15174).

## DECLARATIONS

### Acknowledgements

The country South Tyrol is acknowledged for financial support for the excavations, the staff of the GeoCenter Aldein (Maria Pichler, Herta Obertegger, Christian Weber) for organizational support, and family Kalser (Aldein, South Tyrol) for their kind hospitality during our visits. Hiroshi Okada (Botanical Gardens, Osaka City University), the temple Mito Hachimanguu and Yoshiko Mashiko (Ibaraki-ken, Japan), Wolfgang Schwarz (Technische Universität München) and Josef Bogner (Ludwig-Maximilians-Universität München) are acknowledged for providing samples of *O-ha-tsuki* plant material or for their help to do so. We are grateful to Michael Krings (Ludwig-Maximilians-Universität München), Han van Konijnenburg-Van Cittert (Utrecht University) and Hans Kerp (Westfälische Wilhelms-Universität Münster) for critical discussion. Georg Janssen (Bayerische Staatssammlung für Paläontologie und Geologie) kindly provided photographies of the fossils. Stephan Schultka, Manfred Barthel (both Museum für Naturkunde, Leibniz-Institut für Evolutions- und Biodiversitätsforschung an der Humboldt-Universität zu Berlin) and Ralf Werneburg (Naturhistorisches Museum Schloss Bertholdsburg Schleusingen) are acknowledged for making fossils from the Lower Permian of Thuringia (not mentioned in the text) available to us for comparison. We acknowledge the proof-reading by Elizabeth Schroeder-Reiter (Ludwig-Maximilians-Universität München) as a native speaker and the efforts of two reviewers to improve our manuscript.

## REFERENCES

1. Van Tieghem Ph: Anatomie comparée de la fleur femelle et du fruit des Cycadées, des Conifères, et des Gnetacées. Ann Sci Nat Bot. 1869, V (10): 269-304.
2. Fujii K: On the different views hitherto proposed regarding the morphology of the flowers of *Ginkgo biloba*. Bot Mag Tokyo. 1896, 10 (108): 7-8. 13-15, 104-110
3. Wettstein Rv: Die weibliche Blüte von Ginkgo. Oesterr botan Zeitschrift. 1899, XLIX (12): 417-425. 10.1007/BF01794968.
4. Sakisaka M: On the seed-bearing leaves of *Ginkgo*. Jap J Bot. 1929, 4: 219-235.
5. Krassilov VA: Macroevolutionary events and the evolution of higher taxa. Evolutionary Theory and

Processes, Modern Horizons, Papers in honour of EVIATAR NEVO. Edited by: Wasser SP. 2004, Dordrecht: Kluwer Academic Publishers, 265-289.

6. Zhou ZY: An overview of fossil Ginkgoales. Palaeoworld. 2009, 18: 1-22. 10.1016/j.palwor.2009.01.001.
7. Zhou ZY, Zhang B: A Middle Jurassic *Ginkgo* with ovule-bearing organs from Henan, China. Palaeontographica. 1989, B 211 (4-6): 113-133.
8. Zheng S, Zhou ZY: A new Mesozoic *Ginkgo* from western Liaoning, China and its evolutionary significance. Rev Palaeobot Palynol. 2004, 131: 91-103. 10.1016/j.revpalbo.2004.03.002.
9. Denk T, Velitzelos D: First evidence of epidermal structures of *Ginkgo* from the Mediterranean Tertiary. Rev Palaeobot Palynol. 2002, 120: 1-15. 10.1016/S0034-6667(02)00076-3.
10. Gregor HJ: *Ginkgo geissertii* nov. spec. aus dem Pliozän des Elsaß, der Erstnachweis von *Ginkgo*-Samen im mitteleuropäischen Neogen. Doc nat. 1992, 74: 26-31.
11. Kvaček J, Falcon-Lang HJ, Dašková J: A new Late Cretaceous Ginkgolean reproductive structure *Nehvizdyella* gen. nov. from the Czech Republic and its whole-plant reconstruction. Am J Bot. 2005, 92 (12): 1958-1969. 10.3732/ajb.92.12.1958.
12. Zhou ZY, Zheng S: The missing link in *Ginkgo* evolution. Nature. 2003, 423: 821-822. 10.1038/423821a.
13. Florin R: The morphology of *Trichopitys heteromorpha* Saporta, a seed-plant of Palaeozoic age, and the evolution of the female flowers in Ginkgoinae. Acta Horti Bergiani. 1949, 15: 79-109.
14. Taylor TN, Taylor EL, Krings M: Paleobotany - The biology and evolution of fossil plants. 2009, Amsterdam, Boston, Elsevier: Academic Press
15. Archangelsky S: Fossil Ginkgoales from the Ticó flora, Santa Cruz Province, Argentina. Bull Br Mus nat Hist. 1965, 10: 119-137. Geology
16. Kirchner M, van Konijnenburg-van Cittert JHA: *Schmeissneria microstachys* (Presl, 1833) Kirchner et Van Konijnenburg-Van Cittert, comb. nov. and *Karkenia hauptmannii* Kirchner et Van Konijnenburg-Van Cittert, sp. nov., plants with ginkgoalean affinities from the Liassic of Germany. Rev Palaeobot Palynol. 1994, 83: 199-215. 10.1016/0034-6667(94)90069-8.
17. Anderson JM, Anderson H: Heyday of the gymnosperms: systematics and biodiversity of the Late Triassic Molteno fructifications. Strelitzia. 2003, 15: 1-398.
18. Naugolnykh SV: Foliar seed-bearing organs of Paleozoic Ginkgophytes and the early evolution of the Ginkgoales. Paleontol J. 2007, 41 (8): 815-859. 10.1134/S0031030107080011.
19. Virgili C, Cassinis G, Broutin J: Permian to Triassic sequences from selected continental areas of southwestern Europe. Non-Marine Permian Biostratigraphy and Biochronology. 2006, London: Geological Society Special Publication, 265: 231-259.
20. Tralau H: Evolutionary trends in the genus *Ginkgo*. Lethaia. 1968, 1: 63-101. 10.1111/j.1502-3931.1968.tb01728.x.
21. Harris TM, Millington W, Miller J: Ginkgoales and Czekanowskiales. The Yorkshire Jurassic Flora IV. 1974, London: Trustees of The British Museum (Natural History), 1-78.
22. Wang Y, Guignard G, Thévenard F, Dilcher D, Barale G, Mosbrugger V, Yang X, Mei S: Cuticular anatomy of *Sphenobaiera huangii* (Ginkgoales) from the Lower Jurassic of Hubei, China. Am J Bot. 2005, 92 (4): 709-721. 10.3732/ajb.92.4.709.
23. Anderson JM, Anderson HM: Palaeoflora of Southern Africa. MoltenoFormation (Triassic). Vol. 2: Gymnosperms (excluding Dicroidium). 1989, Rotterdam: Balkema
24. Florin R: Die fossilen Ginkgophyten von Franz-Joseph-Land nebst Erörterungen über vermeintliche Cordaitales mesozoischen Alters. II. Allgemeiner Teil. Palaeontographica. 1936, 82 (B): 1-72.
25. Barthlott W, Wollenweber E: Zur Feinstruktur, Chemie und taxonomischen Signifikanz epicuticularer Wachse und ähnlicher Sekrete. Trop subtrop Pflanzenwelt. 1981, 32: 1-67.
26. Schweitzer HJ: Die Makroflora des niederrheinischen Zechsteins. Fortschr Geol Rheinld Westf. 1962, 6: 331-376.
27. Beerling DJ, McElwain JC, Osborne CP: Stomatal responses of the 'living fossil' *Ginkgo biloba* L. to changes in atmospheric $CO_2$ concentrations. J Exp Bot. 1998, 49 (326): 1603-1607. 10.1093/jexbot/49.326.1603.
28. Schweitzer H-J, Kirchner M: Die rhäto-jurassischen Floren des Iran und Afghanistans. 8. Ginkgophyten. Palaeontographica. 1995, 237 (B): 1-58.
29. Wu X, Yang X, Zhou Z: Ginkgoalean ovulate organs and seeds associated with *Baiera furcata*-type leaves from the Middle Jurassic of Qinghai Province, China. Rev Palaeobot Palynol. 2006, 138: 209-225. 10.1016/j.revpalbo.2006.01.009.
30. Zhou Z, Zhang B: *Baiera hallei* and associated ovule-bearing organs from the Middle Jurassic of Henan, China. Palaeontogr B. 1992, B (224): 151-169.
31. Yang XJ, Friis EM, Zhou ZY: Ovule-bearing organs of *Ginkgo ginkgoidea* (Tralau) comb. nov., and associated leaves from the Middle Jurassic of Scania, South Sweden. Rev Palaeobot Palynol. 2008, 149: 1-17. 10.1016/j.revpalbo.2007.09.005.
32. Rothwell GW: The role of development in plant phylogeny: a paleobotanical perspective. Rev Palaeobot Palynol. 1987, 50: 97-114. 10.1016/0034-6667(87)90041-8.

CHAPTER 11

# EST ANALYSIS IN GINKGO BILOBA: AN ASSESSMENT OF CONSERVED DEVELOPMENTAL REGULATORS AND GYMNOSPERM SPECIFIC GENES

Eric D Brenner, Manpreet S Katari, Dennis W Stevenson, Stephen A Rudd, Andrew W Douglas, Walter N Moss, Richard W Twigg, Suzan J Runko, Giulia M Stellari, WR McCombie and Gloria M Coruzzi

## BACKGROUND

*Ginkgo biloba* is a widely popular tree that is native to China and has been cultivated for well over a millennium. In Asia, *G. biloba* is used medicinally and its seeds are also a popular cuisine item. In the West, *Ginkgo* leaf extracts are commonly used for a variety of folk remedies (for review see: [1]) including as a treatment for improving cognitive function [2, 3]. Today's *Ginkgo biloba* is the sole surviving species of an ancient group (Ginkgophytes) of seed plants that may even date from the Permian (approximately 150–200 million years ago) [4]. The genus *Ginkgo* itself goes back to the Jurassic period – approximately 170 million years ago [5]. Although it is widely believed that the survival of *G. biloba* depended upon Buddhist monks, who venerated the tree cultivated in their temple grounds, molecular evidence suggests that some stands in China (Wuchuan, Guizhou) are of natural origin representing vestige populations [6]. As a living fossil, *Ginkgo biloba*has changed little in morphology from its extinct relatives [5]. Along with the Cycadales, Coniferales and Gnetales, the Ginkgoales is one of four orders of non-flowering seed plants (gymnosperms) that form a sister group to the angiosperms (Figure 1).

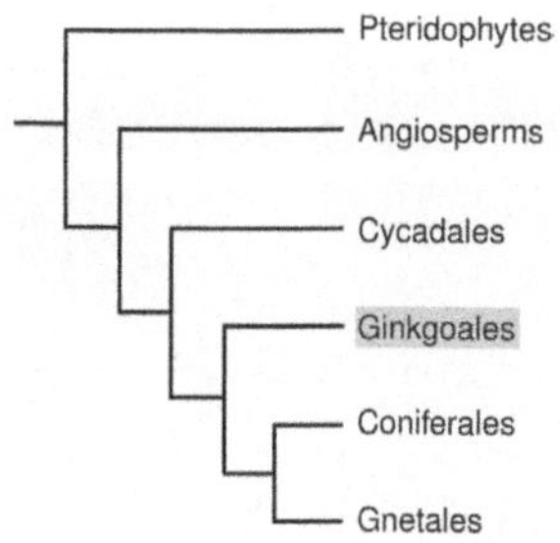


**Figure 1:** Gene tree of extant gymnosperms. *Ginkgo* displays characters suggesting it forms a basal subgroup among the gymnosperms with cycads. Alternately *Ginkgo* is a sister group of the conifers. Here the placement of *Ginkgo* is shown as ambivalent between these two scenarios.

Morphological [7, 8] and molecular analysis have not yet succeeded in defining the precise phylogenetic hierarchy of the four gymnosperm clades [9]. Ginkgo potentially forms a sister group with the Coniferales (partly due to similar characteristics such as axillary branching and simple leaves). Another model, based on molecular sequence data, places *Ginkgo* with the Cycadales [10, 11, 12]. Interestingly cycads and *Ginkgo* both share certain plesiomorphic (ancestral) characters found in early fossil seed plants such as haustorial pollen [13, 14], which release motile male gametes [15] as well as a large four celled opening in the neck of the archegonia [13, 16]. Despite the presence of these and other early seed-plant characteristics, surprisingly little work has been performed on *Ginkgo* and cycads. Some recent molecular [17] and genomic [18] research on cycads have been conducted and molecular studies of *Ginkgo* genes have been initiated as well [19, 20, 21]. However, no genomic work on *Ginkgo biloba* has been completed to date.

**Citation:** Eric D Brenner, Manpreet S Katari, Dennis W Stevenson, Stephen A Rudd, Andrew W Douglas, Walter N Moss, Richard W Twigg, Suzan J Runko, Giulia M Stellari, WR McCombie and Gloria M Coruzzi, EST analysis in Ginkgo biloba: an assessment of conserved developmental regulators and gymnosperm specific genes, DOI: 10.1186/1471-2164-6-143.

To begin our genomic treatment of *Ginkgo biloba*, we focused our initial efforts on developing reproductive and vegetative tissues (Figure 2). Separating *Ginkgo* male and female structures at an early stage is straightforward because *Ginkgo* is strictly a dioecious plant (male and female organs on separate individuals). Organ emergence can generally be pinpointed to a specific time of the year in that both reproductive and vegetative tissues are regularly produced in the beginning of May at our collection site in New York. The reproductive structures, megasporangia bearing ovules (Figure 2A–C) (from female trees) and microsporangia bearing pollen (Figure 2D–F) (from male trees), emerge at the apex of short, determinate (spur) shoots. A discreet flush of leaves are also produced in male and female short-shoots (Figure 2A and 2D) [13, 16, 22]. Long shoots (not shown) exhibit indeterminate growth and yield only vegetative organs. Long-shoots are identifiable by their obvious longer internodes, whereas short shoots (Figure 2A and 2D) have telescoped internodes. Each season, short-shoots might exhibit extensive internode growth and be transformed into long-shoots and vice versa. Consequently, reproductive shoots can become vegetative or vegetative shoots can become reproductive.

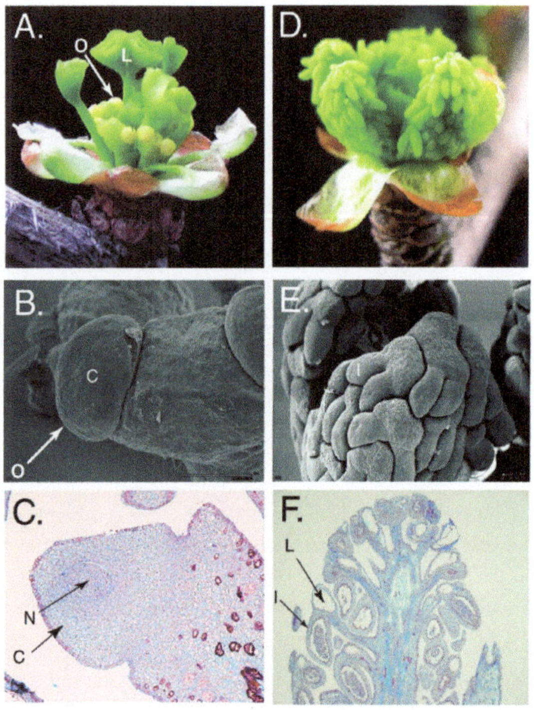


**Figure 2:** Ginkgo male and female short shoots. (A) The fertile female structure (megasporangia) has just emerged from the bud. Two ovules set on a green stalk are visible. Young, unfurled leaves, which have also emerged have extended above the megasporangia. The bracts, which originally enclosed the bud, are now completely opened below the leaves and megasporangia (B) Scanning EM of an ovule, which is completely enclosed by an integument. (C) A longitudinal cross section of the megasporangia reveals the integument enclosing the nucellus. (D) The male reproductive structure is a cluster of microsporangia. In the center of the bud are partly emerged leaves (E) Scanning EM shows two microsporangial lobes containing ripening pollen sacks attached to a stalk. (F) Longitudinal cross section shows a large mucilage containing cavity juxtaposed from the microsporangia filled with immature pollen. C integument; N, nucellus; I, microsporngia; L, mucilage cavity. O, ovule.

Until now, little is known regarding the genetic regulation of development in the oldest living seed plants. In order to uncover the genetic controls directing growth and development in *Ginkgo biloba*, we generated expressed sequence tags (ESTs) from cDNA libraries of very young, recently emerged organs of fertile short-shoots where a large number of regulatory genes are expected to be present. Below is an analysis of these ESTs from *Ginkgo biloba.* In all three tissues examined, vegetative, microsporangia (male), and megasporangia (female), we found a large number of ESTs with similarity to angiosperm developmental genes. Conversely, a certain number of *Ginkgo biloba* ESTs were uncovered with homology to genes only found in gymnosperms and non-seed plants, including a set of *Ginkgo* ESTs that were only common to our cycad EST dataset, further strengthening their classification as gymnosperm specific.

## RESULTS AND DISCUSSION

### Construction of a cDNA Library from *Ginkgo biloba* Fertile and Vegetative Tissue

Young organs (Figure 2A and 2D) were collected during the spring from the opening buds of short shoots immediately after their emergence. At this stage, the megasporangium consists of an axis typically bearing two ovules (Figure 2A–C). The ovule is composed of a single integument surrounding a developing nucellus (Figure2C). The male structure consists of a main axis bearing two or more microsporangia (Figure 2D–F). RNA was extracted from the following organs: megagasporangia, microsporangia and two sets of leaves collected from either male or female trees. mRNA isolated from all four tissues was used to construct four separate cDNA libraries. (Both male and female leaf sequences were pooled during subsequent bioinformatic analysis). Size fractionation was used to enrich for full-length cDNAs during library construction. From this cDNA library, 6,434 sequence reads (Expressed Sequence Tags, ESTs) were generated. All *Ginkgo biloba* EST reads have been deposited in GenBank. It was determined that 3,739 (58%) of the cDNA clones were over 500 bp long. 3618 of the reads were generated from the 5' end of the cDNA, and 2816 were sequenced from the 3' end. Cluster analysis on the EST sequence produced a unigene set of 3,830 contigs consisting of 2,851 singletons and 979 assemblies. Of the clusteredESTs, the longest contig was 2,172 bp. The entire unigene set or complete*Ginkgo* BLAST files can be downloaded at the following website [23]. Each *G. biloba* contig is given a numeric identifier. The constituent ESTs for each contig can be obtained at this website. Additional bioinformatic analysis of the *Ginkgo biloba* dataset can be accessed at the open Sputnik Comparative Genomics Platform at [24]. This site features sequence annotations, peptide sequence predictions, protein domain architectures and putative molecular markers (ISSRs) for the ginkgo EST derived unigenes. The sequence can be downloaded either as a fasta file, a clustered fasta file or as the derived peptide fasta file. In addition, BLAST analysis can be performed with the clustered ESTs from a given ginkgo organ against all genes in *Arabidopsis thaliana* or distinct plant clustered EST datasets using the ViCoGenTa program available at the New York Plant Genomics Consortium website [23].

## Ginkgo Contig Matches To Genes in Angiosperms, Gymnosperms and Non-Seed Plants

TBLASTX (expect < 1eX10-5) was used to compare the *G. biloba* unigenes against all available plant ESTs from TIGR (The Institute for Genomic Research) [25], and the Plant Genome DataBase (Plant GDB) [26]. ESTs from these databases were downloaded and clustered into unigenes, which were used in the comparison. Next the*Ginkgo* unigene set was compared against the *Arabidopsis* and rice genome annotated protein sequences downloaded from TIGR. All genes used in this comparison against *Ginkgo* were divided into one of three taxonomically relevant categories: 1. angiosperms, 2. gymnosperms, and 3. non-seed plants. The angiosperm category encompasses all annotated rice and *Arabidopsis* genes identified from their respective genomic sequences, as well as all higher plant ESTs. The majority of the gymnosperm ESTs came from the conifer groups pine and spruce but also include ESTs generated from the Plant Genomics Consortium containing ESTs from the two other gymnosperm clades, Cycadales and Gnetales. The non-seed plant category consisted of genes from all remaining plant ESTs including ferns, fern allies, bryophytes and algae available with the majority of the sequences originating from *Physcomitrella patens* and *Chlamydamonas reinhardtii.*

A Venn diagram shown in Figure 3A displays the number of *Ginkgo* contigs, which are shared between one or more of the plant EST datasets at low BLAST stringency value (expect < 1eX10-5). From the Venn diagram, it can be seen that a majority of *Ginkgo* unigenes (2749/3830) match genes in other plants, and 1081 have no match to other plant genes. Of those 2749 *Ginkgo biloba* unigenes with matches to other plant genes, a subgroup of 256 unigenes had no corresponding match to genes in the angiosperm dataset. Of these 256*Ginkgo* genes that do not match angiosperms, 4 also match genes in non-seed plants.

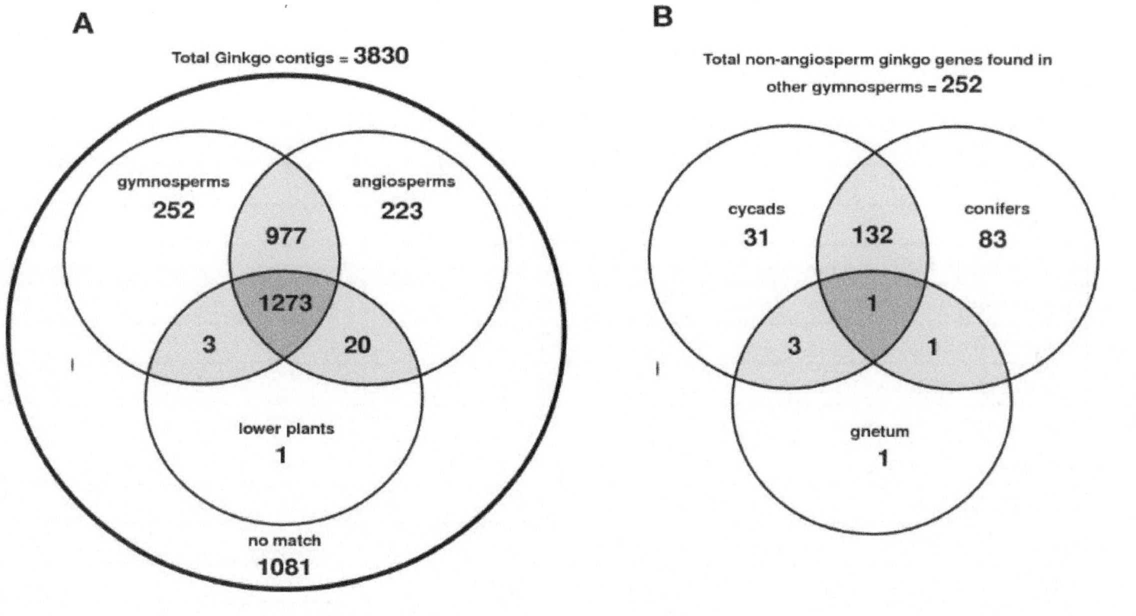


**Figure 3:** A Venn diagram illustrating the number of *Ginkgo* contigs with shared homology to genes found in non-seed plants, gymnosperms and/or angiosperms (A). A BLASTX E value >e 10-5 was used as a cut-off. (B) The *Ginkgo* contigs with similarity to gymnosperms but (no match to angiosperm genes) were further subgrouped according to their BLAST score homology (E value >e 10-5) within gymnosperm taxa.

The 252 *Ginkgo biloba* unigenes that only match gymnosperm genes were next partitioned into matches between the three other gymnosperm orders: Cycadales, Coniferales, and Gnetales (Figure 3B). Since there are significantly more conifer unigenes (>60,000) than cycad unigenes (5459) that were used in this comparison, one would expect that the number of matches between *Ginkgo* and conifers would be significantly greater than matches between *Ginkgo* and cycads. However the actual number of matches between *Ginkgo* and conifers (215) is only slightly more then for *Ginkgo* and cycads (163). In other words despite the fact that there is over 10 times the number of conifer genes then cycad genes used in this comparison, *Ginkgo* matches to conifers are only 1.3 times greater then matches between *Ginkgo* and cycads. Of the matches between *Ginkgo* and gymnosperms, 31 match only cycads, (22 match *Cycas rumphii*, 6 match only unigenes from the cycad species, *Zamia furfuracea*, (753 contigs deposited in Genbank from Brenner et al, unpublished data); and 3 *Ginkgo* contigs match unigenes from both *Cycas rumphii* and *Zamia furfuracea*).

As one might expect, unigenes with matches to other plants are somewhat longer then those that have no match to other plants. Of all Ginkgo unigenes with matches to other plants, 89% are greater than 300 bp, whereas 72% are greater than 300 bp then those *Ginkgo* unigenes with no matches to other plants.

## Common Genes between Cycads and *Ginkgo*

Our comparative analysis of the *Ginkgo* EST dataset builds upon our results from a previous genomic study on the cycad, *Cycas rumphii* [18]. In our current analysis, three

(CB090673:GinkgoA3816, CycadCB089620:GingkoA3730, and CycadCB089926:GinkgoA1532) of the fourteen unigenes from cycads previously found only among gymnosperms (after the full-length clone was sequenced), were also homologues to *Ginkgo* genes found only in gymnosperms. Considering the relatively small number of unigenes from *Ginkgo* (3,830) and cycads (4706) available for our comparative studies, the detection of the same gene match in *Ginkgo* and cycads with homology to only gymnosperms strengthens the argument that these genes are gymnosperms specific.

## *Ginkgo* Matches to Non-Seed Plants But Not Angisosperms

An additional four *Ginkgo* unigenes that are not found in angiosperms were detected in non-seed plants. Three of these *Ginkgo* unigenes (GinkgoA2411, GinkgoA3214 and GinkgoA325) match non-seed plants and other gymnosperm genes whereas the forth *Ginkgo* gene (GinkgoA2273) only matches non-seed plants with similarity to a gene in *Chlamydomonas*.

## Classification of *G. biloba* ESTs by Functional Categories

Each contig from the *Ginkgo* dataset was automatically assigned to a functional category (FunCat) based on its top match against the MIPS FunCat list of functionally annotated gene sequences from *S. cerevisiae* and *A. thaliana* databases using BLASTP. A non-stringent expect value (E-value) of < e-10 was chosen as the threshold. Table 1 below illustrates the relative fraction that each functional category comprises within the entire unigene set compared to our previous study in *Cycas rumphii* [18]. The four largest categories of *Ginkgo*ESTs according to this functional categorization are: "cellular organization" (19%), "metabolism" (11%), "unclassified proteins" (14%), and "protein synthesis" (8%). In general, these same categories are also the highest in the cycad EST library from our previous work, except for "protein synthesis", which appears increased in *Ginkgo*, whereas interestingly, the category of cell growth, cell division and DNA synthesis is reduced in *Ginkgo* compared to cycads.

**Table 1:** Placement of *Ginkgo* unigenes into functional categories (funcats). *Ginkgo* genes with a BLASTP expect value (E-value) of > e-10 were assigned into funcats based on their similarity score. The analysis was performed at the Centre for Biotechnology, Turku, Finland. Previous funcat analysis data from *Cycas rumphii* is shown.

| Functional Category | Ginkgo% | Cycad% |
|---|---|---|
| Metabolism | 11.0 | 10.3 |
| Energy | 5.0 | 5.1 |
| Cell Growth, Cell Division and DNA Synthesis | 5.4 | 9.2 |
| Transcription | 3.6 | 5.0 |
| Protein Synthesis | 8.4 | 5.8 |
| Protein Destination | 5.8 | 8.3 |
| Transport Facilitation | 2.0 | 1.3 |
| Intracellular Transport | 3.4 | 4.3 |
| Cellular Biogenesis | 3.5 | 3.3 |
| Cellular Communication/Signal Transduction | 3.5 | 4.2 |
| Cell Rescue, Defense, Cell Death and Ageing | 7.2 | 6.8 |
| Ionic Homeostasis | 0.5 | 0.1 |
| Cellular Organization | 18.8 | 21.6 |
| Classification Not Yet Clear-Cut | 7.3 | 4.6 |
| Unclassified Proteins | 14.1 | 10.3 |

## *Ginkgo* Genes Involved in Development

Analysis of the *Ginkgo biloba* dataset revealed a number of ESTs with highest BLAST similarity to genes with known roles in higher plant developmental processes. A sampling of some of these genes is shown below (Table 2). These genes included the *Polycomb* gene *CURLY LEAF* [27, 28] as well as *LATERAL ORGAN BOUNDARIES (LOB)* [29], *EARLY FLOWERING 5 (ELF5)* [30], *FLOWERING LOCUS T (FT)* [31], and *CONSTANS* (for review see [32] as well as five ESTs that match *MADS* box genes, some of which appear identical to previously cloned fragments of *MADS* genes from *Ginkgo biloba* including the *G. biloba* ortholog of *AGAMOUS* [19]. Other genes in our EST library have homologies to proteins that regulate development through protein turnover including *SPA-1* [33]*COP1* [34] and *COP9* [35, 36]. This sampling reveals that the EST dataset of *Ginkgo biloba* is a rich source of genes encoding proteins with known roles in development at the transcriptional as well as post-transcriptional stage.

**Table 2:** Similarity match of *Ginkgo* unigenes to genes involved in development. The *G. biloba* unigene set was compared to Genbank using a BLASTP cut-off score < e-5. The top match is listed under subject description. The organ(s) from which the listed ESTs were detected are: G = megagametophyte, I = microgametophyte, and L = leaf.

| Genes in *Ginkgo biloba* with similarity to developmental genes in other plants | | | |
|---|---|---|---|
| **Contig id.** | **organ** | **BLAST homology match** | **E-value** |
| A2725 | G | gi\|1903019\|emb\|CAA71599.1\| curly leaf [*A. thaliana*] | 4.00E-37 |
| A3095 | I | gi\|15231388\|ref\|NP_188001.1\| LOB domain family protein [*A. thaliana*] | 2.00E-21 |
| A2815 | I | gi\|42541156\|gb\|AAS19471.1\| EARLY FLOWERING 5 [*A. thaliana*] | 1.00E-33 |
| A3351 | I | gi\|4903139\|dbj\|BAA77836.1\| extensive homology to FT (FLOWERING LOCUS T, AB027504) | 4.00E-10 |
| A241 | I, L | gi\|41323976\|gb\|AAS00054.1\| CONSTANS-like protein CO1 [*Populus deltoides*] | 1.00E-30 |
| A1591 | L | gb\|AAG43405.1\|AF172931_1 homeobox 1 [*Picea abies*] | 5.00E-23 |
| A1591 | G | gi\|14715183\|emb\|CAC44080.1\| putative MADS-domain transcription factor DEFH7 [*A. majus*] | 2.00E-39 |
| A2737 | G | gi\|30230270\|gb\|AAM76208.1\| AGAMOUS-like MADS-box transcription factor [*Ginkgo biloba*] | 5.00E-28 |
| A2850 | I | gi\|25307918\|pir\|\|S51935 probable MADS-box protein dal1 – Norway spruce dal1 [*Picea abies*] | 5.00E-72 |
| A629 | G, L | gi\|7446554\|pir\|\|T10751 MADS-box protein MADS9 – Monterey pine | 6.00E-17 |
| A352 | G, I | gi\|7446559\|pir\|\|T09571 MADS box protein MADS2 – Monterey pine | 4.00E-75 |
| A914 | G, I | gi\|18401293\|ref\|NP_565632.1\| COP9 / CSN signalosome complex subunit [*A. thaliana*] | 1.00E-45 |
| A2730 | G | gi\|15225760\|ref\|NP_180854.1\| COP1 regulatory protein [*A. thaliana*] | 2.00E-60 |
| A944 | G, I | gi\|30694320\|ref\|NP_849784.1\|argonaute protein(AG01) [*A. thaliana*] | 1.00E-56 |

## CONCLUSION

### The Importance of *Ginkgo* for the study of Plant Evolution

As the sole remaining species of an ancient genus of plants which has survived nearly 170 million years from the Jurassic [5], *Ginkgo biloba* is a taxonomic and geographic relict that may be even older because fossils displaying a "ginkgophyte" vegetative morphology have been found as early as the Permian [4]. *Ginkgo* has a number of plesiomorphic (unspecialized) as well as apomorphic (derived) traits that make it a valuable tool to study the evolution of seed plants. Here we used a genomic approach to investigate the genes involved in regulating development in *Ginkgo* by creating an EST library from both reproductive and vegetative tissues.

Similar to our previous analysis in *Cycas rumphii*, our *Ginkgo* EST study has found significant BLAST homology between *Ginkgo* ESTs with plant genes in gymnosperms and non-seed plants but not in angiosperms. Since ESTs, even when clustered in contiguous genes, may not represent the complete gene [37], often one will find homology to angiosperm genes when the remaining *Ginkgo* gene sequence is revealed. For example in the gymnosperm, *Pinus taeda* EST collection, contigs of increasing length have a higher likelihood then shorter contigs matching a known gene in the *Arabidopsis* genome [38]. However, in this same study a significant subcategory of very long contigs (>1900 bp) have no homology to *Arabidopsis* [38]. It is likely that at least some of these long contigs with no match to angiosperm genes represent full length genes that are specific to the gymnosperm and/or seed-less plant clades. Our strategy to address this question involves screening for these same genes in additional taxa of gymnosperms, in the case of this study, *Ginkgo biloba.* In our analysis three *Ginkgo* genes that were only found in gymnosperms also matched the fourteen ESTs from our previous study of gymnosperm common cycad genes.

Along these same lines, our results suggest the presence of genes common to non-seed plants and gymnosperms that are not present in angiosperms. This non-seed plant/gymnosperm grouping is not surprising considering the fact that gymnosperms have morphologically common characters that are not found in the angiosperms – particularly in their reproductive structures. For example, the megagametophyte is highly reduced both in cell number and in structural organization in angiosperms when compared to gymnosperms. Although these results cannot say for certain that these genes are specific to non-seed plants and gymnosperms, or more specifically that these genes are found in gymnosperm structures that are not found in seed plant, it nonetheless represents an important starting point to correlate the presence or absence of gymnosperm genes in angiosperms and/or lower plants.

## Are cycads and *Ginkgo* Sister Taxa?

One result from our study found that the number of *Ginkgo* contig matches to conifers are only 1.3 times greater then matches between *Ginkgo* and cycads despite the fact that there is over 10 times the number of conifer genes then cycad genes used in this comparison,. Taken together these results might indicate a closer evolutionary association between *Ginkgo* and cycads then between *Ginkgo* and conifers. This bias towards cycad/*Ginkgo* similarity correlates with the fact that the majority of molecular phylogenetic studies place as the cycads sister group to *Ginkgo*. Hopefully, this preliminary data will encourage further phylogenomic studies to fully resolve the hierarchy among extant gymnosperm orders. Until the full genome sequence becomes available for key gymnosperm taxa, EST sampling provides an important initial step for large scale identification of molecular markers to generate robust phylogenetic trees.

## Developmental Regulators in *Ginkgo*

In *Ginkgo biloba* we note here a variety of genes with similarity to developmental regulators in angiosperms. We also note below that homologues to some of these developmental regulators are also present in our *Cycas rumphii* library as either orthologs to those found in higher plants or at least, belonging to the same gene family. An EST from *Ginkgo biloba* that was detected in the megagametophyte library has high similarity to the *Arabidopsis CURLY LEAF* (*CLC*) gene, which belongs to the *Polycomb*- group proteins (PcGs). PcGs epigenetically regulate downstream target genes [28]. PcGs modify chromatin-protein complexes that repress homeotic gene transcription and influence cell proliferation. In *Arabidopsis* PcG genes have been shown to regulate MADS box genes [39]. The *CLC* protein product regulates the expression of *AGAMOUS* [27], a gene controlling floral organ identity [40]. Interestingly, an ortholog for angiosperm *AGAMOUS* was also detected in the *Ginkgo* megagametophyte library (Table 2). *Ginkgo AGAMOUS*, (previously named *GBM5*) was identified in a study where the MADS domains were examined in *Ginkgo* [19]. In this work *Ginkgo AGAMOUS* was shown via RT-PCR to be expressed in not only female but also in male and vegetative tissue. In our analysis, five total MADS box homologues were also detected in the *Ginkgo* EST dataset. Three of the *Ginkgo* ESTs from our library, GinkgoA2340, GinkgoA2730 and GinkgoA2850, are perfectly identical to the MADS domain gene fragments previously cloned by [19] as degenerate PCR products. The other two unigenes from our dataset have homologies to MADS genes (GinkgoA629 and GinkgoA352), but do not specifically match any of the PCR fragments isolated in their study. These two MADS box unigenes either do not include the small region amplified in their degenerate PCR screen or could alternative be unique MADS genes not isolated in their study. Unlike the degenerate primer approach used to isolate MADS genes, our EST approach offers the additional advantage of cloning entire genes or at least substantially large gene fragments. Among the few developmental genes examined in gymnosperms, most attention has focused on the expression of MADS homologs [41, 42].

Other developmental genes found in the *Ginkgo* EST library include those with homology to regulators of flowering such as *EARLY FLOWERING 5 (ELF5)*, which controls the levels of the gene *FLC*, which itself is a central regulator of flowering [30]. Another *Ginkgo* EST includes *FLOWERING LOCUS T (FT)*, which belongs to a small family of genes (*FT/TFL1)* that act to promote flowering as a downstream component from *CONSTANS* [43]. *CONSTANS* is a transcription factor that has a critical role integrating circadian rhythms and light signals (for review see [32]). As one would expect an EST homolog for the *CONSTANS* gene family was found in *Ginkgo*. *CONSTANS* belongs to a large gene family, which may have redundant roles in plants [44]. Not surprisingly, we also found homologs to *CONSTANS* in our previous study on cycad leaf ESTs [18]. In that flowering plants are believed to have evolved from gymnosperms, a survey of *CONSTANS*, *ELF*, and *FT* in gymnosperms, particularly in very young reproductive tissue might help define the origins of reproductive induction in non-flowering plants. Among the other genes related to developmental regulators includes a homologue to *LATEROL ORGAN BOUNDARIES (LOB)* domain gene family which in *Arabidopsis* has over 40 members [29]. The molecular mechanism of *LOB* domain containing genes is unknown, but one gene in *Arabidopsis, ASYMMETRIC LEAVES2*, is required for normal leaf development, by potentially acting as a regulatory repressor of *KNOX* genes [45]. A *KNOX* homolog is also present in our EST library and was found in male reproductive tissues and *HOX* genes were also detected in our previous analysis in *C. rumphii*

Another important component regulating development occurs at the level of protein degradation. A gene recognized in our EST library includes *COP1*. COP1, serves as an E3 ubiquitin targeting photomorphogenic factors such as *HY5* for degradation [33]. Another *Ginkgo* EST from the library has highest similarity to *COP9*. In our previous EST analysis in *Cycas rumphii* an EST was also isolated with similarity to COP9 [18]. COP9 is a subunit of the COP9 signalosome complex that controls multiple signaling pathways that regulate development in all eukaryotes [35, 36]. In *Arabidopsis*, the *cop9* and *cop1* mutants are constitutively photomorphogenic in dark grown seedlings [46]. Unlike angiosperms, seedlings from conifers are constitutively photomorphogenic when grown in the dark [47, 48]. In *Ginkgo*, chlorophyll and chloroplast development is completely dependent on light, however this process proceeds at a markedly slower pace then in flowering plants. That is, photomorphogenic development in *Ginkgo* seedlings is strongly delayed after transfer from dark grown conditions to light grown conditions when compared to seed plants [20, 21]. The dark grown phenotype of cycads is unreported. Considering this variability in photomorphic development among and between the gymnosperms and the angiosperms, the discovery of genes encoding photomorphogenic regulators in gymnosperms will help understand the evolution of photomorphogenesis in seed plants. Taken together, our genomics analysis of *Ginkgo biloba* is an important additional step to analyze the role of molecular development of early seed plants. Thus the stage is set to further determine the role of these genes during the development of ancillary structures found between *Ginkgo*, cycads and other gymnosperms with higher plants as well as the role of those in structures that are unique to gymnosperms and/or the non-seed plants as a step to understand the evolution of the seed plant habit.

## METHODS

### Tissue Collection and Library Construction and DNA purification

Newly emerged microsporangia from accession 76163B, megasporangia from accession 76163D, and immature leaves from both accessions were collected from newly opened buds of *Ginkgo biloba* growing in the New York Botanical Garden outdoor collection on April 12, 2002. Organs were snap frozen in liquid nitrogen. RNA was collected from each organ and a cDNA library was constructed from fractionated cDNA according to [49].

### Microscopy

Ginkgo apices were collected on April 19. Bract tissues were removed from the apex leaving the leaves and reproductive tissue, which was fixed in FAA (50% ethanol, 5% glacial acetic Acid, 3.7% formaldehyde) under vacuum (20 In. Hg) at room temperature. Fresh FAA was vacuum infiltrated two additional times. Tissue was stored in 70% ethanol at 4°C.

For histology, tissue was prepared by sequential (overnight 4 C incubation at each alcohol grade) dehydration in 80, 90, 95 and finally 100% ethanol plus Eosin Y (National Medicinal Products) followed by two treatments in 100% ethanol for 2 hours at room temperature. The tissue was next placed in a 1:1 solution of ethanol and toluene, then twice in toluene alone, each time for 2 hours at room temperature. The tissue was then placed in toluene with a quarter-volume of paraffin (PARAPLAST X-TRA® (Fisher)) chips at 60°C overnight. The tissue was then embedded in melted paraffin with six wax changes over the course of three days at 55°C. Apices were sectioned on a MICROM HM 355 microtome. 8 μm thick sections were taken using a blade angle of 9°. The tissue was stained with Astra Blue and Safranin. After mounting on slides, sections were imaged using a Nikon DXM1200F digital microscope camera.

For scanning electron microscopy, fixed materials were dissected, dehydrated in ethanol and critical point dried in a Denton critical point dryer. Dried materials were affixed to aluminium E. M. stubs and coated with between 80–240 A of palladium in a Hummer II Sputter Coater. Coated materials were then observed using a Jeol scanning electron microscope at 15 or 20 kV. Images were digitally recorded and evaluated using Adobe Photoshop 9.0.

### EST Sequencing and Gene Analysis

Plasmid DNA was collected as described in the manual (Stratagene), catalogue number 200450 in the in vivo mass excision section. Sequence analysis was performed at CSHL using an ABI 3700 Capillary sequencer for separation and nucleotide detection. Reactions were performed using a 1/16 Big Dye Terminator. Sequencing was performed with either the -21 M13 forward and/or reverse primer. ESTs were assembled using Phrap [50,51] and clustered into contigs using the CAP3 program [52]

### Peptide Extraction

Peptide sequences were derived for all unigenes using the ESTScan application [53] run with the default parameters. Prior to the ESTScan predictions, a *Ginkgo* species-specific ESTScan model was created. ESTScan was trained with *Ginkgo* ORFs identified from the best match of BLASTX analyses performed on the unigene sequence against the Swissprot protein database. All BLASTX matches were filtered using the arbitrary expectation value of 1e-10.

### Sequence Annotation

Sequence annotation on each of the *Ginkgo* cluster consensus sequences and derived peptides were performed within the openSputnik application [54]. Results were assessed for possible contamination by searching for homology to the *E. coli* and human genomes and were scored for homology to a wide range of non-coding RNAs and plant chloroplast and mitochondrial genomes. Homology searches were performed using the BLAST application [55] and results were filtered using the expectation value < 1e-10. Functional assignment was performed on both cluster consensus sequence and the peptide sequence. Assignments were made using BLASTX and BLASTP respectively against the MIPS catalogue of functionally assigned proteins (funcat) [50, 51], tentative functional assignments were filtered using the expectation value < 1e-10.

### Categorization of *Ginkgo* Contigs

All *Ginkgo* contigs sequences were aligned against a PlantEST database using TBLASTX [55] and BLASTX against the NR (aa) database. The PlantEST database was created by downloading all plant ESTs in GenBank and assembling them using Phrap [50, 51]. Todd Wood from Clemson University provided the PERL script that creates the PlantEST databases as described above. The NR (aa) database is a non-redundant database of protein sequences from GenBank.

### Determination of Gymnosperm Specific Genes

All available plant ESTs were downloaded from GenBank and separated into three datasets consisting of angiosperms (monocots and dicots), gymnosperms, or non-seed plants (ferns, mosses and algae). Downloaded ESTs were assembled using Phrap [50, 51]. All matches with an expect value < 1e 10-5 are considered significant.

## DECLARATIONS

### Acknowledgements

We would like to thank Vivekanand Balija for sequence generation. Funding for this work comes from the Plant Genomics Consortium. The plant Genomics Consortium is made possible by the generosity of the Altria Group, Inc., The Mary Flagler Cary Charitable Trust, The Eppley Foundation for Research, Inc., The Ambrose Monell Foundation, The Wallace Genetic Foundation, Inc., and the National Science Foundation Plant Genome Group grant number DBI-0421604.

## REFERENCES


1. Hori T, Ridge RW, Tulecke W, Del Tredici P, Tremouillaux-Guiller J, Tobe H: Ginkgo biloba-A Global Treasure. 1997, Tokyo, Springer
2. Curtis-Prior P, Vere D, Fray P: Therapeutic value of Ginkgo biloba in reducing symptoms of decline in mental function. J Pharm Pharmacol. 1999, 51: 535-541. 10.1211/0022357991772817.
3. Gold PE, Cahill L, Wenk GL: The lowdown on Ginkgo biloba. Sci Am. 2003, 288: 86-91.
4. Rothwell GW, Holt B: Fossils and phenology in the evolution of Ginkgo biloba. Ginkgo-biloba-A Global Treasure. Edited by: Hori T, Ridge RW, Tulecke W, Del Tredici P, Trémouillaux-Guiller J and Tobe H. 1997, Tokyo, Springer-Verlag, 223-230.
5. Zhou Z, Zheng S: The missing link in Ginkgo evolution. Nature. 2003, 423: 821-822. 10.1038/423821a.
6. Fagard M, Boutet S, Morel JB, Bellini C, Vaucheret H: AGO1, QDE-2, and RDE-1 are related proteins required for post-transcriptional gene silencing in plants, quelling in fungi, and RNA interference in animals. Proc Natl Acad Sci USA. 2000, 97: 11650-11654. 10.1073/pnas.200217597.
7. Doyle JA, Donoghue MJ: Seed plant phylogeny and the origin of angiosperms: an experimental cladistic approach. Botanical Reviews. 1986, 52: 321-431.
8. Donoghue MJ, Doyle JA: Seed plant phylogeny: Demise of the anthophyte hypothesis?. Curr Biol. 2000, 10: R106-9. 10.1016/S0960-9822(00)00304-3.
9. Magollón S, Sanderson MJ: Relationships among seed plants inferred from highly conserved genes: sorting confilicting phylogenetic signals among ancient lineages. Am J Bot. 2002, 89: 1991-2006.
10. Chaw SM, Parkinson CL, Cheng Y, Vincent TM, Palmer JD: Seed plant phylogeny inferred from all three plant genomes: monophyly of extant gymnosperms and origin of Gnetales from conifers. Proc Natl Acad Sci USA. 2000, 97: 4086-4091. 10.1073/pnas.97.8.4086.
11. Bowe LM, Coat G, dePamphilis CW: Phylogeny of seed plants based on all three genomic compartments: extant gymnosperms are monophyletic and Gnetales' closest relatives are conifers. Proc Natl Acad Sci USA. 2000, 97: 4092-4097. 10.1073/pnas.97.8.4092.
12. Hasebe M: Molecular phylogeny of Ginkgo biloba: close relationship between Ginkgo biloba and cycads. Ginkgo biloba-a Global Treasure. Edited by: Hori T, Ridge RW, Tulecke W, Del Tredici P, Trémouillaux-Guiller J and Tobe H. 1997, Tokyo, Springer-Verlag, 173-181.
13. Chamberlain CJ: Gymnosperms. Structure and Evolution. 1935, Chicago, University of Chicaco Press
14. Friedman WE: Morphogenesis and experimental aspects of growth and development of the male gametophyte of Ginkgo biloba in vitro. American Journal of Botany. 1987, 74: 1816-1830.
15. Ikeno S, Hirase S: Spermatozoids in gymnosperms. Ann of Bot. 1897, 11: 344-345.
16. Foster AS, Gifford EM: Comparative Morphology of Vascular Plants. Edited by: Kennedy D and Park RB. 1974, San Francisco, W. H. Freeman and Company, second edition
17. Zhang P, Tan HT, Pwee KH, Kumar PP: Conservation of class C function of floral organ development during 300 million years of evolution from gymnosperms to angiosperms. Plant J. 2004, 37: 566-577. 10.1046/j.1365-313X.2003.01983.x.
18. Brenner ED, Stevenson DW, McCombie RW, Katari MS, Rudd SA, Mayer KF, Palenchar PM, Runko SJ, Twigg RW, Dai G, Martienssen RA, Benfey PN, Coruzzi GM: Expressed sequence tag analysis in Cycas, the most primitive living seed plant. Genome Biol. 2003, 4: R78-10.1186/gb-2003-4-12-r78.
19. Jager M, Hassanin A, Manuel M, Guyader HL, Deutsch J: MADS-box genes in Ginkgo biloba and the evolution of the AGAMOUS Family. Mol Biol Evol. 2003, 20: 842-854. 10.1093/molbev/msg089.
20. Chinn E, Silverthorne J: Light-dependent chloroplast development and expression of a light-harvesting chlorophyll a/b-binding protein gene in the gymnosperm Ginkgo biloba. Plant Physiol. 1993, 103: 727-732. 10.1104/pp.103.3.727.
21. Chinn E, Silverthorne J, Hohtola A: Light-regulated and organ-specific expression of types 1, 2, and 3 light-harvesting complex b mRNAs in Ginkgo biloba. Plant Physiol. 1995, 107: 593-602. 10.1104/pp.107.2.593.
22. Bierhorst DW: Morphology of Vascular Plants. 1971, , Macmillan
23. The New York Plant Genomics Consortium. [http://nypgenomics.org]
24. OpenSputnik Comparative Genomics Platform. [http://sputnik.btk.fi/]
25. The Institute for Genomic Research. [http://www.tigr.org/]
26. Plant Genomic Database. [http://www.plantgdb.org/]
27. Goodrich J, Puangsomlee P, Martin M, Long D, Meyerowitz EM, Coupland G: A Polycomb-group gene regulates homeotic gene expression in Arabidopsis. Nature. 1997, 386: 44-51. 10.1038/386044a0.
28. Kohler C, Grossniklaus U: Epigenetic inheritance of expression states in plant development: the role of Polycomb group proteins. Curr Opin Cell Biol. 2002, 14: 773-779. 10.1016/S0955-0674(02)00394-0.
29. Shuai B, Reynaga-Pena CG, Springer PS: The lateral organ boundaries gene defines a novel, plant-specific gene family. Plant Physiol. 2002, 129: 747-761. 10.1104/pp.010926.
30. Noh YS, Bizzell CM, Noh B, Schomburg FM, Amasino RM: EARLY FLOWERING 5 acts as a floral repressor in Arabidopsis. Plant J. 2004, 38: 664-672. 10.1111/j.1365-313X.2004.02072.x.
31. Weigel D, Ahn JH, Blazquez MA, Borevitz JO,

Christensen SK, Fankhauser C, Ferrandiz C, Kardailsky I, Malancharuvil EJ, Neff MM, Nguyen JT, Sato S, Wang ZY, Xia Y, Dixon RA, Harrison MJ, Lamb CJ, Yanofsky MF, Chory J: Activation tagging in Arabidopsis. Plant Physiol. 2000, 122: 1003-1013. 10.1104/pp.122.4.1003.

32. Hayama R, Coupland G: The molecular basis of diversity in the photoperiodic flowering responses of Arabidopsis and rice. Plant Physiol. 2004, 135: 677-684. 10.1104/pp.104.042614

33. Saijo Y, Sullivan JA, Wang H, Yang J, Shen Y, Rubio V, Ma L, Hoecker U, Deng XW: The COP1-SPA1 interaction defines a critical step in phytochrome A-mediated regulation of HY5 activity. Genes Dev. 2003, 17: 2642-2647. 10.1101/gad.1122903.

34. Seo HS, Yang JY, Ishikawa M, Bolle C, Ballesteros ML, Chua NH: LAF1 ubiquitination by COP1 controls photomorphogenesis and is stimulated by SPA1. Nature. 2003, 423: 995-999. 10.1038/nature01696.

35. Chamovitz DA, Glickman M: The COP9 signalosome. Curr Biol. 2002, 12: R232-10.1016/S0960-9822(02)00775-3.

36. Hellmann H, Estelle M: Plant development: regulation by protein degradation. Science. 2002, 297: 793-797. 10.1126/science.1072831.

37. Rudd S: Expressed sequence tags: alternative or complement to whole genome sequences?. Trends Plant Sci. 2003, 8: 321-329. 10.1016/S1360-1385(03)00131-6.

38. Kirst M, Johnson AF, Baucom C, Ulrich E, Hubbard K, Staggs R, Paule C, Retzel E, Whetten R, Sederoff R: Apparent homology of expressed genes from wood-forming tissues of loblolly pine (Pinus taeda L.) with Arabidopsis thaliana. Proc Natl Acad Sci U S A. 2003, 100: 7383-7388. 10.1073/pnas.1132171100.

39. Katz A, Oliva M, Mosquna A, Hakim O, Ohad N: FIE and CURLY LEAF polycomb proteins interact in the regulation of homeobox gene expression during sporophyte development. Plant J. 2004, 37: 707-719. 10.1111/j.1365-313X.2003.01996.x.

40. Bowman JL, Smyth DR, Meyerowitz EM: Genetic interactions among floral homeotic genes of Arabidopsis. Development. 1991, 112: 1-20.

41. Brenner ED, Stevenson DW: Genomic Approaches to Understand the Origins of Seeds. Landscapes, Genomics and Transgenic Conifers. Edited by: Williams CG. 2005, Amsterdam, Publisher: Kluwer-Springer

42. Theissen G, Becker A, Di Rosa A, Kanno A, Kim JT, Munster T, Winter KU, Saedler H: A short history of MADS-box genes in plants. Plant Mol Biol. 2000, 42: 115-149. 10.1023/A:1006332105728.

43. Kardailsky I, Shukla VK, Ahn JH, Dagenais N, Christensen SK, Nguyen JT, Chory J, Harrison MJ, Weigel D: Activation tagging of the floral inducer FT. Science. 1999, 286: 1962-1965. 10.1126/science.286.5446.1962.

44. Lagercrantz U, Axelsson T: Rapid evolution of the family of CONSTANS LIKE genes in plants. Mol Biol Evol. 2000, 17: 1499-1507.

45. Lin WC, Shuai B, Springer PS: The Arabidopsis LATERAL ORGAN BOUNDARIES-domain gene ASYMMETRIC LEAVES2 functions in the repression of KNOX gene expression and in adaxial-abaxial patterning. Plant Cell. 2003, 15: 2241-2252. 10.1105/tpc.014969.

46. Suzuki G, Yanagawa Y, Kwok SF, Matsui M, Deng XW: Arabidopsis COP10 is a ubiquitin-conjugating enzyme variant that acts together with COP1 and the COP9 signalosome in repressing photomorphogenesis. Genes Dev. 2002, 16: 554-559. 10.1101/gad.964602.

47. Bogdanovic M: Chlorophyll formation in the dark. Physiol Plant. 1973, 29: 17-18.

48. Peer W, Silverthorne J, Peters JL: Developmental and light-regulated expression of individual members of the light-harvesting complex b gene family in Pinus palustris. Plant Physiol. 1996, 111: 627-634. 10.1104/pp.111.2.627.

49. Brenner ED, Stevenson DW, Twigg RW: Cycads: evolutionary innovations and the role of plant-derived neurotoxins. Trends Plant Sci. 2003, 8: 446-452. 10.1016/S1360-1385(03)00190-0.

50. Ewing B, Green P: Base-calling of automated sequencer traces using phred. II. Error probabilities. Genome Res. 1998, 8: 186-194.

51. Ewing B, Hillier L, Wendl MC, Green P: Base-calling of automated sequencer traces using phred. I. Accuracy assessment. Genome Res. 1998, 8: 175-185.

52. Huang X, Madan A: CAP3: A DNA sequence assembly program. Genome Res. 1999, 9: 868-877. 10.1101/gr.9.9.868.

53. Iseli C, Jongeneel CV, Bucher P: ESTScan: a program for detecting, evaluating, and reconstructing potential coding regions in EST sequences. Proc Int Conf Intell Syst Mol Biol. 1999, 138-148.

54. Rudd S: openSputnik--a database to ESTablish comparative plant genomics using unsaturated sequence collections. Nucleic Acids Res. 2005, 33 (Database Issue): D622-7.

55. Altschul SF, Gish W, Miller W, Myers EW, Lipman DJ: Basic local alignment search tool. J Mol Biol. 1990, 215: 403-410. 10.1006/jmbi.1990.9999.

CHAPTER 12

# FEMALE SHORT SHOOT AND OVULE DEVELOPMENT IN GINKGO BILOBA L. WITH EMPHASIS ON STRUCTURES ASSOCIATED WITH WIND POLLINATION

Biao Jin, Di Wang, Yan Lu, Xiao Xue Jiang, Min Zhang, Lei Zhang, and Li Wang

## INTRODUCTION

Pollination is an important process in the life cycle of seed plants. During the evolution of angiosperms, diverse forms of pollination with complex mechanisms have evolved. Among the gymnosperm species, the majorities are wind pollinated (anemophilous) and their pollen grains are carried from their initial location to female parts via air flow [1]. Thus, wind pollination in gymnosperms is a random and inefficient process with high pollen losses. There have been many studies documenting how gymnosperms achieve pollination via such relatively passive processes. For example, the pollination process has been studied in several conifer families, such as the Podocarpaceae, Cupressaceae, Taxaceae, and Pinaceae, and the proposed evolution of different pollination mechanisms has been widely discussed [2–4]. Based on the results of such studies, it was concluded that adaptive characteristics to wind pollination that contribute to the pollination mechanisms of gymnosperms involve (1) orientation of the ovulate cone, as well as ovule structure and position at the time of pollination [2, 5]; (2) pollen characteristics, especially the presence or absence of sacci [6, 7]; (3) whether a pollination drop appears at pollination or not [8, 9]. Among these characteristics, the ovule as the female reproductive organ varies widely in its morphology among gymnosperms. Studies on ovule development in gymnosperms have revealed various ovule structures that have evolved to increase the chances of pollination success. For instance, the tip of the integument in the ovule shows a variety of shapes for pollen collection. Owens et al. [2] reported that in the Cupressaceae, Taxodiaceae, and Cephalotaxaceae, the ovules are flask-shaped, variable in number, and attached in the axil of the bract scale, and the integument tip has a narrow neck and a small, unspecialized micropyle. However, in Abies species, the integument tip forms a short funnel, often with fluted edges, around a large micropyle [10]. In addition, the orientation and morphology of the megastrobilus (female cone) are important features for wind pollination. For example, in some conifer species, ovulate cones may enhance the probability of pollen entrapment by aerodynamically predetermining airflow patterns around scale-bract complexes [11]. Together, the results of such studies showed that the diverse morphologies and structures of female reproductive organs among gymnosperms are an evolutionary consequence of adaptation for effective pollen reception. Thus, it is important to study ovule development of gymnosperms to further clarify the roles of various ovule characteristics in the wind pollination mechanism.

Ginkgo biloba L. is the unique living species of the ancient lineage of Ginkgophyta, which originated in the Paleozoic Carboniferous period; thus, it is regarded as a "living fossil" [12]. The female reproductive organ of G. biloba, particularly the ovule, has been described and studied extensively because of several unique and primitive characteristics [13, 14]. The ovule differentiates into the integument, nucellus, and collar [15]. Primarily, the integument develops in a circumferential manner and becomes slightly lobed during the formation of the

**Citation:** Biao Jin, Di Wang, Yan Lu, et al., "Female Short Shoot and Ovule Development in Ginkgo biloba L. with Emphasis on Structures Associated with Wind Pollination," ISRN Botany, vol. 2012, Article ID 230685, 9 pages, 2012. doi:10.5402/2012/230685.

micropyle [16]. After that, the nucellar tissues are derived from periclinal divisions in the hypodermal layer, and the tip cells of the nucellus die, resulting in formation of the pollen chamber [17]. Almost simultaneously, one cell in the sporangial tissues differentiates to form a spore mother cell, which further develops to form the female gametophyte, while the other cells differentiate to form the tapetum [18]. During the pollination period, the ovule produces a pollination drop [19, 20]. Although these studies have described the ovule developmental process in G. biloba, most have focused on either the morphological structures or the floristic features. Since G. biloba is a wind-pollinated dioecious plant and this lineage has existed for more than 200 million years, the Ginkgo ovule is obviously a survivor of evolutionary consequences related to wind pollination. However, so far, there have been no systematic studies of how ovule structure has differentiated to adapt to wind pollination. Moreover, while previous studies have provided information about the formation of the female gametophyte, the developmental conditions of the female gametophyte before and during the pollination period remain unclear [21, 22]. In addition, the erect ovules surrounding by several fan-shaped leaves constitute the overall female short shoot system, which clearly works as a whole during wind pollination. However, the specific spatial arrangement and architecture of the short shoot have received little attention, despite the fact that these are important features for comprehensively understanding the female reproductive system associated with wind pollination in G. biloba.

Based on our previous studies [21, 23], in this paper we investigated the morphological and structural development of female short shoots and ovules in G. biloba, focusing on ontogenetic features and their relationships with pollen capture. We used continuous image acquisition, semithin sectioning, and SEM observation techniques to observe the morphological and anatomical changes. We present morphological data that clarify the relationships between female reproductive structures and wind pollination in G. biloba, providing further insights into the wind pollination mechanisms in gymnosperms.

## MATERIALS AND METHODS

### Plant Material

Healthy female G. biloba L. trees were selected from the Ginkgo Experimental Station at Yangzhou University, Yangzhou, China (32°20′N, 119°30′E). Ovules were collected weekly from August 2008 to April 2009. Between mid-March and early April, the period for female gametophyte and pollen chamber development, ovules were collected daily. Ovules in bud were dissected out using a razor blade and a steel needle, and the nucellus together with the pollen chamber was manually isolated from ovules under a dissecting microscope.

### Digital Camera Observation Studies

Potential female buds were randomly sampled and tagged to follow their development. Photographs were taken against a black background each week from August 2008 to mid-March 2009 using a DSC-H7 Sony digital camera. Between mid-March and late April 2009, female buds and ovules were photographed daily until pollination ended.

### Semithin Section Studies

The material was prepared for semi-thin sectioning as follows: at each sampling, 30 fresh ovules were prefixed in 2.5% (v/v) glutaraldehyde in 0.1 mol/L phosphate buffer (pH 7.2) for 20 h at room temperature, postfixed in 1% (w/v) osmium tetroxide for 6 h at 4°C, rinsed twice, dehydrated in an alcohol series (30, 50, 70, 80, 90, 95, and 100%, 15 min each step), treated twice for 10 min with propylene oxide, infiltrated with 1 : 1 propylene oxide/resin in embedding capsules overnight, and then finally embedded in Spurr's resin. Semi-thin sections (1 $\mu$m) were cut with an Leica EM UC6 ultramicrotome (Leica Microsystems GmbH, Wetzlar, Germany) and collected onto glass slides, then stained with 1% (w/v) toluidine blue O dissolved in 1% (w/v) sodium borate prior to examination [24]. Observations and photography were performed under a microscope (Zeiss Axioskop 40: Carl Zeiss Shanghai Company Ltd., Shanghai, China).

### SEM Observation Studies

Samples were prepared for SEM observations as follows: at each developmental stage, 30 intact ovules were dissected from buds under a dissecting microscope. The samples were fixed in 2.5% glutaraldehyde (in 0.1 mol/L phosphate buffer, pH 7.0) for 4 h at room temperature and then kept for 12 h at 4°C. Then, they were dehydrated in a graded ethanol series (30, 50, 70, 80, 90, 95 and 100%, 15 min each step). After drying at the critical point, the specimens were coated with gold using an EMITECH K550 sputter coater for 150 s, and observed under a Hitachi H-300 scanning electron microscope (Tokyo, Japan) at 15 kv accelerating voltage [25].

## RESULTS

### Female Cone Initiation and Sprouting Process

The female buds of G. biloba, borne at the apex of the shortshoots, were conical-shaped with the apex slightly acuminate and nearly enclosed by seven to nine thick-bud scales from September to the next February (Figure1(a)). In late March, the bud scales began to unfold, and over several days, the outer brown scales opened to expose the inner green ones (Figure 1(b)). Continued unfolding of the bud scales resulted in the appearance of the leaves and ovules (Figures 1(c)–1(e)). Most of the ovules were concealed in the bud because both the leaf stalks and ovular stalks were very short. At this stage, the leaves were young and curved inward from the two lateral sides (Figures 1(d) and 1(e)). As the leaves developed, all of the ovules in the buds became visible (Figures 1(f) and 1(g)). At this stage, the ovules were yellow. Afterwards, the leaves began to spread gradually and the leaf stalks rapidly elongated. At the same time, the ovular stalks began to extend rapidly, pushing the ovules outward from the bud (Figure 1(h)). This resulted in a spiral arrangement of the ovules and leaf tufts on the female short shoot. The fan-shaped leaves spread out and bent outwards and downwards (Figure 1(i)). During this period, the ovules were still yellow, erect, and inclined obliquely at various angles. During the pollination period, in mid-April, almost all of the female short shoots were open with several ovules in the center (Figures 1(j) and 1(k)), and each ovule produced a pollination drop on the micropyle (Figure 1(l)). After pollination, the color of ovules changed from yellow to green as the pollination drop disappeared (Figure1(m)). Usually, a single general stalk bore two ovules that were directly attached to the stalk via ovular collars (Figure 1(n)). However, the general stalk of ovules occasionally differentiated into dichotomous separated stalks (ovule stalk), which connected the two ovular collars with the general stalk (Figure 1(o)).

**Figure 1:** Female cone development of G. biloba. (a) Female cone enclosed by thick bud scales during winter. ((b)–(d)) Bud scales opened gradually in late March. (e) Ovules began to appear in early April. ((f)–(h)) Ovular stalks elongated and leaves grew rapidly. (i) Leaves and ovules spirally arranged on the short shoot apex just before pollination. (j) Leaves and ovules during pollination stage (mid-April). (k) Long shoot with several female short shoots during pollination. (l) Pollination drop secreted from micropyle. (m) Ovules changed from yellow to green after pollination. (n) Ovules attached directly to general stalk. (o) Ovule with ovular stalks. bs, bud scale; gs, general stalk; le, leaf; o, ovule; os, ovular stalk; pd, pollination drop; ss, short shoot. Bars: ((a)–(f)) = 50 mm; ((i)–(k)) = 1 cm; ((l)–(m)) = 5 mm; ((n)–(o)) = 2 mm.

## Differentiation of Ovule Tissues

The ovule differentiation process was observed using scanning electron microscopy (SEM) and semi-thin sections. Our materials collected from Yangzhou in January 2009 had ovules in the early stage of development (Figure 2(a)), although we did not know the exact time of ovule initiation (maybe in December or earlier). The general stalk primordia of ovules were initiated at and around the boundary between bud scales and leaves in the first year, and they remained in this stage for a long time (about 1 month). Until the next January, the tip of the general stalk was swollen, giving the stalk an ellipse-like appearance (Figure 2(b)). Afterwards, the swollen primordium stretched on both lateral sides to differentiate into the integument primordium (Figure 2(c)). The cells of the integument primordium were very small and tightly arranged, and the nuclei could not be distinguished easily (Figure 2(d)). The integument tissues showed active cell division, in contrast to the center part of the general stalk. The cell division between the integument tissues led to the formation of a dorsal raphe between them (Figures 2(c) and 2(d)). Further differentiation became visible in mid-March. As the integument developed rapidly, the dorsal raphe developed further, causing the integument to separate into two independent parts. Then, the nucellus was initiated on the convex region of each integument, forming a faint ridge (Figure2(e)). At this stage, the nucellar tissue was rounded with a flat tip, and its cells were similar in size and appearance to those in the integument (Figure 2(f)). Afterwards, the nucellus continued to differentiate and became free from the integument, causing the latter to show a cup-like shape (Figure 2(g)). Then, rapid cell division resulted in rapid growth of the integument, which grew bigger than, and eventually engulfed, the nucellus (Figure 2(h)). The integument completely surrounded the nucellus in late March (Figure 2(i)). Meanwhile, the ovule collar differentiated from cells between the base of the integument and the top of the ovular stalk (Figure 2(i)). After that, the integument proceeded to differentiate upwards, instead of ceasing growth, giving rise to the micropyle and micropylar canal (Figure 2(j)). The mature ovule had a well-defined integument and ovular collar (Figure 2(k)). The micropyle was situated on the opposite side of the collar, and the column-like nucellus was in the center of the integument (Figure 2(l)).

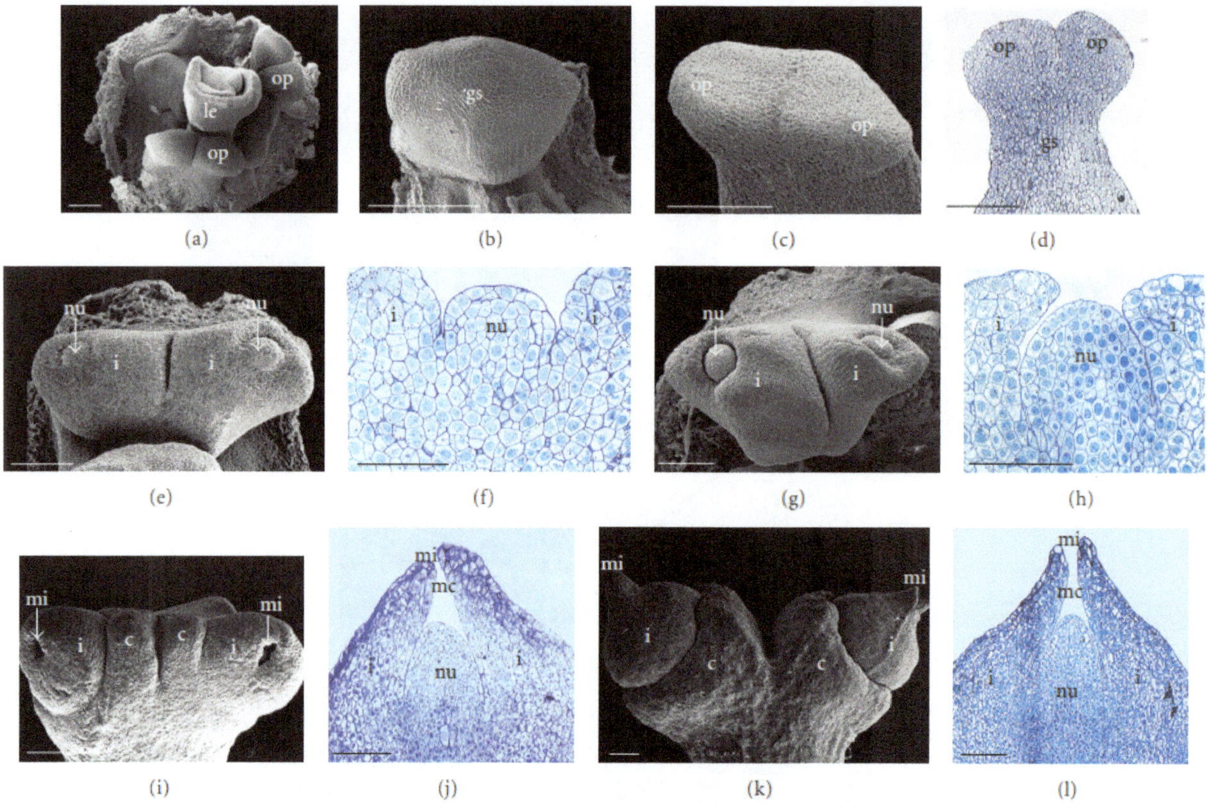


**Figure 2:** Tissues during ovule differentiation process in G. biloba. (a) Position of ovules and leaf in female bud. (b) General stalk differentiated and its tip extended and swelled. ((c)-(d)) Formation of integument primordium on tip general stalk. ((e)-(f)) Center of integument differentiated into nucellar tissue. ((g)–(j)) Integument tissue grew to encircle nucellus. ((k)-(l)) Mature ovules with well-developed collar, integument, nucellus, micropyle, and micropyle tunnel. co, collar; gs, general stalk; i, integument; le, leaf; mi, micropyle; nu, nucellus; op, ovular primordium. Bars: ((a)–(e)) = 100 $\mu$m; ((f), (h), (j), (l)) = 100 $\mu$m; ((g), (i), (k)) = 200 $\mu$m.

## Formation of Pollen Capturing and Transporting Tissues

In G. biloba, the pollen capturing and transporting tissues include the micropyle, micropylar canal, and pollen chamber. Among them, the micropyle and micropylar canal are formed by the integument, while the pollen chamber is formed by the tip of the nucellus. In about mid-March, after the nucellus had differentiated from the center of the integument primordium, it continued to grow upwards by rapid cell division (Figure 3(a)). Because of the more frequent cell divisions of the integument, the nucellus was gradually engulfed by the surrounding integument (Figure 3(b)). When the nucellus was buried deeply in the ovule, the tip of the integument fused toward the very center at the top of the nucellus (Figure 3(c)). Then, the nucellus ceased to extend and became completely concealed by the integument (Figure 3(d)), while the latter continued to grow upwards (Figures 3(e)–3(h)). Therefore, the integument encircled the top of the nucellus to emerge as the micropyle (Figure 3(i)). Some of the cells under the micropyle formed a tunnel-like structure, the micropylar canal (Figure 3(j)). During its differentiation process, the micropyle showed a wavy shape, rather than a ring-shape (Figures 3(f)–3(i)). This was probably because of the differential growth rate of different regions of the upper integument. The micropylar canal cells were almost rectangular and had smooth surfaces (Figures 3(k) and 3(k-1)). As the micropyle and micropylar canal further developed, the nucellus became column-shaped (Figure 3(l)) and slightly acute on the tip side (Figures 3(m) and 3(n)). Some cells near the micropyle elongated longitudinally and collapsed (Figure 2(l)), resulting in a cavity (the pollen chamber) at the nucellus tip (Figures3(o) and 3(p)). During the pollination period, all tissues were specialized for pollen capture and transport. The micropyle slightly curled outwards to form a funnel-like shape, and had a diameter of approximately 200 $\mu$m (Figures 3(q) and 3(r)). The micropylar canal was straight and smooth, and was approximately 100 $\mu$m wide and 500 $\mu$m long (Figures 3(k) and 3(q)). The pollen chamber was well developed with its orifice directly oriented towards the micropyle and micropylar canal, and the diameter was approximately 100 $\mu$m (Figures3(p) and 3(r)).

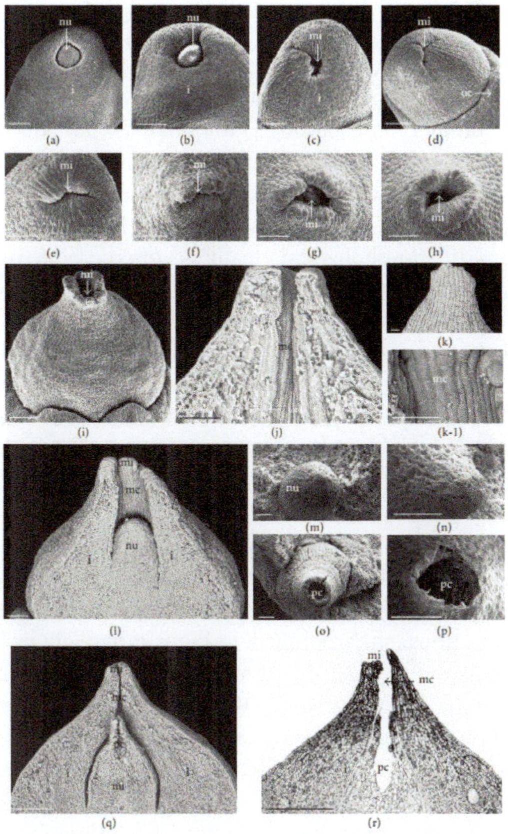

**Figure 3:** Development of tissues for pollen capture and transport in G. biloba. ((a)-(b)) After nucellus differentiation, integument continued to grow upward, surrounding nucellus. ((c)-(d)) Tip of integument gradually fused toward the center on top of nucellus. ((e)–(h)) Tip of integument continued growing upwards. (i) Formation of micropyle. (j) Micropylar canal. ((k), (k-1)) Cells of micropylar canal with smooth surfaces. (l) Column-shaped nucellus and developing micropylar canal. ((m)-(n)) Tip of nucellus became slightly acute. ((o)-(p)) Emergence of pollen chamber. ((q)-(r)) Ovules during pollination period with well-developed micropyle, micropylar canal, and pollen chamber. i, integument; mi, micropyle; mc, micropylar canal; nu, nucellus; oc, ovular collar; pc, pollen chamber. Bars: ((a)–(j), (l)–(p)) = 100 $\mu$m; (k) = 50 $\mu$m; ((q)-(r)) = 500 $\mu$m.

## Formation of the Female Gametophyte

After the micropyle was enclosed by the integument in late March (Figure 2(j)), the megasporocyte (megaspore mother cell) initiated from the center of the nucellus, and was tightly surrounded by spongy tissue. The megasporocyte was very large relative to surrounding cells with a central nucleus (Figure 4(a)). In early April, when the tip of the nucellus began to degenerate to form the pollen chamber, the megasporocyte underwent meiosis. It divided along the longitudinal axis of the ovule, giving rise to the two cells of the dyad (Figure 4(b)). Upon formation of the dyad, the two cells immediately divided again to yield the linear tetrad of megaspores (Figure 4(c)). Among all the megaspores, the one near the chalazal end was typically the largest. The other three near the micropylar end were relatively small, and degenerated soon after (Figure 4(c)). Therefore, the chalazal megaspore enlarged and developed into the functional megaspore (Figure 4(d)). During the pollination period, when the pollen chamber reached its largest volume, the nucleus of the functional megaspore began to undergo series of karyokinetic divisions, forming many free nuclei (Figures 4(e) and 4(g)), indicating the beginning of the female gametophyte phase. At this stage, the surrounding spongy tissue was distinguishable from other cells as they contained vacuoles (Figure 4(f)).

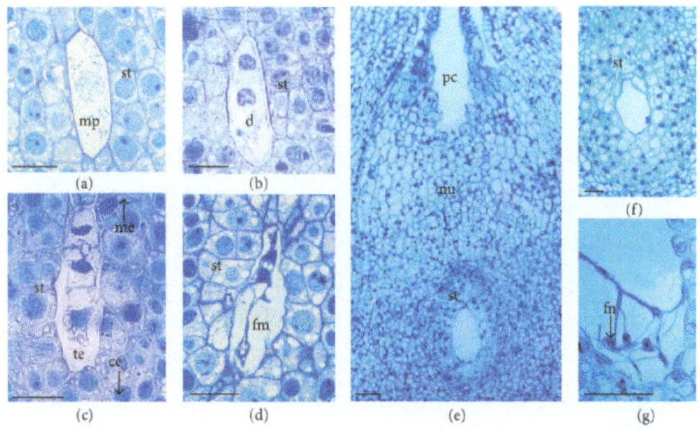

**Figure 4:** Formation of female gametophyte in G. biloba. (a) Megasporocyte differentiated from inner nucellar tissue in late March. (b) Dyad. (c) Tetrad. (d) Functional megaspore. ((e)–(g)) Approaching pollination stage, pollen chamber formed at apex of nucellus, and free nuclear stage of female gametocyte occurred. ce, chalazal end; d, dyad; fm, functional megaspore; fn, free nucleus; me, micropylar end; mp, megasporocyte; nu, nucellus; pc, pollen chamber; st, spongy tissue; te, tetrad. Bars: ((a)–(d), (g)) = 50 $\mu$m; ((e)-(f)) = 100 $\mu$m.

# DISCUSSION

## Differential Stages of the G. biloba Ovule

The structure of the mature G. biloba ovule is similar to that of many other gymnosperms. The mature ovule has a general stalk bearing two ovules at its tip, a single integument with the micropyle positioned opposite the stalk, some pollen grains in the pollen chamber, and many free nuclei [26]. However, some of the developmental stages of the ovule between initiation and maturation remained unclear. In our study, we observed the ovule differentiation process in G. biloba and divided it into six stages: (1) undifferentiated stage (from July to November), the leaf primordium had differentiated, but the ovule primordium was not visible; (2) general stalk differentiation stage (in December), the general stalk primordium appeared in axils of leaves and/or bracts; (3) integument differentiation stage (from January to early March), the tip of the general stalk was extended and swollen (Figure 2(b)), and later, the shoulder of the tip protruded upwards to form the integument primordium (Figures 2(c) and 2(d)), whose upper end then developed into the micropyle and the micropylar canal; (4) nucellus differentiation stage (from mid-March to early April), the nucellus differentiated from the center of the integument (Figure 2(e)); (5) collar differentiation stage (in late March), the ovulate collar differentiated quickly at the base of the integument (Figure 2(i)); and (6) mature stage (in mid-April), a pollination drop appeared on each micropyle (Figure 1(l)), signaling that the ovule was well prepared, morphologically and structurally, for pollen entrapment. The nucellus differentiation stage was the key stage in ovule development, because the nucellus gives rise to the pollen chamber, and later, the female gametophyte.

## Functions of Leaves during Pollination

Many plant lineages show a fossil history of consolidating

individual reproductive organs into compound structures with emergent aerodynamic properties [26]. This appears to have increased the probability of pollination of the reproductive units. For example, wind-pollinated plants tend to show a reduction in the size of branches, leaves, and other structures surrounding pollen-receptive spaces that could catch pollen grains, thereby reducing the ultimate number of pollen grains reaching receptive surfaces [27]. Moreover, the reduction in the size of these structures has been associated with modification of air currents towards receptive surfaces [27]. In addition, in some, but not all species of pine, leaf fascicles emerge around ovulate cones before pollination. As a result, pollen capture by the cone involves the aerodynamics of subtending leaf clusters [28]. In our study, we examined the development of the female short shoot and the orientation of leaves and ovules of G. biloba during pollination. We found before the bud scales unfolded, and ovules and leaves were borne at the apex of the short shoots and were surrounded by thick bud scales (Figure 1(a)). The ovules began to appear as the bud scales and leaves continued to unfold (Figures 1(b)–1(h)). Eventually, all scales, leaves, and ovules formed a spiral arrangement on the short shoot. The ovules were upright, and closer to the vertical axis of the shoot, while the leaves were fan-shaped, and extended further from the center, bending outwards and downwards (Figure 1(i)). Considering this architecture of the leaves and ovules during pollination, we supposed that the shapes and angles of leaves of G. biloba changed the aerodynamics around the ovule, which could increase the probability of pollen capture by the ovule. However, further experiments are required to determine whether or not the leaves of the female short shoot in G. biloba can produce favorable aerodynamics for pollen deposition.

## The Ovulate Stalk

The general evolutionary trends of ovules in G. biloba include a reduction in the number of ovules, an increase in ovule size [29], and shortening or disappearance of the ovulate stalk [17]. Zhou [30] insisted that under the k-strategy, to balance weight and sustainability, the ovulate stalk became shorter, and then ultimately disappeared. Indeed, the single living species of Ginkgo has no ovulate stalk between the collar and general stalk. However, in our study, we found a low frequency of ovulate stalks that had differentiated on the top of the general stalk (Figure 1(o)). Considering that this phenomenon appeared more frequently in Ginkgo fossils [31], we believe the emergence of the ovulate stalk in living Ginkgo might reflect the vestigial expression of an underlying ancestral trait. From the view of the spatial arrangement of the ovules and leaves, the ovulate stalk represents advantageous architecture for ovules to trap pollen grains from the air. However, on the other hand, a long stalk also increases the risk of the pollination drop falling from the micropyle as a result of wind action during the pollination period. In addition, it is unfavorable to have a long developmental period after pollination, owing to the longer distance for nutrient transport and the increased energy demands of ovule bearing. These factors may have contributed to the disappearance of the furcated ovulate stalk during long-term evolution.

## Ovule Morphology and Structures Associated with Wind Pollination

Ovule orientation and morphology are vital features for wind pollination in gymnosperms. Traditionally, wind pollination is viewed as a simple matter of "chance," a mode of reproduction dominated by stochastic processes that predominate in the biology of long-distance dispersal [28]. However, a certain number of morphological features of reproductive units have evolved in divergent land plants that may contribute to the success of wind pollination [7]. For instance, female reproductive organs of gymnosperms showed considerable developmental flexibility, which would significantly influence the behavior of pollen grains in the air around the ovulate cones [11]. Furthermore, ovule geometry and position in relation to floral parts seem to play an essential role in affecting pollination efficiency [28]. In our study, we observed the G. biloba female cone and ovule structure and found many morphological and structural characteristics adapted to wind pollination. First, different from conifers in which ovules are often engulfed by other structures, such as bracts in Pinus monticola or lobes in Cunninghamia [3, 32], the organization of ovulate structures in G. biloba is very simple: the erect ovule is fully naked and is not enclosed by seed scales (Figure 1(j)). In addition, there is no distinctly organized lobe surrounding the ovule; this reduces boundary layer effects near ovules and increases the pollen-scavenging area (Figure 2(k)). Therefore, during pollination, the micropyle orifice enlarges and forms a funnel-like shape, and a pollination drop is produced on the micropyle (Figures 1(l) and 3(q)). It is interesting that the integument tip of the ovule, which includes the micropyle, does not emerge as strictly ring-shaped, but splits to several pieces (Figures 3(i) and 3(p)). Therefore, we propose that the architecture of the micropyle in G. biloba is adapted to produce a pollination drop with its size optimized for attachment of pollen grains, similar to other modifications of the integument tip in some conifers [3, 33–35]. Also, these adaptations could increase the contact area between the ovule and the pollination drop, decreasing the possibility of the pollination drop falling from the micropyle owing to the unexpected motion of the long shoot. Third, before pollination, the ovule differentiates into tissues adapted for pollen transport, including the micropyle, the micropylar canal, and the pollen chamber, of which the micropylar canal is straight and the cells on the inner surface are smooth (Figures 3(j) and 3(k)). The pollen chamber is well-developed, and the orifice reaches its largest size as it faces the micropylar canal (Figure 3(r)). Thus, we assume that, at this moment, the structure of the ovule enables it to transport and accept pollen grains, and facilitates the movement of the pollen from the pollination drop to the pollen chamber.

## Ovule Developmental State during Pollination

At the time of pollination, the developmental state of ovules varies among species, owing to wide variations in the period between pollination and fertilization, or the progamic phase [36]. Ordinarily, the angiosperm ovule, which contains an egg cell, two synergids, and a binucleated central cell at the micropylar pole, is already present during the pollination period, because angiosperms often have a relatively short progamic phase [37–39]. In contrast, since the time interval from pollination to fertilization varies among families, the ovules of gymnosperms have a variable developmental state at the time of pollination. For instance, pollination inPodocarpus totara [40] and Larix decidua [41] occurred when the functional megaspore inside the ovule developed and became the mother cell of the female gametophyte, whereas the female gametophyte of interior spruce was cellular at pollination [42]. In the present study, the female gametophyte of the G. biloba ovule was in the free nuclei stage (Figures 4(e) and 4(g)) during the pollination period. According to Wang et al. [22], there were approximately 130 d between the initiation of karyomitosis in the female gametophyte and the formation of the mature archegonium. Meanwhile, there were also approximately 130 d between entry of the pollen grain into the pollen chamber at the time of pollination and development of the male gametophyte into the mature spermatozoid at the time of fertilization. Thus, we can conclude that the female gametophyte phase occurring inside the ovule during the pollination period might be in preparation for fertilization at the later period. Therefore, the developmental processes of the male and female gametophyte are coordinated after pollination, thereby ensuring that the male and female gametes meet at the correct developmental stage during the short fertilization period.

## Acknowledgments

The authors thank Prof. Peng Chen and Prof. Zhong Wang of Yangzhou University, China, for technical advice. This work was supported by the Natural Science Fund of Jiangsu Province (no. BK2011444), Science and Technology Innovation Fund of Xuzhou City (no. XF11C001), and the National Natural Science Foundation of China (no. 30870436).

## REFERENCES

1. M. Proctor, P. Yeo, and A. Lack, The Natural History of Pollination, Timber Press, Portland, Ore, USA, 1996.
2. J. N. Owens, T. Takaso, and C. John Runions, "Pollination in conifers," Trends in Plant Science, vol. 3, no. 12, pp. 479–485, 1998. · ·
3. J. N. Owens, G. Catalano, and J. S. Bennett, "The pollination mechanism of western white pine,"Canadian Journal of Forest Research, vol. 31, no. 10, pp. 1731–1741, 2001. · ·
4. L. M. Chandler and J. N. Owens, "The pollination mechanism of Abies amabilis," Canadian Journal of Forest Research, vol. 34, no. 5, pp. 1071–1080, 2004. · ·
5. C. J. Runions, K. H. Rensing, T. Takaso, and J. N. Owens, "Pollination of Picea orientalis (Pinaceae): saccus morphology governs pollen buoyancy," American Journal of Botany, vol. 86, no. 2, pp. 190–197, 1999. ·
6. A. B. Schwendemann, G. Wang, M. L. Mertz, R. T. McWilliams, S. L. Thatcher, and J. M. Osborn, "Aerodynamics of saccate pollen and its implications for wind pollination," American Journal of Botany, vol. 94, no. 8, pp. 1371–1381, 2007. · ·
7. Y. Lu, B. Jin, L. Wang et al., "Adaptation of male reproductive structures to wind pollination in gymnosperms: cones and pollen grains," Canadian Journal of Plant Science, vol. 91, no. 5, pp. 897–906, 2011. · ·
8. G. Gelbart and P. Von Aderkas, "Ovular secretions as part of pollination mechanisms in conifers," Annals of Forest Science, vol. 59, no. 4, pp. 345–357, 2002. ·
9. S. Mugnaini, M. Nepi, M. Guarnieri, B. Piotto, and E. Pacini, "Pollination drop in Juniperus communis: response to deposited material," Annals of Botany, vol. 100, no. 7, pp. 1475–1481, 2007. · ·
10. H. Singh and J. N. Owens, "Sexual reproduction in grand fir (Abies grandis)," Canadian Journal of Botany, vol. 60, no. 11, pp. 2197–2214, 1982.
11. K. J. Niklas and K. T. P. U, "Pollination and airflow patterns around conifer ovulate cones," Science, vol. 217, no. 4558, pp. 442–444, 1982. ·
12. L. Wang, D. Wang, M.-M. Lin, Y. Lu, X.-X. Jiang, and B. Jin, "An embryological study and systematic significance of the primitive gymnosperm Ginkgo biloba," Journal of Systematics and Evolution, vol. 49, no. 4, pp. 353–361, 2011. · ·
13. M. Favre-Duchatre, "Ginkgo, an oviparous plant," Phytomorphology, vol. 8, pp. 377–396, 1958.
14. Z. L. Lee, "Recent advances (1949–1959) in morphology, anatomy and cytology of Ginkgo Biloba L," Acta Botanica Sinica, vol. 8, no. 4, pp. 262–270.
15. A. W. Douglas, D. W. Stevenson, and D. P. Little, "Ovule development in Ginkgo biloba L., with emphasis on the collar and nucellus," International Journal of Plant Sciences, vol. 168, no. 9, pp. 1207–1236, 2007. · ·
16. L. Wang, Y. P. Wang, Q. Wang et al., "Anatomical study of development of ovule in Ginkgo biloba L," Acta Botanica Boreali-Occidentalia Sinica, vol. 27, no. 7, pp. 1349–1356, 2007.
17. D. H. Li, X. Yang, X. Cui, K. M. Cui, Z. L. Li, and C. L. Lee, "Early development of pollen chamber inGinkgo biloba Ovule," Acta Botanica Sinica, vol. 44, no. 7, pp. 757–763, 2002. ·
18. C. J. Ji, X. Yang, and Z. L. Li, "Morphological studies on megaspore formation in Ginkgo biloba," Acta Botanica Sinica, vol. 41, no. 7, pp. 219–221, 1999.
19. W. E. Friedman, "Growth and development of the male gametophyte of Ginkgo biloba within the ovule (in Vivo)," American Journal of Botany, vol. 74, no. 12, pp. 1797–1815.
20. B. Jin, L. Zhang, Y. Lu et al., "The mechanism of pollination drop withdrawal in Ginkgo biloba L," BMC Plant Biology, vol. 12, no. 59, 2012.
21. L. Wang, B. Jin, M. M. Lin, Y. Lu, N. J. Teng, and P. Chen, "Studies of the development of female reproductive organs in Ginkgo biloba L," Chinese Bulletin of Botany, vol. 44, no. 6, pp. 673–681, 2009.
22. L. Wang, Y. Lu, B. Jin, M. M. Lin, and P. Chen, "Gametophyte development and embryogenesis in Ginkgo biloba: a current view," Chinese Bulletin of Botany, vol. 45, no. 1, pp. 119–127, 2010.
23. L. Wang, B. Jin, Y. Lu, X. X. Jin, N. J. Teng, and P. Chen, "Developmental characteristics of the ovule and its biological significance in Ginkgo biloba L," Journal of Beijing Forestry University, vol. 32, no. 2, pp. 79–85, 2010. ·
24. B. Jin, L. Wang, J. Wang et al., "The effect of experimental

warming on leaf functional traits, leaf structure and leaf biochemistry in Arabidopsis thaliana," BMC Plant Biology, vol. 11, article 35, 2011. · ·

25. B. Jin, L. Wang, J. Wang et al., "The structure and roles of sterile flowers in Viburnum macrocephalum f. keteleeri (Adoxaceae)," Plant Biology, vol. 12, no. 6, pp. 853–862, 2010. ·
26. M. W. Frohlich and M. W. Chase, "After a dozen years of progress the origin of angiosperms is still a great mystery," Nature, vol. 450, no. 7173, pp. 1184–1189, 2007. · ·
27. K. J. Niklas, "Wind pollination-a study in controlled chaos," American Scientist, vol. 73, no. 5, pp. 462–470, 1985. ·
28. K. J. Niklas, "The aerodynamics of wind pollination," The Botanical Review, vol. 51, no. 3, pp. 328–386, 1985. · ·
29. S. Zheng and Z. Zhou, "A new Mesozoic Ginkgo from western Liaoning, China and its evolutionary significance," Review of Palaeobotany and Palynology, vol. 131, no. 1-2, pp. 91–103, 2004. · ·
30. Z. Y. Zhou, "Heterochronic origin of Ginkgo biloba-type organs," Acta Palaeontologica Sinica, vol. 33, no. 2, pp. 131–139, 1994.
31. Z. Y. Zhou, "An overview of fossil Ginkgoales," Palaeoworld, vol. 18, no. 1, pp. 1–22, 2009. · ·
32. A. Farjon and S. O. Garcia, "Cone and ovule development in Cunninghamia and Taiwania (Cupressaceae sensu lato) and its significance for conifer evolution," American Journal of Botany, vol. 90, no. 1, pp. 8–16, 2003. ·
33. J. R. McWilliam, "The role of the micropyle in the pollination of Pinus," Botanical Gazette, vol. 120, no. 2, pp. 109–117, 1995.
34. J. N. Owens, S. J. Simpson, and M. Molder, "The pollination mechanism and the optimal time of pollination in Douglas-fir (Pseudotsuga menziesii)," Canadian Journal of Forest Research, vol. 11, no. 1, pp. 36–50, 1981.
35. J. N. Owens, S. J. Simpson, and G. Caron, "The pollination mechanism of Engelmann spruce (Picea engelmannii Parry)," Canadian Journal of Botany, vol. 65, no. 7, pp. 1439–1450, 1987.
36. J. H. Williams, "Novelties of the flowering plant pollen tube underlie diversification of a key life history stage," Proceedings of the National Academy of Sciences of the United States of America, vol. 105, no. 32, pp. 11259–11263, 2008. · ·
37. W. E. Friedman and K. C. Ryerson, "Reconstructing the ancestral female gametophyte of angiosperms: insights from Amborella and other ancient lineages of flowering plants," American Journal of Botany, vol. 96, no. 1, pp. 129–143, 2009. · ·
38. V. Olmedo-Monfil, N. Durán-Figueroa, M. Arteaga-Vázquez et al., "Control of female gamete formation by a small RNA pathway in Arabidopsis," Nature, vol. 464, no. 7288, pp. 628–632, 2010. · ·
39. A. Cisneros, R. Benega Garcia, and N. Tel-Zur, "Ovule morphology, embryogenesis and seed development in three Hylocereus species (Cactaceae)," Flora, vol. 206, pp. 1076–1084, 2011.
40. V. R. Wilson and J. N. Owens, "The reproductive biology of totara (Podocarpus totara) (Podocarpaceae),"Annals of Botany, vol. 83, no. 4, pp. 401–411, 1999. · ·
41. K. Rafińska and E. Bednarska, "Localisation pattern of homogalacturonan and arabinogalactan proteins in developing ovules of the gymnosperm plant Larix decidua Mill," Sexual Plant Reproduction, vol. 24, no. 1, pp. 75–87, 2011. · ·
42. J. J. Runions and J. N. Owens, "Sexual reproduction of interior spruce (Pinaceae). I. Pollen germination to archegonial maturation," International Journal of Plant Sciences, vol. 160, no. 4, pp. 631–640, 1999.

CHAPTER 13

# ORIGIN AND DIVERSIFICATION OF LIVING CYCADS: A CAUTIONARY TALE ON THE IMPACT OF THE BRANCHING PROCESS PRIOR IN BAYESIAN MOLECULAR DATING

Fabien L Condamine, Nathalie S Nagalingum, Charles R Marshall and Hélène Morlon

**Keywords:** Bayesian relaxed-clock, Birth-death process, Branching process prior, Cycadales, Cycadopsida, Speciation tree prior, Yule model

## BACKGROUND

Our understanding of biotic evolution relies heavily on phylogenetic and dating reconstructions that provide insight into the periods of major diversification [1]. In the last decade, the advent of molecular dating approaches has fostered an explosion of studies constructing time-calibrated trees for diverse plant clades like bryophytes [2], ferns [3,4], gymnosperms [5-8] and angiosperms [9-11]. These dated trees have permitted the study of character evolution via the reconstruction of ancestral traits [11], inference of biogeographical history [7], as well as estimates of diversification rates [2,8,10]. Dated trees are thus pivotal to our understanding of the evolution of plants, and of the groups that interact with them, such as herbivores and pollinators [12].

Despite the importance of reliable estimates of divergence times, our understanding of the temporal patterns of diversification remain in flux for many groups, in part because the methods for estimating evolutionary timescales from DNA sequences are being refined [1,13,14]. Since the introduction of relaxed-clock methods, which allow substitution rates to vary across the tree, a range of molecular dating methods has been developed. Bayesian inference has received the most attention because of the flexibility with which different parameters and prior assumptions can be incorporated, and the fact that priors are updated as part of the analysis [15,16]. The use of explicit prior distributions is central to the Bayesian perspective; however, the critical role of prior selection in Bayesian analysis is not always fully appreciated [17].

In Bayesian relaxed-clock (BRC) approaches, there are various types of priors, including priors on calibration points, branch-lengths, clock models, and branching processes. Priors on calibration points have been well studied [18] and, not surprisingly, the choice of these priors can affect estimates of node ages [19]. The effects of branch-length priors on posterior probabilities have been studied; priors assuming long internal branches cause high posterior probabilities [20]. In comparison, the impact of different branching process priors has been relatively under-explored [15,21].

The branching process prior (BPP), also called the 'tree shape' or 'speciation tree' prior, is a prior model on how trees are generated. Phylogenetic trees are the result of speciation and extinction events, and their relative roles can be varied and represented as different models of diversification [22]. These models effectively place a prior on how phylogenetic trees grow. Probability distributions over models of diversification were employed in some of the earliest attempts to use likelihood techniques to reconstruct genealogies [23]. The two most commonly used BPPs are the Yule (also called 'pure-birth') process, which models tree formation with a constant rate of speciation and no extinction, and the birth-death process, which includes speciation as well as a constant rate of lineage extinction. Birth-death priors have been used in Bayesian phylogenetics [24]; however, most published analyses use the Yule prior, perhaps because it was initially the only prior for the diversification process implemented in the widely used

**Citation:** Fabien L Condamine, Nathalie S Nagalingum, Charles R Marshall and Hélène Morlon, Origin and diversification of living cycads: a cautionary tale on the impact of the branching process prior in Bayesian molecular dating, DOI: 10.1186/s12862-015-0347-8.

Bayesian software package BEAST [16]. Although the birth-death prior has recently been integrated into this software [25,26], many phylogenetic analyses still use the Yule prior (probably because the BEAST manual recommend the use of the Yule prior, see p. 10 of the BEAST manual, version 1.4). Few studies have used both priors in Bayesian dating, and even fewer have compared the impact of prior choice, even though recent studies have started to do so (e.g. [27]). One of the first to use both priors is the study of Couvreur *et al.* [28], who estimated the evolutionary history of the Brassicaceae using a BRC approach with both priors. Their age estimates did not differ under the Yule or birth–death diversification models, and both models fit the data equally well. While these results could be interpreted to mean that the BPP has little influence on BRC reconstruction in general, the Brassicaceae are a relative young group that originated in the Eocene (credibility interval 24.2-49.4 million years ago, Ma) and likely did not experience major extinction events. However, choice of diversification prior might make a substantial difference for groups that have undergone significant extinction, and for which using a pure-birth Yule prior could give in spurious results.

The cycads (Cycadopsida: Cycadales) are a plant group particularly well suited for testing the influence of BPPs on molecular dating. Today's species (331 species in the tropics and subtropics, [29]) are thought to represent the last remnants of their formidable past, and their evolutionary history extends back to the mid-Permian (~270 Ma) [30-33]. Cycads have witnessed many drastic environmental changes [34-37] and likely suffered major periods of extinction [6,38]. Due to their richness and diversity in Mesozoic fossil records, their current diversity has long been assumed to be of relictual origin dating back to the Late Cretaceous [30,39-42]. This idea has been challenged by molecular time-calibrated phylogenies showing that long branches subtend late Cenozoic (in the Miocene ~15 Ma) and near-simultaneous initiations of diversification of the living genera [6,43].

Despite these recent studies, there remain dating uncertainties in the evolutionary history of cycads. At the generic level, some fossils suggest that the various radiations might be much older than the late Miocene (i.e. Eocene). Indeed, numerous pre-Miocene cycad fossils have been documented (e.g. [44-49]) that are likely early representatives of living genera. For example, recently described fossils from the early Cenozoic of China have been assigned to the crown-group of *Cycas* [50], but that assignment is doubtful given the extensive homoplasy in the characters used to link the fossils to the extant taxon [51]. Even if there is a possibility that fossils instead should be assigned to stem, they cast doubts on the dating. At the level of the whole group, different studies using similar fossil calibrations (but different taxon and molecular sampling) have produced different age estimates, notably for the origin of the cycad crown: ≈200 Ma for Nagalingum *et al.* [6], and≈230 Ma for Salas-Leiva *et al.* [43]. Therefore further dating analyses are needed to estimate the cycad age and validate the recent generic radiations.

In reconstructing the phylogeny of cycads, Nagalingum *et al.* [6] sampled about two-thirds of all known species and used a molecular dataset composed of two chloroplast genes and one nuclear gene (although most of the taxa had the nuclear gene only). In addition to Penalized Likelihood and a strict molecular clock, they inferred the age and divergence times using the BRC approach implemented in BEAST, calibrated with four fossil constraints and a birth-death process as the BPP. The birth-death process used in the Bayesian analyses yielded a high ratio of extinction to speciation (relative death rate, 0.97) [6], suggesting a dominant role of extinction in shaping the phylogeny of cycads. Given our prior knowledge of cycad evolution gleaned from the fossil record, using a prior that includes extinction is realistic. However, Nagalingum *et al.* [6] did not assess the support of the birth-death prior versus other priors.

This study has four objectives: *(i)* investigating the impact of the BPP on the dating of ancient clades using the cycads as an example, *(ii)* assessing whether the birth-death prior is statistically supported, *(iii)* studying the difference between the Yule and birth-death prior on our understanding of cycad diversification, and *(iv)*providing a cycad timetree reconciling fossil and phylogenetic data. We also discuss potential explanations for the differences obtained when using different priors and the consequences of prior choice for Bayesian molecular dating.

## METHODS

### Taxon Sampling and Molecular Dataset

We extended the molecular dataset of Nagalingum *et al.* [6] that initially contained 199 cycad species. We added molecular data for 38 additional species and the same genes retrieved from Genbank [52-57]. Our dataset comprises three genes covering two plastid genes: the maturase K (matK, 2,387 nucleotides, 82 taxa) and the ribulose 1,5-bisphosphate carboxylase large subunit (rbcL, 1,398 nucleotides, 80 taxa), and one nuclear gene: region 1 of phytochrome (PHYP, 1,802 nucleotides for 201 taxa). This resulted in a total of 237 species out of 331 described species (71%), representing all extant genera with the following number of species per genus: 2 of the 2 *Bowenia* (*B. serrulata* and *B. spectabilis*), 65 of the 107 *Cycas*, 24 out of 27*Ceratozamia*, 8 of the 14 *Dioon*, all of the 65 *Encephalartos*, 2 of the 2 *Lepidozamia* (*L. hopei* and *L. peroffskyana*), 26 of the 41 *Macrozamia*, the monotypic *Microcyas calocoma*, the single *Stangeria eriopus*, and 43 out of 71 *Zamia* (including the junior synonym genus *Chigua*) (Figure 1). Following Nagalingum *et al.* [6], the dataset also included six species as outgroups: *Ginkgo biloba* (Ginkgoales), which is recognised as the sister lineage of cycads [58]; plus five conifer species (*Abies firma*, *Araucaria heterophylla*, *Cryptomeria japonica*,*Pinus strobus* and *Pseudotsuga menziezii*). The phylogenetic analyses were thus performed on a dataset containing a total of 243 species (of which 237 are cycads) and 5587 nucleotides.

a

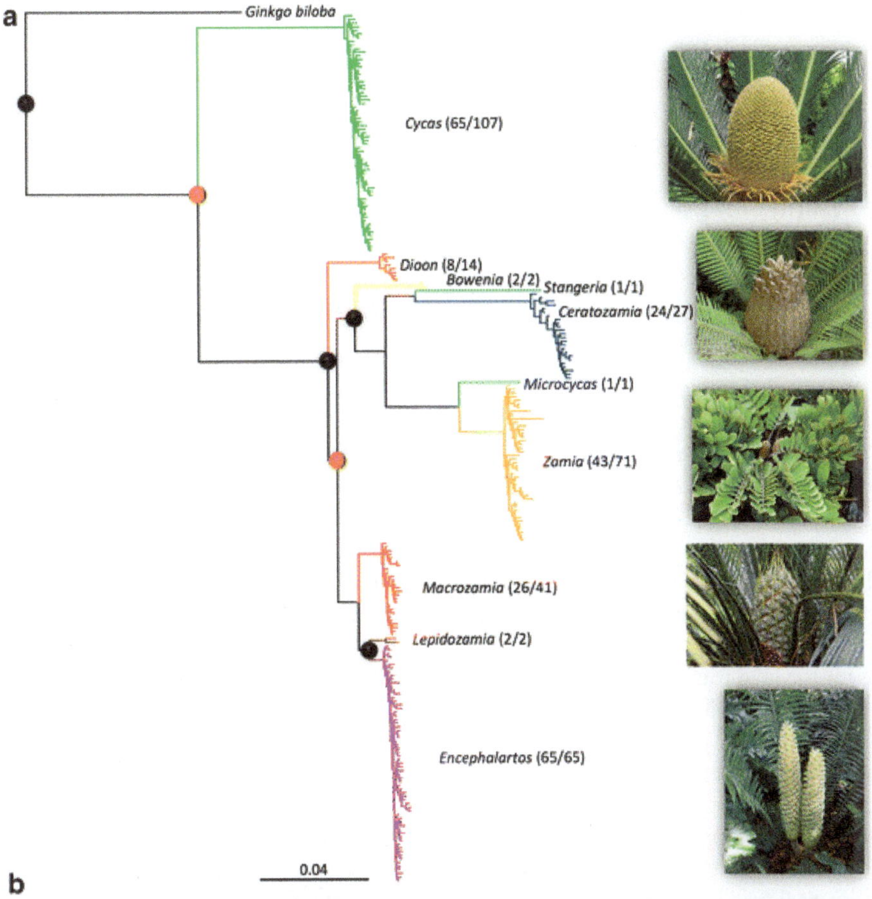


b

| Calibration | Fossil name(s) | Minimum age (geological period) | Calibrated node | Prior distribution | References |
|---|---|---|---|---|---|
| FC ● | †*Crossozamia* (eight species) | 265.1 Ma (middle Permian) | Cycad stem | Uniform 265.1 - 364.7 | Zhifeng & Thomas [67] |
| FC ● | †*Dioon inopinus* and †*D. praespinulosum* | 55.8 Ma (Paleocene) | *Dioon* stem | Uniform 56 - 265.1 | Hollick [44] |
| FC ● | †*Bowenia eocenica* and †*B. papillosa* | 33.9 Ma (Eocene) | *Bowenia* stem | Uniform 33.9 - 265.1 | Hill [69] |
| FC ● | †*Lepidozamia hopeites* and †*L. foveolata* | 33.9 Ma (Eocene) | *Lepidozamia* stem | Uniform 33.9 - 265.1 | Hill [45] |
| FC ● | †*Antarcticycas schopfii* | 235 Ma (Early to Middle Triassic) | Cycad crown | Uniform 235 - 364.7 | Smoot et al. [70] |
| FC ● | †*Wintucycas stevensonii* | 72.1 Ma (middle Campanian to early Maastrichtian) | Encephalarteae stem | Uniform 72.1 - 265.1 | Martinez et al. [60] |

**Figure 1:** The node calibration procedure used for dating the cycads. **(a)** Phylogenetic tree of cycads showing the relationships among the 10 genera. Genera are represented by triangles proportional to their species richness. Numbers in parenthesis are the number of species sampled, and the total number of species, within each genus. Pictures illustrate *Cycas*, *Dioon*, *Zamia*, *Macrozamia*, and *Encephalartos* species. Black dots indicate the four 'traditional' fossil calibrations used for dating, and red dots indicate the two new fossil calibrations evaluated in this study. **(b)**Information related to the four fossil calibrations (FC, see the text for more details). Ma, million years ago.

## Fossil Calibrations

To calibrate the cycad tree, we used two fossil datasets. First, we followed Nagalingum *et al.* [6] and used four cycad fossil constraints (FC1-4), which were retrieved from a careful examination of the cycad fossil record ([32]; Figure 1). These four fossils have also been used by Salas-Leiva *et al.* [43]. Hereafter this fossil dataset is referred to as the 'traditional fossil dataset' (FC1-4). Second we used these four fossils and added two fossils based on the phylogenetic analyses of Hermsen *et al.* [32] and Martínez *et al.* [59]. The fossil record was re-examined after new fossils from the Cretaceous were discovered, which did not affect overall fossil assignments, but allowed the use of new fossils to calibrate the cycad tree [59], two of which are tentatively used here. Hereafter this fossil dataset is referred to as 'new fossil dataset' (FC1-6).

Based on the presence of synapomorphies linking the fossils to the extant clades, all fossils were used as minimum age constraints for specific nodes ([32,59], Figure 1). However, since synapomorphies can evolve anywhere along the stem branch and the fossil may attach anywhere along this branch [60], we used a commonly employed approach whereby the fossils constrain the stem rather than crown node [61,62]. The absolute fossil ages we used (detailed below) are slightly different from those used in Nagalingum *et al.* [6], mainly due to an updated geological timescale [63]. Dating fossils older than ~50,000 years is difficult, which means that those fossils are typically assigned to a stratigraphic interval, for example, the late Miocene. The ages assigned below are the ages of the youngest of the boundaries of stratigraphic interval within which the fossil was found (the fossil will actually be older than this age designation).

## Cycad stem (FC1, Younger = 265.1 Ma and Older = 364.7 Ma)

The oldest possible records of the group Cycadophyta are microstrobili (e.g. [64]) and megasporophylls (e.g. [65,66]), of which †*Crossozamia* (eight known species) is the least equivocal in terms of phylogenetic affinity [32,59,66]. *Crossozamia* consists of megasporophylls with similar morphology to the extant *Cycas* [65,66]. Although Hermsen *et al.* [32] identified *Crossozamia* as being sister to *Cycas*, the two characters supporting this relationship do not provide strong evidence for *Crossozamia* belonging to crown group cycads [59]. In addition, the loosely aggregated cone is most probably an ancestral state within cycads. Therefore the divergence between cycads and *Ginkgo biloba* was constrained using *Crossozamia*. The age for the host rock formation (Shihhotse) that preserves the *Crossozamia* fossils was initially described as lower Permian, but the age of the Shihhotse Formation has been revised to the Roadian-Wordian (265.1-272.3; [63]) in the middle Permian [67]. We used the minimum age of this time interval (265.1 Ma) as a minimum age for the stem of cycads. This age was also used as a conservative maximum possible age for the nodes with internal fossil calibrations (FC2-FC4).

## *Dioon stem (FC2, Younger = 56 Ma and Older = 265.1 Ma)*

The origin of the stem group of *Dioon* was constrained by fossil leaves of †*D. inopinus* and †*D. praespinulosum* [44], which have synapomorphies consistent with the leaflet venation anastomoses and leaflet insertion on the rachis of extant *Dioon* [32]. While originally described as Eocene by Hollick [44], a more recent interpretation of the composition of the Hamilton Bay flora (Kootznahoo Formation), Kupreanof Island, Alaska, from which fossils of both *Dioon* were originally described, suggest they are of Paleocene age [6,32]. Therefore the *Dioon* stem was assigned a minimum age of 56 Ma [63].

## *Bowenia stem (FC3, Younger = 33.9 Ma and Older = 265.1 Ma)*

*Bowenia* leaf fossils of †*B. eocenica* and †*B. papillosa* [68], both found in Australia, were used to constrain the divergence of *Bowenia* from other genera. These fossils were identified as members of the extant genus based on cuticular characters (i.e. number of subsidiary cells) and leaflet morphology (i.e. venation and serrated margin). Both fossil species were found in Australian Eocene deposits (at Anglesea, Victoria for †*B. eocenica*, and at Nerriga, New South Wales for †*B. papillosa*), which confers a minimum age constraint at 33.9 Ma for the *Bowenia* stem [63].

## *Lepidozamia stem (FC4, Younger = 33.9 Ma and Older = 265.1 Ma)*

Fossil *Lepidozamia* leaves of †*L. hopeites* and †*L. foveolata* are also from Australia [45]. Fossils of this genus can be identified by cuticular characters (orientation of the epidermal cells relative to axis of the pinna) that are unique to *Lepidozamia* and support affinity to the extant genus [32,45]. Found in the Australian Eocene (at Nerriga, New South Wales for †*L. foveolata*), this set a minimum age of 33.9 Ma for the *Lepidozamia* stem [63].

## Cycad crown (FC5, Younger = 235 Ma and Older = 364.7 Ma)

The age of modern cycads can be calibrated with †*Antarcticycas schopfii* [69]. The fossil *Antarcticycas* has been extensively studied [32,70]. A reconstruction of the plant habit has even been proposed [70]. This makes*Antarcticycas* the most completely known of the extinct Triassic cycad taxa, if not of all fossil cycads, due to the presence of anatomically preserved organs [70]. Interestingly, phylogenetic studies have also tentatively assigned *Antarcticycas* within modern cycads, and close to the crown of cycads [32,59]. None of the extant genera are phylogenetically sister to *Antarcticycas*, which makes it a valuable fossil calibration for the crown of cycads. *Antarcticycas* was found in the Fremouw Formation of the early Middle Triassic of Antarctica. We used a conservative age of the Middle Triassic, i.e. 235 Ma [63], although there remains uncertainty on the true age of the Fremouw Formation with some authors suggesting an Anisian age (242-247.2 Ma).

## *Encephalarteae stem (FC6, Younger = 72.1 Ma and Older = 265.1 Ma)*

Many fossils have been attributed to the tribe Encephalarteae based on morphological and phylogenetic evidence [32,59]. Among them the recently discovered †*Wintucycas stevensonii* [59] is one of the best-known fossils. *Wintucycas* has features that clearly allow us to assign it to Encephalarteae due to their manoxylic wood, centripetal polyxyly, parenchymatous pith, centrifugal polyxyly and medullary vascular bundles. *Wintucycas* was found in the Allen Formation (middle Campanian to early Maastrichtian) of the Late Cretaceous of Argentina (Patagonia), providing a minimum age for the stem of Encephalarteae at 72.1 Ma [63].

## Tree Root Height

The tree root height is the divergence between Cycadales and Ginkgoales. Following Clarke *et al.* [62], a maximum age can be established with the first records of seeds in the form of preovules that satisfy the criteria of the seed habit. These criteria are the possession of a single functional megaspore that is enveloped in a nucellus (considered equivalent to the megasporangium), which is surrounded (to some extent) by an integument or pre-integument and has mechanisms enabling the capture of pollen before seed dispersal [62,71]. All the criteria are first met with †*Elkinsia polymorpha* found in the VCo spore Biozone, Evieux Formation [71] in the Fammenian (Late Devonian). The VCo spore Biozone spans 364.7-360.7 Ma [63]. A maximum age for gymnosperms is thus 364.7 Ma. We also used this age to set maximum ages for the cycad stem (FC1) and crown (FC5).

## Phylogenetic Reconstructions

To obtain a starting tree for the dating analyses, Bayesian inferences were performed with MrBayes 3.2.3 [72]. To determine the best-fit partitioning scheme of molecular evolution for our dataset, we used PartitionFinder 1.1.1 [73]. For the PartitionFinder analyses, branch lengths were *unlinked* to allow them to be independently estimated for each partition. The searched for best model, among those available in MrBayes, was performed under the *greedy* algorithm based on the Bayesian Information Criterion model metric. We used the partitioning scheme and among-site rate variation suggested by PartitionFinder, but instead of selecting one substitution model *a priori*, we used reversible-jump Markov Chain Monte Carlo (rj-MCMC) to allow sampling across the entire substitution rate model space [74].

MrBayes analyses consisted of four rj-MCMC running for 100 million generations with sampling every 10,000 generations and the first 25% discarded as burn-in. We specified *(i)* a uniform prior probability of phylogenies (i.e. all possible trees are considered *a priori* equally probable), and *(ii)* a uniform prior probability distribution on branch lengths. The convergence of the runs was assessed by checking the potential scale reduction factor (PSRF) values of each parameter in MrBayes and the Effective Sample Size (ESS) values of each parameter in Tracer 1.6 [75]. Values of PSRF close to 1.00 and ESS above 200 were considered as good indicator of convergence. Nodes recovered with posterior probabilities (PP)$\geq$0.95 are considered to be strongly supported.

## Bayesian Relaxed-Clock Analyses

For all dating analyses, we used the BRC approach as implemented in BEAST 1.8 [26]. BEAUti 1.8 was used to create the BEAST input file. We implemented partitioned relaxed-clock models with an uncorrelated lognormal clock model that assumes an underlying lognormal distribution (UCLD) of the evolutionary rates [16], which is more likely to yield accurate estimates than the uncorrelated relaxed clock model that assumes an exponential distribution of the evolutionary rates [76]. We used the same nucleotide substitution model as for the MrBayes analyses. Given that one of the main goals of the study was to investigate the impact of the BPP on molecular dating, we set the tree prior in BEAST to either the Yule or the birth-death process [23]. Following the rigorous approach of Vanneste *et al.* [77], we utilized the following priors: a uniform prior between 0 and 10 with a starting value at 0.1 for the Yule birth rate and birth-death mean growth rate, and a uniform prior between 0 and 1 with a starting value at 0.5 for the birth-death relative death rate. We used an exponential prior with mean one-third on the standard deviation of the UCLD model, and a uniform prior between 0 and 1 on the mean of the UCLD model. Fossil calibrations were conservatively set with a uniform distribution bounded by the minimum and maximum ages as explained above. A starting tree with branch lengths satisfying all of the fossil prior constraints was inserted, as derived from the MrBayes analyses (described above) [77]. The 'tree operators' of the BEAST tree model were disabled to keep the topology fixed so that only the branch lengths were optimized. This procedure ensures better convergence of the tree topology (S. Ho, pers. comm.). Other parameters were kept to their default prior distribution or were indirectly specified through other parameters. We designed four distinct BEAST analyses by mixing the fossil dataset and the BPP as follows *(i)* use of the 'traditional fossil dataset' and the Yule model as BPP; *(ii)* use of the 'traditional fossil dataset' and the birth-death process as BPP; *(iii)* use of the 'new fossil dataset' and the Yule model as BPP; and *(iv)* use of the 'new fossil dataset' and the birth-death process as BPP. For each fossil dataset, we included all fossils for dating the tree, and did not perform cross-validation analyses because Warnock *et al.* [78] questioned the cross-validation approach to measure consistency among calibrations based on minimum constraints [61,79]. The most effective means of establishing the quality of fossil-based calibrations is through *a priori* evaluation of the intrinsic palaeontological, stratigraphic, geochronological and phylogenetic data [78], which we describe above.

The MCMC analyses were run for 100 million generations and sampled every 10,000 generations, resulting in 10,000 trees in the posterior distribution; we discarded the first 2,500 trees as burn-in. We used BEAGLE [80], a library for high-performance statistical phylogenetic inference, with default parameters. This allowed a faster and more accurate computation of likelihood models within our BEAST analyses. All BEAST analyses were performed on the computer cluster CIPRES Science Gateway 3.3 ([81]; http://www.phylo.org/). Tracer was used to assess graphically the convergence of runs, and to check the ESS for all parameters (indicated by ESS above 200). For each analysis, we conducted two independent runs to ensure convergence of the MCMC. Post burn-in trees from the two distinct runs (7,500 trees for each run) were further combined to build the maximum clade credibility tree.

In Bayesian analyses, models have traditionally been compared using the harmonic mean estimate (HME). Other approaches have been developed to calculate accurately the marginal likelihood estimate (MLE) of the specified model, such as path sampling (PS) [82] and stepping-stone sampling (SS) [83]. PS and SS substantially outperform the HME estimator, which overestimates the marginal likelihood and fails to select reliably the best model [76]. We performed MLE using SS with 150 path steps, each with a chain length of one million iterations (G. Baele, pers. comm.), and other parameters were set by default. We directly calculated the log-Bayes factors (hereafter BF) [76,84] from MLEs, and used BF to compare the support of the Yule and the birth-death processes for a given calibration strategy. We considered BF values above 5 to indicate that one model was significantly favoured over another [84].

## Macroevolutionary Rates through Time

We used the Bayesian Analysis of Macroevolutionary Mixture (BAMM, www.bamm-project.org) to estimate speciation and extinction rates through time and among/within clades [85-87]. BAMM is an analytical tool for studying complex evolutionary processes on phylogenetic trees, potentially shaped by a heterogeneous mixture of distinct evolutionary dynamics of speciation and extinction across clades. The method uses rj-MCMC to detect automatically rate shifts and sample distinct evolutionary dynamics that best explain the whole diversification dynamics of the clade. Within a given regime, evolutionary dynamics can involve time-variable diversification rates; in BAMM, the speciation rate is allowed to vary exponentially through time while extinction is maintained constant [86]. Subclades in the tree may diversify faster (or slower) than others, and BAMM detects these diversification rate shifts without *a priori* hypotheses on how many and where these shifts might occur. BAMM provides estimates of marginal probability of speciation and extinction rates at any point in time along any branch of the tree. Marginal probabilities of the number of evolutionary regimes can also be computed, allowing comparisons of models with a given number of shifts with BF.

Estimating extinction rates from reconstructed phylogenies is notoriously difficult [88-90], although not impossible (see Box 3 of Morlon [22] for a discussion of the subject, and ref. [91-94] for recent papers contradicting the widespread idea that extinction rates cannot be estimated from molecular phylogenies). The developer of the approach cautions that BAMM estimates of extinction should be taken with care [87], because they are sensitive to the choice of the prior and have large confidence intervals. Still, these estimates are well correlated with the underlying rates (Figure five B in Rabosky [87]). Biases in extinction rate estimates often come from fitting models based on underlying hypotheses that are violated in nature, such as fitting models with homogeneous rates across clades when major rate shifts occurred (see ref. [90,93], but see [94]). The BAMM approach should in principle help with these issues, because it explicitly accounts for rate heterogeneity across clades while allowing rates to vary over time [85-87]. We share Beaulieu and O'Meara's view [94] that we should use prudent caution, but not abandon all hope, of estimating extinction from molecular phylogenies. Hence, we think that it is crucially important to assess the effect of BPP prior choice on these estimates, and to discuss them in the light of what is known from the cycads fossil record.

BAMM is implemented in a C++ command line program and the BAMMtools R-package [87]. We ran BAMM analyses on the timetree calibrated with the 'new-fossils dataset', and reconstructed with either the Yule or the birth-death prior. We set four rj-MCMC running for 50 million generations and sampled every 50,000 generations. A compound Poisson process is implemented in BAMM for the prior probability of a rate shift along any branch. We used a prior value of 1.0 implying a null hypothesis of no rate shift across the phylogeny, as recommended by Rabosky [87]. We accounted for incomplete taxon sampling using the implemented analytical correction, with a sampling fraction set to 0.716 (i.e. 237 species out of the 331 described species). We performed four independent runs (with a burn-in of 10%) using different seeds, and we used ESS to assess the convergence of the runs, considering values above 200 as indicating convergence. The posterior distribution was used to estimate the configuration of the diversification rate shifts; alternative diversification models were compared using BF. To complement these analyses, we computed pairwise probabilities that any two lineages or clades share a common set of macroevolutionary rate parameters, using the macroevolutionary cohort analysis implemented in BAMM [87].

## RESULTS

### Phylogeny of Cycads

Tracer plots indicated that MrBayes runs reached convergence before 10 million generations (results not shown). Convergence of the analyses was also supported by the average standard deviation of split frequencies (<0.01), all PSRF values close to 1.0 (range between 0.999 and 1.019), and ESS values above 200 (with many >1000) for the post burn-in trees. The resulting Bayesian trees were well resolved: all nodes of the backbone tree and genus crown nodes were recovered with maximal posterior probabilities (PP=1, Figure). Only nodes within genus radiations were recovered with lower PP, which was expected because the genetic divergence among cycad species within genera is low (*Cycas*: [95,96];*Ceratozamia*: [52]; *Dioon*: [56]; *Encephalartos*: [97]; *Macrozamia*: [98]; *Zamia*: [99]). Our MrBayes analyses are congruent with the results of Nagalingum *et al.* [6], but differ from other studies [43,54,55,100]. Notably the genus *Bowenia* is found as sister to the clade ((*Ceratozamia*, *Stangeria*), (*Microcycas*, *Zamia*)) as in Nagalingum*et al.* [6], but an alternative position is that *Bowenia* is sister to all genera, except *Cycas* and *Dioon* [43].

### Divergence Time Estimates and Model Selection for the Branching Process Prior

Convergence of the dating analyses was ensured by ESS values above 200 for all parameters (with many >1000) for the post burn-in trees (Table 1) and Tracer plots indicated that BEAST runs reached convergence before the burn-in threshold. The choice of branching process prior had a striking effect on age estimates (Figures 2 and 3; Table 1). Analyses with the Yule prior inferred much older ages for cycad genera (Figure 4). Genus-level crown ages were on average three times older with a Yule than with a birth-death prior (3.2-fold difference with the 'traditional fossil dataset' and 2.9-fold with the 'new fossil dataset'). This difference was quite striking, with non-overlapping credibility intervals (95% highest posterior density) for the genus crown ages (Figure 4). Generic stem ages and some ages of deeper nodes were also inferred to be older with the Yule than with the birth-death prior, although the difference was not as marked as in the case of the genus crown ages (Figure 5). However, the birth-death and Yule priors usually inferred similar ages for the cycad stem and crown ages (Figure 5). Analyses with the birth-death process inferred a very high estimate of the relative extinction rate (ratio of extinction to speciation, or turnover) with a median=0.966 for the dating with four FC (95% HPD 0.9255-0.9972), and a median=0.962 for the dating with six FC (95% HPD 0.9153-0.9947).

**Table 1:** Age estimates for the six nodes subtending a fossil calibration (FC)

| Fossil calibration strategy | Prior | HME | MLE (SS) | ESS (logL) | BF | Cycad stem (FC1) | *Dioon*stem (FC2) | *Bowenia*stem (FC3) | *Lepidozamia*stem (FC4) | Cycad crown (FC5) | Encephalarteae stem (FC6) |
|---|---|---|---|---|---|---|---|---|---|---|---|
| | | | | | | min. age 265.1 Ma | min. age 56 Ma | min. age 33.9 Ma | min. age 33.9 Ma | min. age 235 Ma | min. age 72.1 Ma |
| Traditional fossil dataset | Birth-death | −26343.61 | −26828.83 | 2451 | 26.7 | 316.6 [269.5-363.3] | 140 [90-192.9] | 116.3 [76.7-160.8] | 38.1 [33.9-50.9] | 257.2 [184.3-337.5] | 129.2 [84.3-180.2] |
| | Yule | −26330.62 | −26855.49 | 3835 | | 295.7 [265.1-350.1] | 206.4 [161.2-258.2] | 170.3 [130.8-216.3] | 68 [41.7-100.4] | 277.2 [231-344.9] | 190.4 [146.9-240.2] |
| New fossil dataset | Birth-death | −26343.36 | −26825.04 | 2076 | 36.3 | 323.2 [273.9-364.6] | 156.1 [107-207.9] | 129.7 [88.7-174.3] | 39.1 [33.9-55] | 274.5 [235-332.4] | 144.5 [97.7-192.5] |
| | Yule | −26331.16 | −26861.34 | 1361 | | 296.9 [265.1-351.7] | 203.5 [158.6-253.6] | 169.5 [129.4-214.2] | 67.8 [40.6-102] | 280.1 [235.6-334.8] | 188 [143.6-235.6] |

Absolute ages and their credibility intervals were obtained for the two pairs of Bayesian calibration strategies, using either the Yule or the birth-death model as branching process prior.

Values are the median ages and 95% Highest Posterior Density (HPD). We denote HME: harmonic mean estimate; MLE (SS): marginal likelihood, estimated with the stepping-stone sampling; ESS: effective sample size; BF: Bayes factor (BF), computed as the difference between the MLE of the birth-death and the Yule prior. A positive value indicates support for the birth-death prior.

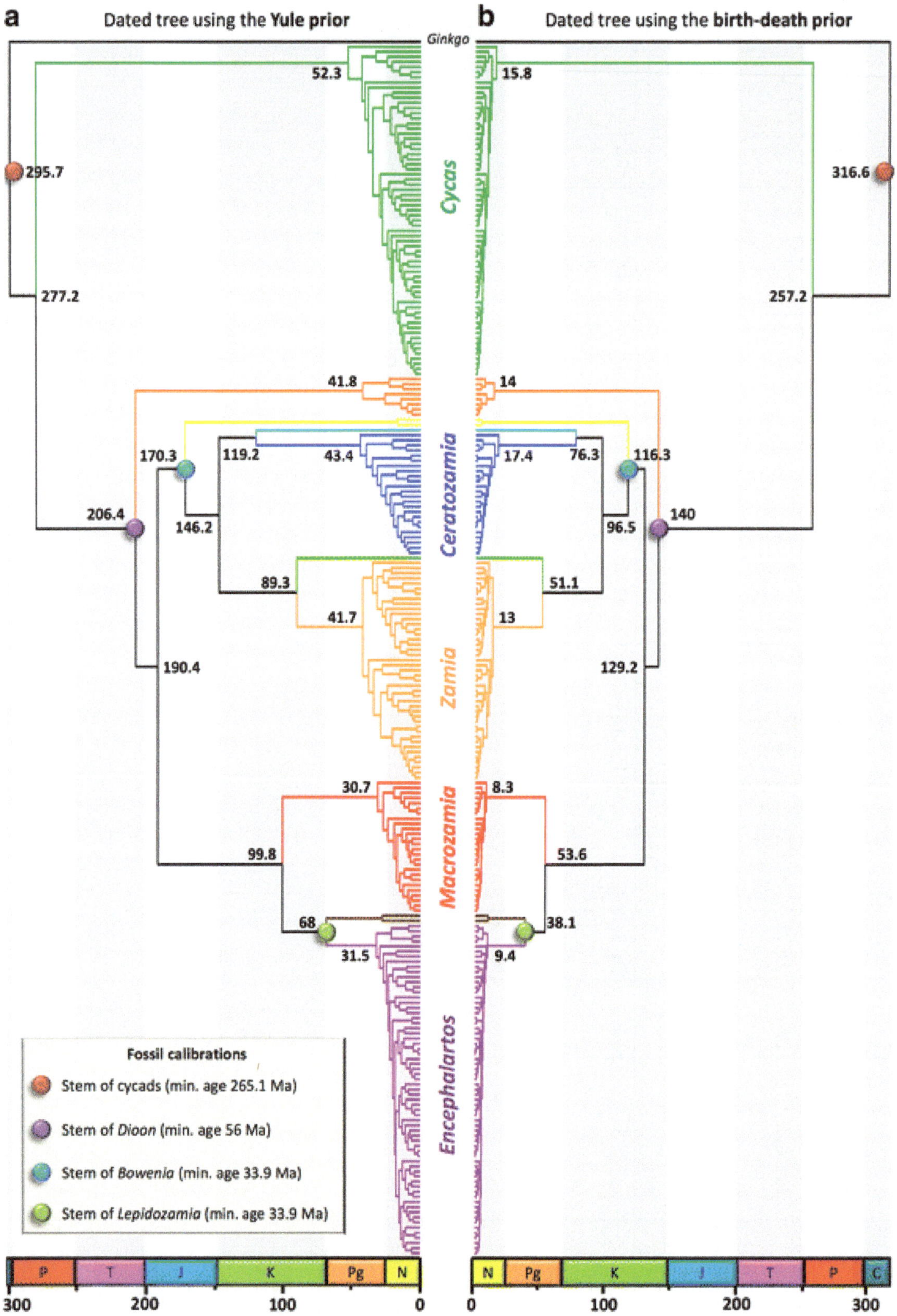


**Figure 2:** Time-calibrated phylogeny of Cycadales obtained with the four fossil calibrations. Timetree obtained with the Yule**(a)** or the birth-death **(b)** model as branching process prior. Each tree is the maximum clade credibility tree with median ages from the Bayesian analyses. The coloured dots highlight nodes on which fossil age constraints were applied. Values are median age estimates for the main nodes, in million years. C, Carboniferous; P, Permian; T, Triassic; J, Jurassic; K, Cretaceous; Pg, Paleogene; N, Neogene. The last geological period, the Quaternary, is missing.

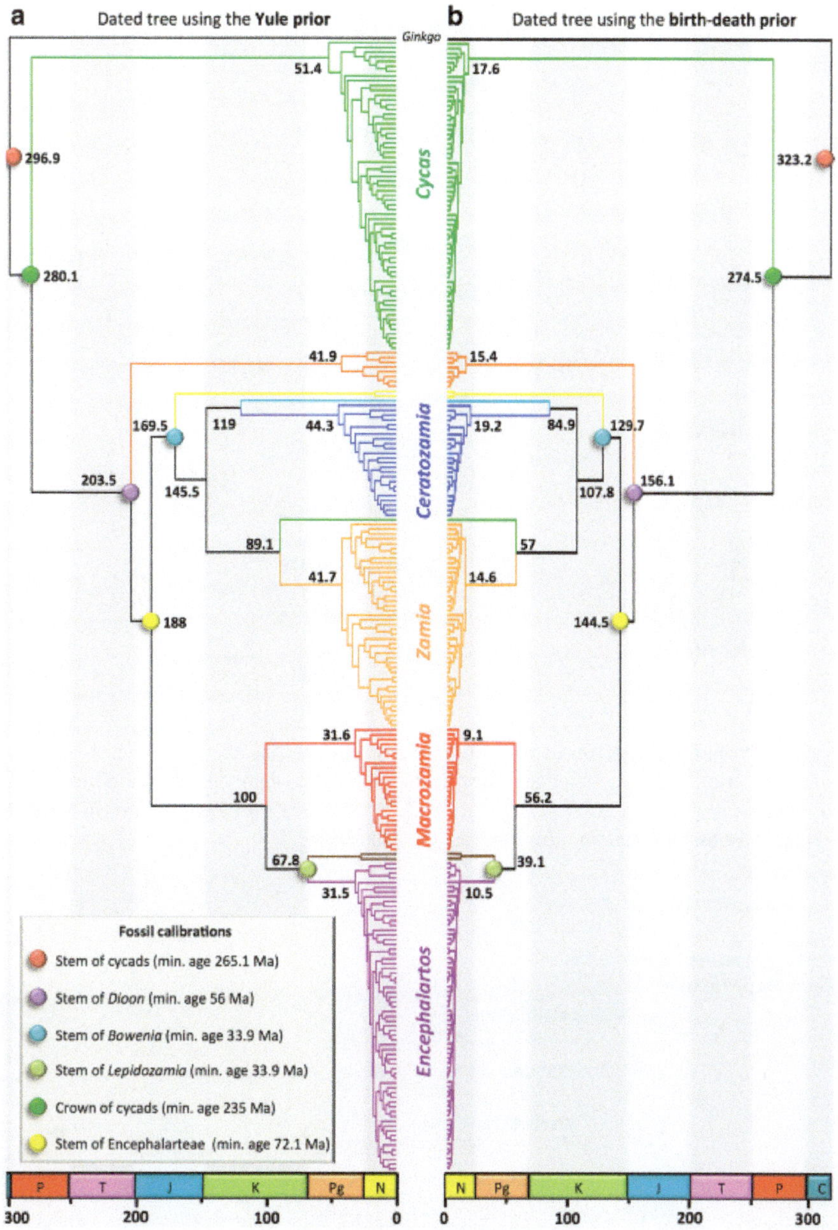


**Figure 3:** Time-calibrated phylogeny of Cycadales obtained with the six fossil calibrations. Timetree obtained with Yule **(a)** or the birth-death **(b)** model as branching process prior. Each tree is the maximum clade credibility tree with median ages from the Bayesian analyses. The coloured dots highlight nodes on which fossil age constraints were applied. Values are median age estimates for the main nodes, in million years. C, Carboniferous; P, Permian; T, Triassic; J, Jurassic; K, Cretaceous; Pg, Paleogene; N, Neogene. The last geological period, the Quaternary, is missing.

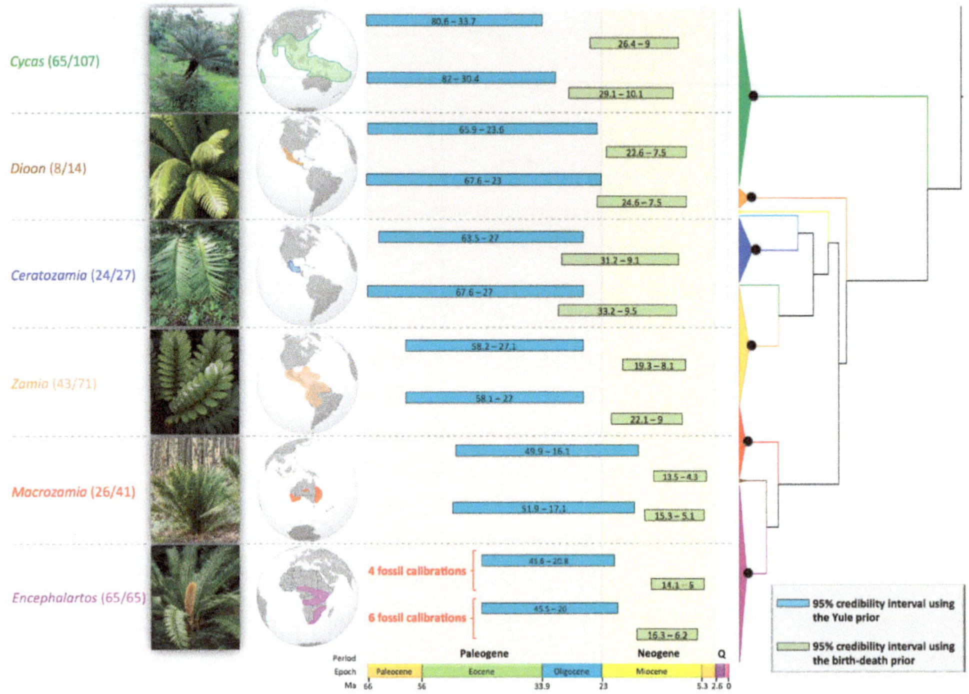


**Figure 4:** Credibility intervals (95% highest posterior density) for the crown ages of the six most species-rich genera. The blue bars depict the age estimates when using the Yule prior, and the green ones show the age estimates obtained with the birth-death prior, shown for the five dating analyses using most fossil calibration points. Analyses with the Yule prior indicate a Paleogene origin, while analyses with the birth-death prior indicate a Neogene origin. Absolute ages are in million years. Q, Quaternary.

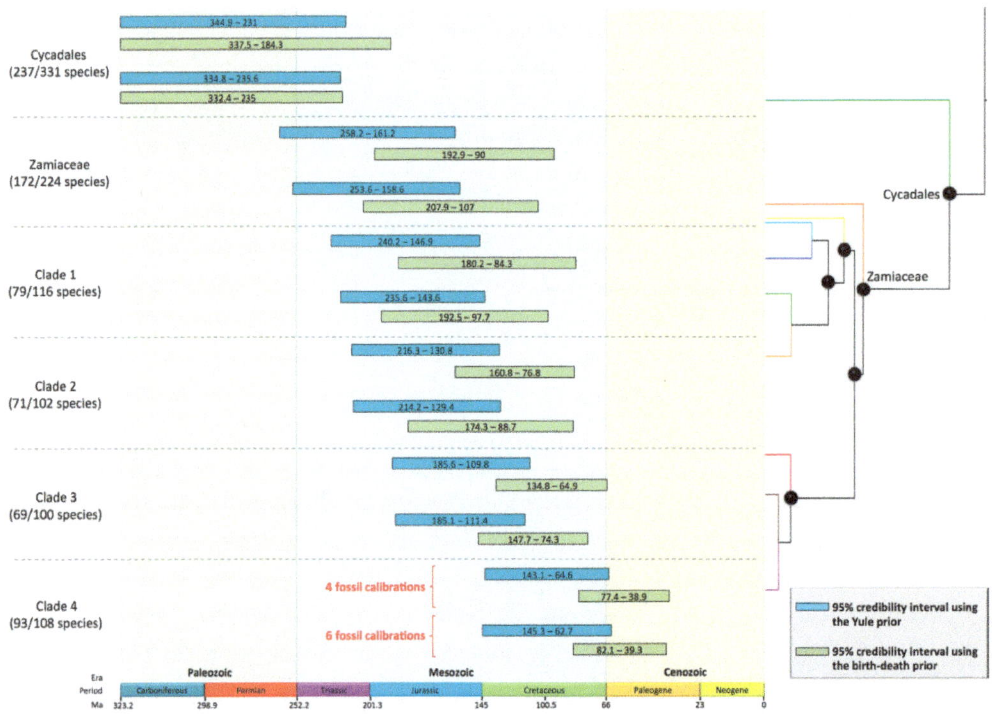


**Figure 5:** Credibility intervals (95% highest posterior density) for the ages of the six deepest nodes. The blue bars depict the age estimates when using the Yule prior, and the green ones show the age estimates obtained with the birth-death prior, shown for the five dating analyses using most fossil calibration points. The numbers associated to each clade correspond to the numbers on the phylogeny. Analyses with the Yule prior consistently indicate older ages. Absolute ages are in million years. The last geological period, the Quaternary, is missing.

Bayes factor values, calculated with the marginal likelihood estimates of the stepping-stone analyses, support the birth-death process as the best tree prior (Table 1). Of the two pairs of BEAST analyses (four FC and six FC), BF values are above the standard threshold: $BF_{4FC}=26.66$ between the Yule and birth-death prior, and $BF_{6FC}=36.3$ between the Yule and birth-death prior. On the contrary, the HME found the opposite, that is the Yule model is the best prior for both datasets for all analyses (Table 1). The results for the cycads thus support previous results showing that the HME estimator is not reliable [76].

## Macroevolutionary Rates through Time

The BAMM analyses converged for both trees ($ESS_{Yule}=719.6$; $ESS_{birth\text{-}death}=750.6$). The choice of the BPP had a major influence on the number of different evolutionary regimes detected in the history of the cycads, as well as on the estimation of speciation and extinction rates across the tree (Figure 6). Analyses with the Yule prior supported a model with three evolutionary regimes (i.e. two rate shifts) located at the crown of the genera *Cycas*and *Encephalartos* (according to BF values>10, and BF=154.5 over the null model). On the other hand, analyses with the birth-death prior supported a model with five evolutionary regimes (i.e. four rate shifts) located at (or near) the crown of the four most species-rich genera (according to BF values>10, and BF=29.3 over the null model). Moreover, both speciation and extinction rates estimated with the birth-death prior are twice as high as those estimated with the Yule prior (see scales on Figure 6). The macroevolutionary cohort analyses showed distinct evolutionary trajectories for the four richest cycad genera with the birth-death prior, but not with the Yule prior.

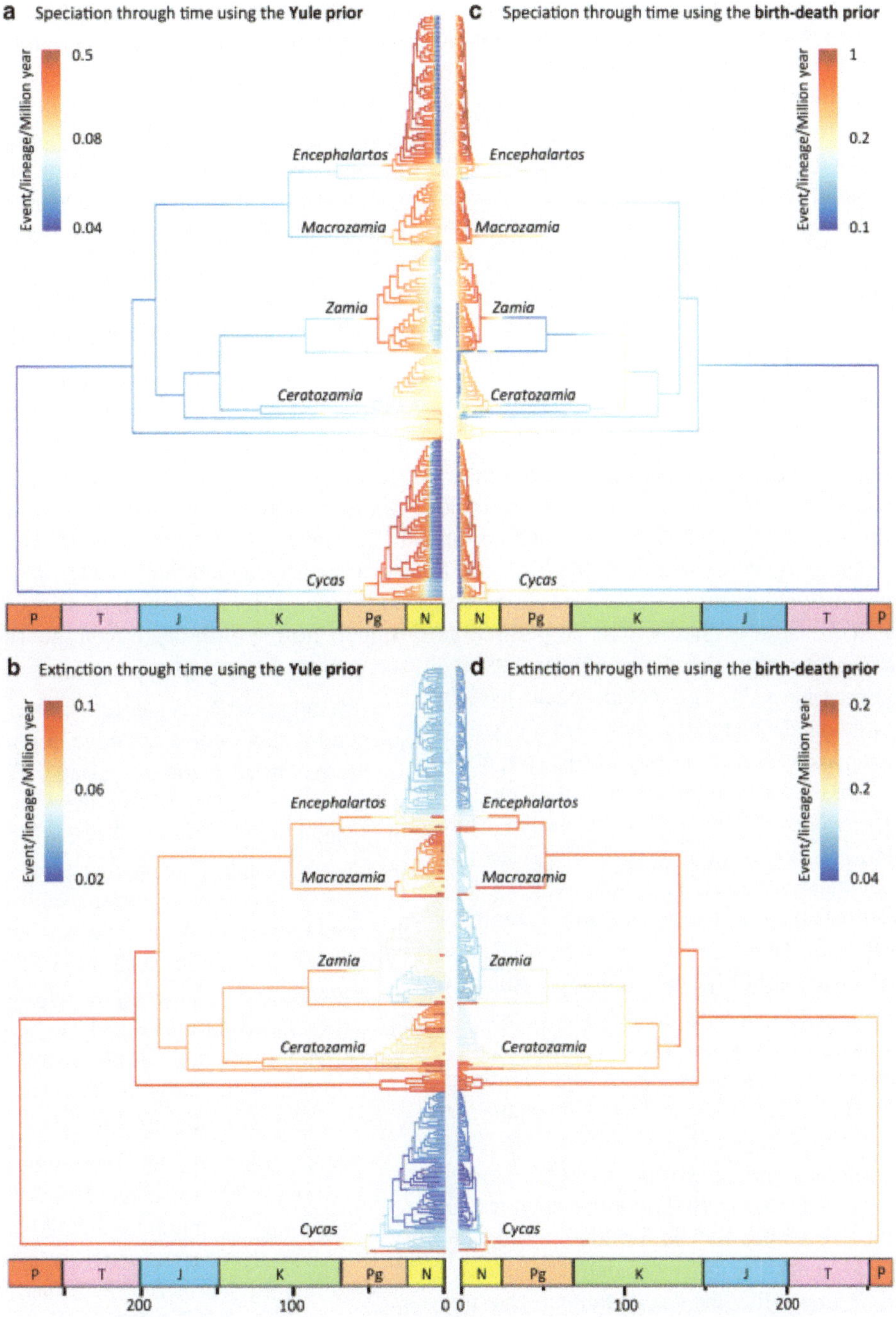


**Figure 6:** Diversification pattern of Cycadales. Estimates of speciation **(a,c)** and extinction **(b,d)** rates along the cycad phylogeny obtained from BAMM analyses, when considering the chronogram reconstructed with the Yule **(a,b)** or the birth-death **(c,d)** prior. Colours at each point in time along branches denote instantaneous rates of speciation or extinction inferred as the mean scenario, with colours indicating mean rates across all the shift configurations sampled in the Bayesian posterior. The diversification scenarios obtained with one versus the other prior are strikingly different. Note the differences for the estimated speciation and extinction rates with each tree prior (rates are twice higher with the tree constrained with a birth-death process). P, Permian; T, Triassic; J, Jurassic; K, Cretaceous; Pg, Paleogene; N, Neogene. The last geological period, the Quaternary, is missing.

# DISCUSSION

## The Branching Process Prior Impacts Bayesian Divergence Time Estimates

Our Bayesian dating analyses highlight an important effect of the branching process prior on the divergence times of cycads. Dating with the Yule prior consistently inferred ages for genus crown groups that were about three times older than those obtained with the birth-death prior. The cycad tree dated with the birth-death process suggested the initiation of crown radiations of the species-rich genera in the Neogene, whereas the Yule model indicated that the radiations began in the Paleogene. These differences in crown ages are noteworthy because the 95% credibility intervals for these nodes did not overlap. It is important to note that the effect of BPP is reduced towards the origin of cycads since, for both priors, genus radiations are subtended by long stem branches originating in the Triassic or Permian. This pattern is recovered for both calibration strategies. Therefore, the BPP is an important influence on the Bayesian dating analyses.

## Potential Causes of Age Differences between Branching Process Priors

It is unclear why differing tree priors have a major impact on divergence time estimates. The priors differ principally in the inclusion of an extinction rate. In the Yule prior, there is no extinction, but in the birth-death prior there is a parameter for the extinction rate. The effect of extinction in the birth-death prior is that there are relatively greater nodes toward the present compared to the past, because extinction has not yet had an effect on the more recent nodes, this phenomenon is known as the "pull of the present" [25]. This pattern could explain the contrasting results found when dating the cycad phylogeny with a birth-death versus a Yule prior. Support for the birth-death prior includes fossil evidence indicating that extinctions occurred and were important in shaping cycad evolution. More generally, extinction is a dominant feature of life given that 99.9% of all species that ever existed are now extinct [101]. But the role and effect of extinction could also potentially explain why the effect of BPPs appears to be taxon specific (i.e. little effect in Brassicaceae [28]). Indeed, it is possible that older clades might be more sensitive to the BPP choice because extinction is likely to be a more important component of their evolutionary histories. Earlier simulations and empirical studies have indicated that our understanding of the statistical properties of the diversification prior combined with prior distribution of calibrations is incomplete, particularly when the tree topology is considered [78,102,103]. It is an open question as to how important BPP choice will be for other groups, but further analyses of additional taxa as well as simulation studies will be valuable in understanding the detailed effects that tree priors have on age estimates.

## Other Potential Biases in the Branching Process Priors

While accounting for extinction in BPP priors is an important first step, a birth-death model with homogeneous rates across time and across the various clades within large groups is still an oversimplified model of their evolution. In particular, the hypothesis of rate homogeneity is thought to strongly bias extinction rate estimates [90,93]. We suggest that more complex priors should be developed for implementation in dating software; indeed evidence for age- or time-dependent diversification in numerous clades have resulted in development of methods to take into account rate-heterogeneity through time and clades [93,104].

Another potential bias in BPP priors is the use of models assuming complete or uniform sampling. Phylogenies are often sparsely sampled, particularly in long-branched outgroups. Sparsely sampled outgroups violate the basic assumption of most (if not all) the tree priors implemented in Bayesian dating, namely that the taxon sampling is random and consistent across all clades in the tree (see p. 98 of Drummond and Bouckaert [105]). In this case, it is difficult to find appropriate priors describing the tree: neither the single-parameter Yule model nor the two-parameter birth-death model fit a situation in which some parts of the tree are densely sampled, while others consist of a single long-branched species (i.e. *Ginkgo* versus the cycads).

## Implications for Bayesian Dating Analyses

The vast majority of molecular dating studies using BEAST have relied on the Yule model, although some recent studies have used both the Yule and birth-death process [27,28,106]. However these studies did not find any differences in age estimates between the two models, nor they did not perform MLE and Bayes factors analyses. Here we tested competing evolutionary hypotheses using a Bayesian relaxed-clock under two different branching priors, and we found stronger support for the birth-death than the Yule process as a prior to reconstruct the cycad tree.

Model selection is a necessary step, because as in the case of cycads, prior choice can have important implications if one simply reconstructs the tree using a Yule model as commonly performed. Conclusions on the age, diversification and/or biogeographic patterns might be different regarding the BPP (see below). Supposedly ancient plant clades might be affected by prior choice as for cycads; specifically, clades with an old and well-documented fossil record that probably experienced periods of extinction such as conifers [7,8], ferns [3,4], gnetophytes [5], and early diverging angiosperm lineages [107,108]. One solution might be to replace the commonly used Yule model by the birth-death process, which is more realistic since all clades are likely to have experienced extinction, even young clades currently diversifying.

As a cautionary tale, we suggest that users routinely test between priors to select the prior that best fit their data. Estimators of the marginal likelihood [82,83] are included in BEAST; therefore, model selection can be relatively straightforward. Furthermore, it has been demonstrated that these marginal likelihood estimators accurately compute BF for BRC studies [76], indicating that the results of our empirical study showing the impact of the BPP are reliable.

## A late Permian Origin for Cycads

According to the dating scenario supported by BF (i.e. timetree reconstructed with six fossil calibrations and a birth-death process), cycads and *Ginkgo* diverged in the late Carboniferous

– early Permian, and the most recent common ancestor of the living cycads is from the late Permian (274.5 Ma). Compared to other dating studies, our timetree fits well with recent works on the genus ages and radiations, but our tree pushes back the origin of cycads more than 50 million years earlier (≈200 Ma for Nagalingum *et al.* [6]; ≈230 Ma for Salas-Leiva *et al.* [43]). For the latter, our results are in agreement with other dating studies that used age constraints on the cycad crown (≈280 Ma for Won & Renner [5]; ≈270 Ma for Burleigh *et al.* [109]). Our cycad timeframe is also congruent with the cycad insect pollinator of the order Thysanoptera originating and diversifying in the late Permian (286 Ma, [110,111]). Finally our results are in agreement with paleobotanical evidence indicating that cycads first appeared in the late Permian and became more abundant and diverse in the Triassic [33,39,40].

Cycads proliferated during the Mesozoic, an era characterized by an increased occurrence and diversity of fossil cycads, broadly distributed throughout the relatively uniform climate of the supercontinent Pangaea [42]. Notably, the Triassic record of Cycadales is among the most well-known of cycad history, not only because of the marked increase in the number of definitive cycad taxa, but also because some of the most well-preserved and informative taxa are from this period [70]. It is generally accepted that cycads first became significant components of terrestrial floras during the Late Triassic as part of a global floral turnover with primitive seed-plant floras being replaced by conifers and cycads after the Permian-Triassic mass extinction (252.2 Ma) [40,112,113]. Cycads are generally believed to have reached a diversity peak during the Jurassic, and remained relatively stable in terms of diversity during the Cretaceous [34,112]. Our results further show that major lineages of extant cycads were established during the Early Cretaceous through to the Late Cretaceous, particularly in the Zamiaceae. The current diversity of Zamiaceae stems from surviving lineages of the Cretaceous, a period of rich fossil diversity for the clade [30,33,114-116].

## Global Diversification Pattern of Cycads

The cycad timetree is remarkable for its long branches subtending the genus radiations. The most remarkable example is the branch leading to the genus *Cycas* that is more than 200 million years long. Our results indicate that extant diversification of cycad species occurred in the Neogene, in agreement with other studies [6,38,43]. No crown node of any modern species-rich genus of cycad has an age >19 Ma (credibility intervals extend back to 33 and 29 Ma for *Ceratozamia* and *Cycas*). Therefore, the ancient stem divergences of genera are decoupled from their recent crown group radiations. This decoupling can be explained by three hypotheses: *(i)*the long branches represent periods of low diversity, *(ii)* the long branches may be the result of periods of extinction [6,38,107,117], but it is unknown whether these were few and sudden or highly protracted, or *(iii)*the presence of ghost lineages, which are lineages that have temporal and/or geographic gaps in their fossil record (e.g. the Cenozoic fossil record of coelacanths). Our results are consistent with the second hypothesis because we show *(i)* quasi-synchronous and independent shifts in diversification at the crown (or near) of genus radiations that may be interpreted as the recovery period after an extinction event [107,117], and *(ii)*high rates of extinction along the long branches of the genus stems that may have extirpated entire lineages [32,33,59]. Our results, as well as fossil evidence [30-33,39,40,42,114-116], support the role of a mass extinction in shaping the cycad timetree, perhaps in the Late Cretaceous when BAMM extinction rates are elevated.

We found periods of extinction along the long branches of the tree. It is important to mention that these inferred periods of extinction are congruent with fossil data also indicating significant extinctions for the clade. From the Jurassic through the Paleogene, the rate of extinction was high for all lineages. The monotypic genera (*Microcycas* and *Stangeria*) and species-poor genera (*Bowenia*, *Dioon*, and *Lepidozamia*) were particularly impacted by extinction (Figure 6). Fossil data and cladistic analyses suggest that, for instance, the lineage today represented by *Stangeria* might have been more diverse in the past [32,59,118]. The surviving cycads re-diversified from the mid- to late Miocene when five lineages, isolated on different continents, underwent successful independent increases in diversification rate. Therefore our results favour the hypothesis of few and sudden extinction events. As potential explanations, this diversification pattern may be attributed to *(i)* competition with angiosperms (e.g. [34,35,112,113,119]), *(ii)* changes in, and final extinction of, non-avian dinosaur faunas that may have acted as seed dispersers (for Mesozoic extinction) [41], and/or *(iii)*global climate change that led to latitudinal shifts of the main vegetation belts and the elimination of cycads from higher latitudes of Eurasia, North America, and Patagonia [6,110,120], perhaps around the Eocene-Oligocene boundary that marked a significant shift from greenhouse to icehouse climate [38,121].

Discussions on cycad diversity through time are partly hampered due to the difficulty of distinguishing cycad foliage and pollen from that of the extinct (and phylogenetically distinct) clade Bennettitales [39,42]; most previous published diversity curves for Mesozoic plants do not separate Cycadales and Bennettitales, with two clades being grouped together as 'cycadophytes' (e.g. [34,39,112]). Also, the Cretaceous and Cenozoic cycad fossil record is probably too limited to enable understanding of the role of extinction, notably around the Cretaceous-Paleogene boundary. Although their apparent post mid-Cretaceous decline is recognized, we do not have quantitative evidence of the diversity decline nor do we know whether Cenozoic cycad diversity remained low until the Neogene radiations, or whether substantial early to mid-Cenozoic diversity existed but was affected by major extinctions. Besides illustrating the potential effect of BPP choice in Bayesian molecular dating, our study re-examines the dating of the cycads, proving a robust phylogenetic framework for future studies aimed at estimating how the diversity of cycads varied through geological time.

## CONCLUSIONS

Our study highlights that the branching process prior (or tree prior) in Bayesian molecular dating has an important effect on age estimates for cycads. The birth-death process had a better fit than the traditionally used Yule process as determined by marginal likelihood estimates and Bayes factors. It is likely that cycads are not an isolated case and we advocate for a closer investigation of the branching process prior, BPP, in all future molecular phylogenetic dating studies. At minimum we suggest conducting a similar model selection as we did for cycads to select the best-fitting prior and thus the best timetree. However, the birth-death process will not necessarily always better fit the data just because it seems more biologically realistic than a simpler Yule model, especially for shallower divergence times (i.e. recent clades). For instance, two recent studies on Australian diving beetles [106] and St. John's wort [122] did not show any age differences using different BPPs, and none of the BPP was supported with Bayes factors; both clades originated in the Oligocene (*ca.* 34-28 Ma). On the contrary for deeper divergence times, age differences might be revealed when comparing a birth-death and a Yule process as exemplified in a dating study of darkling beetles that originated in the Jurassic (≈180 Ma) [27], but those differences were not as marked as in cycads. Based on these studies and our results, we expect that age differences deriving from different BPPs increase when clade age is very old (e.g. Triassic and backward).

As we gain greater understanding of the priors and parameters associated with molecular dating, there have been accordingly advances in our methodology. We note that other avenues for molecular dating are being developed because the standard dating procedure results in overlaying two prior distributions for a calibration node: one from the tree prior and one from the calibration prior [19,21,102]. In the last years, there are new methods that allow the inclusion of fossils in divergence time estimation as non-contemporaneous terminal tips rather than as node calibration points, also known as tip dating [123]. Recently, Heath *et al.* [103] introduced the fossilized birth–death process, a model for calibrating divergence time estimates in a Bayesian framework, explicitly acknowledging that extant species and fossils are part of the same macroevolutionary process. They argued that a single model that acts as a prior on the speciation times for both calibrated and uncalibrated nodes is a better representation of the lineage diversification process. The approach seems promising and will increasingly be used in future studies, like the application on the royal ferns (family Osmundaceae) [124]. The fossilized birth–death process best fitted the Osmundaceae fossil record and also provided speciation and extinction rates associated with the dated phylogeny.

## DECLARATIONS

### Acknowledgements

We thank the Associate Editor Rafael Zardoya, and the two referees Susanne Renner and Diego San Mauro who provided insightful comments on the study. We also thank Frédéric Delsuc and Gael Kergoat for discussions, and Simon Ho and Guy Baele for technical comments on the Bayesian analyses. We also thank Daniel Rabosky for the workshop on the BAMM approach, and Mario Coiro for discussions and feedbacks on the fossil record of cycads. Funding was provided by grant ANR CHEX ECOEVOBIO awarded to HM.

### Competing interests

The authors declare that they have no competing interests.

## REFERENCES

1. Hedges SB, Kumar S. The Timetree of Life. Oxford: Oxford University Press; 2009.
2. Laenen B, Shaw B, Schneider H, Goffinet B, Paradis E, Désamoré A, et al. Extant diversity of bryophytes emerged from successive post-Mesozoic diversification bursts. Nature Comm. 2014;5:5134.
3. Schneider H, Schuettpelz E, Pryer KM, Cranfill R, Magallon S, Lupia R. Ferns diversified in the shadow of angiosperms. Nature. 2004;428:553–7.
4. Schuettpelz E, Pryer KM. Evidence for a Cenozoic radiation of ferns in an angiosperm-dominated canopy. Proc Natl Acad Sci U S A. 2009;106:11200–5.
5. Won H, Renner SS. Dating dispersal and radiation in the gymnosperm *Gnetum* (Gnetales) - Clock calibration when outgroup relationships are uncertain. Syst Biol. 2006;55:610–22.
6. Nagalingum NS, Marshall CR, Quental TB, Rai HS, Little DP, Mathews S. Recent synchronous radiation of a living fossil. Science. 2011;334:796–9.
7. Mao K, Milne RI, Zhang L, Peng Y, Liu J, Thomas P, et al. Distribution of living Cupressaceae reflects the breakup of Pangea. Proc Natl Acad Sci U S A. 2012;109:7793–8.
8. Leslie AB, Beaulieu JM, Rai HS, Crane PR, Donoghue MJ, Mathews S. Hemisphere-scale differences in conifer evolutionary dynamics. Proc Natl Acad Sci U S A. 2012;109:16217–21.
9. Magallón SA, Castillo A. Angiosperm diversification through time. Am J Bot. 2009;96:349–65.
10. Fiz-Palacios O, Schneider H, Heinrichs J, Savolainen V. Diversification of land plants: insights from a family-level phylogenetic analysis. BMC Evol Biol. 2011;11:341.
11. Zanne AE, Tank DC, Cornwell WK, Eastman JM, Smith SA, FitzJohn RG, et al. Three keys to the radiation of angiosperms into freezing environments. Nature. 2014;506:89–92.
12. Percy DM, Page RDM, Cronk QCB. Plant-insect interactions: double-dating associated insect and plant lineages reveals asynchronous radiations. Syst Biol. 2004;53:120–7.
13. Kumar S. Molecular clocks: four decades of evolution. Nature Rev Genet. 2005;6:654–62.
14. Donoghue PCJ, Benton MJ. Rocks and clocks: calibrating the Tree of Life using fossils and molecules. Trends Ecol Evol. 2007;22:424–31.
15. Alfaro ME, Holder MT. The posterior and the prior in Bayesian phylogenetics. Annu Rev Ecol Evol Syst. 2006;37:19–42.
16. Drummond AJ, Ho SYW, Phillips MJ, Rambaut A. Relaxed phylogenetics and dating with confidence. PLoS Biol. 2006;4:e88.
17. Inoue J, Donoghue PCJ, Yang Z. The impact of the representation of fossil calibrations on Bayesian estimation of species divergence times. Syst Biol. 2010;59:74–89.
18. Yang Z, Rannala B. Bayesian estimation of species

divergence times under a molecular clock using fossil calibrations with soft bounds. Mol Biol Evol. 2006;23:212–26.

19. Ho SYW, Phillips MJ. Accounting for calibration uncertainty in phylogenetic estimation of evolutionary divergence times. Syst Biol. 2009;58:367–80.

20. Yang Z, Rannala B. Branch-length prior influences Bayesian posterior probability of phylogeny. Syst Biol. 2005;54:455–70.

21. Heled J, Drummond AJ. Calibrated birth-death phylogenetic time-tree priors for Bayesian inference. Syst Biol. 2015;64:369–383.

22. Morlon H. Phylogenetic approaches for studying diversification. Ecol Lett. 2014;17:508–25.

23. Yang Z, Goldman N, Friday A. Maximum likelihood tree from DNA sequences: a peculiar statistical estimation problem. Syst Biol. 1995;44:384–99.

24. Rannala B, Yang Z. Probability distribution of molecular evolutionary trees: a new method of phylogenetic inference. J Mol Evol. 1996;43:304–11.

25. Gernhard T. The conditioned reconstructed process. J Theoret Biol. 2008;253:769–78.

26. Drummond AJ, Suchard MA, Xie D, Rambaut A. Bayesian phylogenetics with BEAUti and the BEAST 1.7. Mol Biol Evol. 2012;29:1969–73.

27. Kergoat GJ, Bouchard P, Clamens A-L, Abbate JL, Jourdan H, Zahab R, et al. Cretaceous environmental changes led to high turnovers and decreases in diversification rates in a hyperdiverse non-phytophagous beetle family. BMC Evol Biol. 2014;14:220.

28. Couvreur TLP, Franzke A, Al-Shehbaz IA, Bakker F, Koch M, Mummenhoff K. Molecular phylogenetics, temporal diversification and principles of evolution in the mustard family (Brassicaceae). Mol Biol Evol. 2010;27:55–71.

29. Osborne R, Calonje MA, Hill KD, Stanberg L, Stevenson DW. The world list of Cycads. Mem New York Botl Gard. 2012;106:480–510.

30. Pant DD. The fossil history and phylogeny of the cycadales. Geophytol. 1987;17:125–62.

31. Axsmith BJ, Serbet R, Krings M, Taylor TN, Taylor EL, Mamay SH. The enigmatic Paleozoic plants *Spermopteris* and*Phasmatocycas* reconsidered. Am J Bot. 2003;90:1585–95.

32. Hermsen EJ, Taylor TN, Taylor EL, Stevenson DW. Cataphylls of the Middle Triassic cycad *Antarcticycas schopfii* and new insights into cycad evolution. Am J Bot. 2006;93:724–38.

33. Cúneo NR, Escapa I, Villar-de-Seoane L, Artabe A, Gnaedinger S. Review of the Cycads and Bennettitaleans from the Mesozoic of Argentina. In: Gee CT, editor. Plants in Mesozoic Time: Morphological Innovations, Phylogeny, Ecosystems. Bloomington: Indiana University Press; 2010. p. 187–212.

34. Crane PR. Vegetational consequences of the angiosperm diversification. In: Friis EM, Chaloner WC, Crane PR, editors. The Origin of Angiosperms and their Biological Consequences. Cambridge: Cambridge University Press; 1987. p. 107–44.

35. Lupia R, Lidgard S, Crane PR. Comparing palynological abundance and diversity: implications for biotic replacement during the Cretaceous angiosperm radiation. Paleobiol. 1999;25:305–40.

36. van de Schootbrugge B, Quan TM, Lindström S, Püttmann W, Heunisch C, Pross J, et al. Floral changes across the Triassic/Jurassic boundary linked to flood basalt volcanism. Nature Geosci. 2009;2:589–94.

37. Coiffard C, Gomez B, Daviero-Gomez V, Dilcher DL. Rise to dominance of angiosperm pioneers in European Cretaceous environments. Proc Natl Acad Sci U S A. 2012;109:20955–9.

38. Crisp MD, Cook LG. Cenozoic extinctions account for the low diversity of extant gymnosperms compared with angiosperms. New Phytol. 2011;192:997–1009.

39. Willis KJ, McElwain JC. The Evolution of Plants. Oxford: Oxford University Press; 2002.

40. McElwain JC, Punyasena SW. Mass extinction events and the fossil record. Trends Ecol Evol. 2007;22:548–57.

41. Butler RJ, Barrett PM, Kenrick P, Penn MG. Testing co-evolutionary hypotheses over geological timescales: interactions between Mesozoic non-avian dinosaurs and cycads. Biological Rev. 2009;84:73–89.

42. Taylor TN, Taylor EL, Krings M. Paleobotany: The Biology and Evolution of Fossil Plants. 2nd ed. Burlington: Academic Press; 2009.

43. Salas-Leiva DE, Meerow AW, Calonje M, Griffith MP, Francisco-Ortega J, Nakamura K, et al. Phylogeny of the cycads based on multiple single-copy nuclear genes: congruence of concatenated parsimony, likelihood and species tree inference methods. Ann Bot. 2013;112:1263–78.

44. Hollick A. Descriptions of new species of Tertiary cycads, with a review of those previously recorded. Bull Torrey Bot Club. 1932;59:169–89.

45. Hill RS. Three new Eocene cycads from eastern Australia. Austral J Bot. 1980;28:105–22.

46. Horiuchi J, Kimura T. *Dioonopsis nipponica* gen. et sp. nov., a new cycad from the Palaeogene of Japan. Rev Palaeobot Palynol. 1987;51:213–25.

47. Kvaček Z. A new Tertiary *Ceratozamia* (Zamiaceae, Cycadopsida) from the European Oligocene. Flora – Morphology, Distribution Functional Ecology of Plants. 2002;197:303–16.

48. Erdei B, Manchester SR, Kvaček Z. *Dioonopsis* Horiuchi et Kimura leaves from the Eocene of Western North America: a cycad shared with the Paleogene of Japan. Intl J Plant Sci. 2012;173:81–95.

49. Kvaček Z. New fossil records of *Ceratozamia* (Zamiaceae, Cycadales) from the European Oligocene and lower Miocene. Acta Palaeobot. 2014;54:231–47.

50. Su K, Quan C, Liu Y-S. *Cycas fushunensis* sp. nov. (Cycadaceae) from the Eocene of northeast China. Rev Palaeobot Palynol. 2014;204:43–9.

51. Griffith MP, Magellan TM, Tomlinson PB. Variation in leaflet structure in *Cycas* (Cycadales: Cycadaceae):

Does anatomy follow phylogeny and geography? Intl J Plant Sci. 2014;175:241–55.

52. González D, Vovides AP. Low intralineage divergence in *Ceratozamia* (Zamiaceae) detected with nuclear ribosomal DNA ITS and chloroplast DNA trnL-F non-coding region. Syst Bot. 2002;27:654–61.

53. Hill KD, Chase MW, Stevenson DW, Hill HG, Schutzman B. The families and genera of cycads: a molecular phylogenetic analysis of Cycadophyta based on nuclear and plastid DNA sequences. Intl J Plant Sci. 2003;164:933–48.

54. Rai HS, O'Brien HE, Reeves PA, Olmstead RG, Graham SW. Inference of higher-order relationships in the cycads from a large chloroplast data set. Mol Phylogenet Evol. 2003;29:350–9.

55. Chaw S-M, Walters TW, Chang C-C, Hu S-H, Chen S-H. A phylogeny of cycads (Cycadales) inferred from chloroplast matK gene, trnK intron, and nuclear rDNA ITS region. Mol Phylogenet Evol. 2005;37:214–34.

56. González D, Vovides AP, Barcenas C. Phylogenetic relationships of the Neotropical genus *Dioon* (Cycadales, Zamiaceae) based on nuclear and chloroplast DNA sequence data. Syst Bot. 2008;33:229–36.

57. Yessoufou K, Bamigboye SO, Daru BH, van der Bank M. Evidence of constant diversification punctuated by a mass extinction in the African cycads. Ecol Evol. 2013;4:50–8.

58. Wu CH, Chaw SM, Huang YY. Chloroplast phylogenomics indicates that *Ginkgo biloba* is sister to cycads. Genome Biol Evol. 2013;5:243–54.

59. Martínez LCA, Artabe AEE, Bodnar J. A new cycad stem from the Cretaceous in Argentina and its phylogenetic relationships with other Cycadales. Bot J Linn Soc. 2012;170:436–58.

60. Sauquet H, Ho SYW, Gandolfo MA, Jordan GJ, Wilf P, Cantrill DJ, et al. Testing the impact of calibration on molecular divergence times using a fossil-rich group: the case of *Nothofagus* (Fagales). Syst Biol. 2012;61:289–313.

61. Near TJ, Meylan PA, Shaffer HB. Assessing concordance of fossil calibration points in molecular clock studies: an example using turtles. Am Nat. 2005;165:137–46.

62. Clarke JT, Warnock RCM, Donoghue PCJ. Establishing a time-scale for plant evolution. New Phytol. 2011;192:266–301.

63. Gradstein FM, Ogg JG, Schmitz MD, Ogg GM. The Geologic Time Scale 2012, Volume Set. Boston, USA: Elsevier; 2012.

64. Zhu JN, Zhang SS, Ma J. A new genus and species – *Cycadostrobus paleozoicus* Zhu of Cycadaceae from the Permian of China. Acta Phytotaxo Sinica. 1994;32:340–4.

65. Zhu JN, Du XM. A new cycad – *Primocycas chinensis* gen. et sp. nov. discovers from the Lower Permian in Sha-nxi, China and its significance. Acta Bot Sinica. 1981;23:401–4.

66. Zhifeng G, Thomas BA. A review of fossil cycad megasporophylls, with new evidence of *Crossozamia* Pomel and its associated leaves from the Lower Permian of Taiyuan, China. Rev Palaeobot Palynol. 1989;60:205–23.

67. Wang J. Late Paleozoic macrofloral assemblages from Weibei Coalfield, with reference to vegetational change through the Late Paleozoic Ice-age in the North China Block. Intl J Coal Geol. 2010;83:292–317.

68. Hill RS. Two new species of *Bowenia* Hook, ex Hook, f. from the Eocene of eastern Australia. Austral J Bot. 1978;26:837–46.

69. Smoot EL, Taylor TN, Delevoryas T. Structurally preserved fossil plants from Antarctica. I. *Antarcticycas*, gen. nov., a Triassic cycad stem from the Beardmore Glacier area. Am J Bot. 1985;72:1410–23.

70. Hermsen EJ, Taylor EL, Taylor TN. Morphology and ecology of the *Antarcticycas* plant. Rev Palaeobot Palynol. 2009;153:108–23.

71. Rothwell GW, Scheckler SE, Gillespie WH. *Elkinsia* gen nov, a Late Devonian gymnospermn with cupulate ovules. Bot Gazette. 1989;150:170–89.

72. Ronquist F, Teslenko M, Van der Mark P, Ayres DL, Darling A, Höhna S, et al. MrBayes 3.2: efficient Bayesian phylogenetic inference and model choice across a large model space. Syst Biol. 2012;61:539–42.

73. Lanfear R, Calcott B, Ho SYW, Guindon S. PartitionFinder: combined selection of partitioning schemes and substitution models for phylogenetic analyses. Mol Biol Evol. 2012;29:1695–701.

74. Huelsenbeck JP, Larget B, Alfaro ME. Bayesian phylogenetic model selection using reversible jump Markov chain Monte Carlo. Mol Biol Evol. 2004;21:1123–33.

75. Rambaut A, Suchard MA, Xie D, Drummond AJ. Tracer v1.6. Available at: http://beast.bio.ed.ac.uk/Tracer; 2015

76. Baele G, Li WLS, Drummond AJ, Suchard MA, Lemey P. Accurate model selection of relaxed molecular clocks in Bayesian phylogenetics. Mol Biol Evol. 2013;30:239–43.

77. Vanneste K, Baele G, Maere S, Van de Peer Y. Analysis of 41 plant genomes supports a wave of successful genome duplications in association with the Cretaceous–Paleogene boundary. Genome Res. 2014;24:1334–47.

78. Warnock RCM, Parham JF, Joyce WG, Lyson TR, Donoghue PCJ. Calibration uncertainty in molecular dating analyses: there is no substitute for the prior evaluation of time priors. Proc R Soc B. 2015;282:20141013.

79. Near TJ, Sanderson MJ. Assessing the quality of molecular divergence time estimates by fossil calibrations and fossil-based model selection. Philos T R Soc B. 2004;359:1477–83.

80. Ayres DL, Darling A, Zwickl DJ, Beerli P, Holder MT, Lewis PO, et al. BEAGLE: an application programming interface and high-performance computing library for statistical phylogenetics. Syst Biol. 2012;61:170–3.

81. Miller MA, Schwartz T, Pickett BE, He S, Klem EB, Scheuermann RH, et al. A RESTful API for Access to Phylogenetic Tools via the CIPRES Science Gateway.

Evol Bioinformatics. 2015;11:43–8.

82. Lartillot N, Philippe H. Computing Bayes factors using thermodynamic integration. Syst Biol. 2006;55:195–207.

83. Xie W, Lewis PO, Fan Y, Kuo L, Chen MH. Improving marginal likelihood estimation for Bayesian phylogenetic model selection. Syst Biol. 2011;60:150–60.

84. Kass RE, Raftery AE. Bayes factors. J Am Stat Assoc. 1995;90:773–95.

85. Rabosky DL, Santini F, Eastman JM, Smith SA, Sidlauskas B, Chang J, et al. Rates of speciation and morphological evolution are correlated across the largest vertebrate radiation. Nature Comm. 2013;4:1958.

86. Rabosky DL, Donnellan SC, Grundler M, Lovette IJ. Analysis and visualization of complex macroevolutionary dynamics: an example from Australian scincid lizards. Syst Biol. 2014;63:610–27.

87. Rabosky DL. Automatic detection of key innovations, rate shifts, and diversity dependence on phylogenetic trees. PLoS One. 2014;9:e89543.

88. Paradis E. Can extinction rates be estimated without fossils? J Theor Biol. 2004;229:19–30.

89. Quental TB, Marshall CR. Diversity dynamics: molecular phylogenies need the fossil record. Trends Ecol Evol. 2010;25:434–41.

90. Rabosky DL. Extinction rates should not be estimated from molecular phylogenies. Evolution. 2010;64:1816–24.

91. Nee S, Holmes EC, May RM, Harvey PH. Extinction rates can be estimated from molecular phylogenies. Philos Trans R Soc B. 1994;344:77–82.

92. Purvis A. Phylogenetic approaches to the study of extinction. Annu Rev Ecol Evol Syst. 2008;39:301–19.

93. Morlon H, Parsons TL, Plotkin JB. Reconciling molecular phylogenies with the fossil record. Proc Natl Acad Sci U S A. 2011;108:16327–32.

94. Beaulieu JM, O'Meara BC. Extinction can be estimated from moderately sized molecular phylogenies. Evolution. 2015, doi.org/10.1111/evo.12614.

95. Huang S, Chiang YC, Schaal BA, Chou CH, Chiang TY. Organelle DNA phylogeography of *Cycas taitungensis*, a relict species in Taiwan. Mol Ecol. 2001;10:2669–81.

96. Xiao LQ, Möller M. Nuclear ribosomal its functional paralogs resolve the phylogenetic relationships of a late-Miocene radiation cycad *Cycas* (Cycadaceae). PLoS One. 2015;10:e0117971.

97. Treutlein J, Vorster P, Wink M. Molecular relationships in *Encephalartos* (Zamiaceae, Cycadales) based on nucleotide sequences of nuclear ITS 1&2, rbcL, and genomic ISSR fingerprinting. Plant Biol. 2005;7:79–90.

98. Sharma IK, Jones DL, Forster PI. Genetic differentiation and phenetic relatedness among seven species of the*Macrozamia plurinervia* complex (Zamiaceae). Biochem Syst Ecol. 2004;32:313–27.

99. Caputo P, Cozzolino S, Gaudio L, Moretti A, Stevenson D. Karyology and phylogeny of some mesoamerican species of*Zamia* (Zamiaceae). Am J Bot. 1996;83:1513–20.

100. Zgurski JM, Rai HS, Fai QM, Bogler DJ, Francisco-Ortega J, Graham SW. How well do we understand the overall backbone of cycad phylogeny? New insights from a large, multigene plastid data set. Mol Phylogenet Evol. 2008;47:1232–7.

101. Raup DM. Extinction: Bad genes or bad luck? New York: W.W. Norton and Company; 1991.

102. Heled J, Drummond AJ. Calibrated tree priors for relaxed phylogenetics and divergence time estimation. Syst Biol. 2012;61:138–49.

103. Heath TA, Huelsenbeck JP, Stadler T. The fossilized birth-death process for coherent calibration of divergence-time estimates. Proc Natl Acad Sci U S A. 2014;111:E2957–66.

104. Hagen O, Hartmann K, Steel M, Stadler T. Age-dependent speciation can explain the shape of empirical phylogenies. Syst Biol. 2015;64:432–440.

105. Drummond AJ, Bouckaert RR. Bayesian evolutionary analysis with BEAST. Cambridge: Cambridge University Press; 2015.

106. Toussaint EFA, Condamine FL, Hawlitschek O, Watts CHS, Porch N, Hendrich L, et al. Unveiling the diversification dynamics of Australasian predaceous diving beetles in the Cenozoic. Syst Biol. 2015;64:3–24.

107. Antonelli A, Sanmartín I. Mass extinction, gradual cooling, or rapid radiation? Reconstructing the spatiotemporal evolution of the ancient angiosperm genus *Hedyosmum* (Chloranthaceae) using empirical and simulated approaches. Syst Biol. 2011;60:596–615.

108. Thomas N, Bruhl JJ, Ford A, Weston PH. Molecular dating of Winteraceae reveals a complex biogeographical history involving both ancient Gondwanan vicariance and long-distance dispersal. J Biogeogr. 2014;41:894–904.

109. Burleigh JG, Brabazuk WD, Davis JM, Morse AM, Soltis PS. Exploring diversification and genome size evolution in extant gymnosperms through phylogenetic synthesis. J Bot. 2012;2012:1–6.

110. Peñalver E, Labandeira CC, Barrón E, Delclòs X, Nel P, Nel A, et al. Thrips pollination of Mesozoic gymnosperms. Proc Natl Acad Sci U S A. 2012;109:8623–8.

111. Rainford JL, Hofreiter M, Nicholson DB, Mayhew PJ. Phylogenetic distribution of extant richness suggests metamorphosis is a key innovation driving diversification in insects. PLoS One. 2014;9:e109085.

112. Niklas KJ, Tiffney BH, Knoll AH. Patterns in vascular land plant diversification. Nature. 1983;303:614–6.

113. Silvestro D, Cascales-Miñana B, Bacon CD, Antonelli A. Revisiting the origin and diversification of vascular plants through a comprehensive Bayesian analysis of the fossil record. New Phytol. in press, doi.org/10.1111/nph.13247.

114. Artabe A, Archangelsky S. Las Cycadales *Mesodescolea plicata* Archangelsky emend. Archangelsky y Petriella 1971 (Cretacio) y *Stangeria* Moore (Actual). Estudio comparativo de epidermis foliar con microscopia electr6nica de barrido y transmisi6n. Ameghiniana.

1992;28:115–23.

115. Artabe A, Stevenson DW. Fossil Cycadales of Argentina. Botanical Rev. 1999;65:219–38.

116. Passalia MG, Del Fueyo G, Archangelsky S. An Early Cretaceous zamiaceous cycad of South West Gondwana:*Restrepophyllum* nov. gen. from Patagonia, Argentina. Rev Palaeobot Palynol. 2010;161:137–50.

117. Crisp MD, Cook LG. Explosive radiation or cryptic mass extinction? Interpreting signatures in molecular phylogenies. Evolution. 2009;63:2257–65.

118. Uzunova K, Palamarev E, Kvaček Z. *Eostangeria ruzinciniana* (Zamiaceae) from the Middle Miocene of Bulgaria and its relationship to similar taxa of fossil *Eostangeria*, and extant *Chigua* and *Stangeria* (Cycadales). Acta Palaeobot. 2001;41:177–93.

119. Nagalingum NS, Drinnan AN, Lupia R, McLoughlin S. Fern spore diversity and abundance in Australia during the Cretaceous. Rev Palaeobot Palynol. 2002;119:69–92.

120. Tang W. The evolutionary history of North American cycads. Cycad Newslett. 2012;35:7–13.

121. Cramer BS, Miller KG, Barrett PJ, Wright JD. Late Cretaceous–Neogene trends in deep ocean temperature and continental ice volume: Reconciling records of benthic foraminiferal geochemistry ($\delta^{18}O$ and Mg/Ca) with sea level history. J Geophys Res: Oceans. 2011;116:1978–2012.

122. Meseguer AS, Lobo JM, Ree R, Beerling DJ, Sanmartín I. Integrating fossils, phylogenies, and niche models into biogeography to reveal ancient evolutionary history: the case of *Hypericum* (Hypericaceae). Syst Biol. 2015;64:215–32.

123. Ronquist F, Klopfstein S, Vilhelmsen L, Schulmeister S, Murray DL, Rasnitsyn AP. A total-evidence approach to dating with fossils, applied to the early radiation of the Hymenoptera. Syst Biol. 2012;61:973–99.

124. Grimm GW, Kapli P, Bomfleur B, McLoughlin S, Renner SS. Using more than the oldest fossils: Dating Osmundaceae with three Bayesian clock approaches. Syst Biol. 2015;64:96–405.

125. Condamine FL, Nagalingum NS, Marshall CR, Morlon H. Data from: Origin and diversification of living cycads: A cautionary tale on the impact of the branching process prior in Bayesian molecular dating. BMC Evol Biol. 2015,http://dx.doi.org/10.5061/dryad.20k7m.

CHAPTER 14

# DISCOVERY OF A NEW SUB-POPULATION, MAPPING AND UPDATED RED LIST ASSESSMENT OF THE ENDANGERED CYCAS BEDDOMEI DYER (CYCADALES: CYCADACEAE)

B. Ravi Prasad Rao, M. Chennakesavulu Naik, S. Salamma, M.V. Suresh Babu & L. Rasingam

**Keywords:** Beddome's Cycad, Endangered, endemism, restricted distribution, updated IUCN Red List assessment.

## INTRODUCTION

Cycads, often referred to as 'living fossils', comprising 11 genera and 307 species, of which 98 species are classified under the family Cycadaceae, represent a single genus *Cycas*. About 63% of cycads are threatened (IUCN 2014). In India, the genus *Cycas* is represented by nine species (Singh & Khuraijam 2014), of which *Cycas beddomei* Dyer (Thiselton-Dyer 1883) is reported to be confined to the Seshachalam Hills (Tirupati-Kadapa Hills) of Andhra Pradesh State, India. Its conservation status has been variously assessed as Vulnerable (Nayar & Sastry 1987), Endangered (Rao et al. 2003) and Critically Endangered (Jadhav et al. 2001; Hill et al. 2003). Based on direct field observation the threat status of the species has been evaluated as 'Endangered' through grid sampling (Rao et al. 2010) and this work has been reviewed and adopted by the IUCN Red List (IUCN 2014). It is the only Indian cycad and the only species of *Cycas* that is listed on Convention on International Trade in Endangered Species of Wild Fauna and Flora (CITES) Appendix I (Inskipp & Gillett 2005) and this categorization has also been supported by Rao et al. (2010).

Rao et al. (2010) assessed its conservation status as 'Endangered' by laying 15 transects in the Seshachalam Hills. In the present study, the population was monitored in those transects and further sampling was carried out in 11 new transects, four in the Seshachalam Hills and seven in the Velikonda Hills. The latter is a new area discovered with a new subpopulation of the species through ecological niche modelling. Based on earlier and current datasets, species distribution is updated, current population size is estimated and conservation status is evaluated.

**Citation:** Rao, B.R.P., M.C. Naik, S. Salamma, M.V.S. Babu & L. Rasingam (2015). Discovery of a new sub-population, mapping and updated Red List assessment of the Endangered Cycas beddomei Dyer (Cycadales: Cycadaceae). Journal of Threatened Taxa 7(12): 7902–7909; http://dx.doi.org/10.11609/JoTT.o4377.7902-9.

## MATERIALS AND METHODS

The study area is confined to the Seshachalam and Velikonda hills of Andhra Pradesh (Figure 1). From 2011 to 2014, the *Cycas beddomei* population was monitored in 15 transects of 1kmx5m size and inventoried the species in four new transects of size 500x5 m in Seshachalam hills all laid in 13 grids each of 40km$^2$ size. In addition, based on the findings through ecological niche modelling the population was assessed in the Velikonda Hills bordering Nellore and Kadapa districts. Due to sparse and scattered distribution of the species in the Velikonda hills, the area is sampled through seven transects of 500×5 m size each laid in seven grids of 40km$^2$ size. The number of mature individuals (>2cm stem height) were enumerated and the area of occupancy was estimated in individual grids. The number of mature individuals was extrapolated taking into consideration the topographic limitations of the species occurrence based on field observations. All the transect sites are geo-coded. In both the areas, the numbers of male and female plants were counted transect-wise based on Whitelock's (2002) observations that male plants tend to form clumps and the females grow as single stems.

The area of occupancy was calculated in 20 grids; 13 in the Seshachalam Hills and seven in the Velikonda Hills, by combining the observations from the present study as well as the earlier one. The locations and subpopulations were revised for the species throughout its terrain following IUCN guidelines (IUCN 2001; IUCN Standards and Petitions Subcommittee 2014). Accordingly, subpopulations are considered as geographically or otherwise distinct groups in the population between which there is little demographic exchange. Taking into consideration the topographic limitations of species occurrence in its native habitat, the area of occupancy was calculated and extrapolation of the number of mature individuals was done on the basis of the grids in the study area. The whole geographic range of the species is mapped based on field observations. The conservation status of *Cycas beddomei* has been reassessed based on the observations from the newly sampled transects. For determining the three threatened categories, namely Critically Endangered, Endangered and Vulnerable, five criteria are used in the IUCN Species Survival Commission guidelines (IUCN 2001). However, for *C. beddomei*, only criteria B, C and D are applied as A and E criteria require long-term observations. For the purpose of measuring the Extent of Occurrence (EOO) and the Area of Occupancy (AOO), satellite Figurery of IRS-1C has been stratified with 40km$^2$ grids. The AOO is defined as the area within its 'EOO' which is occupied by the taxon, excluding cases of vagrancy. The measure reflects the fact that a taxon will not usually occur throughout the area of its 'EOO', which may contain unsuitable or unoccupied habitats (IUCN Standards and Petitions Subcommittee 2014).

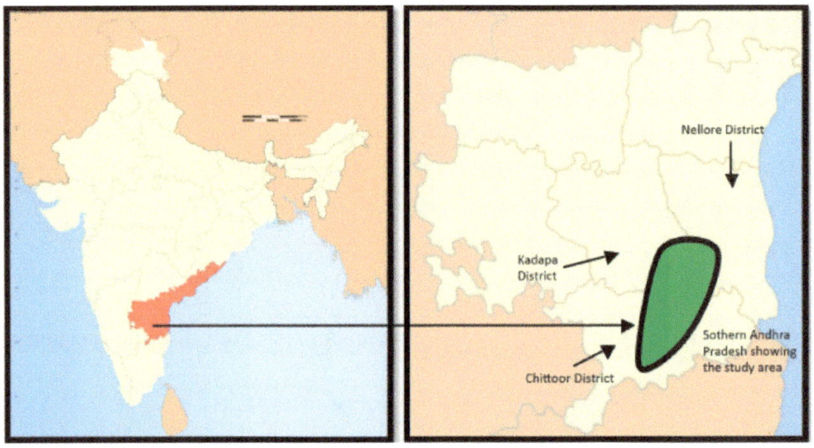


**Figure 1:** Study area.

## RESULTS AND DISCUSSION

### Nomenclature

*Cycas beddomei* Dyer, Trans. Linn. Soc. London, Bot. 2(5): 85, pl. 17. 1883. Lindstrom & Hill in Telopea 11(4): 470. 2007. *Cycas circinalis* var. *beddomei*(Dyer) J. Schust., Pflanzenr. IV, 1: 67.1932. *Cycas pluma* W. Bull, Retail List New Beautiful Rare Pl. 142: 478. 1877 (Figure 2).

### Botanical description

Stems arborescent, appear like a small palm with a distinct trunk of up to 1.5m high. The stem is covered with the remnants of leaf bases. Leaves up to 30 in number form a crown at top of the stem, 70–120 cm long. Petioles up to 20cm long, with minute spines. Leaflets narrow, linear, 12–18 × 2–0.4 cm, with revolute margins. The plants are dioecious. Male cones oblong-ovoid, up to 32×16 cm, orange, with a short peduncle; microsporophyll with an apical spine up to 3cm long. Megasporophylls grow up to 4×2 cm and with pectinate margins. Ovules 2–4, occasionally 6–8, inserted above the middle of the stalk, up to 4cm across. Seeds globose-ovoid, with fibrous sarcotesta, up to 5×5 cm, green, turning yellow on maturity.

**Figure 2:** Cycas beddomei: A- Habit; B-Male plant; C-Female plant; D-Germinating seeds; E-Diseased plants; F- Fire affected plant.

### Ecology

Found on dry, open hill slopes, in woodland or grass dominated forested areas at an altitude between 500–1165 m. The species is mostly restricted to the top slopes and predominantly in black soils. Common associate species are: *Phoenix humilis, Terminalia pallida, Syzygium alternifolium, Pimpinella tirupatiensis, Curculigo orchioides, Cymbopogon* spp. and *Dechaschistia* spp.

Reproductive period: April–August.

## Common name

Beddome's Cycas (English); Peritha, Madanakamakshi and Kondaeetha (Telugu).

## Uses

The seeds are harvested for preparing flour by the local people. The, male cone extracts are used to prepare a health tonic by the local Yanadi tribes and rural communities. The pith is harvested often for use in abortion.

## Sub-populations and locations of the species

Totally, 3962 mature individuals of *Cycas beddomei* were enumerated in 26 transects laid in 20 grids (Table 1). Of these, 2834 individuals were recorded in the Seshachalam hills and 1128 in the Velikonda Hills. In both the areas, the male: female ratio was about 1:3. Since all of the individuals in the Seshachalam Hills were inferred to have some minimum demographic exchange throughout the terrain, these were considered together as 'one sub-population' since fire is only threat factor affecting the population all are considered under one location. In the Velikonda Hills, all the individuals are considered as one sub-population distributed in one location as the whole sub-population is affected by pest. This discovery of the species in the Velikonda Hills is significant as till date, no published data is available on the species occurrence in the area. Although there is no overall estimate of decline, the population is considered to be experiencing decline because of habitat modification and continuing threats from forest fires. Hence with the current field observations in the native habitat of the species, we could demarcate two subpopulations separated by about 40km; one entirely falling in the Seshachalam Hills and the other in the Velikonda Hills with a total of eight locations.

## Distribution pattern of the species

The species is now recorded from the Seshachalam and the Velikonda Hills covering Chittoor, Kadapa and Nellore districts. The species has patchy distribution within the grids and substantial areas in the individual grids (about 80% in many cases) did not have plants. Although we located isolated trees at 500–600 m altitude, most of the population lies at an altitudinal range of 600–1150 m. We found in the past four years the AOO is reduced by 5–10 % in different locations and the number of mature individuals has come down in some grids (ranging from 100 to 300) which is attributed to intensive and recurrent forest fires.

## Population size

The species population is estimated to comprise over 3.9 lakh individuals (Table 1); 97% of it is in the Seshachalam Hills.

## Threats to the species

*Cycas beddomei* is experiencing population decline because of habitat modification and continuing threats especially from forest fires. These fires have become more prominent and recurrent in the past few years. The natural threats are in the form of hemipteran scale *Saissetia coffeae* and lepidopteran butterfly *Chilades pandava* in its natural habitat; persistent occurrence of the former and massive infestation by the latter may prove fatal in the future (Srivastava 2011). Harvesting of male cones for medicinal purpose; uprooting plants for ornamental purpose; occasional collection of individuals for pith extraction for flour making are other major threats.

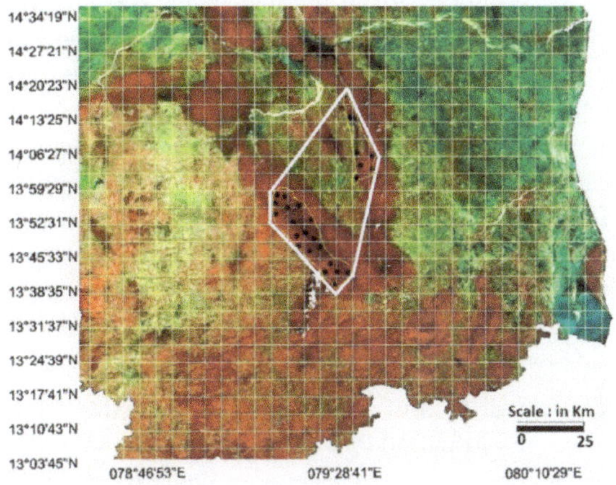


**Figure 3:** Cycas beddomei – Distribution sites and ECO/AOO map.

**Table 1:** Sampled inventory of cycas beddomei population

| Grid No. | Transect No. | Transect size (in m) | Altitude (above MSL) | No. of individuals (male+female) | Estimated AOO in km² (extrapolated individuals) |
|---|---|---|---|---|---|
| A. Seshachalam Hills sub-population | | | | | |
| 1 | 1 | 1000×5 | 800 | 141 (42+99) | 2.0 (40,533) |
| | 2 | 1000×5 | 790 | 132 (29+103) | |
| | 3 | 1000×5 | 750 | 31 (7+24) | |
| 2 | 4 | 1000×5 | 860 | 120 (35+85) | 0.9 (21,600) |
| 3 | 5 | 1000×5 | 650 | 103 (24+79) | 2.4 (34,920) |
| | 6 | 1000×5 | 792 | 55 (17+38) | |
| 4 | 7 | 1000×5 | 980 | 143 (36+107) | 1.2 (34,320) |
| 5 | 8 | 1000×5 | 1010 | 170 (45+125) | 1.1 (37,400) |
| 6 | 9 | 1000×5 | 960 | 190 (48+142) | 1.2 (40,560) |
| | 10 | 1000×5 | 950 | 148 (30+118) | |
| 7 | 11 | 1000×5 | 920 | 161 (40+121) | 1.0 (32,000) |
| 8 | 12 | 1000×5 | 850 | 76 (21+55) | 1.9 (28,880) |
| 9 | 13 | 1000×5 | 1150 | 137(34+103) | 1.2 (32,880) |
| 10 | 14 | 1000×5 | 820 | 124 (32+92) | 1.2 (29,760) |
| 11 | 15 | 1000×5 | 1108 | 194 (50+144) | 0.8 (16,586) |
| | 16 | 500×5 | 872 | 58 (16+42) | |
| 12 | 17 | 500×5 | 838 | 142 (37+105) | 0.6 (14,340) |
| | 18 | 500×5 | 814 | 97 (23+74) | |
| 13 | 19 | 500×5 | 832 | 163 (46+117) | 0.5 (15,750) |
| ubtotal | | 8.5 ha | - | 2834 (734+2100) | 16.2 (3,79,529) |
| B. Velikonda Hills sub-population | | | | | |
| 14 | 20 | 500×5 | 694 | 102(24+78) | 0.05 (1020) |
| 15 | 21 | 500×5 | 731 | 78(21+57) | 0.05 (780) |
| 16 | 22 | 500×5 | 742 | 202(46+156) | 0.05 (2020) |
| 17 | 23 | 500×5 | 680 | 250(72+178) | 0.05 (2500) |
| 18 | 24 | 500×5 | 809 | 83(18+65) | 0.05 (830) |
| 19 | 25 | 500×5 | 670 | 193(42+151) | 0.03 (1158) |
| 20 | 26 | 500×5 | 684 | 220(64+156) | 0.04 (1760) |
| Subtotal | | 1.75 ha | - | 1128 (287+841) | 0.32 (11,088) |
| Total | | 10.25 ha | - | 3962 (2834+1128) | 16.52 (3,90,617) |

## EVALUATION OF CURRENT CONSERVATION STATUS

The whole population of the species is considered under '2 locations' comprising two subpopulations. The Extent of Occurrence (EOO) of the species is calculated as 1713km$^2$ (Figure 3). Taking the topographic limitations throughout its geographic range, the Area of Occupancy (AOO) was measured as 16.52km$^2$ (1652 ha) (Table 1; Figure 3) and the number of mature individuals was extrapolated to 3.90 lakhs.

**Criterion B: B1**: The EOO of the species is estimated at 1713km$^2$ and hence categorized as 'Endangered'. Of the three conditions to be fulfilled under B1, the species is found to qualify two: (a) no more than five locations and (b) Continuing decline observed under the sub criteria: (ii) area of occupancy (iii) area, extent and or quality of habitat and (v) no. of mature individuals. B2: The AOO is 16.52km$^2$and since this estimate is less than 500km$^2$, the species is qualified under 'Endangered' category. The species also fulfils two of the three conditions: (a) no more than five locations and (b) continuing decline observed under the sub criteria: (ii) area of occupancy (iii) area, extent and or quality of habitat and (v) no. of mature individuals.

**Criterion C and D:** Since the estimated population comprise more than 10,000 individuals it is not qualified for any of the three threatened categories under these criteria.

**Final assessment:** EN B1ab(ii,iii,v)+B2ab(ii,iii,v)

## CONCLUSIONS

The global distribution of the species is redefined as 'Endemic to the Seshachalam and the Velikonda hills of the southern Eastern Ghats of Andhra Pradesh, India' due to the discovery of a new sub-population in the Velikonda Hills. Despite its extended geographic distribution, its 'Endangered' status has been retained following the IUCN Red List guidelines. Owing to its restricted distribution and slow growth its current status of Appendix I of CITES is also to be retained.


## ACKNOWLEDGEMENTS

We thank the Department of Biotechnology for supporting the ongoing multi-centred project on 'Prevention of extinction......... Biotechnological Tools' (BT/Env/BC/01/2010:23-3-2012). We acknowledge the constructive support received from Dr. P.C. Panda, RPRC, Bhubaneswar and Coordinator of the multi-centred project at the Eastern Ghats level and from Prof. S.K. Barik, NEHU, Shillong, National level coordinator. Thanks are also due to Dr. T.R. Kiran Chand, Forestry and Ecology Division, NRSC, Hyderabad for kind help. We are also grateful to the support the Andhra Pradesh Forest Department provided during our field visits from time to time.


## REFERENCES


1. Hill, K.D., C.J. Chen & P.K. Loc (2003).Regional Overview: Asia. In Donaldson, J.S. (ed.). *Cycads: Status Survey and Conservation Action Plan*. IUCN/ SSC Cycad Specialist Group, IUCN, Gland, Switzerland and Cambridge, UK, ix+86pp
2. Inskipp, T. & H.J. Gillett. (eds.) (2005). *Checklist of CITES species and Annotated CITES Appendices and reservations*. Compiled by UNEPWCMC. CITES Secretariat, Geneva, Switzerland and UNEP-WCMC Cambridge, UK, 339pp & CD-ROM
3. IUCN (2014). The IUCN Red List of Threatened Species. *Version 2014.3*. <http://www.iucnredlist.org>. Downloaded on 03-03-2015.
4. IUCN Species Survival Commission (2001). IUCN Red List Categories, Version 3.1. www.iucn.org/themes/ssc/redlists/ssc-rl-c.htm.
5. IUCN Standards & Petitions Subcommittee (2014). Guidelines for using the IUCN Red List Categories and Criteria. Version 11. Prepared by standards & petitions subcommittee Downloadable from <http://www.iucnredlist.org.documents/Red List Guidelines.pdf.
6. Jadhav, S.N., D.K. Ved, U. Ghate, K.N. Reddy & S. Reddy (2001). *Proceedings of the Workshop on Conservation Assessment and Management Planning for Medicinal Plants of Andhra Pradesh* (CAMP) MPCC. Hyderabad, 97pp.
7. Nayar, M.P & A.R.K. Sastry (1987). *Red Data Book of Indian Plants - Volume 1.* Botanical Survey of India, Calcutta, xiii+367pp.
8. Rao, C.K., B.L. Geetha & G. Suresh (2003). *Red List of Threatened Vascular Plant Species in India.* BSI, Calcutta, xxiv+144pp.
9. Rao, B.R.P., M.V.S. Babu & J. Donaldson (2010). A Reassessment of the conservation status of *Cycas beddomei* Dyer (Cycadaceae), an endemic of the Tirupati-Kadapa Hills, Andhra Pradesh, India, and comments on its CITES Status. *Encephalartos* 102: 19–24.
10. Rao, B.R.P. (2010).*Cycas beddomei*. In: IUCN 2011. IUCN Red List of Threatened Species. Version 2011.2. <www.iucnredlist.org>. Accessed on 03 January 2015.
11. Singh, R. & J.S. Khuraijam (2014). *Cycad Diversity in India.* In: Cycads of India: Online Database. www.cycadsofindia.in Downloaded on 07 April 2015.
12. Srivastava, P.C. (2011). Observation on serious threats to *Cycas beddomei* Thiselton-Dyer, endemic and critically endangered species in its natural habitat Tirupathi-Cudapa hills of Andhra Pradesh, India. *Journal of Fairylake Botanical Garden* 10: 15.
13. Thiselton-Dyer, W.T. (1883). On a new species of *Cycas* from southern India. *Transactions of the Linnaean Society London (Botany Series 2)*2: 85–86, pl. 17.
14. Whitelock, L.M. (2002). *The Cycads.* Timber Press, Oregon, 374pp.

CHAPTER 15

# PARAPHYLY OF ORGANELLE DNAS IN CYCAS SECT. ASIORIENTALES DUE TO ANCIENT ANCESTRAL POLYMORPHISMS

Yu-Chung Chiang, Kuo-Hsiang Hung, Shann-Jye Moore, Xue-Jun Ge, Shong Huang, Tsai-Wen Hsu, Barbara A Schaal and TY Chiang

## BACKGROUND

The analysis of geographical structure and phylogeographical pattern has been a major focus in population genetics and molecular ecology [1]. Populations descending from a common ancestral population are expected to differentiate from each other and eventually show reciprocal monophyly. Reciprocal monophyly is a result of the accumulated effects of genetic drift and selection when little gene exchange occurs after divergence. In contrast, populations may be genetically similar due to recent isolation, continuous gene exchange, or secondary contact [2]. Plant species on islands provide ideal subjects for teasing apart the contrasting effects of geographical isolation and gene flow on population differentiation. Island populations are often genetically isolated from each other and most island populations are small, thus facilitating random genetic drift.

Coalescence studies provide information on the genealogical processes (backwards in time) that have led to the current apportionment of genetic variation within populations or species. Coalescence-based phylogeography uses haplotype networks, which trace the mutational relationships among alleles within or among species in a geographical context [3], to infer processes such as migration, divergence, and isolation [4,5, 6]. Noncoding spacer regions of organelle DNAs are often used as genetic markers because they are nearly neutral with low functional constraints and are free of the complexities associated with recombination for nuclear markers [6].

Cycads are a unique and ancient lineage of land plants that have attracted much attention from evolutionary biologists [7]. *Cycas revoluta* Thunb. and *C. taitungensis* Shen *et al.* constitute the section *Asiorientales* which is a basal lineage of this Laurasian genus [8]. *Cycas revoluta* occurs in a scattered distribution pattern from the southernmost part of Kyushu and throughout the Ryukyu Islands in Japan, and is also distributed in Fukien Province of China [9]; while *C. taitungensis* is the only extant and endemic species with two endangered populations along the eastern coast of Taiwan. The two species resemble each other morphologically; a few characters such as revolute vs. plane leaflet margins distinguish the species from each other. An Eocene fossil,*C. fujiana* Yokoyama, is much like *C. revoluta*; both the morphological similarity and the occurrence of both taxa in Japan suggest that these species are the same [10]. The divergence of the species in section *Asiorientales* is thought to have occurred some 30 million years before present [10]. *Asiorientales* is allopatric from other cycads.

Both species are agriculturally important. Each year thousands of seeds and plants have been exported for use in the nursery trade and in floral arrangements [11]. In the early 20$^{th}$ Century, cycad seeds and stems were used for making cake and soup. Cycad flour is toxic and causes neurological disorder [12]. Fermented seeds are used for producing alcoholic drinks [13]. Unfortunately, human overexploitation has led to the extinction of populations of *C. revoluta* in Fukien Province of China [9], and only two remaining populations of *C. taitungensis* occur in Taiwan [14, 15]. Large specimens of *C. revoluta* are prized

**Citation:** Yu-Chung Chiang, Kuo-Hsiang Hung, Shann-Jye Moore, Xue-Jun Ge, Shong Huang, Tsai-Wen Hsu, Barbara A Schaal and TY Chiang, Paraphyly of organelle DNAs in Cycas Sect. Asiorientales due to ancient ancestral polymorphisms, DOI: 10.1186/1471-2148-9-161.

by the nursery industry and natural populations of *C. revoluta* in mainland China have been decimated by poachers [9]. In contrast, populations of *C. revoluta* are numerous across the Ryukyu Archipelago and Kyushu [13, 16]. On the Island Amami-O-Shima, protected cycad forests contain thousands of plants [17]. Despite the high morphological similarity, the two species differ in habitat preference. *Cycas revoluta* usually grows along coasts, often subject to salt spray. In contrast, *C. taitungensis* grows on cliffs along riverbanks [18].

Island systems have long served as natural laboratories for evolutionary studies [e.g. [19, 20]]. Despite their often small geographic size, islands can contain a diversity of habitats suitable for the colonization of an array of plants and animals. Once an island is colonized, population differentiation between islands can be facilitated by isolation, vicariance events and local extinction-recolonization [21]. Taiwan and the Ryukyu islands lie at the boundary between the Eurasian and Pacific plates and are a constituent of the island-arc system along the western edge of the Pacific Ocean [22]. In contrast to sequentially emerging oceanic islands of volcanic origin, these continental islands emerged almost simultaneously via collision between the continental and oceanic tectonic plates and are geologically young, estimated at 5–6 million years old [23, 24]. Since emerging from the Pacific Ocean, these islands gradually acquired their floras and faunas from the Eurasian mainland [25, 26]. Most island species/populations thus have close phylogenetic links to their mainland relatives.

Today, islands of the Ryukyu archipelago are isolated from each other. In the past, sea level oscillated during periodic glacial cycles (Milankovitch cycles) [27, 28, 29]. During glacial maxima, the global sea level dropped by some 100–120 meters, exposing land bridges that connected islands [30] and thus providing migration corridors between today's isolated islands. Repeated stepwise migrations, plus occasional, long-distance colonization [cf. [31, 32, 33]], would allow various plants to fill newly available habitats on islands left by the expanding or retreating glaciers. The glacial cycles of the Quaternary played an important role in the evolution of species [cf. [34, 35]].

In the present study, phylogenies of the rITS region of mtDNA and the intergenic spacer between *atp*B and*rbc*L genes of cpDNA were used to trace phylogeographical relationships in *C. revoluta* and *C. taitungensis*. Organelle DNAs are maternally inherited in cycads; they disperse only via seeds [cf. [15]]. Most cycad seeds, including those of *C. revoluta* and *C. taitungensis*, are heavy and sink reducing the possibility of water/current dispersal over long distances [36, 37]. The fleshy outer seed coat of cycads attracts animals, mainly rodents, small fruit-eating bats, and birds which serve as dispersal agents [38]. In natural habitats, seeds drop and occasionally disperse for a limited distance via gravity. Migration between islands is thought to be very limited and such restricted dispersal has been documented in other island species [20, 39]. Ocean straits between large geographical regions (Fukien, Taiwan, and Ryukyu), and channels between islands make migration between islands unlikely in these cycads. Thus, genetic differentiation between populations for maternally inherited markers would be expected. Nevertheless, multiple colonization and recent isolation can counter forces leading to genetic differentiation between islands and monophyly [20]. Many plants distributed on the island arcs of the western Pacific Rim, including *Cunninghamia* [40], *Michelia* [41], *Kandelia* [42], *Trema* [43] and *Lithocarpus* [32, 44], show little differentiation. Coalescence in island populations can be a complicated process, determined by counteracting effects of isolation (i.e., historical connection), gene flow, drift, selection and repeated colonization.

Given the similar systems of mating and demography of both cycad species, non-coding regions of organelle DNA's should provide information on species/population histories and demographic processes as well as gene flow and colonization. In this study, we examine the genetic diversity within and between species and populations, determine geographical associations and analyze phylogenetic relationships to infer the historical processes that have lead to the current pattern of genetic structure. Several specific questions are addressed:

1. Given the long time that these species have been isolated, has reciprocal monophyly of sections and species been attained?
2. Are populations and geographical regions genetically differentiated?
3. Is the spatial apportionment of genetic variation consistent with a stepwise model of colonization?
4. Is erosion of genetic diversity associated with population decline? Are levels of genetic diversity in populations of Fukien and Taiwan lower than those of Ryukyu?

## RESULTS

### Genetic Diversity within Cycas Revoluta and C. taitungensis

The rITS region of mtDNA varies in length from 411 to 547 bp with a consensus length of 598 bp. The sequence length of *atp*B-*rbc*L intergenic spacer of cpDNA varies from 762 to 830 bp with a 913 bp consensus length. In total, 105 and 97 haplotypes of cpDNA as well as 170 and 55 haplotypes of mtDNA were detected in populations of *Cycas revoluta* (307 individuals) and *C. taitungensis* (102 individuals), respectively (Table 1). Haplotype sequences of mt- and cpDNAs were deposited in the GenBank database with accession numbers of FN397910-FN397954 and FM999745-FM999775, respectively.

**Table 1:** Population localities, and number of samples (N) used for a phylogeographical analysis of Cycas revoluta and C. taitungensis in Ryukyu, Fukien, and Taiwan

| Population | Area | Sampling size (N) | Symbol | Coordinate |
|---|---|---|---|---|
| INGROUPS: | | | | |
| *Cycas revoluta* | | | | |
| Japan: Kagoshima | Kyushu, Japan | 37 | KO | 31°05' N, 130°70' E |

| | | | | |
|---|---|---|---|---|
| Amami-O-Shima | Ryukyu, Japan | 124 | AM | 26°90' N, 128°30' E |
| Okinawa | Ryukyu, Japan | 26 | OK | 28°15' N, 130°40' E |
| Iriomote | Ryukyu, Japan | 36 | IR | 24°19' N, 123°54' E |
| Ishigaki | Ryukyu, Japan | 43 | IS | 24°50' N, 123°30' E |
| Yonaguni | Ryukyu, Japan | 15 | YO | 24°45' N, 123°05' E |
| China: Fukien | Fukien, China* | 26 | FU | 24°26' N, 118°04' E |
| *Cycas taitungensis* | | | | |
| Taiwan: Red Leaf Conservation Area | Taitung, Taiwan | 81 | RL | 22°85' N, 120°95' E |
| Coastal Conservation Area | Taitung, Taiwan | 21 | C | 23°05' N, 121°30' E |
| **OUTGROUPS:** | | | | |
| *Cycas taiwaniana* | Fukien, China** | 1 | CT | 23°15' N, 113°35' E |
| *Cycas thouarsii* | Pingtung, Taiwan*** | 1 | CK | 21°95' N, 120°85' E |

* Sampled from the South China Botanic Garden, transplanted from the Fukien wild population before extinction, which located in the Lienchiahg County, Fukien.

** Sampled from the South China Botanic Garden, transplanted from Fukien, China.

*** Sampled from the Taiwan Forestry Research Institute, transplanted form Pingtung, Taiwan.

Higher levels of genetic diversity of mtDNA are detected in *C. revoluta*, $\varphi$ = 0.0500, than in *C. taitungensis*, $\varphi$ = 0.0038 (Table 2). Among the Ryukyu Islands, populations from Yonaguni and Ishigaki have the highest levels of mtDNA diversity while the diversity in populations on Okinawa and Amami-O-Shima is relatively low. Despite its extinct status in the wild, the Fukien sample contains much higher levels of mtDNA diversity than any of the Ryukyu populations. The existence of the phylogenetically most distant mitotype, G, that occurs only in the Fukien population contributes to the high level of genetic diversity (Fig. 1). In *C. taitungensis* the level of genetic diversity is mostly associated with population size. Genetic diversity of the Red-Leaf (RL) population is about twice that of the Coastal (C) population. In contrast, the levels of genetic diversity of cpDNA are much lower in *C. revoluta*, $\varphi$ = 0.008, than in *C. taitungensis*, $\varphi$ = 0.018 (Table 3). Fukien has the lowest level of diversity among all populations, while the populations of Ishigaki ($\varphi$ = 0.055) and Iriomote ($\varphi$ = 0.022) have higher levels of genetic diversity.

**Table 2:** Haplotype diversity (h), the number of segregating sites (S), the average number of pairwise difference (k), and nucleotide diversity ($\pi$), and neutrality tests at mtDNA within cycad populations

| | ***h*** | **S** | **k** | $\pi$ | **Fu and Li's D*** | **Tajima's D** | **Fu's Fs** |
|---|---|---|---|---|---|---|---|
| *C. revoluta*: | | | | | | | |
| Overall: | 0.932 | 240 | 11.9 | 0.0500 ± 0.00600 | | | |
| Japan: AM | 0.923 | 340 | 22.4 | 0.0392 ± 0.00382 | -4.6531** | -2.4548*** | -58.223 *** |
| IR | 0.962 | 199 | 24.3 | 0.0423 ± 0.00533 | -3.2490 | -2.0509* | -5.908 *** |
| IS | 0.908 | 399 | 43.3 | 0.0876 ± 0.02082 | -3.9556** | -2.4527*** | -0.656 |
| KO | 0.941 | 209 | 24.7 | 0.0429 ± 0.00523 | -2.9419* | -2.0702* | -4.241*** |
| OK | 0.886 | 110 | 19.4 | 0.0335 ± 0.00437 | -1.6804 | -1.4166 | -0.225 |
| YO | 0.933 | 211 | 56.2 | 0.1050 ± 0.02041 | -1.0720 | -1.0007 | 0.462 |
| China: FU | 0.982 | 440 | 108.3 | 0.2597 ± 0.04094 | -1.6974 | -1.0465 | -0.326 |
| *C. taitungensis*: | | | | | | | |
| Overall: | 0.978 | 293 | 21.8 | 0.0383 ± 0.00470 | | | |
| Taiwan: C | 0.967 | 64 | 9.9 | 0.0197 ± 0.00598 | -0.4979 | -1.6527 | -1.482 |
| RL | 0.975 | 271 | 23.8 | 0.0419 ± 0.00553 | -0.0325 | -2.1674* | -13.788*** |

*P < 0.05, **P < 0.02; ***P < 0.01

**Table 3:** Haplotype diversity (h), the number of segregating sites (S), the average number of pairwise difference (k), nucleotide diversity (π), and neutrality tests at cpDNA within cycad populations

| | *h* | S | k | *π* | Fu and Li's D* | Tajima's D | Fu's Fs |
|---|---|---|---|---|---|---|---|
| *C. revoluta*: | | | | | | | |
| Overall: | 0.959 | 245 | 5.2 | 0.0581 ± 0.00316 | | | |
| Japan: AM | 0.912 | 165 | 13.4 | 0.0154 ± 0.00328 | -4.5405** | -1.8676* | -41.080*** |
| IR | 0.955 | 105 | 19 | 0.0216 ± 0.00520 | -0.1257 | -1.0012 | -1.879 |
| IS | 0.833 | 328 | 42 | 0.0551 ± 0.01876 | -1.7192** | -1.7929*** | 4.265** |
| KO | 0.951 | 69 | 9 | 0.0101 ± 0.00359 | -0.3475 | -1.6622*** | -7.981*** |
| OK | 0.797 | 85 | 12.3 | 0.0142 ± 0.00717 | -0.9808 | -1.7768*** | 0.674 |
| YO | 0.971 | 59 | 8.9 | 0.0100 ± 0.00558 | -2.9824** | -2.1975*** | -3.635** |
| China: FU | 0.939 | 17 | 2.3 | 0.0026 ± 0.00030 | -2.4588*** | -1.4785** | -5.673*** |
| *C. taitungensis*: | | | | | | | |
| Overall: | 0.998 | 336 | 16.5 | 0.0181 ± 0.00279 | | | |
| Taiwan: C | 0.999 | 127 | 13 | 0.0148 ± 0.00646 | -4.0436** | -2.5904*** | -16.787*** |
| RL | 0.999 | 307 | 17.4 | 0.0196 ± 0.00309 | -5.2128** | -2.5911*** | -113.774*** |

*P < 0.05, **P < 0.02; ***P < 0.01

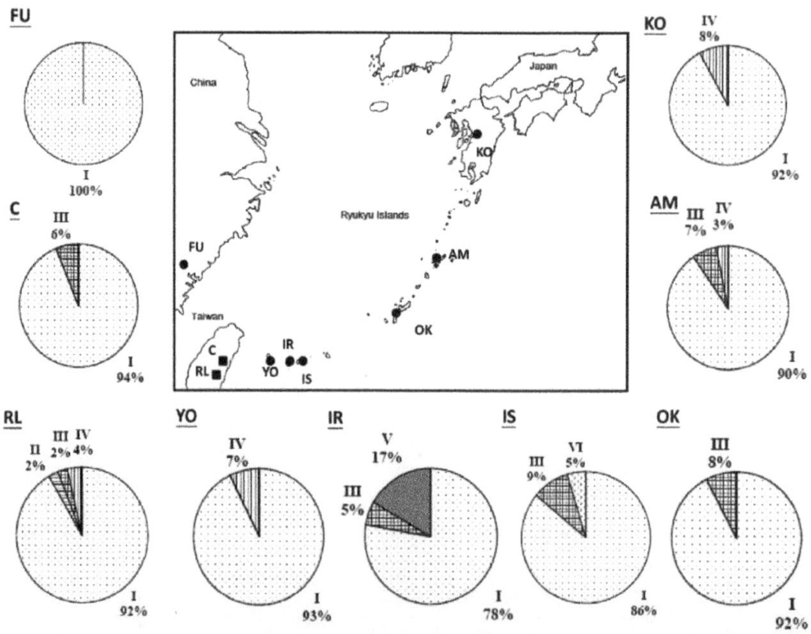


**Figure 1:** Map showing the spatial distribution of genetic polymorphisms of mtDNA in populations of *Cycas revoluta*(circles), and *C. taitungensis* (squares).

## Phylogenies of organelle DNAs in C. revoluta and C. taitungensis

Knowledge of the process and pattern of nucleotide substitution is important for estimating the number of substitutional events between DNA sequences and phylogenetic reconstruction that incorporates models of DNA sequence evolution. Transition bias results in a substitution pattern that is characterized by transition/transversion (ti/tv) ratios typically ranging between 2 and 10 [45, 46, 47]. In contrast, higher rates of tranversional substitutions resulting in ti/tv ratios ~1 have often been equated with transitional saturation, in which the same nucleotide position undergoes multiple transitions, or considered indicative of high levels of homoplasy [48]. In this study, the transition/transversion ratios were detected to be 1.808 and 1.707 for mtDNA ITS and cpDNA, respectively, suggesting low likelihood of saturation, although relatively low. The phylogeny of the mtDNA ITS region was reconstructed by maximum-likelihood (ML) and maximum-parsimony (MP) analyses. Both analyses recover phylogenetic trees with consistent topologies. The gene tree identifies seven mitotypes, including haplotypes A to G (Fig. 2A). Reciprocal monophyly is not supported in either *Cycas*species. Geographically, mitotypes A and B are widespread (Fig. 1), while four mitotypes are restricted to single populations, including C (YO), D (AM), E (IS), and G (FU). Mitotypes C-G are absent in *C. taitungensis* while mitotype G does not occur in Ryukyus. Sequences from the two outgroups are mitotype A and do not form a monophyletic group, indicating that sections of *Cycas* are also paraphyletic for mtDNA (Fig. 2A and 3A). An ML tree based on nucleotide variation of the cpDNA *atp*B-*rbc*L intergenic spacer identifies six chlorotypes I-VI (Fig.2B). Chlorotypes I, II, and IV are geographically widespread across islands and Fukien, while chlorotype II is restricted to *C. taitungensis*, and chlorotypes V and VI are found exclusively in Iriomote and Ishigaki (Fig. 4). The hypothesis of a molecular clock was not rejected at either locus based on the relative rate test (data not shown).

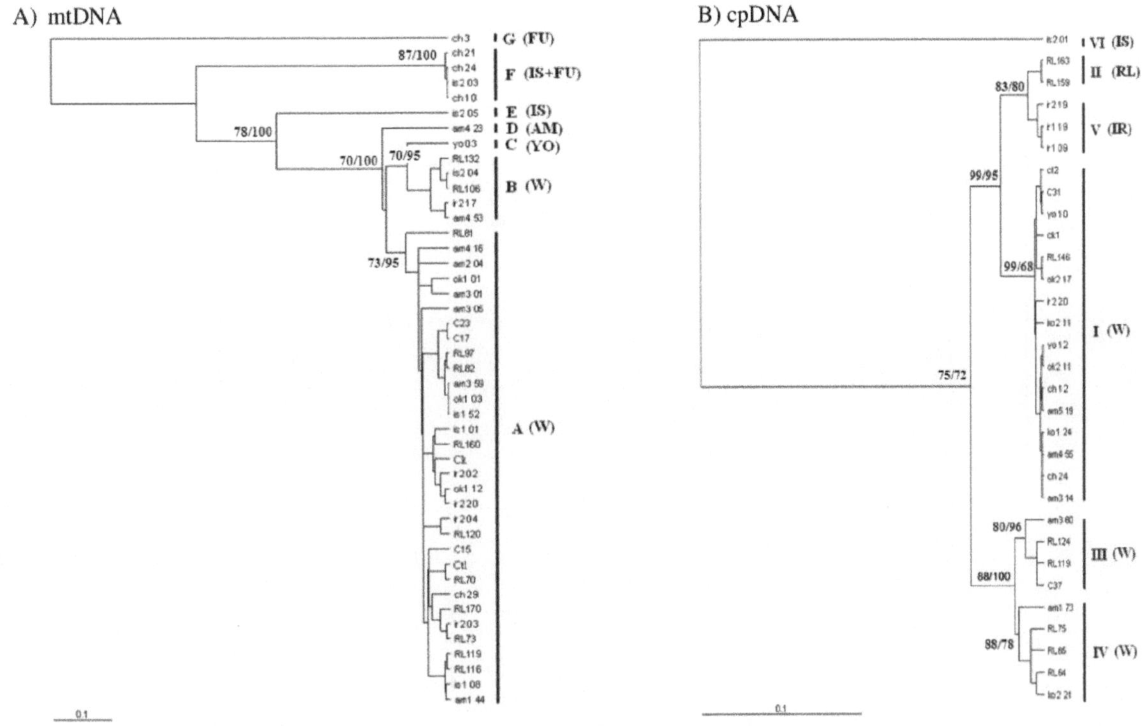


**Figure 2:** The maximum likelihood tree of haplotypes of *Cycas revoluta* and *C. taitungensis* with outgroup sequences (O). Numbers at nodes indicate the bootstrap values (ML/MP) in percentage. Symbols in parentheses indicate the geographical distribution of each mitotype. W = widespread. A) mtDNA. Clades A and B occur in both species, while clades D-G are restricted to *C. revoluta*. B) cpDNA. Clades I, III and IV occur in both species, while other are restricted to *C. revoluta*.

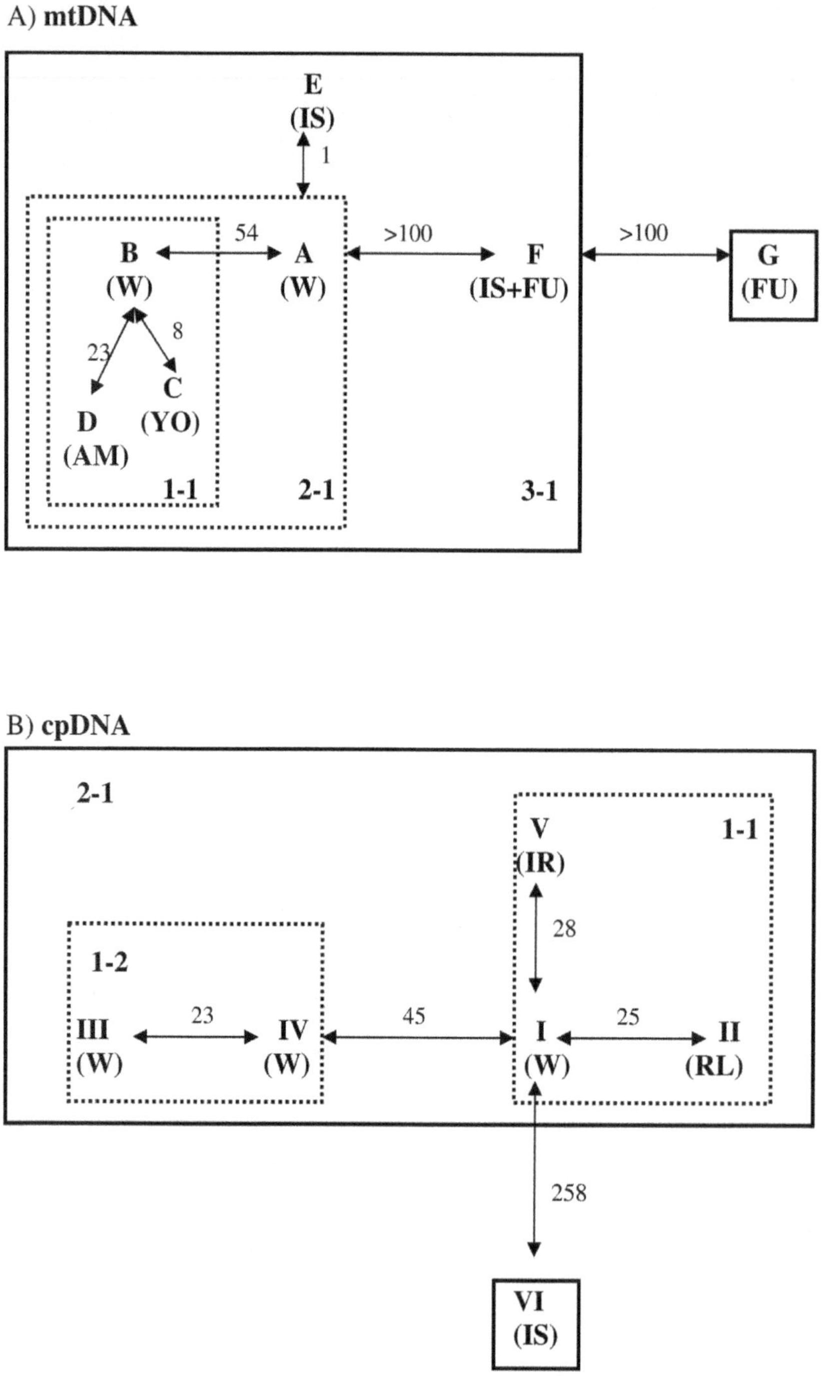


**Figure 3:** Network of haplotypes of *Cycas revoluta* and *C. taitungensis*. Numbers at arrows indicate the mutational changes between mitotypes. A) mtDNA, B) cpDNA, C) combined data.

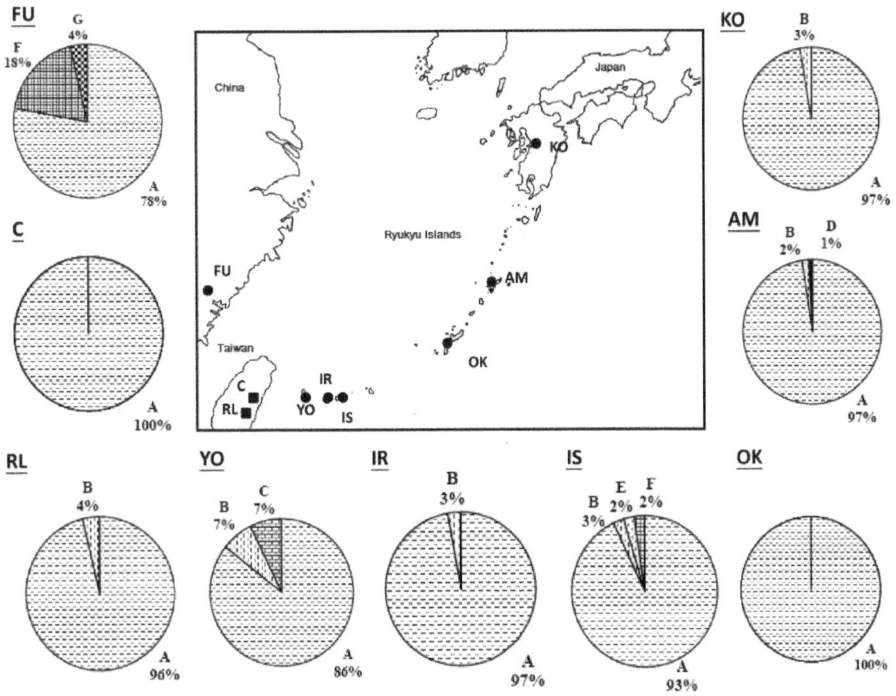


**Figure 4:** Map showing the spatial distribution of genetic polymorphisms of cpDNA in populations of *Cycas revoluta*(circles), and *C. taitungensis* (suqares).

*Cycas* species are not monophyletic for cpDNA. Haplotypes of outgroups belong to chlorotype I and are also found in *C. revoluta* and *C. taitungensis* (Fig. 2B and 3B), again reflecting paraphyly of the section *Asiorientales*. In both phylogenies numerous mutations are located at tips of the network, and are derived from ancestral haplotypes at the interior nodes in the network. This pattern results in a star-like phylogeny. Furthermore, tip haplotypes are mostly confined to a single population while the interior haplotypes are geographically widespread.

## Genetic Differentiation and Population Demography

Almost all dominant cytotypes of both organelle genes are shared among cycad populations. For mtDNA, mitotype A (95.3%) is common and widely distributed across populations, while rare mitotypes C, D, E, and G occur in single populations (Fig. 1). Likewise, for the cpDNA intergenic spacer, chlorotype I (90.1%) is the most common and widespread in all populations, while rare chlorotypes II, V, and VI are restricted to single populations (Fig. 4). Despite the lack of physical linkage between the two organelle genomes, high levels of linkage disequilibrium are detected in both *Cycas* species ($P < 0.05$). All rare mitotypes (4.7%), including B-G, are associated with chlorotype I; while all rare chlorotypes II–VI (9.9%) are associated with the mitotypes A, in the pooled data reflecting a nonequilibrium state.

Since common haplotypes are shared among populations, cycads show little genetic differentiation at the two organelle DNA markers. *Nm* values obtained from Fst estimates in *C. revoluta* vary among pairwise comparisons between populations (Tables 4 and 5). High levels of differentiation between Japan and China populations are detected for all comparisons, e.g., Fst of 0.115 ($P < 0.01$) between OK and FU based on mtDNA, and Fst = 0.159 between IR and FU ($P < 0.01$) based on cpDNA (Tables 4 and 5). For example the estimated number of migrants per generation is high between populations C and R of *C. taitungensis*, *Nm* = 90.41 and 23.18 based on cp- and mtDNAs, respectively (60). The *Nm* values based on Fst estimates from mtDNA average 106.10 which is greater than the average from cpDNA data, 10.95 (Tables 4 and 5).

**Table 4:** Pairwise comparisons of Fst (below diagonal) and Nm values (above diagonal) between populations of Cycas revoluta deduced from mtDNA sequences.

| Fst\Nm | AM | IR | IS | KO | OK | YO | FU |
|---|---|---|---|---|---|---|---|
| AM | | 44.16 | 89.16 | 90.17 | 45.15 | 12.29 | 4.19 |
| IR | 0.011 | | 105.02 | 111.27 | 1030.15 | 18.48 | 4.00 |
| IS | 0.006** | 0.005 | | 475.91 | 70.97 | 24.17 | 7.31 |
| KO | 0.006 | 0.005 | 0.001 | | 51.75 | 17.8 | 4.13 |
| OK | 0.011* | 0.001 | 0.007 | 0.010** | | 11.7 | 3.87 |

| | | | | | | | |
|---|---|---|---|---|---|---|---|
| YO | 0.039* | 0.026 | 0.020 | 0.027 | 0.041 | | 6.52 |
| FU | 0.107** | 0.111** | 0.064** | 0.108** | 0.115** | 0.071 | |

Probability value associated with the Fst is shown. The Fst values significantly greater than zero are noted by adding *P < 0.05 and **P < 0.01

**Table 5:** Pairwise comparisons of Fst (below diagonal) and Nm values (above diagonal) between populations of Cycas revoluta deduced from cpDNA sequences.

| Fst\Nm | AM | IR | IS | KO | OK | YO | FU |
|---|---|---|---|---|---|---|---|
| AM | | 6.02 | 13.58 | 17.86 | 7.63 | 3.71 | 7.13 |
| IR | 0.077* | | 38.08 | 3.07 | 7.38 | 3.89 | 2.64 |
| IS | 0.036* | 0.013 | | 8.06 | 12.57 | 8.58 | 6.49 |
| KO | 0.027 | 0.140** | 0.058 | | 3.59 | 2.34 | 10.25 |
| OK | 0.061 | 0.063 | 0.038 | 0.122 | | 62.62 | 2.66 |
| YO | 0.119 | 0.114* | 0.050 | 0.176 | 0.008 | | 1.87 |
| FU | 0.066 | 0.159** | 0.072* | 0.047 | 0.158 | 0.211 | |

Probability value associated with the Fst is shown. The Fst values significantly greater than zero are noted by adding *P < 0.05 and **P < 0.01

The program IM defined six demographic parameters estimated by the MCMC (Markov chain Monte Carlo) multiple loci simulations and tested for an 'isolation with migration' model in cycads. Of these demographic parameters, the distribution of migration parameters reveals a peak at the lower limit of resolution in one direction from Japanese populations to the Chinese population in *C. revoluta* (Table 6). When the MLE (maximum likelihood estimates) for the migration parameters are transformed into the population migration rates, $M_1 = 2N_1m_1 = m_1*(\theta_1/2)$ estimated from Japanese populations to the Chinese population in *C. revoluta* is higher than the estimate, $M2 = 2N_2m_2 = m_2*(\theta_2/2)$, from China to Japan, i.e., 8.0134 vs. 0.5431 based on the combined data (Table 6). The Neτ (the product of effective population size and the generation time in years) value is also assessed; a higher effective population size is detected for Japan's populations than China population, i.e., $Ne\tau = 4.753*10^6$ vs. $1.437*10^5$ based on the combined data (Table 6).

**Table 6:** Scaled parameter estimates for IM model as inferred from the combined data of cpDNA and mtDNA.

| | $\theta_1$ [a] | $\theta_2$ [a] | $\theta_A$ [a] | $m_1$ [b] | $m_2$ [b] | $N_1\tau$ [c] | $N_2\tau$ [c] | $N_A\tau$ [c] | $2N_1m_1$ | $2N_2m_2$ |
|---|---|---|---|---|---|---|---|---|---|---|
| A. HiPt[d] | 8.9914 | 0.2426 | 10.9320 | 1.4500 | 3.5500 | $4.517*10^6$ | $1.210*10^5$ | $5.492*10^6$ | 6.5188 | 0.4306 |
| 95%HPDLo[e] | 6.4040 | 0.2426 | 5.4983 | 0.6500 | 1.3500 | | | | | |
| 95%HPDHi[e] | 51.8138 | 2.6683 | 70.9610 | 14.0500 | 7.4500 | | | | | |
| B. HiPt[d] | 10.0668 | 0.2860 | 10.3903 | 1.9978 | 4.1023 | $5.057*10^6$ | $1.426*10^5$ | $5.219*10^6$ | 10.0557 | 0.5867 |
| 95%HPDLo[e] | 7.8028 | 0.2860 | 5.2962 | 0.8820 | 1.5548 | | | | | |
| 95%HPDHi[e] | 79.2004 | 2.5066 | 36.1031 | 16.1243 | 7.8965 | | | | | |
| C. HiPt[d] | 9.3231 | 0.3356 | 10.6652 | 1.6345 | 3.6577 | $4.683*10^6$ | $1.674*10^5$ | $5.358*10^6$ | 7.6193 | 0.6138 |
| 95%HPDLo[e] | 7.2336 | 0.3356 | 5.3365 | 0.7586 | 1.4723 | | | | | |
| 95%HPDHi[e] | 68.5891 | 2.4523 | 65.2581 | 15.4572 | 7.6894 | | | | | |
| Average | 9.4604 | 0.2881 | 10.6634 | 1.6941 | 3.7700 | $4.753*10^6$ | $1.437*10^5$ | $5.358*10^6$ | 8.0134 | 0.5431 |

Note: Values are presented for each of the three runs with different seed numbers (A, B and C).

[a] The population size parameters for the Japan population ($\theta_1$) and Fukien population ($\theta_2$) *of C. revoluta*, and ancestral population ($\theta_A$).

[b]Migration rate estimate ($m_1$ – from Japan to Fukien population; $m_2$ – from Fukien to Japan population).

[c] Effective population size estimate (Nτ: population size × generation time) for Japan population ($N_1\tau$) and Fukien population ($N_2\tau$) *of C. revoluta*, and ancestral population ($N_A\tau$).

[d]The value of the bin with the highest count (HiPt).

[e]The lower (HPD95Lo) and upper (HPD95Hi) bound of the estimated 95% highest posterior density (HPD) interval.

In the study, the demographic scenarios for the two species and populations of Japan and China of *C. revoluta*are represented by the Bayesian skyline plot (Fig. 5 and 6). Based on the pattern of variation in mtDNA, a long history of constant population size of *C. revoluta*, followed by a slight decline (bottleneck) with subsequent demographic expansion, is recovered (Fig. 5A). At the population level, in *C. revolute*, the Japanese and Fukien populations also display similar patterns (Fig. 5C and 5D). In contrast, slow population growth is detected in *C. taitungensis* based on the Bayesian skyline plot (Fig. 5B). Based on cpDNA, a scenario of population expansion is suggested for both *C. revoluta* and *C. taitungensis*, while a weak bottleneck event may have occurred in *C. revoluta* (Fig. 6A and 6B). Likewise, Japan's populations of *C. revoluta* almost remain nearly constant before a recent population expansion, whereas, the Fukien's population did not undergo large fluctuations in population size until its recent extinction (Fig. 6C and 6D).

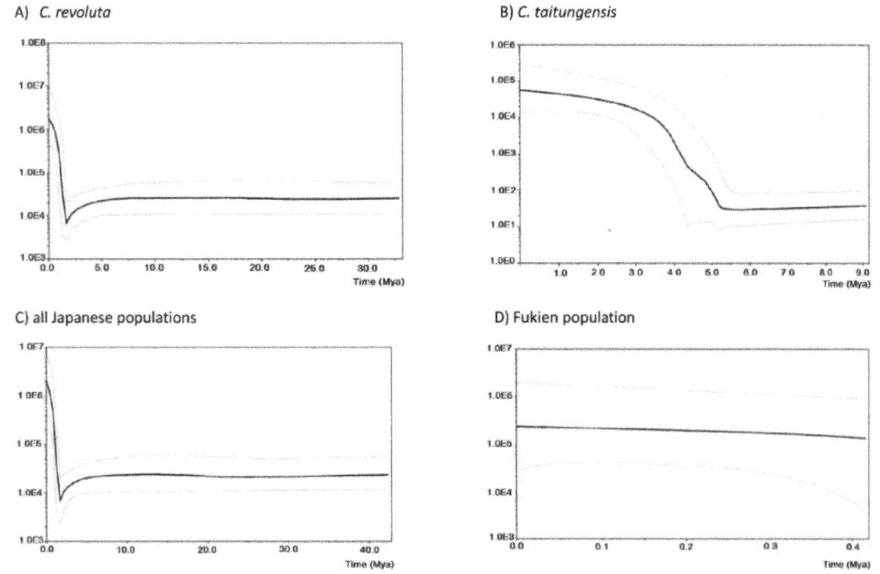


**Figure 5:** Bayesian skyline plot based on the mtDNA for the effective population size fluctuation throughout time. X axis: population size × generation time; Y axis: time (million years ago). A) *Cycas revoluta*; B) *C. taitungensis*; C) all Japanese populations of *C. revoluta*; D) the Fukien population of *C. revoluta* (black line, median estimations; area between gray lines, 95%.

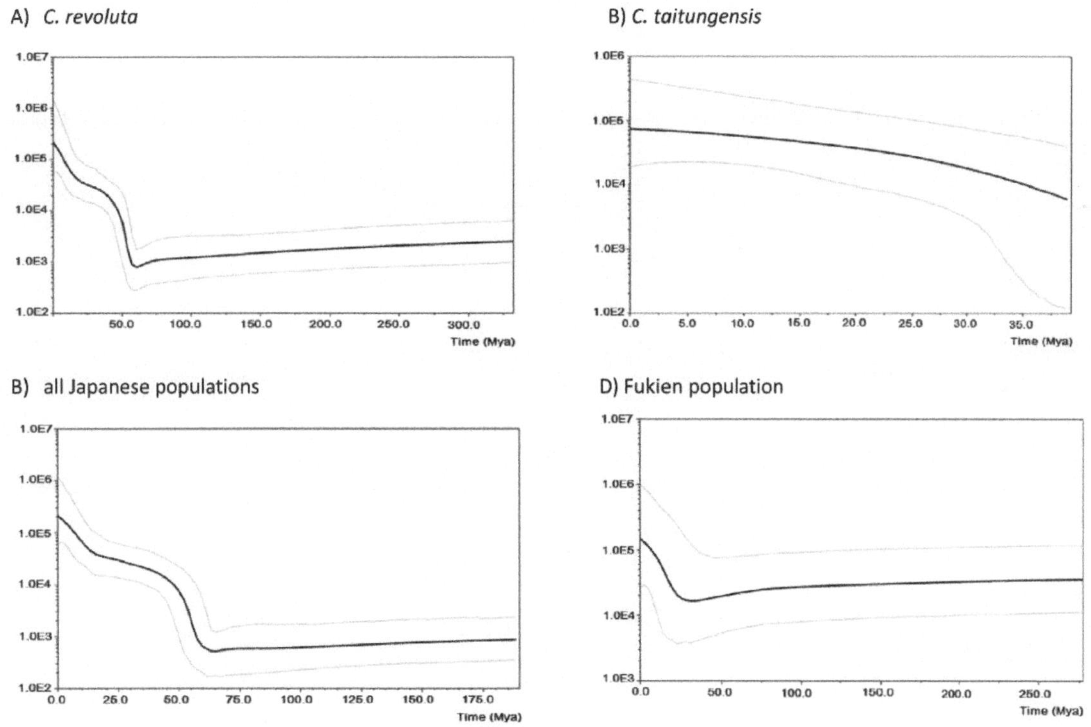


**Figure 6:** Bayesian skyline plot based on the cpDNA for the effective population size fluctuations throughout time. X axis: population size × generation time; Y axis: time (million years ago). A) *Cycas revoluta*; B) *C. taitungensis*; C) all Japanese populations of *C. revoluta*; D) the Fukien population of *C. revoluta* (black line, median estimations; area between gray lines, 95% confidence interval.

We also use mismatch distribution analyses to infer the long-term demographic history of populations. Mismatch distribution analyses (Fig. 7 and 8) show different patterns between the cp- and mtDNA genomes. In most cycad populations, cpDNA has a unimodal mismatch distribution (Fig. 8), while multimodal and very ragged distributions are detected in mtDNA within all populations (Fig. 7). Nevertheless, when sequences from all populations are pooled together, both cp- and mtDNAs have unimodal mismatch distributions in *C. revoluta*, while only cpDNA displays a unimodal mismatch distribution in *C. taitungensis*. Unimodal mismatch distributions are often the result of past demographic expansions [50, 51]; whereas, multimodal mismatch distributions usually indicate stationary population structure. Since the data used to produce mismatch distributions are not independent [52], Tajima's *D*, Fu and Li's D* and Fu's Fs statistics can be used to detect departure from population equilibrium. In this study, values of Fu and Li's D* based on cpDNA sequences for*C. revoluta* and *C. taitungensis* are significantly negative, except for populations IR, KO, and OK (Table 3); while Tajima's D statistics are all significantly negative, except for the population IR. Fu's Fs statistics are significantly negative, except for populations IS, IR, and OK (Table 3). Likewise, Fu and Li's D*, Tajima's D, and Fu's Fs statistics are significantly or marginally significantly negative based on mtDNA data in most populations (Table2).

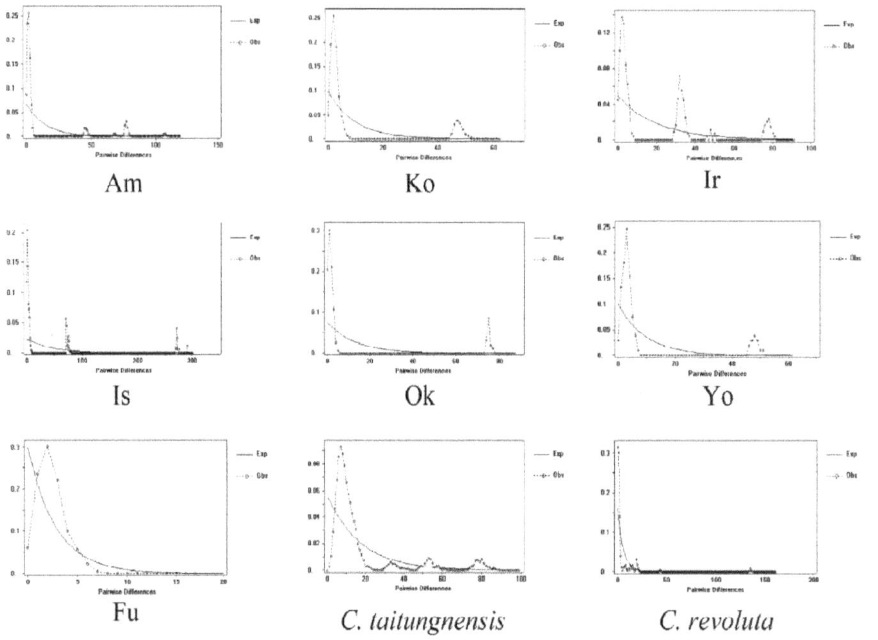


**Figure 7:** Mismatch distributions of mtDNA haplotypes based on pairwise sequence differences against the frequencies of occurrence for cycad species and populations.

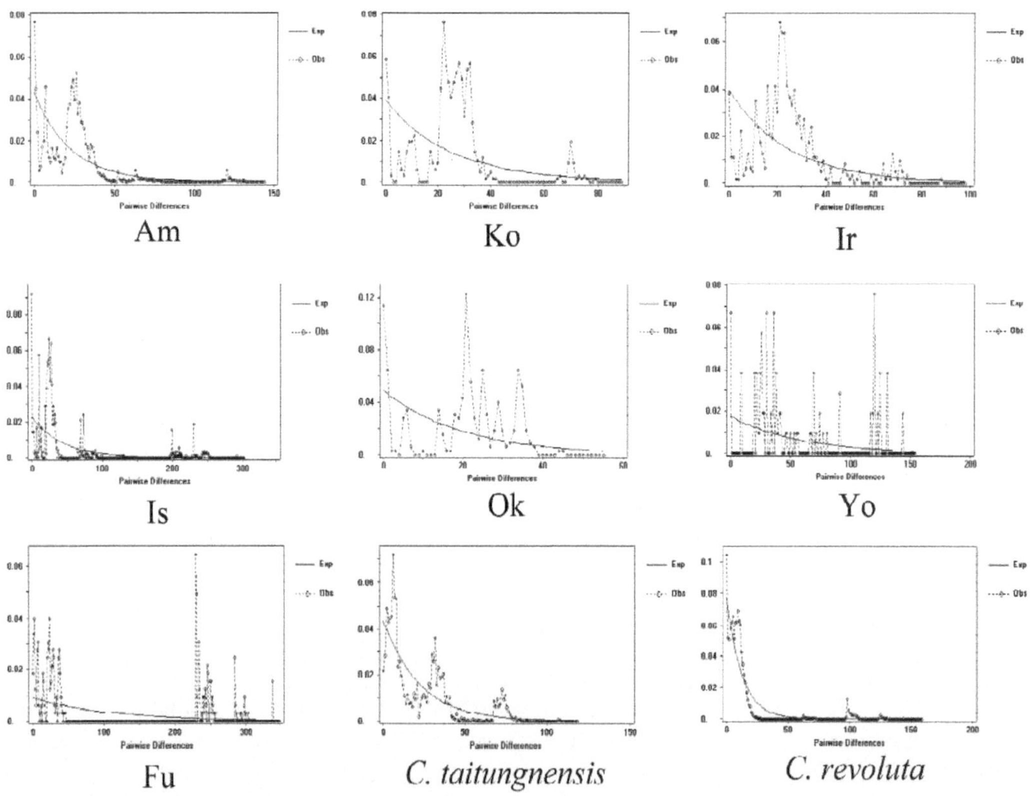


**Figure 8:** Mismatch distributions of cpDNA haplotypes based on pairwise sequence differences against the frequencies of occurrence for cycad species and populations.

## Coalescence of Lineages and Molecular Dating

Bayesian estimates of the mutation rates and the age of the most recent common ancestor (TMRCA) of the cycad sequences were obtained using BEAST v. 1.3. For mtDNA, all the haplotypes coalesced at about 327.3 MYA, while all chlorotypes could be traced back to a common ancestor about 204.0 MYA (Fig. 9). In the mtDNA haplotype tree, clusters (A, B, C) vs. D split at some 74.1 MYA; and the clusters A and (B, C) coalesced at about 30.6 MYA. Likewise, in the cpDNA tree, clusters I and (II, V) split some 25.6 MYA, while cluster (I, II, V) diverged from its sisters (III, IV) an estimated 43.9 MYA. Almost all coalescences of the major lineages predate the formation of the section *Asiorientales*. Coalescent events occurred much earlier in mtDNA than in cpDNA. On the other hand, most tip haplotypes in both gene trees coalesced recently to their common ancestors.

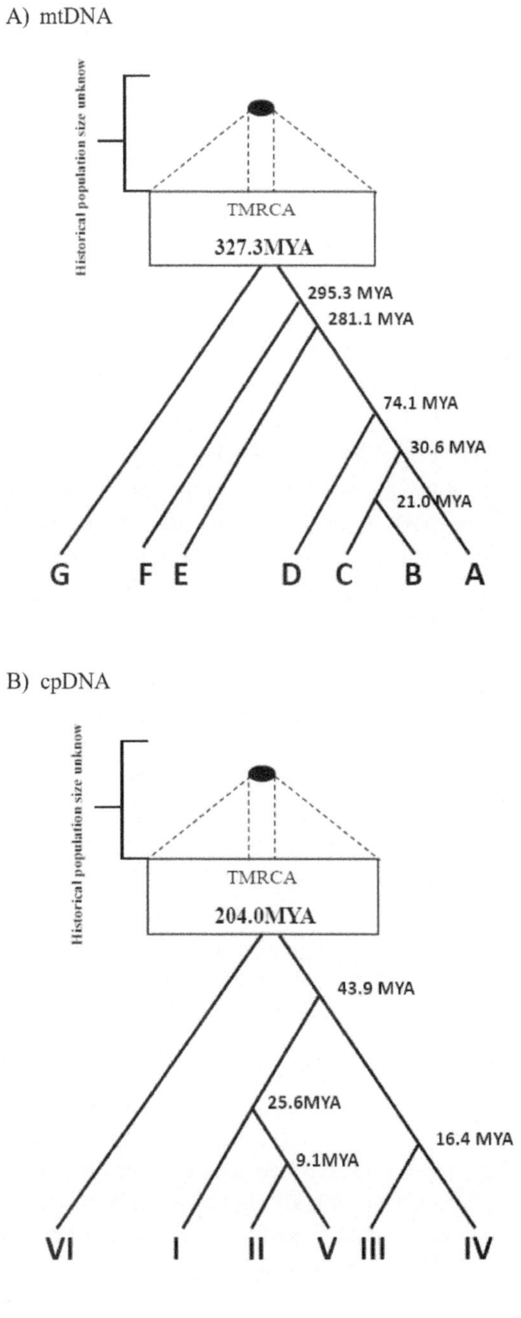


**Figure 9:** Coalescence-based analyses of lineages. Time coalesced to the most recent common ancestor (TMRCA) of major lineages are indicated. A) mtDNA, B) cpDNA.

# DISCUSSION

## Highly polymorphic organelle DNAs in Cycas revoluta and C. taitungensis

Genetic variation in the two cycad species is high, comparable to or higher than many tree species (such as*Quercus* [31, 49]; *Fagus* [50]; *Lithocarpus* [32, 44, 51]. Such high levels of genetic variation within mt- and cpDNAs suggest that these are ancient lineages with enough time for mutations to accumulate within lineages, given the slow rate of organelle evolution [52]. Evidence of the deep splits in both phylogenetic trees (Fig. 2) and the estimated dates of divergence support this scenario. Moreover, star-like phylogenies with a mixture of long and short branches, which represent recently evolved haplotypes, illustrate the high levels of variation.

Although the two cycad species have high levels of genetic diversity at both loci, the loci do not vary in concert. For example, the *C. revoluta* population YO possesses the highest level of mtDNA variation while having the lowest level of cpDNA diversity. Similarly, the Fukien China *C. revoluta* population is highly diverse for mtDNA haplotypes while containing lower than average of cpDNA haplotypes. The large native cycad population on the island of Amami-O-Shima (AM) has genetic diversity at both loci comparable to or even lower than most populations.

In this study, genetic diversity tends to be higher in the south than in the north. A cline-like distribution of genetic diversity from south to north occurs in many continental plants of the Northern Hemisphere and is thought to be related to colonization history over glacial cycles [cf. [34, 53]]. Considering the geological history of the Taiwan-Ryukyus Archipelago, one expects a similar pattern for these cycad species where colonization occurs in a stepwise manner from China northwards through the Archipelago. Genetic diversity should decrease to the north, assuming that a small number of individuals colonize each subsequent island in a stepwise manner. However, the occurrence of highly diverse organelle DNAs across the range of *Cycas revoluta* and *C. taitungensis*, and many other tree species of the Taiwan-Ryukyus Archipelago suggests that single stepwise colonization events may not be the sole determinant of the overall pattern of genetic diversity [cf. [33]].

Austerlitz *et al.* [51] suggest that the level of diversity in tree species that recolonize glaciated regions is affected by restricted gene flow but is also tempered by their demography. In tree populations, the genetic effects of founder events are reduced due to overlapping generations and delayed reproduction associated with long life spans. These are traits found in most long lived trees including these cycads, which are characterized by great longevity, overlapping generations, age structured populations and a juvenile phase of 30 to 50 years. During the juvenile period when cycads grow vegetatively, newly founded cycad potentially populations can increase in number for many years only through the arrival of new migrants, especially when migration corridors across islands for migration are available. This pattern of colonization and reproduction effectively acts to increase in the number of initial colonizers of a given population and in turn decreases any founder effects. Even given a high likelihood of stepwise colonization along the Taiwan-Ryukyu Archipelago, as observed in the *Pinus luchuensis* complex [52, 33], multiple colonization events could compensate for potential reduction in genetic diversity during each founder event.

Secondary contact of ancestral lineages can be another factor that contributes to high genetic diversity. The coalescence-based analysis reveals that mitotypes F and G diverged some 327.3 and 295.3 million years ago, respectively. Today these haplotypes exist exclusively in the mainland Fukien (mitotype G), and in the Fukien and Ishigaki (mitotype F) populations. Mitotype F is common in the Fukien population, suggesting an origin most likely on the Asian mainland with subsequent dispersal northward onto Ishigaki. Such secondary contact of divergent lineages is consistent with a U-shaped site-frequency distribution according to Wakeley and Aliacar's simulations (Fig. 8) [54].

## Isolation and Lack of Geographical Subdivision

Current gene flow between islands is greatly restricted due to their geographical isolation and the low dispersability of cycad seeds in water [36]. Thus, one expects that island populations should be significantly differentiated from each other. But, like many other tree species of the Taiwan-Ryukyu Archipelago [33], no significant geographical differentiation is detected between cycad populations among islands. Fst values are small and all pairwise *Nm* values, deduced from Fst values, are much greater than one (Tables 4 and 5). The*Nm* values based on mtDNA data are about ten-fold greater than those based on cpDNA sequences, 106.10 vs. 10.95. While both organelles are transmitted only via seeds, the difference in *Nm* values determined from cp- and mtDNAs coupled with deviation from an isolation-by-distance model, and the lack of geographical subdivision and genetic differentiation, suggest that the two loci are responding to different evolutionary forces.

Clearly, *Nm* values cannot be interpreted as current gene flow between populations. These high *Nm* values likely represent either ancient migration events or shared ancestral polymorphisms [55, 56]. For example, the*Nm* values between Chinese and Japanese populations range from 3.87 to 7.31 (mtDNA) and 1.87 to 10.25 (cpDNA) (Tables 4 and 5). Testing of 'isolation with migration' (IM), nevertheless, reveals asymmetrical, historical gene flow with more migrants of *C. revoluta* from Japan to China than vice versa, 8.0134 vs. 0.5431 based on combined data. These results suggest long periods of isolation with limited genetic exchange between the two geographical regions. *C. taitungensis* shows a similar asymmetric pattern of ancient gene flow. Migration is estimated to be more frequent from population C to population RL (3.86445 vs. 41.10288, cpDNA, and 2.25430 vs. 48.99630, mtDNA), contributing to the lack of genetic differentiation between populations, as indicated by Fst values of 0.0056 (cpDNA) and 0.021 (mtDNA) [cf. [55]].

Like other island species, cycad populations were colonized via founders from neighboring islands or the mainland. Today, deep water channels hinder seed dispersal among islands and most seed dispersal occurs only within islands. However, during the glacial maxima [cf. [29]], sea levels were much lower and migration could have occurred among islands. The rise of sea levels during glacial retreats terminated between-island seed exchange and isolated cycad populations of the Ryukyu Archipelago from each other [57]. The time since isolation approximates 200–400 cycad generations. Given the very short time that populations have been isolated and the low rate of molecular evolution for both organelle loci, it is not surprising that populations show genetic coherence. A recent simulation study [51] indicates that tree populations can remain genetically similar even with an isolation-by-distance pattern of colonization and can yield a phylogeographical pattern with little differentiation, here for the cycads of the Taiwan-Ryukyu Archipelago. The high levels of nucleotide diversity in both organelle DNAs and the existence of the highest number of private cytotypes suggest that population IS of the south Ryukyu and population RL of Taiwan may have acted as glacial refugia for *C. revoluta* and *C. taitungensis*, respectively.

## Paraphyly of Loci in Cycas Species and Sections

Although cycad populations are undifferentiated from each other due to recent historical gene flow (Table 6), given that the generic sections of cycads sections have been isolated for very long periods of time, one would expect monophyly at the sectional level. Nevertheless, both mt- and cpDNAs are paraphyletic or polyphyletic between section *Asiorientales* and its sisters. Such lack of reciprocal monophyly between taxa may result from selection, e.g., the S locus of *Brassica* [58], and α-tubulin genes of *Miscanthus* [59], lineage sorting [60], or hybridization/introgression [61]. Genetic exchange among species via hybridization and introgression introduces foreign alleles into a species or population and can result in a genetic admixture. In contrast, lineage sorting occurs when populations or species are recently diverged and retain ancestral polymorphisms [cf. [5, 52]].

Hybridization and/or lineage sorting are likely causes of shared polymorphisms in *Cycas* given that these are neutral loci and thus not under strong selection. Discerning between hybridization and lineage sorting can be difficult [61, 62, 63, 64], since most statistical analyses have limited power. However, in this study neither hypothesis can by itself explain the current patterns of genetic variation among sections. First, all four *Cycas*species are allopatric which greatly reduce the possibility of recent recurrent interspecific hybridization. Lineage sorting can result in shared polymorphism but requires a relatively short time of isolation. The fossil record indicates that the sections of *Cycas* have been diverged for 30 million years, making such a scenario unlikely [10].

The pattern of coalescing DNA segments provides information on the genealogical processes that occur within populations [4, 65, 66]. The star-like phylogeny of the organelle trees shows that recently derived haplotypes are numerous, external on the networks and are geographically restricted. Ancestral haplotypes are found at interior nodes, and are geographically more widespread than the recently derived tip haplotypes. This pattern is consistent with a birth-and-death model where most mutations go extinct with only very few ancestral polymorphisms surviving to become dominant [67]. If the dynamics of mutations and fixation do not follow a birth and death model, one expects topologies with deep splits for most haplotypes.

The pattern of mutations in cycads follows a 'surfing' model [68], which predicts that a majority of new mutations do not spread geographically and either remain at low frequencies or are lost by genetic drift. Only successful 'surfing mutations' can reach high frequencies and eventually occupy a large geographical region [69]. In this model both population size and migration rate determine the success of new mutations. Deme size not only affects the probability of a mutation to 'surf' but also determines its final frequency and spatial distribution. When the effective population size is small, a mutation within a population has a high probability of extinction due to genetic drift and it is able to spread only if it reaches a very high frequency by chance. In contrast, large effective population sizes counter genetic drift while a large number of exchanged migrants between neighboring demes prevents a mutation from reaching high local frequencies [68]. Accordingly, as most cycad populations were probably large (Fig. 5 and 6), one expects to observe numerous recently derived mutations that were derived from a single ancestral haplotype, and the observation that few haplotypes appear to survive over long periods (Fig. 2). Most of the new mutations appear to be ephemeral and can be lost from a population or species during population contraction (the glacial maximum). As a consequence, very few cytotypes, *e.g.*, mitotypes A, and B, and chlorotypes I, and III, survived and became dominant across the cycad populations. Such a birth-and death process corresponding to demographic fluctuations may have occurred regularly in cycads over repeated glacial cycles. As a result, the existence of rare and geographically restricted cytotypes contributes to the high levels of genetic diversity within populations and within species, whereas, they played very minor roles in differentiating populations or species from each other due to their rarity.

While the surfing phenomenon explains the dominance of widespread haplotypes and low genetic differentiation between cycad populations, it does not explain shared ancestral polymorphisms between sections. Both genetic drift and changes in population size have most likely influenced polymorphism as well. Population contraction-expansion, which has occurred regularly in the Quaternary, provides an opportunity for rapid stochastic loss or fixation of alleles (haplotypes). Rapid population growth and expansion in cycads is suggested by the star-shape phylogenies found in most lineages of both organelle genomes, or the Bayesian skyline plot and by the observation that DNA sequences are saturated with singletons at both the population and species levels (Fig. 3 and 4) [cf. [70, 71]], as well as the Bayesian skyline plotting. A mixture of long and short branch lengths in the star-like phylogenies could also be caused by heterogeneous mutation rates. The significant values for Tajima's D and Fu and Li's D* statistics, nevertheless, favor a demographic expansion hypothesis [72].

Mismatch distribution analyses are also used to infer the long-term demographic history of populations. In this study, contrasting patterns of unimodal vs. multimodal mismatch are found for mt- and cpDNAs (Fig. 7and 8). Usually, unimodal mismatch distributions are interpreted as the result of past demographic expansions [73, 74], whereas multimodal mismatch distributions indicate stationary population sizes. These contrasting patterns in cycads are difficult to interpret although a mismatch inconsistency between different data sets is not unprecedented. A well-known example of contrasting mismatch patterns occurs between expanding human Neolithic and hunter-gatherer populations [75]; the unimodal distribution for Neolithic populations is interpreted as the result of greater *Nm* values [76]. Even with a multimodal distribution, demographic expansion cannot be rejected, especially if the *Nm* values are small. In this study, the contrasting mismatch patterns between unlinked organelle genomes seem to be uncorrelated with *Nm* values. The ragged distributions in mtDNA are more likely attributable to the low $\tau$ (fig. 5). In cycads, a very recent population expansion about 200–400 generations, a slow evolutionary rate, and putative intramolecular recombination [data not shown; cf. [55]], may preclude mismatch distribution analyses from assessing recent population expansion. Nevertheless, Tajima's D, Fu and Li's D*, and Fu's Fs statistics provide independent estimates for spatial expansion, especially considering that the DNA sequences are from neutral noncoding spacer regions.

Coalescence-based analyses measure the time of coalescence to the most recent common ancestor (TMRCA). In both organelle DNAs, the divergence time is estimated at 327.3 MYA in mtDNA and 204.0 MYA in cpDNA, agreeing with the divergence of major sections of cycads in the Mesozoic [77]. Moreover, the split of the common haplotypes, mitotype B and chlorotype I, from their sister haplotypes occurred about 21.0 and 25.6 MYA, respectively in the Tertiary period when cycads were diversifying and when divergence of section *Asiorientales* (30 MYA) occurred. Since then the two dominant cytotypes have persisted in cycads [cf. [55]]. The sharing of common cytotypes in both genomes is consistent with range contraction-expansion during glacial cycles where common haplotypes could persist, while most rare, newly derived haplotypes were lost. Consequently, both phylogenies show topologies with few ancient coalescence events vs. numerous recent coalescence events.

The evolution of these two cycad species is complex, affected by range expansion and contraction, fluctuations in population size, widespread historical and current restricted gene flow during complex glacial cycles as well as low rates of molecular evolution in organelle DNA. All of these processes result in a pattern of genetic similarity among populations, a lack of reciprocal monophyly, and long coalescent times.

## CONCLUSION

In summary, paraphyly at both organelle DNAs has been long maintained in *Cycas* section *Asiorientales* ever since the split from its sisters some 30 million years ago. Large population sizes of the constituting species *C. revoluta* and *C. taitungensis* contributed to the lack of reciprocal monophyly betwen *Cycas* sections. Providing limited current seed dispersal across oceans, unexpected low levels of geographical subdivision are attributable to the sharing of common haplotypes among island populations. Demographic expansion and contraction with fluctuating population sizes, widespread historical vs. current restricted gene flow during complex glacial cycles, and low rates of molecular evolution in organelle DNA all have to be aroused to explain such population structuring.

## METHODS

### Plant Materials, PCR, and Nucleotide Sequencing

We assessed levels of sequence variation for cp and mt-DNAs among individuals and populations of *Cycas revoluta* and *C. taitungensis* in Taiwan, Fukien, and Ryukyus (Table 1, Fig. 1 and 4). A total of 307 and 102 plants of *C. revoluta* and *C. taitungensis*, respectively, were sampled. *Cycas revoluta* is extinct from the wild in Fukien; only old trees were collected from public gardens or nurseries to avoid sampling from recently introduced material. The samples of *C. revoluta* in China were collected from South China Botanic Garden, which had plants collected from the Fukien wild population before extinction, which located in the Lienchiahg County, Fukien. An additional six native island populations were collected in Ryukyus for *C. revoluta* and two populations of *C. taitungensis*. Two species, *C. taiwaniana* Carruthers (Section *Stangerioides*) and *C. thouarsii*R. Brown (Section *Rumphiae*) [cf. [8]], were chosen as outgroups. *C. taiwaniana* was obtained from the South China Botanic Garden, transplanted from a native population in Fukien, China, and *C. thouarsii* was sampled from the Taiwan Forestry Research Institute, transplanted from a native population in Pingtung, Taiwan. Young and healthy leaves were collected in the field, rinsed with tap water and dried in silica gel. All samples were stored at -70°C until they were processed. The voucher specimens (*Hsu s.n.*) of this research are deposited in the Herbarium, Endemic Species Research Institute, Nantou, Taiwan. Leaf tissue was ground to a powder in liquid nitrogen and stored at -70°C. Genomic DNA was extracted from the powdered tissue following a CTAB procedure [78]. The *atp*B-*rbc*L noncoding spacer of cpDNA and the rDNA internal transcribed spacer (rITS) of mtDNA were amplified and sequenced. PCR amplification was carried out in 100 μL reaction using 10 ng of template DNA, 10 μL of 10× reaction buffer, 10 μL MgCl2 (25 mM), 10 μL dNTP mix (8 mM), 10 pmole of each primer, 10 μL of 10% NP-40, and 2 U of *Taq* polymerase (Promega, Madison, USA). The reaction was programmed on a MJ Thermal Cycler (PTC 100) as one cycle of denaturation at 95°C for 4 min, 30 cycles of 45 s denaturation at 92°C, 1 min 15 s annealing at 52°C, and 1 min 30 s extension at 72°C, followed by 10 min extension at 72°C. A pair of universal primers for cpDNA *atp*B-*rbc*L spacer [79] or mtDNA rITS (mito 18 s F 5'-GTG,AAG,TCG,TAA,CAA, GGT,AGC-3'; mito 5 s R 5'-TCG,AGG,TCG,GAA,TGG,GA T,CGG-3') [cf. [80]], dNTP and *Taq*polymerase were added to the above ice-cold mix. The reaction was restarted at the first annealing at 52°C.

PCR products were purified by electrophoresis in 1.0% agarose gel using 1 × TAE buffer. The gel was stained with ethidium bromide and the desired DNA band was excised and eluted using QIAquick Gel Kit (QIAGEN). For all individuals, purified amplicons were ligated to a pGEM-T easy vector system (Promega). Five clones were randomly selected and purified using the Plasmid Mini Kit (QIAGEN). Purified cloned ampicons were sequenced in both directions using BigDye chemistry (Perkin Elmer) on ABI 377A automated sequencer (Applied Biosystems). Primers for sequence determination were T7-promoter and SP6-promoter located on p-GEM-T easy Vector termination site. As no within-individual variation was detected for either DNA marker, direct sequencing with eluted PCR products was also conducted to all individuals.

### Data Analysis

#### *Sequence Alignment and Phylogenetic Analyses*

Nucleotide sequences were aligned with the program BlastAlign [81]. After the alignment, the indels were coded as missing data. Phylogenetic trees were reconstructed using maximum likelihood (ML) analyses of the nucleotide sequences with software PHYML v2.4.5 [82] and bootstrap consensus values calculated using 1,000 replicates. The general time reversible GTR + I + G model with 6 substitution categories was determined to be the most suitable model by Modeltest v3.6 [83] and was used for all subsequent nucleotide analyses. Maximally parsimonious (MP) trees were also sought using PAUP heuristic search strategies [84]. The length of the shortest trees was obtained by initiating 1,000 heuristic searches, each using random addition starting trees, with tree bisection-reconnection (TBR) branch swapping. The equally MP trees were then used as starting trees for TBR branch swapping. In all analyses, the maximum number of trees to be saved was set at 5000. Bootstrap values [85] were calculated from 1,000 replicate analyses using a heuristic search strategy, simple addition sequence of the taxa, and TBR branch swapping. The values of random number seed set as 64238 and 70189 for mtDNA and cpDNA data sets, respectively, were used to start the random number generator. Relationships among haplotypes were determined and displayed as nested phylogenies. In order to visualize the phylogeographical relationships, a network of each locus based on statistical parsimony was constructed with the computer program TCS [86]. A distance matrix was composed of all pairwise comparisons of haplotypes, for which the maximum number of mutations was determined that did not exceed the probability of parsimony by 0.95 [as defined in [87]]. All haplotypes satisfying this parsimony criterion were then connected to a single network, while those of which the probability exceeded 0.95 were resolved as a separate network. According to coalescent theory, tip nodes of a network are likely to represent descendents derived from ancestral, interior nodes [cf. [88, 89]].

The hypothesis of a molecular clock was tested by relative rate tests [90, 91]. The null hypothesis of a molecular clock suggests that the number of nucleotide substitutions between two lineages would be the same. Based on the assumption of a normal distribution of nucleotide substitutions [91], the hypothesis of a molecular clock will be rejected with 95% significance, when the difference of substitution rates between two lineages is greater than 1.96 times the standard error.

#### *Population Genetic Analysis*

Summary statistics such as the number of segregating sites (S), and the average number of pairwise difference (k) between haplotypes were determined. The level of genetic diversity within populations was quantified by measures of nucleotide divergence, $\theta$ [92] and $\varphi$ [93], using DnaSP Version 4.10 [94]. In order to make inferences about demographic changes in the cycads, we employed both mismatch distributions and statistical tests of neutrality. We calculated Tajima's D [91], Fu and Li's D* statistic [95], and Fu's Fs statistics [96], which is powerful for detecting population growth [97], in the noncoding DNA fragments as indicators of demographic expansion in DnaSP. We also investigated the historical demographics of populations by plotting mismatch distributions [73] and comparing them to Poisson distributions. Mismatch distributions for each sample were used to distinguish between models that invoke past exponential growth and historical population stasis. Parameters of demographic expansion were estimated using the methods of Schneider and Excoffier [98]. The validity of the demographic expansion hypothesis was tested using a parametric bootstrap approach, in which the sum of squared deviation (SSD) between the observed distribution and the expected distribution was compared to the SSD between the simulated distributions and the expected distribution (Arlequin version 3.0 [99]). Patterns of geographical subdivision and genetic differentiation were estimated hierarchically in DnaSP. Statistical significance of Fst estimates was assessed using software Arlequin with 10,000 permutations.

## Estimation of Coalescence Times of cp and mtDNA Lineages

We estimated the coalescence time of the the sister lineages of *C. taitungensis* and *C. revoluta* split for both mt- and cpDNAs. To estimate divergence between lineages a well-supported rate of evolution is required. In seed plants, evolutionary rates are estimated at $1.01 \times 10^{-9}$ ahd $4.50 \times 10^{-10}$ substitutions per site per year for synonymous sites of cp- and mt-DNAs, respectively [cf. [52, 100]]. These values approximate the evolutionary rates of introns and noncoding spacers of organelle DNAs. Bayesian estimates of the mutation rates and the ages of the most recent common ancestor (TMRCA) of the cycad sequences were obtained using BEAST v. 1.4, available from http://beast.bio.ed.ac.uk[101]. We used the GTR+I+G model of nucleotide substitution with estimated base frequencies, gamma shape distribution (with six categories), proportion of invariant sites, and a strict molecular clock with uncorrelated log normal distribution of branch lengths. Posterior estimates of the mutation rate and age of the TMRCA were obtained by Markov chain Monte Carlo (MCMC) analysis, with samples drawn every 500 steps over a total of 25,000,000 steps. The BEAST program was also used to create a Bayesian skyline plot with 10 steps. The analysis was run for $10^7$ iterations with a burn-in of $10^6$ under the GTR+I+G model with estimated base frequencies, gamma shape distribution (with six categories), proportion of invariant sites, and a strict molecular clock with uncorrelated log normal distribution of branch lengths. Genealogies and model parameters were sampled every 1000 iterations. All operators were automatically optimized. Convergence of parameters and mixing of chains were followed by visual inspection of parameter trend lines and checking of ESS values by three pre-runs. The effective sampling size (ESS) parameter was found to exceed 100, which suggests acceptable mixing and sufficient sampling. Adequate sampling and convergence to the stationary distribution were checked using TRACER v. 1.3 [102]. Posterior estimates of parameters were all distinctly unimodal (although with wide 95% highest posterior densities), and all parameters were identifiable, despite the relatively low information content in the sequences and the small age range of the sequences.

## Isolation by Migration and Estimating Of Ancestral Population Size

We used the simulation program IM [1, 103] to investigate isolation with a migration model. By applying coalescent simulations and Bayesian computation procedures, IM yielded six demographic parameters, including population-split time, effective population size for the ancestral and two current populations, and migration rates. The posterior probability densities of these parameters are generated by MCMC simulations, and simulations were run with individual simulations being updated 50 million times. Within each simulation, we used the procedure to swap among 10 heated chains (Metropolis coupling) and observed sufficient swapping rate while the simulation was running. These simulations were carried out using several independent runs, with each chain started at a different starting point and initiated with a burn-in period of 100,000 updates. Convergence upon the stationary distribution was assessed by estimating the effective sample size (ESS) for the parameters, based on the autocorrelation of parameter values measured over the course of the run. The analysis was considered to have converged upon a stationary distribution if the independent runs generated similar posterior distributions with a minimum ESS of 100. Each simulation yielded a marginal density histogram for the six parameters of interest. The peaks of the resulting distributions were considered as the maximum likelihood estimates (MLE) of the parameter with credibility intervals equaling the 95% highest posterior density (HPD) intervals.

# DECLARATIONS

## Acknowledgements

This study was supported by the National Science Council (NSC) and Council of Agriculture of (COA) of Taiwan. We are grateful to the Tropical Research Centre of the University of Ryukyus for the assistance in the sample collecting.

# REFERENCES

1. Chiang TY, Schaal BA: Phylogeography of plants in Taiwan and the Ryukyu Archipelago. Taxon. 2006, 55: 3-41.
2. Hoelzer GA, Wallman J, Melnick DJ: The effects of social structure, geographical structure, and population size on the evolution of mitochondrial DNA: II. Molecular clocks and the lineage sorting period. J Mol Evol. 1998, 47: 21-31.
3. Schaal BA, Olsen K: Gene genealogies and population variation in plants. Proc Natl Acad Sci. 2000, 97: 7024-7029.
4. Donnelly P, Tavaré S: Coalescents and genealogical structure under neutrality. Ann Rev Genet. 1995, 29: 401-421.
5. Chiang TY: Lineage sorting accounting for the disassociations between chloroplast and mitochondrial lineages in oaks of southern France. Genome. 2000, 43: 1090-1094.
6. Matos JA, Schaal BA: Chloroplast evolution in the *Pinus montezumae* complex: a coalescent approach to

hybridization between *P. hartwegii* and *P. montezumae*. Evolution. 2000, 54: 1218-1233.

7. Willis KJ, McElwain JC: The evolution of plants. 2002, Oxford: Oxford University Press
8. Hill KD: *Cycas* – an evolutionary perspective. Biology and conservation of cycads. Proceedings of the Fourth International Conference on Cycad Biology. Edited by: Chen CJ. 1999, Beijing: International Academic Publishers, 98-115.
9. Ye ZY: The investigation into the past and the current state of *Cycas revoluta* in Fujian. Biology and conservation of cycads. Proceedings of the Fourth International Conference on Cycad Biology. Edited by: Chen CJ. 1999, Beijing: International Academic Publishers, 73-74.
10. Pant DD: A study in contrasts: the present and past distribution and form of certain cycads. Biology and conservation of cycads. Proceedings of the Fourth International Conference on Cycad Biology. Edited by: Chen CJ. 1999, Beijing: International Academic Publishers, 1-23.
11. Zhou GS, Jian SX, Zhou W: Cycad cultivation and trade in Fujian of China. Biology and conservation of cycads. Proceedings of the Fourth International Conference on Cycad Biology. Edited by: Chen CJ. 1999, Beijing: International Academic Publishers, 368-370.
12. Miller G: From cycad flour, a new emerges. Science. 2006, 313: 431-
13. Whitelock LM: The cycads. 2002, Portland: Timber Press
14. Osborne R, Stevenson DW, Hill KD: The world list of cycads. Biology and conservation of cycads. Proceedings of the Fourth International Conference on Cycad Biology. Edited by: Chen CJ. 1999, Beijing: International Academic Publishers, 224-239.
15. Huang S, Hsieh HT, Fang K, Chiang YC: Patterns of genetic variation and demography of *Cycas taitungensis* in Taiwan. Bot Rev. 2004, 70: 86-92.
16. Shimabuku K: Checklist of vascular flora of the the Ryukyu Islands. 1997, Kyushu University Press
17. Uehara K: Illustrations of trees. Cycadaceae. 1970, Tokyo: Erushima Press, 1:
18. Hsieh HT: Studies on population ecology and genetic variability of *Cycas taitungensis*. Master Thesis. 1999, Taipei: Department of Biology, National Taiwan Normal University
19. Darwin C: On the Origin of Species. 1859, London: John Murray
20. Emerson BC: Evolution on oceanic islands: molecular phylogenetic approaches to understanding pattern and process. Mol Ecol. 2002, 11: 951-966.
21. Juan II, Emerson BC, Orom II, Hewitt GM: Colonization and diversification: towards a phylogeographic synthesis for the Canary Islands. Trends Ecol Evol. 2000, 15: 104-109.
22. Vita-Finzi C: Deformation and seismicity of Taiwan. Proc Natl Acad Sci. 2000, 97: 11176-80.
23. Sibuet JC, Hsu SK: Geodynamics of the Taiwan arc-arc collision. Tectonophysics. 1997, 274: 221-251.
24. Sibuet JC, Hsu SK: How was Taiwan created?. Tectonophysics. 2004, 379: 159-181.
25. Wang JM: The Fenwei rift and its recent periodic activity. Tectonophysics. 1987, 133: 257-275.
26. Hikida T, Ota H: Biogeography of reptiles in the subtropical East Asian Islands. The Symposium on the Phylogeny, Biogeography and Conservation of Fauna and Flora of East Region. 1997, Taipei, Taiwan: National Taiwan Normal University, 11-18.
27. Bennett KD: Milankovitch cycles and their effects on species in ecological and evolutionary time. Paleobiology. 1990, 16: 11-21.
28. Lambeck K, Chappell J: Sea level change through the last glacial cycle. Science. 2001, 292: 679-685.
29. Hewitt GM: A climate for colonization. Heredity. 2004, 92: 1-2.
30. Kizaki K, Oshiro I: Paleogeography of the Ryukyu Islands. Marine Sci Month. 1977, 542-549.
31. Petit E, Balloux F, Goudet J: Sex-biased dispersal in a migratory bat: a characterization using sex-specific demographic parameters. Evolution. 2001, 55: 635-640.
32. Hung KH, Hsu TW, Schaal BA, Chiang TY: Loss of genetic diversity and erroneous phylogeographical inferences in*Lithocarpus konishii* (Fagaceae) of Taiwan caused by the Chi-Chi Earthquake: Implications for Conservation. Ann Missouri Bot Gard. 2005, 92: 52-65.
33. Chiang TY, Schaal BA: Phylogeography of plants in Taiwan and the Ryukyu Archipelago. Taxon. 2006, 55: 31-41.
34. Hewitt GM: The genetic legacy of the Quaternary ice ages. Nature. 2000, 405: 907-913.
35. Zachos J, Pagani M, Sloan L, Thomas E, Billups K: Trends, rhythms, and aberrations in global climate 65 Ma to present. Science. 2001, 292: 686-693.
36. Dehgan B, Yuen CKH: Seed morphology in relation to dispersal, evolution, and propagation of *Cycas*. Bot Gaz. 1983, 144: 412-418.
37. Hill KD: Evolution and biogeography of the genus *Cycas*. Proceedings of the 3rd International Conference of Cycas Biology. Edited by: Vorster P. 1996, Stellenbosch: The Cycad Society of South Africa
38. Tang W: Seed dispersal in the cycad *Zamia pumila* in Florida. Can J Bot. 1989, 67: 2066-2070.
39. Keppel G, Lee SW, Hodgskiss PD: Evidence for the long isolation among populations of a Pacific cycad: genetic diversity and differentiation in *Cycas seemannii* A. Br. (Cycadaceae). J Heredity. 2002, 93: 133-139.
40. Lu SY, Peng CI, Cheng YP, Hong KH, Chiang TY: Chloroplast DNA phylogeography of *Cunninghamia konishii*(Cupressaceae), an endemic conifer of Taiwan. Genome. 2001, 44: 797-807.
41. Lu SY, Hong KH, Liu SL, Cheng YP, Wu WL, Chiang TY: Genetic variation and population differentiation of *Michelia formosana* (Magnoliaceae) based on cpDNA variation and RAPD fingerprints: relevance to post-

Pleistocene recolonization. J Plant Res. 2002, 115: 203-216.

42. Chiang TY, Chiang YC, Chen YJ, Chou CH, Havanond S, Hong TN, Huang S: Phylogeography of *Kandelia candel* in East Asiatic mangroves based on nucleotide variation of chloroplast and mitochondrial DNAs. Mol Ecol. 2001, 10: 2697-2710.
43. Chiang YC, Schaal BA, Ge XJ, Chiang TY: Range expansion leading to departures from neutrality in the nonsymbiotic hemoglobin gene and the cp DNA *trn* L-*trn*F intergenic spacer in *Trema dielsiana* (Ulmaceae). Mol Phyl Evol. 2004, 31: 929-942.
44. Chiang TY, Hung KH, Hsu TW, Wu WL: Lineage sorting and phylogeography in *Lithocarpus formosanus* and *L. dodonaeifolius* (Fagaceae) from Taiwan. Ann Missouri Bot Gard. 2004, 91: 207-222.
45. Purvis A, Bromham L: Estimating the transition/transversion ratio from independent pairwise compaisons with an assumed phylogeny. J Mol Evol. 1997, 44 (1): 112-119.
46. Ina Y: Estimation of the transition/transversion ratio. J Mol Evol. 1998, 46 (5): 521-533.
47. Bakker FT, Culham A, Gomez-Martinez R, Carvalho J, Compton J, Dawtrey R, Gibby M: Patterns of nucleotide substitution in angiosperm cpDNA trnL (UAA)-trnF (GAA) regions. Mol Biol Evol. 2000, 17: 1146-1155.
48. Hillis DM, Allard MW, Miyamoto MM: Analysis of DNA sequence data: phylogenetic inference. Methods Enzymol. 1993, 224: 456-490.
49. Dumolin-Lapeue S, Pemonge MH, Petit RJ: An enlarged set of consensus primers for the study of organelle DNA in plants. Mol Ecol. 1997, 6: 393-397.
50. Demesure B, Comps B, Petit RJ: Chloroplast DNA phylogeography of the common beech (*Fagus sylvatica* L.) in Europe. Evolution. 1996, 50: 2515-2520.
51. Austerlitz F, Mariette S, Machon N, Gouyon PH, Godelle B: Effects of colonization processes on genetic diversity: differences between annual plants and tree species. Genetics. 2000, 154: 1309-1321.
52. Chiang YC, Hung KH, Schaal BA, Ge XJ, Hsu TW, Chiang TY: Contrasting phylogeographical patterns between mainland and island taxa of the *Pinus luchuensis* complex. Mol Ecol. 2006, 765-779.
53. Petit RJ, Duminil J, Funeschi S, Hampe A, Salvini DA, Vendramin GG: Comparative organization of chloroplast, mitochondrial and nuclear diversity in plant populations. Mol Ecol. 2005, 14: 689-701.
54. Wakeley J, Aliacar M: Gene genealogies in a metapopulation. Genetics. 2001, 159: 893-905.
55. Huang S, Chiang YC, Schaal BA, Chou CH, Chiang TY: Organelle DNA phylogeography of *Cycas taitungensis*, a relict species in Taiwna. Mol Ecol. 2001, 10: 2669-2681.
56. Whitlock MC, McCauley DE: Indirect measures of gene flow and migration: FST not equal to $1/(4Nm + 1)$. Heredity. 1999, 82: 117-125.
57. Lin CC: An outline of Taiwan's Quaternary geohistory with a special discussion of the relation between natural history and cultural history in Taiwan. Bull Dept Archaeol Anthropol. 1966, 23: 7-44.
58. Miege C, Ruffio-Chable V, Schierup MH, Cabrillac D, Dumas D, Gaude T, Cock JM: Intrahaplotype polymorphism at the*Brassica* S locus. Genetics. 2001, 159: 811-822.
59. Wu WL, Schaal BA, Hwang CY, Hwang MD, Chiang YC, Chiang TY: Characterization and adaptive evolution of α-tubulin genes in *Miscanthus sinensis* complex (Poaceae). Amer J Bot. 2003, 90: 1513-1521.
60. Jacob S, Blattner FR: A chloroplast genealogy of *Hordeum* (Poaceae): long-term persisting haplotypes, incomplete lineage sorting, regional extinction, and the consequences for phylogenetic inference. Mol Biol Evol. 2006, 23 (8): 1602-1612.
61. Chiang TY, Hung KH, Peng CI: Experimental hybridization reveals biased inheritance of the internal transcribed spacer of the nuclear ribosomal DNA in *Begonia* × *taipeiensis*. J Plant Res. 2001, 114: 343-351.
62. Donnelly MJ, Pinto J, Girod R, Besansky NJ, Lehmann T: Revisiting the role of introgression vs shared ancestral polymorphisms as key processes shaping genetic diversity in the recently separated sibling species of the *Anopheles gambiae* complex. Heredity. 2004, 92: 61-68.
63. Holder MT, Anderson JA, Holloway AK: Difficulties in detecting hybridization. Syst Biol. 2001, 50: 978-982.
64. Osada N, Wu CI: Inferring the mode of speciation from genomic data: a study of the great apes. Genetics. 2005, 169: 259-264.
65. Hudson RR: Island models and the coalescent process. Mol Ecol. 1998, 7: 413-418.
66. Templeton AR: Gene tree: A powerful tool for exploring the evolutionary biology species and speciation. Plant Species Biol. 2000, 15: 211-222.
67. Graham J, Thompson EA: A coalescent model of ancestry for a rare allele. Genetics. 2000, 156: 375-384.
68. Klopfstein S, Currat M, Excoffier L: The fate of mutations surfing on the wave of a range expansion. Mol Biol Evol. 2006, 23: 482-490.
69. Edmonds CA, Lillie AS, Cavalli-Sforza LL: Mutations arising in the wave front of an expanding population. Proc Natl Acad Sci. 2004, 101: 975-979.
70. Rosenberg NA, Hirsh AE: On the use of star-shaped genealogies in inference of coalescence times. Genetics. 2003, 164: 1677-1682.
71. Joy DA, Feng X, Mu J, Furuya T, Chotivanich K, Krettli AU, Ho M, Wang A, White NJ, Suh E, Beerli P, Su XZ: Early origin and recent expansion of *Plasmodium falciparum*. Science. 2003, 300: 318-321.
72. Aris-Brosou S, Excoffier L: The impact of population expansion and mutation rate heterogeneity on DNA sequence polymorphism. Mol Biol Evol. 1996, 13: 494-504.
73. Slatkin M, Hudson RR: Pairwise comparisons of mitochondrial DNA sequences in stable and exponentially growing populations. Genetics. 1991, 129: 555-562.
74. Rogers AR, Harpending H: Population growth makes

waves in the distribution of pairwise genetic differences. Mol Biol Evol. 1992, 9: 552-569.

75. Excoffier L, Schneider S: Why hunter-gatherer populations do not show signs of Pleistocence demographic expansions. Proc Natl Acad Sci. 1999, 96: 10597-10602.
76. Ray N, Currat M, Excoffier L: Intra-deme molecular diversity in spatially expanding populations. Mol Biol Evol. 2003, 20: 76-86.
77. Treutlein J, Wink M: Molecular phylogeny of cycads inferred from *rbc* L sequences. Naturwissenschaften. 2002, 89: 221-225.
78. Murray MG, Thomposon WF: Rapid isolation of high molecular weight DNA. Nucleic Acids Res. 1980, 8: 4321-4325.
79. Chiang TY, Schaal BA, Peng CI: Universal primers for amplification and sequencing a noncoding spacer between atpB and rbcL genes of chloroplast DNA. Bot Bull Acad Sin. 1998, 39: 245-250.
80. Brennicke A, Möller S, Blanz PA: The 18S and 5S ribosomal RNA genes in Oenothera mitochondria: Sequence rearrangments in the 18S and 5S rRNA genes of higher plants. Mol Gen Gent. 1985, 198: 404-410.
81. Belshaw R, Katzourakis A: BlastAlign: a program that uses blast to align problematic nucleotide sequences. Bioinformatics. 2005, 21 (1): 122-123.
82. Guindon S, Gascuel O: A simple, fast, and accurate algorithm to estimate large phylogenies by maximum likelihood. Syst Biol. 2003, 52: 696-704.
83. Posada D, Crandall KA: MODELTEST: testing the model of DNA substitution. Bioinformatics. 1998, 14 (9): 817-818.
84. Swofford DL: PAUP*: Phylogenetic Analysis Using Parsimony (and Other Methods), Version 4.0b10. 2002, Sinauer Associates, Inc, Sunderland, Massachusetts
85. Felsenstein J: Confidence limits on phylogenies: an approach using the bootstrap. Evolution. 1985, 39: 783-791.
86. Clement M, Posada D, Crandall K: TCS, a computer program to estimate gene genealogies. Mol Ecol. 2000, 9: 1657-1660.
87. Templeton AR, Crandall KA, Sing CF: A cladistic analysis of phenotypic associations with haplotypes inferred from restriction endonuclease mapping and DNA sequence data. III. Cladogram estimation. Genetics. 1992, 132: 619-633.
88. Donnelly P, Tavare S: The ages of alleles and a coalescent. Adv Appl Probab. 1986, 18: 1-19.
89. Crandall KA, Templeton AR: Empirical tests of some predictions from coalescent theory with applications to intraspecific phylogeny reconstruction. Genetics. 1993, 134: 959-969.
90. Sarich VM, Wilson AC: Generation time and genomic evolution in primates. Science. 1973, 179: 1144-1147.
91. Wu CI, Li WH: Evidence for higher rates of nucleotide substitution in rodents than in man. Proc Natl Acad Sci USA. 1985, 82: 1741-1745.
92. Watterson GA: On the number of segregating sites in genetical models without recombination. Theor Popul Biol. 1975, 7: 256-275.
93. Tajima F: Statistical method for testing the neutral mutation hypothesis by DNA polymorphism. Genetics. 1989, 123: 585-595.
94. Rozas J, Sanchez-DelBarrio JC, Messeguer X, Rozas R: DnaSP, DNA polymorphism analyses by the coalescent and other methods. Bioinformatics. 2003, 19: 2496-24.
95. Fu YX, Li W: Statistical tests of neutrality of mutations. Genetics. 1993, 133: 693-709.
96. Fu YX: Statistical tests of neutrality of mutations against population growth, hitchhiking and background selection. Genetics. 1997, 147: 915-925.
97. Ramos-Onsins SE, Rozas J: Statistical properties of new neutrality tests against population growth. Mol Biol Evol. 2002, 19: 2092-2100.
98. Schneider S, Excoffier L: Estimation of past demographic parameters from the distribution of pairwise differences when mutation rates vary among sites: application to human mitochondrial DNA. Genetics. 1999, 152: 1079-1089.
99. Excoffier L, Laval G, Schneider S: Arlequin ver. 3.0: An integrated software package for population genetics data analysis. Evol Bioinform Online. 2005, 1: 47-50.
100. Graur D, Li WH: Fundamentals of molecular evolution. 2000, Sunderland: Sinauer Associates
101. Drummond AJ, Ho SY, Phillips MJ, Rambaut A: Relaxed phylogenetics and dating with confidence. PLOS Biology. 2006, 4: e88-
102. Rambaut A, Drummond A: Tracer – MCMC trace analysis tool. 2004, Oxford, UK: University of Oxford
103. Hey J, Nielsen R: Multilocus methods for estimating population sizes, migration rates, and divergence time, with applications to the divergence of *Drosophila pseudoobscura* and *D. persimilis*. Genetics. 2004, 167: 747-760.

CHAPTER 16

# MIDDLE-UPPER PLEISTOCENE CLIMATE CHANGES SHAPED THE DIVERGENCE AND DEMOGRAPHY OF CYCAS GUIZHOUENSIS (CYCADACEAE): EVIDENCE FROM DNA SEQUENCES AND MICROSATELLITE MARKERS

Xiuyan Feng, Ying Zheng & XunGong

## INTRODUCTION

Different regions have experienced different scales and frequencies of glaciations in the Pleistocene[1]. Although China was not glaciated during the Pleistocene[2] and had a relatively mild Quaternary climate[3], climatic oscillations in the Pleistocene have also had profound effects on the demography and genetic diversity of many extant species[4,5,6,7]. To respond to climatic oscillations, some species contracted into glacial refugia, became extinct and possibly locally adapted[6,8]. Nevertheless, historical processes leave imprints on the genetic structure of existing populations, especially those of long-lived and sessile organisms[9]. As an ancient and perennial gymnosperm, cycads play an important role for tracking the evolutionary history of seed plants. An increasing number of studies are examining the evolutionary history of cycads and how they have responded to historic climate change in recent years[9,10,11].

Cycads are generally considered to be the oldest group of living seed plants. Cycads reached their greatest levels of diversity during the Jurassic-Cretaceous (approximately 199.6 to 65.5 million years ago, MYA). Their diversity dwindled to the present number of species (approximately 300 species) as a result of the emergence and dominance of flowering plants[12,13,14,15,16]. Although the cycad lineage is ancient, research based on a fossil-calibrated molecular phylogeny showed that cycads underwent a recent synchronous global rediversification in the late Miocene, indicating that extant cycad species are not much older than 12 million years[14]. Currently, cycads contain two families (Cycadaceae, Zamiaceae) with ten genera, primarily distributed in Africa, Asia, Australia and South and Central America[17]. Unfortunately, cycad species are threatened with extinction because of over exploitation and habitat loss. In China, there is only one cycad genus, *Cycas*, belonging to Cycadaceae[18]. All of the *Cycas* species are given 'First Grade' conservation status in China[19].

Recent research which has combined ancestral area reconstructions with fossil evidence proposed that the origin of *Cycas* was in South China, with the deepest divergence occurring via a vicariant barrier: the Red River Fault[20]. In South China, *Cycas* species generally grow on low-altitude slopes of ridges and cliffs along river valleys. Their fertile seeds are large, heavy and sink in water, which precludes water dispersal over long distances. In addition, cycad seeds contain virulent cycasin, preventing their dispersal by animals for long distances. Thus, in this region, seed dispersal was likely limited to short distances[21,22,23], resulting in the expected high levels

**Citation:** Feng, X. et al. Middle-Upper Pleistocene climate changes shaped the divergence and demography of Cycas guizhouensis (Cycadaceae): Evidence from DNA sequences and microsatellite markers. Sci. Rep. 6, 27368; doi: 10.1038/srep27368 (2016).

of genetic differentiation as observed in maternally inherited DNA. *Cycas* species are dioecious and allogamous. In plant species, both the mating system and dispersal mechanism have important effects on genetic differentiation among populations24·25. For neutral markers, genetic differentiation is typically influenced by genetic drift, mutation of novel alleles and gene flow between populations26. If levels of gene flow between populations are sufficient, there may be no relationship between geographic distance and genetic differentiation.

*Cycas guizhouensis* Lan & R.F. Zou was described in 1983 and can be identified by its morphological characteristics of a cylindrical trunk up to one meter in height, fusiform male cone, loose and open female cone, densely hairy sporophylls, nearly round apical lobe (the margins with numerous tapered lobes) and yellow with reddish brown mucro of seeds (nearly globose)13. *Cycas guizhouensis* is endemic to Southwest China, occurring primarily in the valleys of the Nanpan River basin of southwestern Guizhou, northwestern Guangxi and the eastern Yunnan provinces. The species mainly prefers sand or limestone habitats where the climate is characterized by mild with cool, moist summers and moist, frost-free winters13. In recent decades, *C. guizhouensis* has been severely threatened by rapid habitat destruction for the cultivation of economic plants and over-collection because of its edible stem and ornamental attributes. This species has dramatically decreased in numbers and is becoming endangered. Field surveys have shown that there are two populations with fewer than 20 individuals. Thus, the need to develop effective protection measures for this species is urgent.

Understanding the genetic variation, genetic structure and demographic history in a rare species is essential for the establishment of effective and efficient conservation practices. Protection measures can be proposed by relying on detailed population genetic studies. Several population genetic studies on cycads based on different molecular markers have been recently published9·10·11·19·27·28. These publications provide references for the study of other cycad species. Previous studies on the genetic diversity of *C. guizhouensis* were by Yang and Meerow who used allozymes to investigate three populations of *C. guizhouensis*, and by Xiao who studied 12 populations of *C. guizhouensis* using ISSR markers19·29. These two studies were not consistent with each other in the levels of genetic differentiation reported between populations. Meanwhile, both of these markers in previous studies have their shortcomings: the number of allozyme loci that can be probed is limited and allozyme variation may not be able to accurately or completely measure nucleotide variation in the genome30. ISSR markers are dominant and are not able to distinguish homozygosity and heterozygosity directly; in addition, errors in counting bands may also arise. However, the nuclear DNA (nDNA) of cycads is biparentally inherited and is dispersed by both seeds and pollen. The organelle DNA (cpDNA) is maternally inherited and is dispersed only through seeds31. Microsatellite markers (SSRs) are known to be codominant and possess more genetic variation than other molecular markers. To comprehensively investigate genetic variation in*C. guizhouensis* and compare these results with previous studies on this species, we used four cpDNAs (*psb*A-*trn*H32, *psb*M-*trn*D33, *trn*L-*trn*T34and *trn*L-*trn*F34), three nDNAs (*GTP*, GTP genes35; *PHYP*, phytochrome P gene and *F3H*, flavanone 3-hydroxyrase gene (unpublished)) and 13 microsatellite markers36·37·38·39·40·41·42 in this study, aiming to estimate the genetic variation, genetic structure and demographic history of *C. guizhouensis* in an further effort to investigate the causes of its endangerment and provide basic guidelines for its conservation.

## RESULTS

### DNA Sequence Diversity and Haplotype Fixation and Distribution

The four cpDNA regions surveyed across the 110 individuals (11 populations, Table S1) of *C. guizhouensis* identified 12 haplotypes (guiH1-guiH12) in total. Of these, eight haplotypes were specific to one population, whereas haplotypes guiH1, guiH2, guiH4 and guiH6 were shared by five, four, two and two populations, respectively. Detection of recombination in nuclear genes showed that no recombination occurred in *GTP* and *PHYP*. In contrast, eight recombination events were occurred in *F3H*. Most software used in our analysis didn't need the data to remove recombination. Thus, we did not remove recombination from gene *F3H* because some individuals with recombination were also removed. A total of seven haplotypes (guiG1-guiG7) were derived from the nuclear gene *GTP*, of which haplotype guiG1 was the most abundant and predominant in all 11 populations. The remaining six haplotypes were private. Ten haplotypes (guiP1-guiP10) were derived from the nuclear gene *PHYP*. Of these, haplotypes guiP1 and guiP2 were most frequent and were widely shared by all 11 and ten populations, respectively. Haplotype guiP4 was distributed in three populations and the remaining haplotypes were specific to single populations. The nuclear gene *F3H* identified 32 haplotypes (guiF1-guiF32) in total. guiF1 and guiF4 were the most widely distributed haplotypes. Information on cpDNA and three nDNA haplotypes and their distribution in populations are shown in Table S2 and Fig. 1, respectively.

Genetic diversity indices *H*d and *P*i for each population were summarized in Table S2 and were highly variable among cpDNA and nDNA. In sum, population SP had the highest genetic diversity. Total genetic diversity $H_T$ was higher than the average intrapopulation diversity *H*s (Table S3), which meant that there were high levels of genetic differentiation ($N_{ST}$, ranging from 0.192 to 0.698; $G_{ST}$, ranging from 0.233 to 0.676). $N_{ST}$ was not significantly greater than $G_{ST}$ ($P>0.05$), suggesting that there was no correspondence between haplotype similarities and their geographic distribution in *C. guizhouensis*.

The AMOVA revealed that there was more variation (69.82%) partitioned among populations rather than within (30.18%) populations based on cpDNA data, whereas there was less variation (24.11%, 30.75% and 19.20%) partitioned among populations than variation (75.89%, 69.25% and 80.80%) within populations based on nuclear genes *GTP*, *PHYP*, and*F3H*, respectively. $F_{ST}$ for these genes thus ranged from 0.192–0.69843, with all indicating highly significant genetic differentiation among *C. guizhouensis* populations (Table 1).

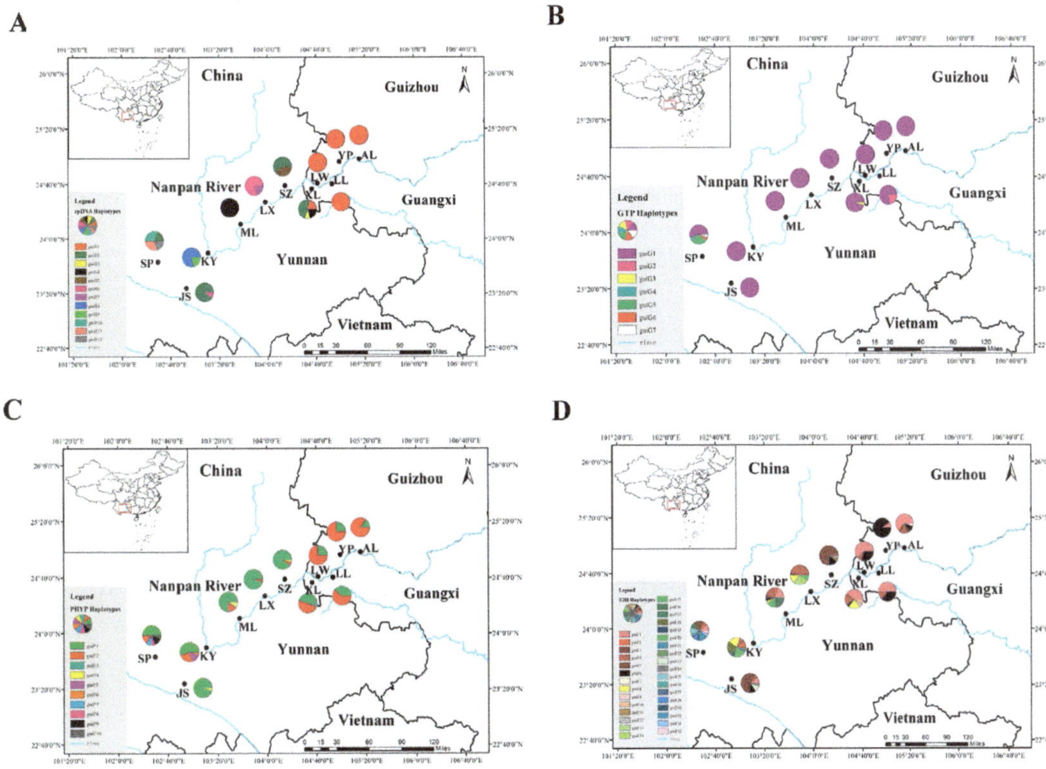


**Figure 1.** Geographical distribution of 11 populations of *C. guizhouensis* and distribution of its haplotypes detected from cpDNA (**A**), *GTP* (**B**), *PHYP* (**C**) and *F3H* (**D**). Population codes refer to Table S1. Maps were drawn using the software ArcGIS version 10.2 (http://desktop.arcgis.com) and modified using Photoshop (Adobe Corporation, California, America).

**Table 1:** Analysis of molecular variance (AMOVA) based on DNA sequences and microsatellites for populations of *C. guizhouensis*, note: $^{***}P<0.001$.

| Marker | Source of variation | d.f. | Sum of squares | Variance components | Percentage of variation (%) | $F_{ST}$ |
|---|---|---|---|---|---|---|
| cpDNA | Among populations | 10 | 105.545 | 1.01172 | 69.82 | 0.698*** |
| | Within populations | 99 | 43.300 | 0.43737 | 30.18 | |
| *GTP* | Among populations | 10 | 6.809 | 0.02942 | 24.11 | 0.241*** |
| | Within populations | 209 | 19.350 | 0.09258 | 75.89 | |
| *PHYP* | Among populations | 10 | 21.818 | 0.09805 | 30.75 | 0.308*** |
| | Within populations | 209 | 46.150 | 0.22081 | 69.25 | |
| *F3H* | Among populations | 10 | 88.264 | 0.36458 | 19.20 | 0.192*** |
| | Within populations | 209 | 320.750 | 1.53469 | 80.80 | |
| SSR | Among populations | 10 | 197.779 | 0.44772 | 13.80 | 0.138*** |
| | Within populations | 407 | 1138.037 | 2.79616 | 86.20 | |

## Phylogeny and Network of Haplotypes

Phylogenetic relationships among cpDNA and nDNA haplotypes were constructed using Bayesian inference (Fig. S1). The four phylogenies were strongly supported with Bayesian probabilities, revealing a haplotype monophyly for *C. guizhouensis* with *Cycas edentata* and *Cycas rumphii* as outgroups. The 12 cpDNA haplotypes were grouped into two clades, showing that haplotype guiH7 was the firstly diverged haplotype, and the others were grouped into a clade. Among this clade, haplotypes guiH2, guiH3, guiH4, guiH5, guiH8, guiH9 and guiH10 were clustered into a subclade, indicating that they were more closely related than others, although the phylogenetic relationship within the subclade was not resolved (Fig. S1A). The *GTP* haplotype phylogenetic tree showed that haplotypes guiG1, guiG2 and guiG4 were grouped together, and haplotypes guiG6 and guiG7 were clustered together, indicating that they shared closer relationships, respectively (Fig. S1B). For the*PHYP* haplotypes in the Bayesian phylogram, guiP2, guiP3, guiP7 were firstly and parallelly diverged haplotypes, while the relationship of the remaining seven haplotypes which formed a subclade was not resolved, either (Fig. S1C). Recombination gave rise to many haplotypes in the nuclear gene *F3H*. For haplotypes guiF13, guiF14 and guiF29 that were the earliest diverged in the *F3H* phylogram, their further relationships had not been figured out for lacking sufficient informative sites. Within the subsequent diverged but weakly supported clade (PP=57) where 26 haplotypes were clustered in, although some genetic relationships of some *F3H* haplotypes were basically resolved such as haplotypes guiF1, guiF5 and guiF10, parallel relationships were also observed in most haplotypes (Fig. S1D).

Genealogies reflecting topology and haplotype frequency were constructed for cpDNA and three nuclear genes (Fig. 2). The haplotype network of cpDNA was concordant with the haplotype network of nuclear genes *GTP* and *PHYP*, with each showing 1–2 centrally located nodes representing hypothetical ancestral haplotypes (Fig. 2A–C). The remaining haplotypes were linked to these central haplotypes by 1–3 steps in a star-like network. A reticulate evolutionary relationship was observed in the *F3H* haplotype network (Fig. 2D).

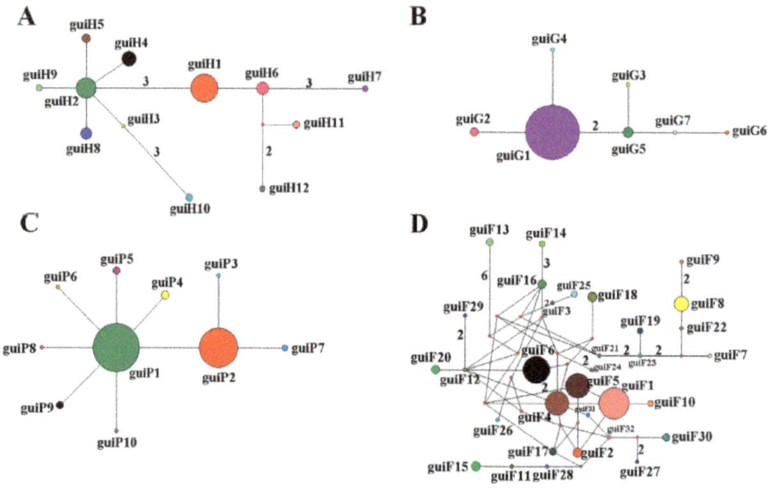


**Figure 2:** Network of haplotypes of *C. guizhouensis* based on cpDNA (**A**), *GTP* (**B**), *PHYP* (**C**) and *F3H* (**D**). The numbers on branches indicate mutational steps. Haplotype distribution in 11 populations refers to Table S2.

## Phylogenetic Divergence Time Estimation from Haplotypes

The BEAST-derived trees for the four DNA sequences revealed similar topologies (Fig. 3). The 12 cpDNA haplotypes clustered into two lineages, one including five haplotypes with guiH1 as the ancestral haplotype and the other including seven haplotypes with guiH2 as the ancestral haplotype. The two lineages were split at approximately 0.460 MYA and haplotypes in the inner lineages were diverged from 0.155 MYA to 0.050 MYA and 0.171 MYA to 0.048 MYA, respectively (Fig. 3A). For the three nDNA haplotype trees, haplotypes were also grouped into two lineages, separately coalescing at 0.453 MYA (*GTP*), 0.181 MYA (*PHYP*) and 1.334 MYA (*F3H*) (Fig. 3B–D). Coalescent events occurred much earlier in *F3H* than in *GTP, PHYP* and cpDNA. In total, most tip haplotypes in the four BEAST-derived trees diverged recently from their common ancestors. These results entirely implied that haplotypes of *C. guizhouensis* diverged in the Middle-Upper Pleistocene.

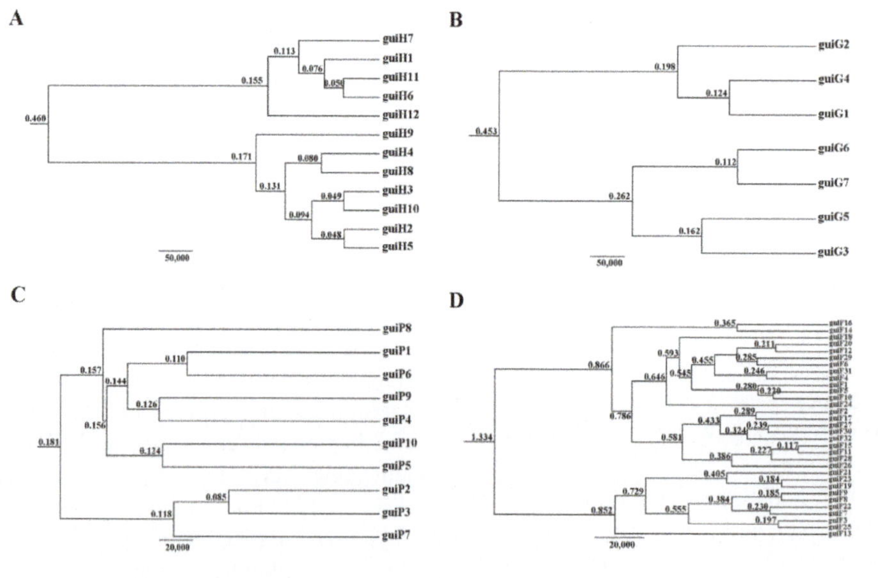


**Figure 3.** BEAST-derived trees based on cpDNA (**A**) and the nuclear genes *GTP* (**B**), *PHYP* (**C**) and *F3H*(**D**). The numbers on branches repre-

sent divergence time (MYA). Haplotype distribution in 11 populations refers to Table S2.

## Neutrality test, Mismatch Analysis and the Bayesian Skyline Plot

Values for Tajima's *D*, Fu and Li's $D^*$ and $F^*$, Fu's *F*s tests as well as SSD and raggedness statistics were listed in Table S4. For cpDNA, all of the values were positive and non-significant, indicating that *C. guizhouensis*had not undergone a recent population expansion. For the nuclear genes, all values were non-significant except for significant negative values of Fu's *F*s for all three genes, which suggested the presence of population growth or recombination44. Mismatch distribution analyses displayed bimodal and multimodal graphs for cpDNA and *F3H*, respectively (Fig. S2A, D), but unimodal graphs for *GTP* and *PHYP* (Fig. S2B, C). A multimodal curve is indicative of a population at demographic equilibrium, whereas a unimodal curve is indicative of recent population expansion. More detailed evidence of demographic came from the Bayesian Skyline Plots. Based on the pattern of variation in cpDNA, *C. guizhouensis* had a long history of a constant population size, followed by a slight population expansion (over 25-4 Ka years) and a subsequent slight decline (approximately 2 Ka-present) (Fig. 4A). During the recent 10 Ka years, rapid population growth was detected in *C. guizhouensis* based on the gene *GTP* (Fig. 4B), while slow population growth with subsequent slight declines in demography was detected in the Bayesian Skyline Plot of the gene *PHYP* (Fig. 4C). In contrast, coalescence of the gene *F3H* dated back to much longer ago, showing that *C. guizhouensis* experienced a long period of continuous growth over 600–80 Ka years before undergoing a decline from approximately 50 Ka years ago to the present (Fig. 4D).

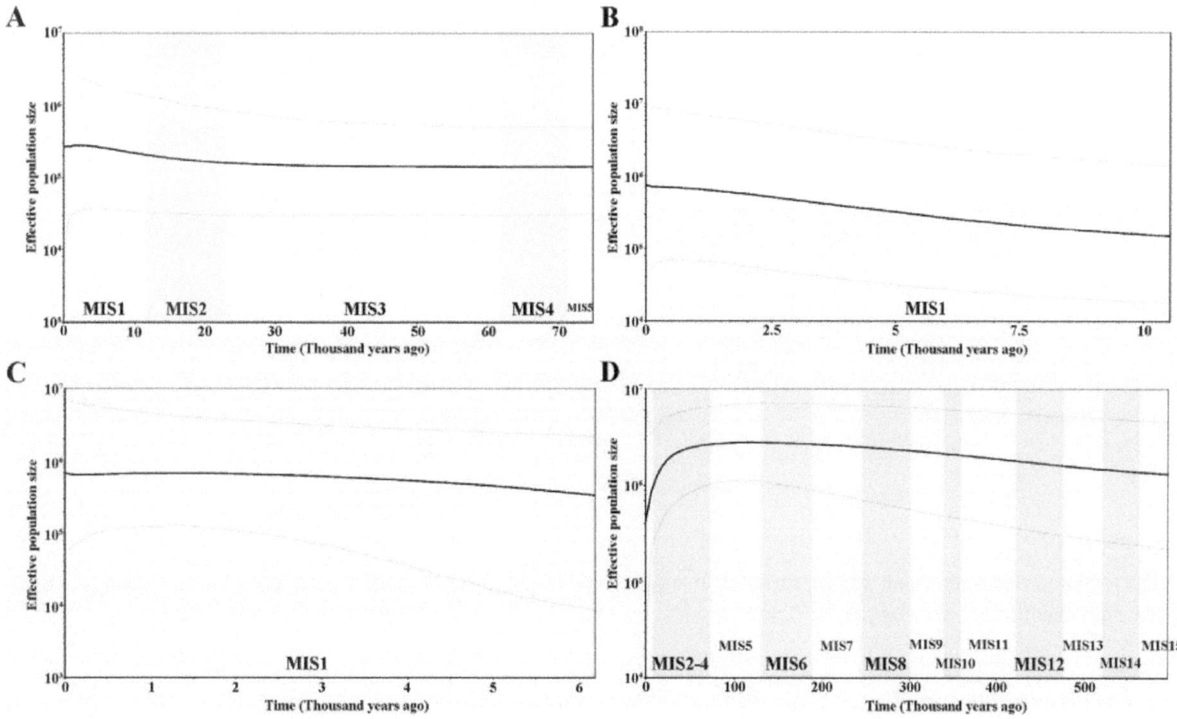


**Figure 4:** Bayesian skyline plots based on cpDNA (**A**) and the nuclear genes *GTP* (**B**), *PHYP* (**C**) and *F3H*(**D**) for the estimate of fluctuations in effective population size over time. Black line: median estimation; area between gray lines: 95% confidence interval. MIS: Marine Isotope Stage.

## Nuclear Microsatellite Genotyping

Diversity estimates from 13 microsatellites varied among populations (Table 2), with the lowest and highest measures of diversity consistently found in populations YP and SP, respectively. Fixation indices (*F*) were positive for all populations, with a mean value $F=0.204$, indicating that there was a high level of inbreeding within each population. This inbreeding resulted in a deficiency of heterozygotes and significant deviations from Hardy-Weinberg equilibrium at most loci and in most populations (Table S5). Estimates of effective population sizes (Ne) with the lowest allele frequency (0.05) were listed in Table 2. There were only two populations, LL and SP, whose Nes were greater than 50. The AMOVA revealed that 13.80% of the genetic variation was partitioned among populations and 86.20% was partitioned within populations with the significant genetic differentiation coefficient $F_{ST}=0.138$ (Table 1). The correlation between genetic and geographic distances was not significant (P=0.18), indicating that there was no significant effect of isolation by distance (IBD) (Fig. S3). Estimates of gene flow (*N*m) between pairs of populations were shown in Table S6. Most of these estimates was greater than 1.

**Table 2:** Genetic diversity within populations of *C. guizhouensis*, note: $N_T$, total number of alleles; $N_P$, private alleles; $A_R$, allelic richness;$N_A$, number of alleles; $A_E$, the effective number of alleles; *I*, Shannon's information index; $H_O$, observed heterozygosity; $H_E$, expected heterozygosity; UHE, Nei's unbiased heterozygosity; *F*, fixation index; *PPB*, the percentage of polymorphic loci; Ne, effective population size.

| Population | $N_T$ | $N_P$ | $A_R$ | $N_A$ | $A_E$ | *I* | $H_O$ | $H_E$ | UHE | *F* | *PPB* (%) | Ne |
|---|---|---|---|---|---|---|---|---|---|---|---|---|
| AL | 43 | 1 | 2.963 | 3.308 | 2.248 | 0.717 | 0.285 | 0.372 | 0.382 | 0.159 | 84.62 | 8 |
| YP | 33 | 0 | 2.255 | 2.538 | 1.680 | 0.543 | 0.250 | 0.328 | 0.336 | 0.163 | 84.62 | 13.8 |
| LW | 51 | 6 | 3.305 | 3.923 | 2.390 | 0.786 | 0.300 | 0.402 | 0.412 | 0.204 | 92.31 | 9.8 |
| LL | 57 | 6 | 3.530 | 4.385 | 2.753 | 0.770 | 0.281 | 0.362 | 0.371 | 0.246 | 76.92 | 981 |
| XL | 44 | 5 | 3.299 | 3.385 | 2.054 | 0.739 | 0.273 | 0.394 | 0.413 | 0.249 | 84.62 | 12.2 |
| SZ | 54 | 3 | 3.662 | 4.154 | 2.560 | 0.720 | 0.204 | 0.331 | 0.339 | 0.203 | 84.62 | 48.8 |
| LX | 67 | 3 | 4.292 | 5.154 | 3.118 | 1.003 | 0.331 | 0.463 | 0.475 | 0.243 | 92.31 | 21.7 |
| ML | 65 | 5 | 3.985 | 5.000 | 2.758 | 0.984 | 0.427 | 0.483 | 0.495 | 0.102 | 92.31 | 4.9 |
| KY | 65 | 6 | 4.001 | 5.000 | 2.674 | 0.947 | 0.323 | 0.453 | 0.464 | 0.212 | 92.31 | 34.2 |
| JS | 71 | 7 | 4.420 | 5.462 | 2.872 | 1.023 | 0.333 | 0.474 | 0.488 | 0.251 | 92.31 | 33.5 |
| SP | 76 | 7 | 4.799 | 5.846 | 3.757 | 1.175 | 0.415 | 0.546 | 0.560 | 0.214 | 92.31 | 51.9 |
| Mean | 57 | 4.46 | 3.656 | 4.378 | 2.624 | 0.855 | 0.311 | 0.419 | 0.431 | 0.204 | 88.11 | 110.9 |

The clustering analyses resulted in population groupings that were generally consistent with the above haplotype aggregation analyses. The UPGMA clustering dendrogram, Structure analysis, and principal coordinate analysis (PCoA), all indicated the same general grouping: individuals belonging to populations AL, YP, LW and LL clustered into one group (Clade I), while the remaining individuals clustered into a second group (Clade II) (Fig. S4; Fig. 5).

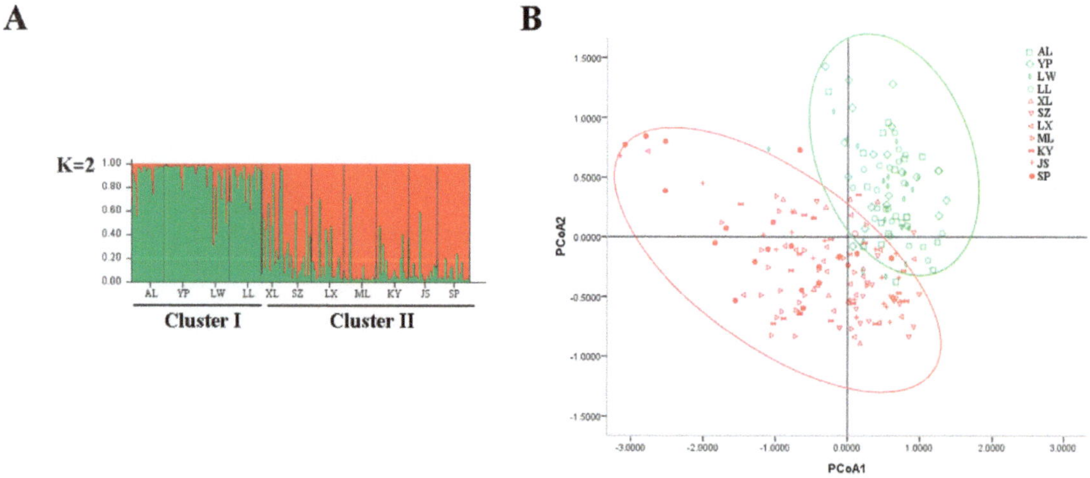


**Figure 5:** (**A**) Bayesian inference using STRUCTURE (K=2) and (**B**) Principal coordinate analysis (PCoA) of SSR phenotype from 11 populations of 209 individuals of *C. guizhouensis*.

The bottleneck analysis showed that no populations had a significant excess of heterozygosity in the two methods under two models, indicating that none of them deviated from mutation-drift equilibrium (Table 3). Mode shift models showed that all of the populations had a normal L-shaped distribution, which suggested that *C. guizhouensis* had not experienced a recent severe bottleneck. However, *C. guizhouensis*apparently experienced a population decline (bottleneck) in history as inferred from the calculation of Garza-Williamson indices which were lower than the critical *M*c value of 0.68 (Table 3).

**Table 3:** Bottleneck analysis using microsatellite loci from 11 populations of *C. guizhouensis*, note: *P* **is test for** heterozygosity excess, $^{*}P<0.05$, significantly different; $^{**}P<0.01$, highly significantly different.

| Population | T.P.M | | S.M.M | | Mode shift | Garza-Williamson |
|---|---|---|---|---|---|---|
| | Sign test | Wilcoxon test | Sign test | Wilcoxon test | | index |
| AL | $0.007^{**}$ | $0.021^{*}$ | 0.122 | 0.080 | L | 0.433 |
| YP | 0.124 | $0.021^{*}$ | 0.297 | 0.376 | L | 0.359 |
| LW | $0.048^{*}$ | 0.058 | 0.460 | 0.787 | L | 0.414 |
| LL | 0.260 | 0.216 | 0.529 | 0.946 | L | 0.478 |
| XL | 0.141 | 0.146 | 0.473 | 0.787 | L | 0.455 |
| SZ | $0.038^{*}$ | $0.007^{**}$ | 0.298 | 0.191 | L | 0.424 |
| LX | 0.318 | 0.736 | 0.115 | 0.244 | L | 0.374 |
| ML | 0.308 | 0.168 | 0.524 | 0.946 | L | 0.426 |
| KY | 0.322 | 0.273 | 0.276 | 0.588 | L | 0.417 |
| JS | 0.556 | 0.636 | 0.119 | 0.340 | L | 0.436 |
| SP | 0.055 | $0.001^{**}$ | 0.302 | 0.273 | L | 0.363 |

## DISCUSSION

Previous study on genetic patterns of *C. guizhouensis* using ISSR markers showed that this species has low genetic diversity and a high degree of differentiation among populations with the highest genetic diversity in population JS and the lowest in SZ19. In contrast, via DNA sequences (Tables 1, S2 and S3) and microsatellites (Tables 1 and 2) data analysis, our study revealed that *C. guizhouensis* had high genetic diversity and high genetic differentiation, respectively. However, population SP had the highest genetic diversity and YP had the lowest (Tables S2 and 2). Our results were not completely consistent with the previous study19. The types of molecular markers, the number of populations and sample sizes per population, life history traits and distance between populations can all affect genetic variation estimates within or among species45. Most *Cycas* species in Asia are narrowly distributed and their life history traits, such as dioecy, longer life span and pollination syndromes, are similar46. Thus, we can compare the genetic diversity of *C. guizhouensis* with other Asian *Cycas* species using the similar molecular markers. *Cycas guizhouensis* had higher genetic diversity (Tables 2, S2 and S3) than *C. debaoensis*27, but had slightly lower genetic diversity than *C. simplicipnna*9 and *C. multipinnata*10. cpDNA-based studies on 170 plant species revealed a mean value of $H_T$=0.6747, while $H_T$ estimated from *C. guizhouensis* by four cpDNA was 0.841 which implied that *C. guizhouensis* had a high level of genetic diversity.

All 11 *C. guizhouensis* populations deviated significantly from Hardy-Weinberg equilibrium (Table S5), and the fixation indices (*F*) were also all greater than zero (Table 2). Therefore, we can conclude that *C. guizhouensis* populations are notably deficient in heterozygosity and have experienced severe levels of inbreeding, which have contributed to a high level of genetic differentiation. In general, mating systems and dispersal mechanisms have significant effects on levels of genetic differentiation observed among populations of plant species. *Cycas guizhouensis* is dioecious with both wind and insects as primary pollen dispersal vectors23. This species has evolved large, heavy seeds that sink, limiting their dispersal by water21. These seeds also contain the neurotoxin cycasin, which limits the ability of animals to disperse them over long distances23. Most seeds germinate and grow near the mother plant, increasing the chances of inbreeding within the species. Consistent with restricted seed dispersal21·23, cpDNA data showed a high level of genetic differentiation and restricted seed-mediated gene flow among populations, despite their relative proximity. In contrast, our microsatellite data indicate much higher rates of interpopulation gene flow at nuclear loci with *N*m estimates mostly greater than 1 (Table S6). If $Nm > 1$, gene flow can largely prevent population differentiation due to genetic drift, while if $Nm < 1$, high levels of differentiation will develop48.

The AMOVA results assessed from cpDNA data revealed that more genetic variation existed among populations and less within populations, while the opposite result was obtained from nuclear genes and microsatellites (Table 1). This inconsistency can be explained by the different inheritance patterns and evolutionary rates of organelle DNA and nuclear DNA. In cycads, chloroplast DNA (cpDNA) is haploid, maternally inherited, and slowly evolving, while nuclear DNA (nDNA) is diploid, biparentally inherited, and more rapidly evolving31. These factors combined with typically greater rates of pollen- than seed-mediated gene flow25·47 contribute to the maintenance of higher within- and lower between-population nuclear than chloroplast diversity.

The observation of isolation by distance (IBD) was generally consistent with the previous study19, showing that there was no significant correlation between nuclear genetic and geographic distance in *C. guizhouensis* (Fig. S3). In addition, a U test showed that $N_{ST}$ was not significantly greater than $G_{ST}$, suggesting that there was no distinct phylogeographic structure of cpDNA in *C. guizhouensis*. However, obvious genetic structure was detected in *C. guizhouensis*. The UPGMA clustering of microsatellite data showed the 11 populations to comprise two clades (I and II) (Fig. S4). Populations in Guizhou and Guangxi province (except for population XL) were grouped into Clade I, while six populations in Yunnan province and population XL from Guangxi province were grouped into Clade II. The existence of the two clades was also supported by the Structure analysis (Fig. 5A), PCoA (Fig. 5B) and haplotype distributions (Fig. 1). The population XL is situated at the border of the two clades geographically and contains genetic components from both, suggesting this location as an area of admixture (gene flow) between the two clades.

Phylogenetic relationships of haplotypes of cpDNA and nDNA were reconstructed using Bayesian and Median Joining methods. Although monophyly of *C. guizhouensis* haplotypes from the four phylogenies was strongly supported by Bayesian inference, the phylogenetic relationships of haplotypes were not well resolved (Fig. S1). Nevertheless, network analysis provided a better resolution of the phylogenetic relationships of haplotypes (Fig. 2). Parsimony network analysis (except for the nuclear gene *F3H*) all suggested that there are two star-like evolutionary units separately dominating two lineages. The assumed two ancestral haplotypes were distributed in the inner nodes of the reticulate evolutionary diagrams at a high frequency. The derived haplotypes were distributed around the ancestral haplotypes with 1–3 mutational steps, showing a star-like topology, which indicated a recent rapid divergence. The topology of the haplotypes from the nuclear gene *F3H* was highly reticulate because eight recombination events had occurred (Fig. 2D).

Meanwhile, the BEAST-derived trees were consistent with the network of four DNA sequences (Fig. 3). Haplotypes were separated into two groups with an ancestral haplotype at the inner node of each group. All divergence times within *C. guizhouensis* estimated by BEAST occurred during the Middle-Upper Pleistocene, indicating that this species was the product of a recent rapid divergence, which is consistent with the former conclusion that the extant cycad species are not much older than 12 million years14. Previous studies have proposed that climatic oscillations in the Pleistocene had significant effects on the genetic diversity of many extant species4·5·6·7. However, different scales and frequencies of glaciations occurred in different regions in the Pleistocene1. In Europe, the Pleistocene ice sheet was evident during major glaciations49. In contrast, a relatively mild Quaternary climate occurred in China3. In general, the relatively high altitudes in the southwest mountains of China were deeply influenced by glaciations50·51, while the lower slopes or valleys were not affected during the cooler periods52. Therefore, we suggest that the relatively mild Pleistocene climate in China contributed to the survival and divergence of *C. guizhouensis* during that period.

Additionally, exploring the historical demography of a species can help in understanding its ancient evolutionary environment9. The glacial and interglacial climatic fluctuations have had profound influences on some plant species. To respond to the glaciation, most plants shifted latitude or elevation of their ranges8. Distinct demographics were detected in different plant taxa as a result of glacial and interglacial periods. Gymnosperm species, such as *Taxus wallichiana*53, *Cycas revoluta*, and *Cycas taitungensis*54, have experienced population expansions during the most recent glacial period, while *Cycas debaoensis*27, *Cycas simplicipinna*9 and *Cycas multipinnata*10 have experienced population contractions. In this study, four cpDNA, three nuclear genes and 13 microsatellites generated inconsistent results for the demography of *C. guizhouensis*. For cpDNA, neutrality tests and mismatch analysis showed that *C. guizhouensis* did not experience a recent population expansion (Table S4; Fig. S2A), while the Bayesian Skyline Plot revealed a recent population expansion (over 25–4 Ka years) in this species (Fig. 4A). For the three nuclear genes, *GTP* and *PHYP* indicated that *C. guizhouensis* experienced a recent population expansion; however, a recent population decline was detected by the nuclear gene *F3H* (Table S4; Fig. S2B–D; Fig. 4B–D). In addition, bottleneck analysis (Table 3) based on microsatellites showed that populations of *C. guizhouensis* have not experienced a recent bottleneck event, but these populations experienced a historical reduction in population size. Normally, different genes and markers are subjected to different selective pressures, which may generate the above inconsistency.

Marine Isotope Stages (MIS), based on oscillations in oxygen-18 values of benthic foraminifera, are timescales that are now widely used to express warm and cool periods in the Quaternary55. The odd-numbered troughs are warm or interglacial intervals, while even-numbered peaks are cold or glacial intervals55. Based on the Bayesian Skyline Plot results, *C. guizhouensis* was believed to have experienced an expansion during MIS1, except for the nuclear gene *F3H* (Fig. 4). *Cycas guizhouensis* had a population expansion after the post-glacial period (10 Ka-Present). As a result, Pleistocene climatic oscillations had no noticeable effect on the demography of *C. guizhouensis*. Alternatively, the warmer climate during the post-glacial period (MIS1) promoted the expansion of this species.

Information on the genetic diversity, genetic structure, and demographic history of rare and endangered species can aid the development of appropriate management strategies for their conservation. One goal for the conservation of threatened plants is to preserve their genetic diversity56. Our study detected relatively higher levels of genetic diversity in *C. guizhouensis* than some other *Cycas*species. However, massive illegal digging of *C. guizhouensis* for trading or ornaments has jeopardized the existing populations of this species and its effective population size. For example, some populations that were sampled in Xiao's study, such as LP, have since been extirpated19. An effective population size greater than 100 is suggested to prevent inbreeding depression and a Ne greater than 1,000 is suggested to maintain the evolutionary potential of a population in perpetuity57. In contrast, the estimated Ne in populations of *C. guizhouensis* was less than 100 (especially less than 50), except for population LL (Table 2). Therefore, we suggest that protection zones or plots in the distribution areas of *C. guizhouensis* be established to protect the habitat for this species. In our study, two genetic clusters were detected in *C. guizhouensis*. Thus, these two genetic clusters could be managed as two evolutionary units and be given the highest priority protection. The demography of this species and network of its haplotypes indicated that there has been a recent expansion. Therefore, protection of the species in its natural habitat is required (*in situ*). Moreover, *ex situ* conservation is a viable auxiliary conservation approach for protecting this species by collecting seeds or seedling individuals from different populations which possess key sources of genetic diversity to remote adaptive areas. In addition, we suggest that more efforts should be made to raise awareness of conservation to local farmers and that a prohibition on deforestation in *Cycas* distribution areas be implemented. Taken together, these measures could protect *C. guizhouensis* from extinction.

# METHODS

## Sampling and Genotyping

A total of 209 individuals were collected from 11 populations of *C. guizhouensis* in southwestern Guizhou, northwestern Guangxi and eastern Yunnan province, China. Young and healthy leaves were dried in silica gel immediately after collection. All samples were stored in Kunming Institute of Botany, Chinese Academy of Sciences, China. Within the 209 samples, 10 individuals from each population were randomly selected for chloroplast and nuclear DNA sequencing, while all of the 209 individuals were used for the microsatellite study. Information on each sampling location and the number of individuals from each population that were used in the DNA sequence and microsatellite analyses are presented in Table S1 and Fig. 1, respectively. Among the 11 populations, seven populations are the same population examined in Xiao's study and the same population codes are used19. Several of the populations examined in Xiao's study are not included in this study because individuals from these populations (LP, MZ, TX, WF and XY) could not be detected. Fortunately, there were other populations of *C. guizhouensis* that were used in this study (AL, LW, XL and SP).

Genomic DNA was extracted from dried leaves using the modified CTAB method58. We chose four cpDNA intergenic spacers, *psb*A-*trn*H32, *trn*L-*trn*F34, *psb*M-*trn*D33 and *trn*L-*trn*T34 and three nuclear genes, *GTP*35, *PHYP* and *F3H* for complete analysis after preliminary screening from universal chloroplast and nuclear primers. The seven pairs of fragments were amplified for the most polymorphic sites of the 110 individuals. PCR amplification procedures were the same as those used in *C. simplicipinna*9. All PCR products were sequenced in both directions with the same primers for the amplification reactions, using an ABI 3770 automated sequencer at Shanghai Meiji Biological Medicine and Technology Co Ltd. All sequences were deposited in GenBank with the accession numbers KT824878-KT824962.

We screened microsatellite loci from recently developed nuclear microsatellites in *Cycas*36·37·38·39·40·41·42. PCR amplification was conducted with the same protocol as used in *C. simplicipinna*9. After preliminary screening microsatellite loci for *C. guizhouensis*, the selected microsatellite loci were stained with fluorescent dye at the 5′ end, their PCR products were separated and visualized using an ABI 3770 automated sequencer, and their profiles were read with GeneMapper software. An individual was declared null (nonamplifying) at a locus and was treated as missing data after two or more amplification failures. Finally, we chose 13 polymorphic microsatellite loci (Cha0236, Cha0836, Cy-TaiEST-SSR1137, E00138, E00438, Cpz2639, HL0840, CY23241, Cha-estssr0142, Cha-estssr0242, Cha-estssr0442, Cha-estssr0342, Cha0536) for *C. guizhouensis* after calculating polymorphism indices.

## Data Analysis

### *Data Analysis of DNA Sequences*

Sequences were edited and assembled using SeqMan. Multiple alignments of DNA sequences were performed manually and subsequently adjusted in Bioedit, version 7.0.4.159. We combined the four cpDNA regions and performed a congruency test in PAUP* 4.0b1060. The testing results showed a significant rate of homogeneity (P=1, >0.5), suggesting that there was a high degree of homogeneity between the four cpDNA regions. The combined cpDNA sequences were therefore used in following analyses. For three nuclear genes, heterozygous sites were identified by overlapping peaks in chromatograms. We resolved the nuclear sequences by applying the algorithms of PHASE61·62 in the software package DnaSP, version 5.063. The phased nuclear sequences were used in following analyses. Maps were drawn using the software ArcGIS version 10.2 (http://desktop.arcgis.com).

We first used DnaSP, version 5.063 to detect recombination in nuclear genes. Haplotypes were inferred from aligned DNA sequences by DnaSP, version 5.063. Genetic diversity was estimated by calculating *Nei's*nucleotide diversity (*P*i) and haplotype diversity (*H*d) indices using DnaSP, version 5.063. We calculated within-population gene diversity ($H_S$), gene diversity in total populations ($H_T$)64, and two measures of population differentiation, $G_{ST}$ and $N_{ST}$, according to the methods described by Pons & Petit65 using the Permut, 1.0 (http://www.pierroton.inra.fr/genetics/labo/Software/Permut). We used the program Arlequin, version 3.1166 to conduct an analysis of molecular variance (AMOVA) 67 and to estimate the genetic variation that was assigned within and among populations.

Phylogenetic relationships among cpDNA and nDNA haplotypes of *C. guizhouensis* were inferred using Bayesian methods implemented in MrBayes, version 3.1.268. We chose *Cycas edentata* and *Cycas rumphii* as outgroups. The degree of relatedness among cpDNA and nDNA haplotypes was also estimated using Network, version 4.2.0.169. In network analysis, indels were treated as single mutational events.

We used the evolutionary rates that had previously been estimated for seed plants to be $1.01 \times 10^{-9}$ and $5.1–7.0 \times 10^{-9}$70 mutations per site per year for synonymous sites to estimate the coalescent time of haplotypes for cpDNA and nDNA, respectively. Substitution models for cpDNA and nDNA were separately tested with MEGA, version 571. We used BEAST, version 1.6.172, to estimate the time of divergence using a strict molecular clock and the HKY model. The HKY+I model was used for the gene *F3H*. The BEAST program was also used to create a Bayesian Skyline Plot to infer the historical demography of *C. guizhouensis*. Posterior estimates of the mutation rate and time of divergence were obtained by Markov Chain Monte Carlo (MCMC) analysis. The analysis was run for $10^8$ iterations with a burn-in of $10^4$under a strict clock and the HKY model or HKY+I model. Log parameters were sampled every 10,000 iterations. The analysis ran three times under the condition that all parameters were unchanged. The ESS parameter was found to exceed 200 by TRACER, version 1.573, which suggested that there was acceptable mixing and sufficient sampling. Next, the three times running log files and tree files were combined, respectively. The combined log and tree files were used to construct a Bayesian skyline plot in TRACER, version 1.573. We used a pairwise mismatch distribution to test for population expansion in DnaSP, version 5.063, to further investigate the demography of the species. In addition, the signatures of demographic change were examined with neutrality tests, including Tajima's *D*, Fu and Li's $D^*$ and$F^*$ and Fu's $F_S$44, using DnaSP, version 5.063. The sum-of-squared deviations (SSD) and raggedness index as well as their *P*-values were calculated with the software Arlequin, version 3.1166.

### *Data Analysis of SSR Markers*

Dataset editing and formatting were performed in GenAlEx, version 6.374. Genetic diversity indices, including the number of alleles ($N_A$), private alleles ($A_P$), effective number of alleles ($A_E$), expected heterozygosity ($H_E$), observed heterozygosity ($H_O$), information index (*I*), fixation index (*F*) and percentage of polymorphic loci (*PPB*), were calculated using GenAlEx, version 6.374, and POPGENE, version 1.3275, with mutual correction. Allelic richness ($A_R$) was estimated with FSTAT, version 1.276. The differentiation index $F_{ST}$ between pairs of populations was computed with Arlequin, version 3.1166. Isolation by distance (IBD) was tested by performing Mantel tests in GenAlEx, version 6.374 on the correlation of $F_{ST}/(1-F_{ST})$ with geographic distance for all pairs of populations. $F_{ST}/(1-F_{ST})$ was calculated with Genepop, version 4.1.477. Gene flow between pairs of populations was estimated according to Wright's principles $Nm=(1-F_{ST})/4F_{ST}$26. Tests for departure from Hardy-Weinberg equilibrium (HWE) were performed in each locus and each population as well as a globally unified population using Genepop, version 4.1.477.

The genetic structure of sampled populations was estimated by unweighted pair group mean analysis (UPGMA) using TEPGA, version 1.378, with 5,000 permutations. A Bayesian analysis of population structure on the SSR data was conducted with STRUCTURE, version 2.279. The combination of an admixture and a correlated-allele frequencies model was used for the analysis. The simulation was run with values of K from 1 to 20 and repeated 20 times for each set. Each run included a burn-in of $1 \times 10^5$ iterations and $1 \times 10^5$ subsequent MCMC steps. The best-fit number of groupings was evaluated using $\Delta K$ and the log-likelihood value by STRUCTURE HARVESTER, version 0.6.880. Based on genetic distances among microsatellite phenotypes, an individual-based principal coordinate analysis (PCoA) was visualized with

MVSP, version 3.1281. The effective population sizes of each population were estimated in the program LDNe at three levels of the lowest allele frequency (0.01, 0.02, 0.05) with a 95% confidence interval82. We tested the bottleneck effect at the population level to explore population demography using different models and testing methods implemented in BOTTLENECK, version 1.2.0283. A stepwise mutation model (SMM) and a two-phased model (TPM) were chosen. Under the two models, the standardized differences test was removed from this study because this test is typically only used when at least twenty polymorphic loci are available. Two other methods (Sign tests and Wilcoxon tests) were applied in the analyses. We also used a mode shift model84 to test for bottlenecks in each population. These methods implemented in BOTTLENECK are most powerful when bottlenecks are severe and recent. These methods have low power unless the decline is greater than 90%84. Moreover, we further investigated a genetic bottleneck using the Garza-Williamson index (also called M-ratio, the ratio of number of alleles to range in allele size)85. The index was calculated by Arlequin, version 3.1166. When seven or more loci are analysed, the Garza-Williamson index is lower than the critical *M*c value of 0.68, a value obtained by simulations based on the empirical data in bottlenecked populations, which suggests a reduction in population size66·85. The Garza-Williamson index is more powerful for detecting genetic bottlenecks if the bottleneck lasted several generations or if the population made a rapid demographic recovery86. All figures were modified using Photoshop (Adobe Corporation, California, America). All statistical analyses were conducted under an open source environment. Methods were conducted in accordance with approved guidelines.

## SUPPLEMENTARY INFORMATION

**Table S1.** Details of sample locations, sample sizes (n) surveyed for DNA sequences and microsatellites of C. guizhouensis.

| Population Code | Population | Latitude N° | Longitude E° | Altitude m | Individuals for DNA sequences/microsatellites |
|---|---|---|---|---|---|
| AL | Anlong, Guizhou | 24.965 | 105.261 | 500 | 10/20 |
| YP | Yangping, Xingyi, Guizhou | 24.938 | 104.994 | 930 | 10/20 |
| LW | Luowan, Xingyi, Guizhou | 24.681 | 104.686 | 880 | 10/20 |
| LL | Longlin, Guangxi | 24.665 | 104.889 | 970 | 10/20 |
| XL | Xilin, Guangxi | 24.606 | 104.611 | 860 | 10/11 |
| SZ | Shizong, Yunnan | 24.652 | 104.248 | 1360 | 10/20 |
| LX | Luxi, Yunnan | 24.452 | 103.975 | 1160 | 10/20 |
| ML | Mile, Yunnan | 24.181 | 103.632 | 1430 | 10/20 |
| KY | Kaiyuan, Yunnan | 23.829 | 103.180 | 1400 | 10/20 |
| JS | Jianshui, Yunnan | 23.400 | 102.880 | 1320 | 10/18 |
| SP | Shiping, Yunnan | 23.720 | 102.500 | 1400 | 10/20 |
| Total | 11 | | | | 110/209 |

**Table S2.** The composition of haplotypes, haplotype diversity (Hd) and nucleotide diversity (Pi) surveyed for combined cpDNA and nuclear genes of C. guizhouensis.

| Population code | cpDNA (10) | | | *GTP* (20) | | | *PHYP* (20) | | | *F3H* (20) | | |
|---|---|---|---|---|---|---|---|---|---|---|---|---|
| | Haplotypes (No.) | *H*d | *P*i × $10^3$ | Haplotypes(No.) | *H*d | *P*i × $10^3$ | Haplotypes(No.) | *H*d | *P*i × $10^3$ | Haplotypes(No.) | *H*d | *P*i × $10^3$ |
| AL | guiH1(10) | 0 | 0 | guiG1(20) | 0 | 0 | guiP1(3) guiP2(17) | 0.268 | 0.27 | guiF1(12) F2(1) F3(1) F4(2) F5(1) F6(2) F7(1) | 0.642 | 2.79 |
| YP | guiH1(10) | 0 | 0 | guiG1(20) | 0 | 0 | guiP1(5) guiP2(15) | 0.395 | 0.39 | guiF1(2) F6(18) | 0.189 | 0.75 |
| LW | guiH1(10) | 0 | 0 | guiG1(20) | 0 | 0 | guiP1(5) guiP2(15) | 0.395 | 0.39 | guiF1(12) F4(2) F6(6) | 0.568 | 1.84 |
| LL | guiH1(10) | 0 | 0 | guiG1(16) guiG2(4) | 0.337 | 0.59 | guiP1(8) guiP2(12) | 0.505 | 0.50 | guiF1(10) F4(3) F6(7) | 0.637 | 1.96 |
| XL | guiH1(2) guiH2(5) guiH3(1) guiH4(2) | 0.733 | 0.54 | guiG1(19) guiG3(1) | 0.100 | 0.52 | guiP1(9) guiP2(10) guiP3(1) | 0.574 | 0.62 | guiF1(8) F4(4) F6(1) F8(4) F9(1) F10(2) | 0.784 | 5.68 |
| SZ | guiH2(6) guiH5(4) | 0.533 | 0.18 | guiG1(20) | 0 | 0 | guiP1(18) guiP2(1) guiP4(1) | 0.195 | 0.20 | guiF4(1) F5(14) F6(3) F11(1) F12(1) | 0.505 | 2.43 |
| LX | guiH6(8) guiH7(2) | 0.356 | 0.36 | guiG1(20) | 0 | 0 | guiP1(19) guiP2(1) | 0.100 | 0.10 | guiF4(9) F5(1) F8(3) F13(3) F14(2) F15(2) | 0.768 | 7.84 |
| ML | guiH4(10) | 0 | 0 | guiG1(20) | 0 | 0 | guiP1(15) guiP2(3) guiP4(2) | 0.426 | 0.46 | guiF1(4) F2(1) F4(4) F5(1) F6(1) F8(1) F15(3) F16(3) F17(2) | 0.900 | 4.96 |
| KY | guiH8(8) guiH9(2) | 0.356 | 0.24 | guiG1(20) | 0 | 0 | guiP1(11) guiP2(5) guiP5(3) guiP6(1) | 0.642 | 0.76 | guiF2(2) F4(2) F8(4) F18(5) F19(1) F20(4) F21(1) F22(1) | 0.874 | 6.99 |
| JS | guiH2(9) guiH6(1) | 0.2 | 0.27 | guiG1(20) | 0 | 0 | guiP1(19) guiP4(1) | 0.100 | 0.10 | guiF1(2) F4(1) F5(12) F6(2) F23(2) F24(1) | 0.637 | 3.29 |
| SP | guiH2(2) guiH10(3) guiH11(3) guiH12(2) | 0.822 | 1.63 | guiG1(11) guiG4(1) guiG5(6) guiG6(1) guiG7(1) | 0.632 | 2.44 | guiP1(10) guiP2(3) guiP7(2) guiP8(1) guiP9(3) guiP10(1) | 0.726 | 1.05 | guiF1(2) F2(1) F4(1) F5(3) F16(1) F17(1) F25(2) F26-F29(1) F30(3) F31(1) F32(1) | 0.958 | 6.06 |
| Total | 12 | 0.794 | 0.91 | 7 | 0.123 | 0.42 | 10 | 0.555 | 0.62 | 32 | 0.870 | 4.93 |

**Table S3.** Genetic diversity, differentiation parameters for the combined cpDNA sequences and nuclear genes in all populations of C. guizhouensis.

| Marker | *Hs* | $H_T$ | $G_{ST}$ | $N_{ST}$ |
|---|---|---|---|---|
| cpDNA | 0.273 | 0.841 | 0.676 | 0.698 |
| *GTP* | 0.097 | 0.125 | 0.223 | 0.241 |
| *PHYP* | 0.393 | 0.571 | 0.311 | 0.307 |
| *F3H* | 0.678 | 0.902 | 0.248 | 0.192 |

**Table S4.** Parameters of neutrality tests and mismatch analyses based on cpDNA and nuclear genes of C. guizhouensis, note: ** , P < 0.01; *** , P < 0.001.

| Marker | Tajima'*D* | Fu and Li'$D^*$ | Fu and Li'$F^*$ | Fu's *Fs* | SSD | raggedness |
|---|---|---|---|---|---|---|
| cpDNA | 0.74126 | 1.18632 | 1.22583 | 1.659 | 0.04352 | 0.23852 |
| *GTP* | -1.67165 | -1.75178 | -2.05712 | -6.269$^{**}$ | 0.00753 | 0.11759 |
| *PHYP* | -1.32224 | -2.07520 | -2.15827 | -5.505$^{**}$ | 0.03506 | 0.30727 |
| *F3H* | -0.19361 | 1.40253 | 0.90810 | -10.910$^{***}$ | 0.07958 | 0.21545 |

**Table S5.** P-value of Hardy-Weinberg equilibrium tests for 11 populations of C. guizhouensis, note: -, Monomorphic; $^*$ , P < 0.05; $^{**}$ , P < 0.01; $^{***}$ , P < 0.001.

| Population | Cha02 | Cha08 | Cy-TaiEST-SSR11 | E001 | E004 | Cpz26 | HL08 | CY232 | Cha-estssr01 | Cha-estssr02 | Cha-estssr04 | Cha-estssr03 | Cha05 | All loci |
|---|---|---|---|---|---|---|---|---|---|---|---|---|---|---|
| AL | 0.002$^{**}$ | 0.000$^{***}$ | 1.000$^{ns}$ | - | 1.000$^{ns}$ | 0.544$^{ns}$ | - | 0.129$^{ns}$ | 0.000$^{***}$ | - | 0.214$^{ns}$ | 0.280$^{ns}$ | - | 0.000$^{***}$ |
| YP | 0.000$^{***}$ | 0.000$^{***}$ | 1.000$^{ns}$ | - | 1.000$^{ns}$ | 0.039$^{*}$ | - | 0.002$^{**}$ | 0.093 | - | 0.149$^{ns}$ | 1.000$^{ns}$ | - | 0.000$^{***}$ |
| LW | 0.034$^{*}$ | 0.000$^{***}$ | 1.000$^{ns}$ | 0.026$^{*}$ | 0.800$^{ns}$ | 0.009$^{**}$ | - | 0.654$^{ns}$ | 0.000$^{***}$ | 1.000$^{ns}$ | 1.000$^{ns}$ | 0.647$^{ns}$ | - | 0.000$^{***}$ |
| LL | 0.017$^{*}$ | 0.001$^{**}$ | - | - | 1.000$^{ns}$ | 0.182$^{ns}$ | - | 1.000$^{ns}$ | 0.000$^{***}$ | 0.026$^{*}$ | 1.000$^{ns}$ | 0.528$^{ns}$ | 0.026$^{*}$ | 0.000$^{***}$ |
| XL | 0.173$^{ns}$ | 0.022$^{*}$ | - | - | 0.406$^{ns}$ | 0.008$^{**}$ | - | 0.143$^{ns}$ | 0.000$^{***}$ | 0.023$^{*}$ | 1.000$^{ns}$ | 0.215$^{ns}$ | - | 0.000$^{***}$ |
| SZ | 0.013$^{*}$ | 0.000$^{***}$ | - | 1.000$^{ns}$ | 0.468$^{ns}$ | 0.041$^{*}$ | 1.000$^{ns}$ | 0.004$^{**}$ | 0.000$^{***}$ | 1.000$^{ns}$ | 1.000$^{ns}$ | - | - | 0.000$^{***}$ |
| LX | 0.035$^{*}$ | 0.000$^{***}$ | - | 0.003$^{**}$ | 0.440$^{ns}$ | 0.052$^{ns}$ | 0.105$^{ns}$ | 0.198$^{ns}$ | 0.000$^{***}$ | 0.000$^{***}$ | 0.136$^{ns}$ | - | - | 0.000$^{***}$ |
| ML | 0.001$^{**}$ | 0.224$^{ns}$ | 0.633$^{ns}$ | 0.012$^{*}$ | 0.156$^{ns}$ | 0.004$^{**}$ | 1.000$^{ns}$ | 0.339$^{ns}$ | 0.000$^{***}$ | 0.000$^{***}$ | 0.018$^{*}$ | - | - | 0.000$^{***}$ |
| KY | 0.225$^{ns}$ | 0.000$^{***}$ | - | 1.000$^{ns}$ | 0.076$^{ns}$ | 0.284$^{ns}$ | 1.000$^{ns}$ | 0.032$^{*}$ | 0.000$^{***}$ | 0.000$^{***}$ | 0.619$^{ns}$ | 1.000$^{ns}$ | - | 0.000$^{***}$ |
| JS | 0.000$^{***}$ | 0.001$^{**}$ | 0.169$^{ns}$ | 0.124$^{ns}$ | 0.000$^{***}$ | 0.371$^{ns}$ | 1.000$^{ns}$ | 0.004$^{**}$ | 0.000$^{***}$ | 0.267$^{ns}$ | 0.036$^{*}$ | - | - | 0.000$^{***}$ |
| SP | 0.434$^{ns}$ | 0.017$^{*}$ | 0.041$^{*}$ | 0.208$^{ns}$ | 0.000$^{***}$ | 0.656$^{ns}$ | 1.000$^{ns}$ | 0.054$^{ns}$ | 0.000$^{***}$ | 0.026$^{*}$ | 1.000$^{ns}$ | 0.544$^{ns}$ | - | 0.000$^{***}$ |
| All pop. | 0.000$^{***}$ | 0.000$^{***}$ | 0.541$^{ns}$ | 0.001$^{**}$ | 0.000$^{***}$ | 0.000$^{***}$ | 0.972$^{ns}$ | 0.000$^{***}$ | 0.000$^{***}$ | 0.000$^{***}$ | 0.233$^{ns}$ | 0.832$^{ns}$ | 0.000$^{***}$ | 0.000$^{***}$ |

**Table S6.** Estimates of gene flow (Nm) between pairs of C. guizhouensis populations.

| Population | AL | YP | LW | LL | XL | SZ | LX | ML | KY | JS | SP |
|---|---|---|---|---|---|---|---|---|---|---|---|
| AL | 0.000 | | | | | | | | | | |
| YP | 1.620 | 0.000 | | | | | | | | | |
| LW | 2.354 | 1.187 | 0.000 | | | | | | | | |
| LL | 2.328 | 0.893 | 12.096 | 0.000 | | | | | | | |
| XL | 0.937 | 0.591 | 1.345 | 1.224 | 0.000 | | | | | | |
| SZ | 1.758 | 0.833 | 1.150 | 1.052 | 0.786 | 0.000 | | | | | |
| LX | 2.176 | 1.048 | 1.755 | 1.321 | 1.464 | 6.716 | 0.000 | | | | |
| ML | 1.124 | 0.805 | 1.046 | 0.824 | 1.578 | 1.675 | 4.050 | 0.000 | | | |
| KY | 1.638 | 0.965 | 1.785 | 1.537 | 2.163 | 2.534 | 4.966 | 4.250 | 0.000 | | |
| JS | 1.676 | 0.956 | 1.532 | 1.245 | 1.484 | 2.578 | 13.486 | 2.746 | 3.955 | 0.000 | |
| SP | 1.143 | 0.910 | 1.532 | 1.038 | 1.986 | 1.335 | 3.740 | 4.478 | 3.895 | 3.043 | 0.000 |

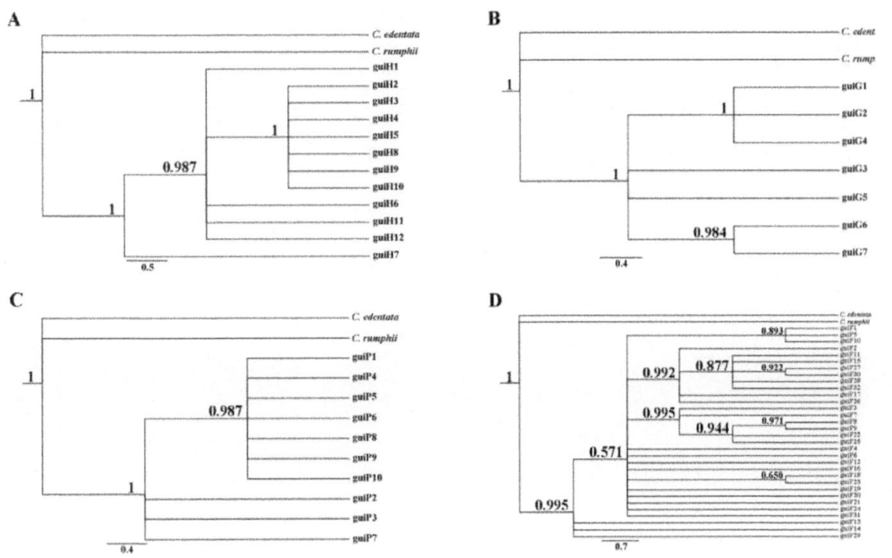


**Figure S1.** Bayesian tree based on the haplotypes of cpDNA (A), GTP (B), PHYP (C) and F3H (D). Cycas edentata and Cycas rumphii were used as outgroups. The number on each branch indicates the posterior probability (PP).

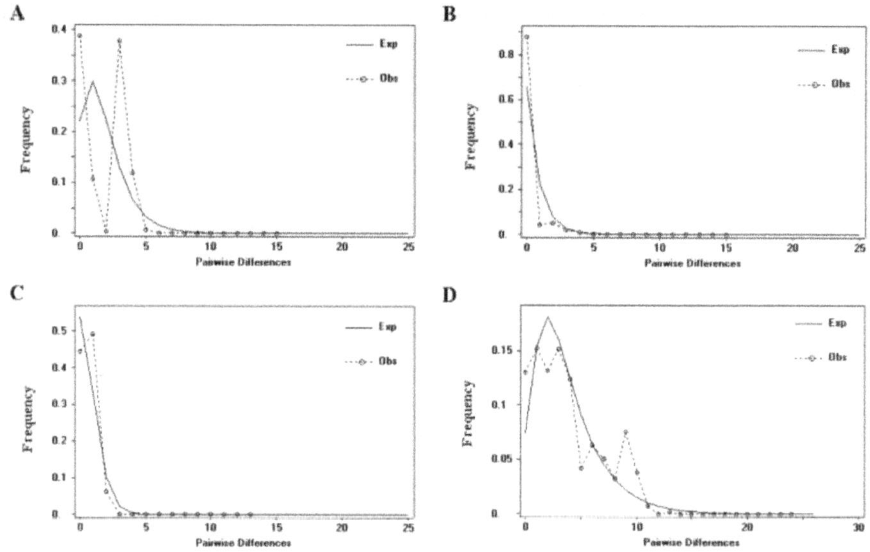

**Figure S2.** Mismatch distribution of cpDNA (A) and the nuclear genes GTP (B), PHYP (C) and F3H (D) haplotypes based on pairwise sequence differences against the frequency of occurrence for C. guizhouensis.

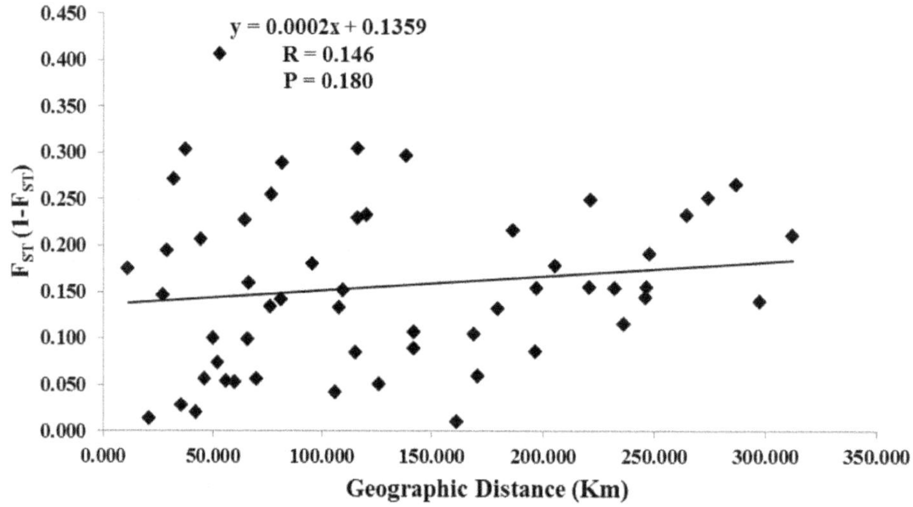


**Figure S3.** Figure plot of geographical distance against genetic distance for 11 populations of C. guizhouensis.

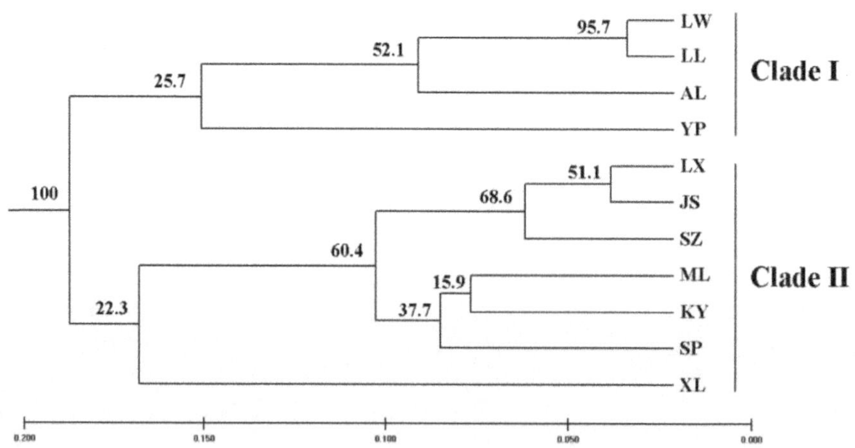


**Figure S4.** An unweighted pair-group method with arithmetic averages (UPGMA) phenogram of 11 populations of C. guizhouensis based on microsatellites. Numbers on branches indicate bootstrap values from 5,000 replicates.

## Acknowledgements

The authors thank Jian Liu and Yiqing Gong for their assistance with field sampling. This research was supported by the United Fund of the NSFC and the Yunnan Natural Science Foundation (U1136602 to X.G.).

## REFERENCES

1. Zhao, N. et al. Pleistocene climate changes shaped the divergence and demography of Asian populations of the great tit Parus major: evidence from phylogeographic analysis and ecological niche models. J. Avian. Biol. 43, 297–310 (2012).
2. Williams, M. A. J., Dunkerley, D. L., De Deckker, P., Kershaw, A. P. & Stokes, T. Quaternary environments. (Arnold Press, London, 1998).
3. Weaver, A. J., Eby, M., Fanning, A. F. & Wiebe, E. C. Simulated influence of carbon dioxide, orbital forcing and ice sheets on the climate of the Last Glacial Maximum. Nature 394, 847–853 (1998).
4. Hewitt, G. M. Some genetic consequences of ice ages, and their role in divergence and speciation. Biol. J. Linn. Soc. 58, 247–276 (1996).
5. Hewitt, G. The genetic legacy of the Quaternary ice ages. Nature 405, 907–913 (2000).
6. Hewitt, G. Genetic consequences of climatic oscillations in the Quaternary. Philos. Trans. R. Soc. Lond. B: Biol. Sci. 359, 183–195 (2004).
7. Avise, J. C. Phylogeography: the history and formation of species. (Harvard University Press, Cambridge, 2000).
8. Davis, M. B. & Shaw, R. G. Range shifts and adaptive responses to Quaternary climate change. Science 292, 673–679 (2001).
9. Feng, X. Y., Wang, Y. H. & Gong, X. Genetic diversity, genetic structure and demographic history of Cycas simplicipinna (Cycadaceae) assessed by DNA sequences and SSR markers. BMC Plant Biol. 14, 187 (2014).
10. Gong, Y. Q. et al. The historical demography and genetic variation of the endangered Cycas multipinnata (Cycadaceae) in the red river region, examined by chloroplast DNA sequences and microsatellite markers. PLos One 10, e0117719 (2015).
11. Yang, Y. Q., Huang, B. H., Yu, Z. X. & Liao, P. C. Inferences of demographic history and fine-scale landscape genetics in Cycas panzhihuaensis and implications for its conservation. Tree Genet. Genomes 11, 1–15 (2015).
12. Taylor, E. L., Taylor, T. N. & Krings, M. Paleobotany: the biology and evolution of fossil plants. (Academic Press, Burlington, 2009).
13. Jones, D. L. Cycads of the World: Ancient Plant in Today's Landscape. (Smithsonian Institution Press, Washington, 1993).
14. Nagalingum, N. S. et al. Recent synchronous radiation of a living fossil. Science 334, 796–799 (2011).
15. Friis, E. M. The origins of angiosperms and their biological consequences. (Cambridge University Press, Archive Cambridge, 1987).
16. Norstog, K. & Nicholls, T. The Biology of the Cycads. (Cornell University Press, New York, 1997).
17. Christenhusz, M. et al. A new classification and linear sequence of extant gymnosperms. Phytotaxa 19, 55–70 (2011).
18. Hill, K. D., Stevenson, D. W. & Osborne, R. The world list of cycads. Bot. Rev. 70, 274–298 (2004).
19. Xiao, L. Q., Ge, X. J., Gong, X., Hao, G. & Zheng, S. X. ISSR variation in the endemic and endangered plant Cycas guizhouensis (Cycadaceae). Ann. Bot. 94, 133–138 (2004).
20. Xiao, L. Q. & Moller, M. Nuclear ribosomal ITS functional paralogs resolve the phylogenetic relationships of a late-miocene radiation cycad Cycas (Cycadaceae). PLos One 10, e0117971 (2015).
21. Dehgan, B. & Yuen, C. Seed morphology in relation to dispersal, evolution, and propagation of Cycas L. Bot. Gaz. 144, 412–418 (1983).
22. Keast, A. & Miller, S. E. The origin and evolution of Pacific island biotas, New Guinea to Eastern Polynesia: patterns and processes. (SPB Academic Publishing, Amsterdam, 1996).
23. Schneider, D., Wink, M., Sporer, F. & Lounibos, P. Cycads: their evolution, toxins, herbivores and insect pollinators. Naturwissenschaften 89, 281–294 (2002).
24. Loveless, M. D. & Hamrick, J. L. Ecological determinants of genetic structure in plant populations. Annu. Rev. Ecol. Evol. S. 15, 65–95 (1984).
25. Duminil, J. et al. Can population genetic structure be predicted from life-history Traits? Am. Nat. 169, 662–672 (2007).
26. Wright, S. Evolution in Mendelian populations. Genetics 16, 97 (1931).
27. Zhan, Q. Q., Wang, J. F., Gong, X. & Peng, H. Patterns of chloroplast DNA variation in Cycas debaoensis (Cycadaceae): conservation implications. Conserv. Genet. 12, 959–970 (2011).
28. CIBRIÁN-JARAMILLO, A., Daly, A. C., Brenner, E., Desalle, R. & Marler, T. E. When North and South don't mix: genetic connectivity of a recently endangered oceanic cycad, Cycas micronesica, in Guam using EST-microsatellites. Mol. Ecol. 19, 2364–2379 (2010).
29. Yang, S. L. & Meerow, A. W. In Biology and conservation of Cycads. (ed. Chen, C. J.) 175–186 (International Academic Publishers, Beijing, 1999).
30. Brown, A. H., Clegg, M. T., Kahler, A. L. & Weir, B. S. Plant population genetics, breeding, and genetic resources. (Sinauer Associates
31. Inc., Sunderland, 1990).
32. Huang, S., Hsieh, H. T., Fang, K. & Chiang, Y. C. Patterns of genetic variation and demography of Cycas taitungensis in Taiwan. Bot. Rev. 70, 86–92 (2004).
33. Chiang, T. Y. & Peng, C. I. In Proceedings of the Symposium of Conservation of Endemic Species. (ed. Yang, S. D.) 148–155 (Research Institute of Taiwan

Endemic Species, Taipei, 1998).

34. Shaw, J. et al. The tortoise and the hare II: relative utility of 21 noncoding chloroplast DNA sequences for phylogenetic analysis. Am. J. Bot. 92, 142–166 (2005).

35. Taberlet, P., Gielly, L., Pautou, G. & Bouvet, J. Universal primers for amplification of three non-coding regions of chloroplast DNA. Plant Mol. Biol. 17, 1105–1109 (1991).

36. Salas-Leiva, D. E. et al. Conserved genetic regions across angiosperms as tools to develop single-copy nuclear markers in gymnosperms: an example using cycads. Mol. Ecol. Resour. 14, 831–845 (2014).

37. Zhang, M. et al. Isolation and characterization of microsatellite markers for Cycas hainanensis CJ Chen (Cycadaceae). Conserv. Genet. 10, 1175–1176 (2009).

38. Ju, L. P. et al. Microsatellite primers in the native perennial cycad Cycas taitungensis (Cycadaceae). Am. J. Bot. 98, e84–e86 (2011).

39. Yang, Y., Li, Y., Li, L. F., Ge, X. J. & Gong, X. Isolation and characterization of microsatellite markers for Cycas debaoensis YC Zhong et CJ Chen (Cycadaceae). Mol. Ecol. Resour. 8, 913–915 (2008).

40. Zhang, F. M. et al. Development of seven novel EST-SSR markers from Cycas panzhihuaensis (Cycadaceae). Am. J. Bot. 97, e159–e161 (2010).

41. Li, L. et al. Isolation and characterization of microsatellite loci in endangered Cycas changjiangensis (Cycadaceae). Conserv. Genet. 10, 793–795 (2009).

42. Cibrián-Jaramillo, A., Marler, T. E., DeSalle, R. & Brenner, E. D. Development of EST-microsatellites from the cycad Cycas rumphii, and their use in the recently endangered Cycas micronesica. Conserv. Genet. 9, 1051–1054 (2008).

43. Wang, Z. F., Ye, W. H., Cao, H. L., Li, Z. C. & Peng, S. L. Identification and characterization of EST-SSRs and cpSSRs in endangered Cycas hainanensis. Conserv. Genet. 9, 1079–1081 (2008).

44. Wright, S. Evolution and the Genetics of Populations. (University of Chicago Press, Chicago, 1978).

45. Fu, Y. X. Statistical tests of neutrality of mutations against population growth, hitchhiking and background selection. Genetics 147, 915–925 (1997).

46. Reisch, C. & Bernhardt-Römermann, M. The impact of study design and life history traits on genetic variation of plants determined with AFLPs. Plant Ecol. 215, 1493–1511 (2014).

47. Wu, Z. Y. & Raven, P. H. Flora of China. (Science Press and Missouri Botanical Garden Press, Beijing and St. Louis, 1999).

48. Petit, R. J. et al. Invited review: Comparative organization of chloroplast, mitochondrial and nuclear diversity in plant populations. Mol. Ecol. 14, 689–701 (2005).

49. Slatkin, M. Gene flow in natural populations. Annu. Rev. Ecol. Systemat. 16, 393–430 (1985).

50. Taberlet, P., Fumagalli, L., Wust-Saucy, A. G. & Cosson, J. F. Comparative phylogeography and postglacial colonization routes in Europe. Mol. Ecol. 7, 453–464 (1998).

51. Li, Z. W., Chen, Z. R. & Wang, M. L. Classification and correlation of the quaternary glacial epoch in the Hengduan (Transverse) Mountains. Geol. Rev. 37, 125–132 (1991).

52. Zhou, S. Z., Wang, X. L., Wang, J. & Xu, L. B. A preliminary study on timing of the oldest Pleistocene glaciation in Qinghai-Tibetan Plateau. Quatern. Int. 154, 44–51 (2006).

53. Qu, Y. H. et al. Lineage diversification and historical demography of a montane bird Garrulax elliotii - implications for the Pleistocene evolutionary history of the eastern Himalayas. BMC Evol. Biol. 11, 174 (2011).

54. Liu, J. et al. Geological and ecological factors drive cryptic speciation of yews in a biodiversity hotspot. New Phytol. 199, 1093–1108 (2013).

55. Chiang, Y. C. et al. Paraphyly of organelle DNAs in Cycas Sect. Asiorientales due to ancient ancestral polymorphisms. BMC Evol. Biol. 9, 161 (2009).

56. Ogg, J. G., Ogg, G. & Gradstein, F. M. The concise geologic time scale. (Cambridge University Press, New York, 2008).

57. Montalvo, A. M. et al. Restoration biology: a population biology perspective. Restor. Ecol. 5, 277–290 (1997).

58. Frankham, R., Bradshaw, C. J. & Brook, B. W. Genetics in conservation management: revised recommendations for the 50/500 rules, Red List criteria and population viability analyses. Biol. Conserv. 170, 56–63 (2014).

59. Doyle, J. J. & Doyle, J. L. A rapid DNA isolation procedure for small quantities of fresh leaf tissue. Phytochemical Bulletin 19, 11–15 (1987).

60. Hall, T. A. BioEdit: a user-friendly biological sequence alignment editor and analysis program for Windows 95/98/NT. Nucleic. Acids. Symp. Ser. 41, 95–98 (1999).

61. Swofford D. L. PAUP*. Phylogenetic Analysis Using Parsimony (*and Other Methods). Version 4 (Sinauer Associates, Sunderland, MA, 2002).

62. Stephens, M., Smith, N. J. & Donnelly, P. A new statistical method for haplotype reconstruction from population data. Am. J. Hum. Genet. 68, 978–989 (2001).

63. Stephens, M. & Donnelly, P. A comparison of bayesian methods for haplotype reconstruction from population genotype data. Am. J. Hum. Genet. 73, 1162–1169 (2003).

64. Librado, P. & Rozas, J. DnaSP v5: a software for comprehensive analysis of DNA polymorphism data. Bioinformatics 25, 1451–1452 (2009).

65. Nei, M. Analysis of gene diversity in subdivided populations. Proc. Nat. Acad. Sci. 70, 3321–3323 (1973).

66. Pons, O. & Petit, R. J. Measuring and testing genetic differentiation with ordered versus unordered alleles. Genetics 144, 1237–1245 (1996).

67. Excoffier, L., Laval, G. & Schneider, S. Arlequin (version 3.0): an integrated software package for population genetics data analysis. Evol. Bioinform Online 1, 47–50 (2005).

68. Excoffier, L., Smouse, P. E. & Quattro, J. M. Analysis of

molecular variance inferred from metric distances among DNA haplotypes: application to human mitochondrial DNA restriction data. Genetics 131, 479–491 (1992).

69. Ronquist, F. & Huelsenbeck, J. P. MrBayes 3: Bayesian phylogenetic inference under mixed models. Bioinformatics 19, 1572–1574 (2003).

70. Bandelt, H. J., Forster, P. & Röhl, A. Median-joining networks for inferring intraspecific phylogenies. Mol. Biol. Evol. 16, 37–48 (1999).

71. Graur, D. & Li, W. H. Fundamentals of molecular evolution. (Sinauer Associates, Sunderland, MA, 2000).

72. Tamura, K. et al. MEGA5: molecular evolutionary genetics analysis using maximum likelihood, evolutionary distance, and maximum parsimony methods. Mol. Biol. Evol. 28, 2731–2739 (2011).

73. Drummond, A. J. & Rambaut, A. BEAST: Bayesian evolutionary analysis by sampling trees. BMC Evol. Biol. 7, 214 (2007).

74. Rambaut, A. & Drummond, A. Tracer: MCMC Trace Analysis Tool, Version 1.5.0. (2009). Available at: http://tree.bio.ed.ac.uk/software/tracer/University of Oxford.

75. Peakall, R. & Smouse, P. E. GENALEX 6: genetic analysis in Excel. Population genetic software for teaching and research. Mol. Ecol.

76. Notes. 6, 288–295 (2006).

77. Yeh, F. C., Yang, R. C., Boyle, T. B. J., Ye, Z. H. & Mao, J. X. POPGENE, the user-friendly shareware for population genetic analysis. Molecular Biology and Biotechnology Centre, University of Alberta, Edmonton, Canada 10 (1997).

78. Goudet, J. F. S. T. A. T. (version 1.2): a computer program to calculate F-statistics. J. Hered. 86, 485–486 (1995).

79. Rousset, F. GENEPOP'007: a complete re-implementation of the GENEPOP software for Windows and Linux. Mol. Ecol. Resour. 8, 103–106 (2008).

80. Miller, M. P. Tools for population genetic analyses (TFPGA) 1.3: A Windows program for the analysis of allozyme and molecular population genetic data. Comput. Softw. 4, 157 (1997).

81. Pritchard, J. K., Stephens, M. & Donnelly, P. Inference of population structure using multilocus genotype data. Genetics 155, 945–959 (2000).

82. Earl, D. A. & vonHoldt, B. M. STRUCTURE HARVESTER: a website and program for visualizing STRUCTURE output and implementing the Evanno method. Conserv. Genet. Resour. 4, 359–361 (2012).

83. Kovach, W. MVSP-A multivariate statistical Package for Windows, ver. 3.1. Kovach Computing Services, Pentraeth, Wales, UK 137, (1999).

84. Waples, R. S. & Do, C. LDNE: a program for estimating effective population size from data on linkage disequilibrium. Mol. Ecol. Resour. 8, 753–756 (2008).

85. Piry, S., Luikart, G. & Cornuet, J. M. BOTTLENECK: a computer program for detecting recent reductions in the effective size using allele frequency data. J. Hered. 90, 502–503 (1999).

86. Cornuet, J. M. & Luikart, G. Description and power analysis of two tests for detecting recent population bottlenecks from allele frequency data. Genetics 144, 2001–2014 (1996).

87. Garza, J. C. & Williamson, E. G. Detection of reduction in population size using data from microsatellite loci. Mol. Ecol. 10, 305–318 (2001).

88. Williamson-Natesan, E. G. Comparison of methods for detecting bottlenecks from microsatellite loci. Conserv. Genet. 6, 551–562 (2005).

# CHAPTER 17 ASYMMETRIC INTROGRESSION IN THE HORTICULTURAL LIVING FOSSIL CYCAS SECT. ASIORIENTALES USING A GENOME-WIDE SCANNING APPROACH

Yu-Chung Chiang, Bing-Hong Huang, Chun-Wen Chang, Yu-Ting Wan, Shih-Jie Lai, Shong Huang and Pei-Chun Liao

**Keywords:** Introgression, AFLP, Cycas revoluta, Cycas taitungensis, horticulture

## INTRODUCTION

Introgression usually happens in contact zones of sympatric or parapatric species even they have diverged for millions of years after speciation [1]. In contrast to the sympatry or parapatry, introgression between allopatrically distributed species is more equivocal as the longer isolation results in deep divergence and lower chance for introgression, but species of recent allopatry have higher opportunity for introgression than species of sympatry in nature [2]. In addition, introgression occurs more easily in species of niche conservation than species of niche specialization when they are secondarily contacting [3]. Horticulturally, the introduction of alien species could accelerate the opportunity of interspecific gene flow, break the reproductive isolation, even create new chimera by hybridization between deep diverged species [4,5]. Therefore, degrees of introgression between horticulturally introduced species and native species provide clues for exploring the mechanism of species divergence in different geographic habitats.

**Citation:** Yu-Chung Chiang, Bing-Hong Huang, Chun-Wen Chang, Yu-Ting Wan, Shih-Jie Lai, Shong Huang and Pei-Chun Liao, Asymmetric Introgression in the Horticultural Living Fossil Cycas Sect. Asiorientales Using a Genome-Wide Scanning Approach, doi:10.3390/ijms14048228.

Cycads are the classic living fossils with long evolutionary history but recently and rapidly diversified since the late Miocene [6]. Geographical isolation and small effective population sizes might be the reason for the rapid diversification [6]. The eastern Asian *Cycas* section *Asiorientales* is composed of two allopatric species *C. revoluta* and *C. taitungensis*, distributed widely in the Ryukyu archipelagos and Fujian Province of China and endemic to Taiwan, respectively. These two eastern Asian cycad species could coalesce to approximate 350 million years ago (mya) as inferred from fluctuations in very recent demography since around 5~3.5 mya [7]. The paraphyletic relationship was also suggested as a consequence of historical demographic fluctuations with past gene flow among ancestral populations [7].

*Cycas revoluta* is distributed around islands of the southern Japan and the eastern China, and widely planted in eastern Asia for horticultural reason. In contrast to the wider distributed and planted *C. revoluta*, *C. taitungensis* is endemic and restrictedly distributed in eastern Taiwan. Plantation of *C. taitungensis* is also found locally in the eastern Taiwan. Earlier unlawful lumbering and habitat destruction limited the population expansion of the native populations of *C. taitungensis* in Taiwan. Invasion of *Aulacaspis* spp. due to the introduction of foreign cycads in recent years severely infects the native *C. taitungensis* and results in high mortality in wild populations. Even though there is an endangered situation of extant populations, the genetic diversity of *C. taitungensis* does not appear low as estimated by either plastid and ribosomal DNA [7,8] or isozymes [9]. Low, even insignificant, genetic differentiation between populations of *C. taitungensis* was also reported, which is unusual for the species constrained by migratory capability of pollinators and seed carriers [8]. Restricted distribution and low migration rates for pollination decreased the probability of interspecific gene flow. However, *C. taitungensis* and *C. revoluta* still share identical genotypes and revealed paraphyletic relationships, which was suggested as a consequence of sharing high degrees of common ancestral polymorphisms [7]. Although Chiang *et al.*[7] indicated that the interspecific gene flow is greatly restricted between the extant wild populations by geographic isolation and low dispersability, gene flow between sympatrically cultivated samples were not examined yet. Whether such ancestral genetic compatibility still persisted between extant species is still unknown.

Hybridization between *C. taitungensis* and phylogenetically distant *C. ferruginea* was also performed in the botanical garden for examining the maternal inherited plastid genome [10]. However, the natural hybridization between these geographical isolated species is not reported. Genetic introgression between related plant species is probably more frequently than what we thought [4] through the process of gene transfer between plastid and nuclear genomes [11]. Sympatric growing of *C. revoluta* and *C. taitungensis* in gardens could enhance interspecific pollen (microspore) flow, and the amount of compatible microspores dictates the rate of introgression [12]. Cycad is mostly considered as entomophily (insect-pollination) [13,14] while the anemophily (wind-pollination) in certain cycad species could be a newly derived trait, *i.e.*, autapomorphy [13]. *Cycas revoluta* is also entomophilic by Coleoptera insects [15]. However, the trait of anemophily is also observed in *C. revoluta* although the amount of airborne pollens drop quickly in male cones distant from >2 [15], which supports Chiang *et al.*'s [7] speculation of restricted gene flow but also implies higher probability of successful pollinating via either insects or wind between individuals growing nearby.

For horticultural reasons, *C. revoluta* is introduced and widely planted in gardens, schools, and as the shade trees in Taiwan, which means that *C. revoluta* is not as distant from its relative *C. taitungensis*. In considering pollination compatibility of interspecific *Cycas* species [10], a prediction of enhanced introgression between *C. revoluta* and *C. taitungensis* by horticultural introduction of *C. revoluta* was made. Examination of genetic composition of individuals both in the wild and in the gardens in Taiwan was performed to clarify the prediction of introgression by horticulture. We hypothesized that interspecific gene flow increases the shared genetic polymorphisms in the cultivated individuals but not in the allopatrically wild populations. Therefore, we compared the genetic composition of the wild and cultivated samples of both *C. revoluta* and *C. taitungensis* using multilocus markers (*i.e.*, the amplified fragment length polymorphisms, AFLPs) for clarifying the interspecific gene flow at the artificially sympatric areas (e.g., gardens).

Neutral genes that experienced different rates of gene flow provide hints for investigating the asymmetric gene flow [16]. For evaluating the genetic impact of the horticultural introduction of *C. revoluta* on the native *C. taitungensis*, population genetic analyses with the multilocus neutral loci were used to address two specific objectives of this study: (1) to reevaluate the genetic diversity of the cultivated and extant wild populations of *C. taitungensis* by multilocus marker and (2) to evaluate the degrees of introgression between *C. revoluta* and *C. taitungensis* in Taiwan. In this study, we also discuss the mechanisms (e.g., geographic isolation or niche specialization) of the divergence of these two phylogenetically related species through an examination of the horticultural introgression.

## RESULTS

### Sampling and Neutrality Test of AFLP Polymorphisms

Twenty one and 62 samples of *C. revoluta* and *C. taitungensis* were collected from wild populations, respectively, and 29 cultivated samples of *C. revoluta* and 151 and 134 cultivated adults and seedlings of *C. taitungensis* were collected in the South and Southeast Taiwan (Figure 1 and Table A1). In total, 311 polymorphic loci were obtained from 397 samples of both*C. revoluta* and *C. taitungensis*. One and 48 of the 311 loci had significantly lower and higher $F_{ST}$ values deviating from 95% confidence intervals, respectively, and were defined as negative and positive outliers, respectively (Figure A1). The rest of the 262 loci located within the 95% confidence intervals were considered as neutral loci. Example panels of species- and population-specific loci detected by the AFLP genotyping are provided in Figure A2.

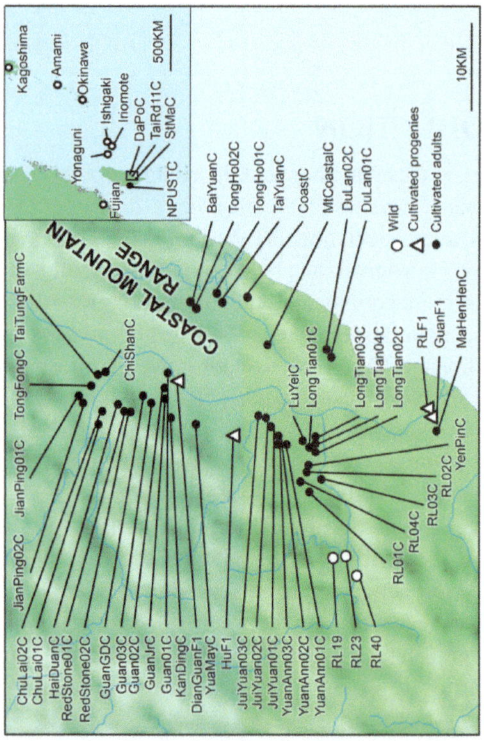


**Figure 1.** Map of sampling sites in this study. The large panel indicates the sampling sites of *C. taitungensis* in the southeast Taiwan and the upper-left panel indicates the sampling sites of *C. revoluta*. Wild populations, cultivated adults and progenies are marked in hollow circles, full circles, and hollow triangles, respectively. Codes of the sampling sites correspond to Table A1.

## Genetic Diversity

Both *C. revoluta* and *C. taitungensis* have 77.86% polymorphic neutral loci (Table 1). *Cycas revoluta* revealed slightly higher genetic diversity than *C. taitungensis* in total samples in Shannon's information index ($I$ = 0.422 ± 0.017 and 0.401 ± 0.016, respectively, $p$ = 0.0140, Student's $t$ test) and the expected heterozygosity ($h$ = 0.286 ± 0.012 and 0.266 ± 0.011, respectively, $p$ = 0.0294) but insignificantly different in indices number of effective alleles ($Ne$ = 1.504 ± 0.024 and 1.446 ± 0.021, respectively, $p$= 0.1261) and unbiased heterozygosity ($uh$ = 0.292 ± 0.012 and 0.267 ± 0.011, respectively, $p$ = 0.1208) even though the sample size of *C. revoluta* is relatively small (Table 1). However, when only comparing the wild population samples between two species, *C. revoluta* has relatively smaller but insignificant genetic diversity than *C. taitungensis* in all indices ($p$ = 0.9051, 0.3934, 0.5477 and 0.8968 in $Ne$, $I$, $h$ and $uh$, respectively). The relatively lower genetic diversity of wild *C. revoluta* than *C. taitungensis* is consistence with estimates of plastid DNA sequences by Huang *et al.*[8] and Kyoda and Setoguchi [17] but inconsistent with Chiang *et al.*[7]. Although the different estimates could be due to sample sizes, the probability of heterogeneous evolutionary rates in different genomic markers cannot be excluded. When comparing with the cultivated adults, *C. taitungensis* has relatively higher genetic diversity than the introduced *C. revoluta* in $I$ and $h$ ($p$ = 0.0195 and 0.0268, respectively) but insignificantly different in the other indices; non-significant differences of diversity indices ($p$ > 0.05) were also estimated between cultivated adults and progenies of *C. taitungensis*, between cultivated progenies of *C. taitungensis* and cultivated adults of *C. revoluta*, and between wild populations and cultivated adults or progenies of both species (Table 1). Detailed genetic diversity of each population is shown in Table 1.

**Table 1.** Genetic diversity of wild populations and cultivated samples of *C. revoluta*and *C. taitungensis* estimated using 262 neutral AFLP loci.

| Species/Population | $N$ | $Ne$ | $I$ | $h$ | $uh$ | $\%P$ |
|---|---|---|---|---|---|---|
| *Cycas revoluta* | 50 | 1.504 ± 0.024 | 0.422 ± 0.017 | 0.286 ± 0.012 | 0.292 ± 0.012 | 77.86% |
| Wild population | 21 | 1.406 ± 0.024 | 0.353 ± 0.017 | 0.236 ± 0.012 | 0.248 ± 0.013 | 69.47% |
| Cultivated-Adults | 29 | 1.390 ± 0.024 | 0.343 ± 0.017 | 0.227 ± 0.012 | 0.235 ± 0.013 | 69.47% |
| *Cycas taitungensis* | 347 | 1.446 ± 0.021 | 0.401 ± 0.016 | 0.266 ± 0.011 | 0.267 ± 0.011 | 77.86% |
| Wild population | 62 | 1.410 ± 0.022 | 0.373 ± 0.016 | 0.246 ± 0.011 | 0.250 ± 0.012 | 77.10% |
| Cultivated-Adults | 151 | 1.446 ± 0.022 | 0.397 ± 0.016 | 0.264 ± 0.011 | 0.266 ± 0.011 | 76.72% |
| Cultivated-Progeny | 134 | 1.410 ± 0.021 | 0.374 ± 0.016 | 0.246 ± 0.011 | 0.248 ± 0.011 | 75.95% |

$N$, sample size; $Ne$, number of effective alleles; $I$, Shannon's Information Index; $h$, expected heterozygosity; $uh$, unbiased expected heterozygosity; $\%P$, percentage of polymorphic loci.

## Species-Associated Loci

Linkage group with the character "species", named species-associated loci, was detected by neutral loci using loose (LOD 3.0) and strict criteria (LOD 6.0). These loci were chosen as the loci diverged after species divergence. Among 262 neutral loci, 91 and 23 loci were detected linked with the character "species" at linkage threshold LOD 3.0 and LOD 6.0, respectively (Figure A3). The abundance of species-associated loci at loose and strict criteria composes about one-third and one-tenth neutral loci. In addition, the linkage-group loci at LOD 6.0 revealed two linkage subgroups, named linkage subgroup 1 (tightly linked with "species") and subgroup 2, composed of 10 and 13 loci, respectively (Figure A3B). Four sets of loci (262 neutral loci, 91 species-associated loci at LOD 3.0, 23 species-associated loci at LOD 6.0, and 10 species-associated loci of linkage subgroup 1 at LOD 6.0) were used for the further PCoA and Bayesian clustering analysis.

## Principle Coordinate Analyses

Genetic homogeneity of wild and cultivated samples between *C. revoluta* and *C. taitungensis* is tested by PCoA. In comparison of wild populations by 262 neutral loci, genetic compositions of two species cannot be distinguished at the first axis (explained 36.1% variations) but are obviously distinct by the first two axes (explained 55.64% variations) (Figure 2A). Similar result was also detected in species-associated loci at linkage threshold LOD 3.0 that the resolution of only first axis (38.48%) is worse but is better in the first two axes (61.27%) (Figure 2B), while genetic distinction is clear at the first axis at the stricter criterion LOD 6.0 (explains 42.27% variations) or its linkage subgroup 1 (explains 40.82% variations) (Figure 2C,D).

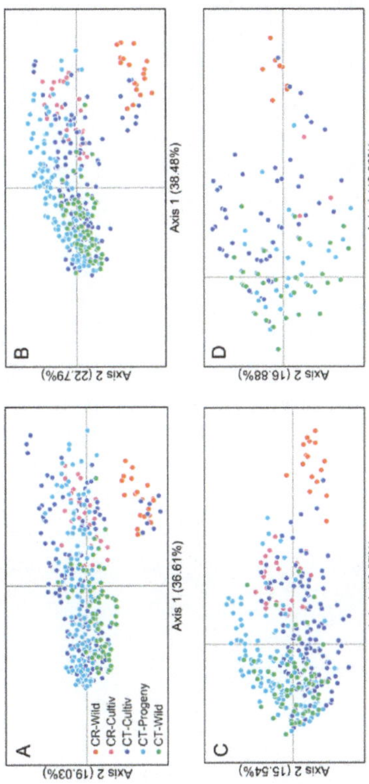


**Figure 2.** The first two axes plots of principle coordinate analysis (PCoA) calculated using (**A**) 262 neutral loci; (**B**) 91 species-associat-

ed loci determined under criteria LOD 3.0; (**C**) 23 species-associated loci determined under criteria LOD 6.0; and (**D**) 10 species-associated loci of the linkage subgroup 1 under criteria LOD 6.0. Abbreviations CR ad CT indicate *C. revoluta* and *C. taitungensis*, respectively; Wild, Cultiv, and Progeny indicate the wild populations, the cultivated adults, and the cultivated seedlings, respectively.

When comparing cultivated samples to the wild populations, genetic compositions of *C. revoluta* are indistinguishable at the first axis by 262 neutral loci or species-associated loci at LOD 3.0 but can be obviously distinguished at the first two axes (Figure 2A,B). However, obvious distinction is shown at the first axis under stricter criterion of species association (Figure 2C,D), which implies the shift of genetic components in cultivated samples of *C. revoluta* from the wild populations. In contrast to *C. revoluta*, both cultivated adults and progenies of *C. taitungensis* have wider ranges of genetic composition, covering the genetic distribution of wild populations of *C. taitungensis* and cultivated samples of *C. revoluta* in four loci sets (Figure 2). This result also indicates shifts of genetic components of partial samples of cultivated *C. taitungensis* from wild populations. Furthermore, a very clear pattern was found of the cultivated samples of *C. revoluta* grouped with the cultivated samples of *C. taitungensis*, especially with the cultivated adults, in all four loci-set analyses (Figure 2). However, the genetic distribution of the cultivated *C. revoluta* is not entirely covered by wild samples but by cultivated samples of *C. taitungensis*. It is also noticeable that certain samples of cultivated adults of *C. taitungensis* are grouped to the wild *C. revoluta*. We are not sure whether it implies a long-distant introgression from Ryukyus or Fujian, but these results clearly indicate more severely genetic admixture among cultivated cycads than among allopatrically wild populations.

## Bayesian Clustering Analysis

In the Bayesian clustering analysis, the best group manner is inferred as two by the $\Delta K$ evaluation ($\Delta K = 1154.354$ when $K = 2$) when using the 262 neutral loci (Figure A4). However, the grouping pattern is not completely consistent with taxonomic grouping, *i.e.*, revealed an admixture genetic composition, especially for the cultivated samples of *C. taitungensis* (Figure 3A). Also, for resolving the paradox of sharing common ancestral polymorphisms or recent introgression, the Bayesian clustering analysis was redone using species-associated loci. The Bayesian clustering analysis still revealed mosaic genetic composition at the species-associated loci at both criteria of species association (*i.e.*, LOD 3.0 and LOD 6.0, Figure 3B,C, respectively). Cultivated samples of both species were apparently composed of higher frequencies of alien genes than the wild populations (Figure 4A–C). However, the degree of genetic mosaicism increases when using the loci of linkage subgroup 1 of LOD 6.0 (Figure 3D), and even cultivated samples of *C. revoluta* were inferred to be composed of relatively higher genetic components belong to *C. taitungensis* than belong to *C. revoluta* itself (Figure 4D). This implied that (1) these ten loci of linkage subgroup 1 of LOD 6.0 could be locally selected or adapted in Taiwan, or (2) these ten loci revealed an opposite direction of introgression from *C. taitungensis* to *C. revoluta*. Because the outlier-loci have been eliminated in clustering analysis, the possibility of local adaptation can be excluded, and the introgression from *C. taitungensis* to *C. revoluta* is more appropriate to explain the high genetic components of *C. taitungensis* in cultivated *C. revoluta* (Figures 3Dand 4D).

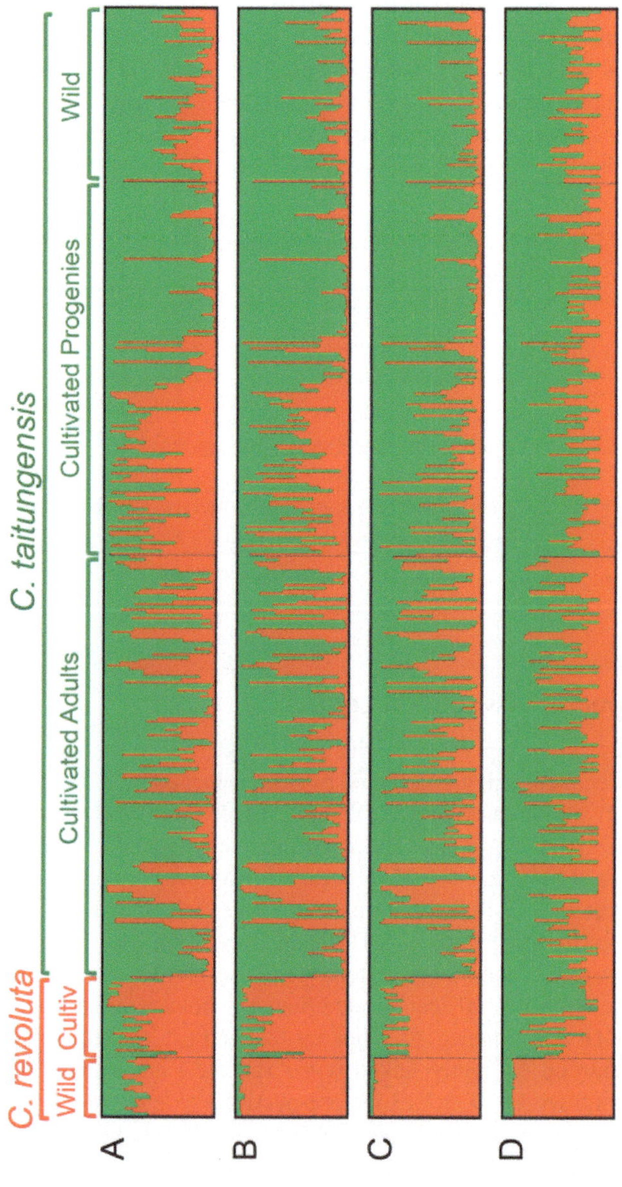


**Figure 3.** Assignment test by Bayesian clustering analysis of wild populations and cultivated samples of *C. revoluta* and *C. taitungensis* at clustering number $K = 2$. Patterns of genetic composition of each individual are detected by (**A**) 262 neutral loci; (**B**) species-associated loci at criteria LOD 3.0; and (**C**) LOD 6.0; and (**D**) the loci of linkage subgroup 1 of LOD 6.0.

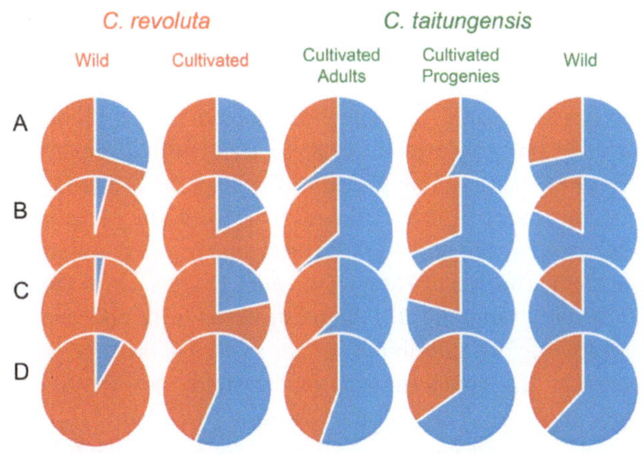


**Figure 4.** Average genetic composition of wild populations and cultivated samples ofC. *revoluta* and *C. taitungensis*, detected by (**A**) 262 neutral loci; (**B**) species-associated loci at criteria LOD 3.0; and (**C**) LOD 6.0; and (**D**) the loci of linkage subgroup 1 of LOD 6.0.

## DISCUSSION

Cycad is commonly planted for horticultural reasons for a long time. Introduction history of *C. revoluta* into Taiwan is probably decades or hundreds years ago, since the Chinese Hans or Japanese colonization. Horticultural introduction spreads *C. revoluta* in Asia and leads this species to secondarily contact with other cycads (e.g., *C. taitungensis* in Taiwan) since they diverged. Our genomic survey by AFLP multilocus scans evidenced that sympatric plantation increases the opportunity of introgression. This study evidences asymmetric introgression among invading and native cycads and suggests their niche conservatism after speciation by geographic isolation.

### Asymmetric Introgression between Cycad Species in Taiwan

Successful pollination of *C. revoluta* is limited by distance [15]. Allopatric distribution of wild populations of *C. revoluta*and *C. taitungensis* restricts the pollen flow from each other, while the horticultural introduction increases the chance of interspecific gene flow. Introgression could be considered as a kind of genetic invasion [1,16], and direction of introgression is commonly considered from native species into the invaders due to population size effect [16,18]. However, the contrast phenomenon of introgression from invaders into natives is also reported in poplar [19], rice [20], and bitter melons of Taiwan [21]. In this case of *Cycas* in Taiwan, severe introgression from invaders (*C. revoluta*) to natives (*C. taitungensis*) is obviously revealed in the undistinguishable patterns of PCoA at the first axis (Figure 2) and revealed in the Bayesian clustering analysis (Figures 3 and 4), while the opposite-directional introgression is also detected, despite being relatively small (Figure 4). Based on the distribution pattern (wider distribution in *C. revoluta vs.* restricted distribution in *C. taitungensis*) and the paraphyletic relationship [7], *C. taitungensis* could be just a unique lineage of the ancestor of *C. revoluta*with more autapomorphies. Therefore, *C. revoluta* could be more incompatible to receive the alien genes from *C. taitungensis*while *C. taitungensis* is more compatible to receive plesiomorphic alleles from *C. revoluta*.

### Paradox of Sharing Ancestral Polymorphisms and Recent Introgression

The shared polymorphisms are usually questioned as a consequence of common ancestral polymorphisms instead of introgression. For resolving this question, we redid the PCoA and Bayesian clustering analysis by the "species-associated loci". If there was no introgression, the species-associated loci would differentiate well without admixture; in contrast, if introgression happens, the species-associated loci would represent admixture pattern. This assumption is similar to the concept of "divergence hitchhiking" [22] but we only considered the neutral loci rather than the "outlier loci" for eliminating the interference of adaptation or speciation genes. The wild populations that allopatrically distributed were used as template for determining the species-associated loci in order to detect the introgression between horticultural *C. revoluta* and *C. taitungensis*. Frequencies of the sharing polymorphisms apparently decreased in wild populations of both species by the reanalysis of Bayesian clustering analysis, but unvaried in cultivated samples (Figure 4B,C, in comparison ofFigure 4A). This result indicated that the introgression occurred in cultivated cycads and evidenced the acceleration of introgression by horticultural introduction of *C. revoluta*.

### Genetic Chimera of Cultivated Cycads

Obvious patterns that a broader and continuous genetic distribution of cultivated *C. taitungensis* covers the wild populations and cultivated *C. revoluta* while the grouping of cultivated *C. revoluta* is distinct from wild populations of both species are shown in PCoA (Figure 2). This is probably because the cultivated samples are composed of genes of both species by hybridizing recombination, *i.e.*, chimeric DNA [23,24]. It also implied that the introgression could only happen between cultivated samples but not between wild populations. However, the cultivated chimeric *C. taitungensis* could probably backcross with wild populations in short geographic distance, which explains (1) continuous and broader genetic distribution of cultivated *C. taitungensis* covering with the wild populations in PCoA (Figure 2) and (2) highly genetic admixture of wild samples of *C. taitungensis* in Bayesian clustering analysis (Figure 3). In contrast, both PCoA and Bayesian clustering analysis showed that the backcross of cultivated samples with long-distant wild populations seems rarer in *C. revoluta*, which is probably because of low dispersability of pollens [15] and seeds [7]. Genetic chimera explain not only the sustention of genetic diversity of horticultural *C. taitungensis* but also the distinction between cultivated and wild samples of*C. revoluta* (Table 1), which is broadly evidenced in microbes [25,26]. Hybridizing recombination would also raise the number of rare alleles [27,28] especially in those newly derived hybrids [29]. However, this speculation of genetic chimera is difficult to test by AFLP markers and we hope to test this further by codominant-marker surveys (e.g., by microsatellite DNAs) like the beautiful case of grapevine [30,31].

### Niche Conservation Accelerates Sympatric Introgression

The two cycad species *C. revoluta* and *C. taitungensis* were geographically isolated with different habitats: *C.*

*revoluta*mostly grows along coasts and is subjected to salt spray in Ryukyus and Fujian Province of China while *C. taitungensis* grows under forests along river valleys in Taiwan Island [7]. Therefore, what mechanism, *i.e.*, geographic isolation or adaptive divergence, results in the species divergence is curious. Since we know that degrees of gene flow decrease between organisms "if adaptation to a particular habitat determines where organisms mate [32]" but would recover between organisms of niche conservatism [33], estimating the interspecific gene flow could be useful for determining the mechanism of geographically or adaptively reproductive isolation. In other words, the interspecific gene flow might be recovered when organisms met (*i.e.*, secondary contact) through niche conservation; in contrast, if species divergence with niche specialization, the reproductive isolation would be retained by the incompatibility of adapted genes between species [34] or by eliminating immigrant alleles [35]. In this case, higher genetic admixture was shown in cultivated samples of two species than the allopatrically distributed wild populations, implying that the species divergence could be mainly affected by geographic isolation rather than adaptive divergence.

Although these two *Cycas* species grow in different environments in the wild, the growing condition of both species is similar, implying their broad adaptability without niche specialization. In addition, multiple extant Asian cycad species, including *C. taitungensis*, are allopatric and restrictedly distributed with small population sizes [36–40] and is considered as relicts from glacial refugia [8]. The small population size and geographic isolation from other populations increase the effect of genetic drift resulting in species divergence. However, the time to geographic isolation seemed not enough to complete the reproductive isolation and was broken off by transplantation. In fact, frequent hybridization could be seen in botanical gardens in several species whether naturally or artificially [29,41]. Artificial hybridization between *C. revoluta* and *C. taitungensis* is also successfully done by horticulturists [42,43]. Hybridization between *C. taitungensis* and *C. ferruginea* (sect.*Stangerioides*) was even performed in the botanical garden [10]. This indicates that the introgression could occur more easily among these living fossil cycads than what we thought when secondarily contacting.

Although the introgression between the *C. revoluta* and *C. taitungensis* has been proved by genetic analyses, morphological characters (leaf traits) that are commonly used to identify these two cycads do not change. The unchanged leaf traits of the cultivated *C. taitungensis*, such as the flat leaves and plane leaflet margins (in contrast to the deep keeled leaves and revolute leaflet margins of *C. revoluta*), reflect the fact of none or rare effects on the leaf character shift after introgression. The unchanged morphotypes of the genetically chimeric individuals have made the introgression an unseen threat to the native cycads.

## EXPERIMENTAL SECTION

### Sampling

The sampling of cycad species included two parts: the wild samples and the cultivated samples. In the sampling of wild populations, because the main purpose of this study focused on the genetic introgression of horticultural cycads in Taiwan, an indicative sampling of three individuals from each wild population of *C. revoluta* were performed; in contrast, *C. taitungensis* is only restricted distributed in the Hong-Yeh valley (the preserve areas of 19th, 23rd and 40th Compartment of Yen-Ping Area, Taitung County) and sparse in the Coastal Mountain Range of the southeastern Taiwan, the sampling of wild*C. taitungensis* only focused on the main wild population at three compartments of Hong-Yeh valley (Figure 1). In the part of cultivated sampling, the sampling areas were concentrated on sympatrically distributed areas of two cultivated cycad species in the southern and southeastern Taiwan. Species identification of horticultural samples was based on two distinguished leaf characters: flatter leaves and plane leaflet margins in *C. taitungensis vs.* deep keeled leaves and revolute leaflet margins in *C. revoluta*. In addition to the adults, the seedlings (progenies) of *C. taitungensis* in nursery gardens were also collected. In total, 397 individuals were collected for genetic analyses. Detailed information of the sampling sites is listed in Table A1.

### DNA Extractions and AFLP Genotyping

Total genomic DNA was extracted with cetyl trimethylammonium bromide (CTAB) method [44]. The AFLP was performed following the method developed by Vos *et al.*[45] with little modification. Two restricted enzymes *Eco*RI (10 Unit) and *Mse*I (10 Unit) (New England Biolabs, Beverly, MA, USA) were used to digest the sample DNA with the following amplification by the pre-selected primers Eco+A (GACTGCGTACCAATTCA) and Mse+C (GATGAGTCCTGAGTAAC) and the selected primer pairs Eco+AGT/Mse+CTA, Eco+ACG/Mse+CTC, and Eco+AAT/Mse+CTG. These primers were labeled with florescence dye (6FAM, JOE, and TAMRA, respectfully), and the genotyping was performed on ABI Prism 3730XL (Applied Biosystems, Foster City, CA, USA). LIZ600 was used as size standard and peak size detection was conducted by Peak Scanner ver. 1.0 (Applied Biosystems, Foster City, CA, USA). Detailed methods are provided in Supplementary Materials.

### Data Scoring and Data Analyses

The present and absent loci of the AFLP bands (peaks) ranged from 50 to 300 bps were scored as 1 and 0, respectively. For evaluation of neutrality of AFLP loci, the Dfdist approach was used by the program McHeza [46]. A strict criterion of 95% confidence interval (CI) was set for defining the neutral-evolving loci. The percentage of polymorphic loci (PPL), number of effective alleles (*Ne*), expected heterozygosity (*h*), unbiased heterozygosity (*He*), and Shannon's information index (*I*) were estimated using the neutral loci by GenAlEx ver. 6.3 [47] in order to reveal the genetic diversity of wild and cultivated populations of *C. taitungensis* and *C. revoluta*. The principle coordinate analysis (PCoA) and the model based Bayesian clustering analysis were performed to evaluate the degrees of genetic admixture and genetic structure. The PCoA and the Bayesian clustering analysis were conducted by GenAlEx ver. 6.3 [47] and STRUCTURE ver. 2.3.3 [48–50], respectively. Simulation results of the best grouping number *K* analyzed by STRUCTURE were evaluated using $\Delta K$[51] by STRUCTURE HARVESTER ver. 0.6.8 [52] (see Figure A4).

Detailed descriptions and program settings were available in Supplementary Materials.

In order to ascertain the paradox of sharing common ancestral polymorphism [7] from introgression, we used the concept of linkage to determine the "species-associated" loci by detecting the samples from allopatrically wild populations of *C. revoluta* and *C. taitungensis*. This hypothesis was made under the premise of the species-associated loci were diverged after species divergence. Therefore, the divergence of species-associated loci would follow the divergence of species and would "link" with the character "species". Wild populations of allopatric distribution were used for looking for the species-associated loci. The "species" was treated as one character to join the 262 neutral loci to determine the linkage group using JoinMap ver. 4.0 [53]. Low and high linkage thresholds were set at the LOD 3.0 and LOD 6.0 to evaluate the loci that associated with "species", respectively. The other para followed the default setting of JoinMap. After determining the species-associated loci, the PCoA and Bayesian clustering analysis were redone for all samples (including the wild populations and cultivated samples) for detecting whether the introgression was occurred after horticultural introduction.

## CONCLUSIONS

Introgression between sympatrically cultivated *C. revoluta* and *C. taitungensis* reveals incomplete reproductive isolation between deep divergent species of cycads. The natural introgression among horticultural individuals from different sources also supports the inference that the selection is not necessary for introgression [16]. Genetic evaluation of the wild populations of both *Cycas* species indicates a more severe impact on population genetic structure of the native *C. taitungensis*than *C. revoluta*. Detection of the divergence pattern of the species-associated loci helps to distinguish the sources of genetic admixture between the recent introgression and sharing common ancestral polymorphisms. Furthermore, asymmetric introgression is probably due to the demographic imbalance of these two species at the wave front for surfing [7,16], which could threat the native species by rapid spread of invasive genes [54,55]. The introgression hence becomes another important conservation issue of cycads beyond the illegal logging, habitat destruction, and the plague of vermin.

## Acknowledgments

Funds are partially supported by National Science Council, Republic of China (NSC 99-2621-B-020-002-MY3) and the Forestry Bureau, Council of Agriculture, Republic of China (100 Forest development-7.1-conservation-88 and 100AS-1.1.8-FB-e1) to P.-C. Liao.

## APPENDIX

## Methods

### Methods for DNA Extractions and AFLP Genotyping

Fresh leaves were dried by silica gel immediately and ground to powder by liquid nitrogen after being carryied to the laboratory. Total genomic DNA was extracted by the cetyl trimethylammonium bromide (CTAB) method [44]. The extracted DNA was dissolved in 1X TE buffer and stored at −20 °C.

For the multilocus genome-scan approach, we adopted the amplified fragment length polymorphism (AFLP) for genetic assessment. The AFLP was performed following the method developed by Vos *et al.*[45] with little modification. A total of 250 ng DNA were digested by restriction enzymes, EcoRI (10 Unit) and MseI (10 Unit) (New England Biolabs, Beverly, MA, USA); in total 25 μL reaction in 37 °C for 3 h, followed by 70 °C for 15 min to inactivate enzyme activity. The 5 μL digested products were added to 15 μL ligation mix with 5 pmol EcoRI adapter, 50 pmol MseI adapter, and 1 Unit T4 DNA ligase (New England Biolabs, Beverly, MA, USA) in 16 °C for 1 h then 37 °C for 3 h. Ligated products were pre-selected by 0.5 mM primers Eco+A (GACTGCGTACCAATTCA) and Mse+C (GATGAGTCCTGAGTAAC), with 2.5 nmole dNTP, 3 nmole MgCl2, 0.2 μL 1% BSA and 1 Unit DNA Taq polymerase. Amplification reactions were performed under 94 °C for 30 s, 56 °C for 1 min, and 72 °C for 1 min with 20 cycles. Pre-selected products were used as template for selective amplification. Selective amplifications were conducted by using 0.5 mM primer pairs Eco+AGT/Mse+CTA, Eco+ACG/Mse+CTC, and Eco+AAT/Mse+CTG. These primers were labeled with florescence dye (6FAM, JOE, and TAMRA, respectfully), with 3 nmole MgCl2, 2.5 nmole dNTP, and 1 Unit DNA Taq polymerase. Selective PCR was set for 94°C for 2 minutes for reaction activation, followed by a total of 25 cycles of 94 °C for 30 s, 65 °C for 30 s (decreasing 1 °C every cycle until it reached 56 °C) and 72 °C for 1 min, with subsequently 72 °C for 30 min for final extension. Concentration of selective amplified products was checked under 1.5% agarose gels. Genotyping was performed on ABI Prism 3730XL (Applied Biosystems, Foster City, CA, USA). LIZ600 was used as size standard and peak size detection was conducted by Peak Scanner ver. 1.0 (Applied Biosystems, Foster City, CA, USA).

### Data Scoring, Genetic Diversity, and Population Structure

AFLP bands with the same migration distances were considered homologous loci and scored manually as present (1) or absent (0). The sizes of the AFLP bands scored ranged from 50 to 300 bps. For evaluation of neutrality of AFLP loci, the Dfdist approach which evaluates the distribution of heterozygosity and genetic differentiation ($F_{ST}$) of each locus [56,57] was used by the program McHeza [46]. Since we wanted to eliminate the interference of adaptive effect for evaluating introgression, a strict criterion of 95% confidence interval (CI) rather than the 99% CI was set for defining the neutral-evolving loci. Two strategies were used for detecting the loci with outlier $F_{ST}$: (1) the mean $F_{ST}$ was calculated by McHeza and forced the simulations according to the mean $F_{ST}$; (2) we ran McHeza as in Strategy 1 but ran the simulation after removing the loci outside the 95% CI, as recommended by Antao *et al.*[58]. One million Markov chain simulations were performed. Each strategy was run 3 times to obtain a converged inference to ensure the accuracy of estimation.

Genetic diversity indices, including the percentage of polymorphic loci (PPL), number of effective alleles [*Ne* = 1/

$(p_2 + q_2)$, where $p$ = band frequency and $q = 1 - p$], expected heterozygosity [$h = 1 - (p_2 + q_2)$], unbiased heterozygosity {$He = [N/(N-1)] \times h$, where $N$ is sample size}, and Shannon's information index [$I = -1 \times (p \times L_n p + q \times L_n q)$], were estimated using the neutral loci by GenAlEx ver. 6.3 [47]. Genetic composition and genetic distinction between the wild populations and cultivated samples of *C. revoluta* and *C. taitungensis* was evaluated by the principle coordinate analysis (PCoA). The Bayesian clustering analysis was also performed to evaluate the degrees of genetic admixture and the genetic structure among populations, using STRUCTURE ver. 2.3.3 [48–50]. The admixture model was used [59]. Posterior probability of the grouping number ($K$= 1~10) was estimated by the Markov chain Monte Carlo (MCMC) method with 10 independent runs to evaluate the consistency of the results, using 3,000,000 steps with a 500,000-step burn-in for each run. The best grouping number was evaluated using $\Delta K$[51] by STRUCTURE HARVESTER ver. 0.6.8 [52]. A final 10,000,000 simulations, with a 1,000,000-step burn-in, were executed based on the best $K$.

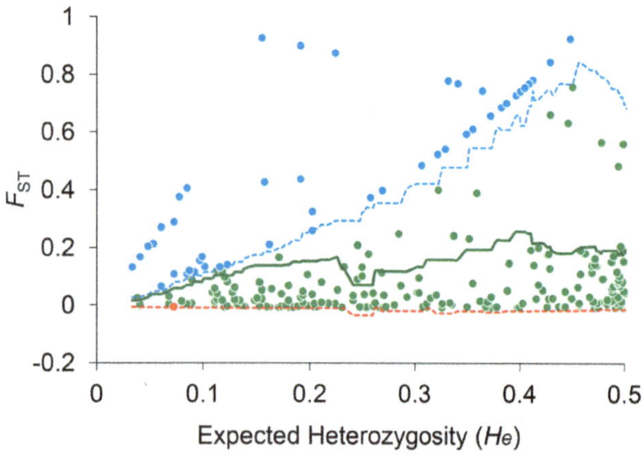


**Figure A1.** Examination of neutrality of AFLP loci based on Beaumont and Nichols's [56] method by program Mcheza [46]. Loci located within the 95% confidence intervals are taken as neutral loci (solid dots) and the positive (open diamonds) and negative outliers (open squares) are taken as candidate loci under positive and balancing selection, respectively.

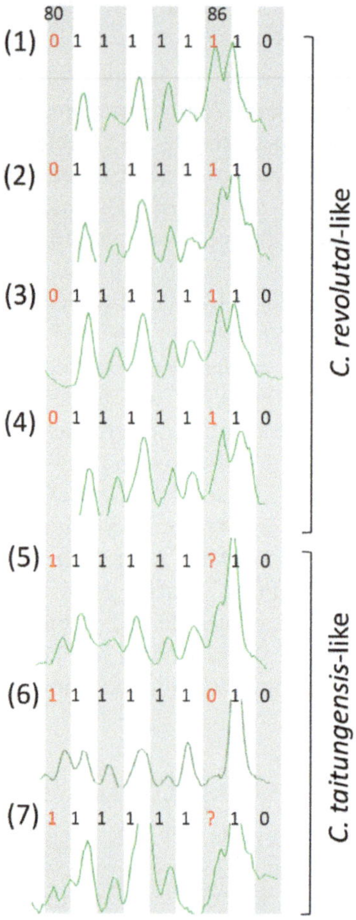


**Figure A2.** Examples of the ABI prism graphs revealing the loci with present (1) and absent (0) peaks. The question mark (?) indicates the missing (ambiguous) allele. Examples were separated to seven populations: (**1**) the wild population of *C. revoluta* in Ryukyus; (**2**) the wild population of *C. revoluta* in Fujian; (**3**) the *C. revoluta*-like cultivated population of *C. revoluta* in Taiwan; (**4**) the *C. revoluta*-like cultivated population of *C. taitungensis* in Taiwan; (**5**) the wild population of *C. taitungensis* in Taiwan; (**6**) the *C. taitungensis*-like cultivated population of *C. taitungensis* in Taiwan; and (**7**) the *C. taitungensis*-like cultivated population of *C. revoluta* in Taiwan. The polymorphic loci represented in this example graph are indicated in red (*i.e.*, the 80 and 86 sites), which separate these individuals to the *C. revoluta*-like and the *C. taitungensis*-like groups in genetic components.

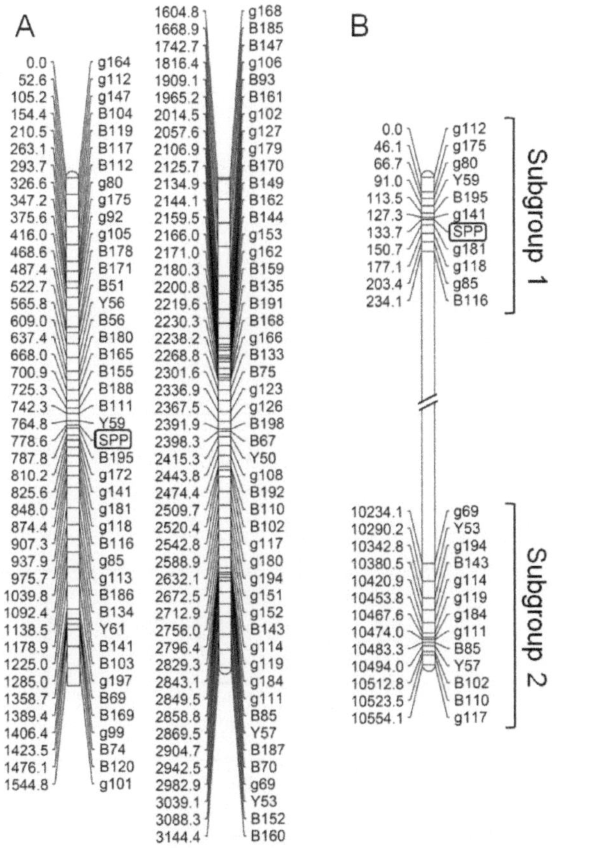


**Figure A3.** Linkage maps show the species-associated loci estimated by the maximum likelihood mapping algorithm using JoinMap ver. 4.0 [53]. Linkage group of species-associated loci estimated by criteria of (**A**) LOD 3.0 and (**B**) LOD 6.0 are shown. Two linkage subgroups at LOD 6.0 were inferred. The code SPP indicates the "character" as species.

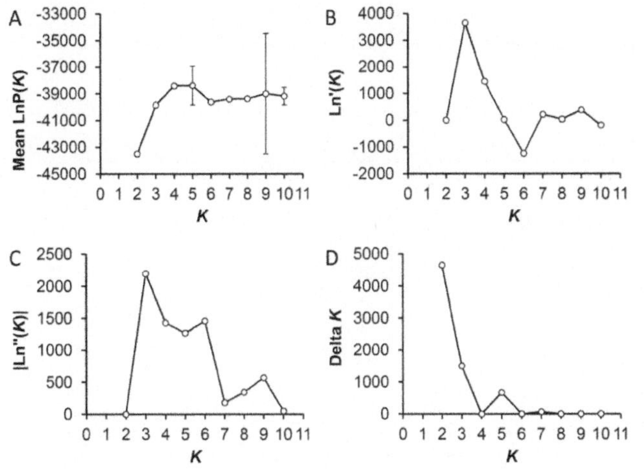


**Figure A4.** Evaluation of the best grouping number ($K$) of the Bayesian clustering analysis using Evanno *et al.*'s [51] methods. (**A**) Mean $L_nP(K)$; (**B**) $L_n'(K)$, equation to $L_n'(K) = L_nP(K) - L_nP(K-1)$; (**C**) $|L_n''(K)|$, equation to $|L_n'(K+1) - L_n'(K)|$; and (**D**) $\Delta K$, equation to $m(|L''(K)|)/s[L(K)]$, where $m$ and $s$ are the mean and standard deviation, respectively.

**Table A1.** Information of sampling sites of *Cycas revoluta* and *C. taitungensis*.

| Species | Wild/ Cultivated | Adult/Progeny | Sampling Site (Population) | Code | Leaf trait | N | Latitude | Longitude |
|---|---|---|---|---|---|---|---|---|
| *Cycas revoluta* | Wild | Adult | Iriomote | Iriomote | A | 3 | 24.333708 | 123.820337 |
| *Cycas revoluta* | Wild | Adult | Ishigaki | Ishigaki | A | 3 | 24.337730 | 124.153348 |
| *Cycas revoluta* | Wild | Adult | Yonaguni | Yonaguni | A | 3 | 24.440478 | 122.985870 |
| *Cycas revoluta* | Wild | Adult | Fujian | Fujian | A | 3 | 26.191753 | 119.537165 |
| *Cycas revoluta* | Wild | Adult | Okinawa | Okinawa | A | 3 | 26.207440 | 127.675227 |
| *Cycas revoluta* | Wild | Adult | Amami | Amami | A | 3 | 28.286530 | 129.385748 |
| *Cycas revoluta* | Wild | Adult | Kagoshima | Kagoshima | A | 3 | 31.591265 | 130.554270 |
| *Cycas revoluta* | Cultivated | Adult | Campus of National Pingtung University of Science and Technology | NPUSTC | A | 10 | 22.638740 | 120.600127 |
| *Cycas revoluta* | Cultivated | Adult | Notre Dame Heath Farm, Taitung County | StMaC | A | 7 | 22.712072 | 121.073585 |
| *Cycas revoluta* | Cultivated | Adult | The roadside of Highway No. 11 at 125K | TaiRd11C | A | 7 | 23.012964 | 121.324124 |
| *Cycas revoluta* | Cultivated | Adult | Taitung Tapo Elementary School | DaPoC | A | 5 | 23.125284 | 121.230655 |

| | | | | | | | | |
|---|---|---|---|---|---|---|---|---|
| *Cycas taitungensis* | Wild | Adult | The preserve area of 40th Compartment of Yen-Ping Area, Taitung County | RL40 | B | 15 | 22.857918 | 120.975018 |
| *Cycas taitungensis* | Wild | Adult | The preserve area of 23rd Compartment of Yen-Ping Area, Taitung County | RL23 | B | 31 | 22.867292 | 121.008930 |
| *Cycas taitungensis* | Wild | Adult | The preserve area of 19th Compartment of Yen-Ping Area, Taitung County | RL19 | B | 16 | 22.870879 | 121.019618 |
| *Cycas taitungensis* | Cultivated | Adult | Mahengheng Blvd., Taitung City | MaHenHenC | B | 18 | 22.771136 | 121.145511 |
| *Cycas taitungensis* | Cultivated | Adult | Taitung Dulan Elementary School | DuLan01C | B | 1 | 22.877599 | 121.227522 |
| *Cycas taitungensis* | Cultivated | Adult | A residence house near the Dulan Bridge | DuLan02C | B | 1 | 22.878884 | 121.230977 |
| *Cycas taitungensis* | Cultivated | Adult | A residence house at Hong-Yeh Village, Taitung County | RL01C | B | 4 | 22.893270 | 121.066878 |
| *Cycas taitungensis* | Cultivated | Adult | Outside of Taitung Hong-Yeh Elementary School | RL03C | B | 13 | 22.893641 | 121.063870 |
| *Cycas taitungensis* | Cultivated | Adult | Naruwan Hong-Yeh Hot Spring, Taitung County | RL04C | B | 1 | 22.899838 | 121.067684 |
| *Cycas taitungensis* | Cultivated | Adult | The intersection of Neighborhoods No. 2 and 3 at at Hong-Yeh Village, Taitung County | RL02C | B | 4 | 22.901864 | 121.082500 |
| *Cycas taitungensis* | Cultivated | Adult | The office of Longtian Old Folk's Club, Taitung County | LongTian02C | B | 4 | 22.903850 | 121.125340 |
| *Cycas taitungensis* | Cultivated | Adult | Taitung Longtian Elementary School | LongTian04C | B | 1 | 22.903850 | 121.124396 |
| *Cycas taitungensis* | Cultivated | Adult | No.400, Guangrong Rd., Luye Township, Taitung County | LongTian03C | B | 1 | 22.904206 | 121.126199 |
| *Cycas taitungensis* | Cultivated | Adult | No.23, Shengping Rd., Yanping Township, Taitung County | YenPinC | B | 1 | 22.904280 | 121.083471 |
| *Cycas taitungensis* | Cultivated | Adult | Longtian Cycad Orchard (private) | LongTian01C | B | 12 | 22.906056 | 121.123083 |
| *Cycas taitungensis* | Cultivated | Adult | Taitung Lu-Ye Junior High School | LuYeiC | B | 2 | 22.907052 | 121.135275 |
| *Cycas taitungensis* | Cultivated | Adult | Fuder House at the Yong'an Village, Luye Township, Taitung County | YuanAnn01C | B | 1 | 22.925728 | 121.124197 |

| | | | | | | | | |
|---|---|---|---|---|---|---|---|---|
| *Cycas taitungensis* | Cultivated | Adult | Community Center of Yong'an Village, Taitung County | YuanAnn02C | B | 1 | 22.930631 | 121.139224 |
| *Cycas taitungensis* | Cultivated | Adult | Taitung Yong'an Elementary School | YuanAnn03C | B | 2 | 22.933397 | 121.128774 |
| *Cycas taitungensis* | Cultivated | Adult | Ruiyuan Station, Luye Township, Taitung County | JuiYuan02C | B | 2 | 22.953617 | 121.155438 |
| *Cycas taitungensis* | Cultivated | Adult | Taitung Ruiyuan Elementary School | JuiYuan03C | B | 3 | 22.954660 | 121.153372 |
| *Cycas taitungensis* | Cultivated | Adult | A residence house at Coastal Range | MtCoastalC | B | 4 | 22.958790 | 121.183673 |
| *Cycas taitungensis* | Cultivated | Adult | A residence house at Ruiyuan Village | JuiYuan01C | B | 5 | 22.972250 | 121.164278 |
| *Cycas taitungensis* | Cultivated | Adult | Taitung Tai-yuan junior high school | TaiYuanC | B | 4 | 23.002525 | 121.289878 |
| *Cycas taitungensis* | Cultivated | Adult | No.45-1, Ganjyulin, Beiyuan, Donghe Township, Taitung County | TongHo01C | B | 1 | 23.004343 | 121.286874 |
| *Cycas taitungensis* | Cultivated | Adult | Taitung Yuemei Elementary School | YuaMayC | B | 1 | 23.009320 | 121.148922 |
| *Cycas taitungensis* | Cultivated | Adult | A residence house at Guanshan Township, Taitung County | Guan01C | B | 5 | 23.009578 | 121.172317 |
| *Cycas taitungensis* | Cultivated | Adult | Visitor Center of the East Coast National Scenic Area Administration | CoastC | B | 1 | 23.025080 | 121.327515 |
| *Cycas taitungensis* | Cultivated | Adult | Donghe Farm, Taitung County | TongHo02C | B | 1 | 23.037402 | 121.278763 |
| *Cycas taitungensis* | Cultivated | Adult | Taitung Beiyuan Elementary School | BaiYuanC | B | 2 | 23.040996 | 121.293182 |
| *Cycas taitungensis* | Cultivated | Adult | Guanshan Junior High School | GuanJrC | B | 7 | 23.044511 | 121.159973 |
| *Cycas taitungensis* | Cultivated | Adult | Taitung Kanding Elementary School | KanDingC | B | 3 | 23.045083 | 121.146476 |
| *Cycas taitungensis* | Cultivated | Adult | Guanshan Nursery garden | GuanGDC | B | 14 | 23.046528 | 121.177194 |
| *Cycas taitungensis* | Cultivated | Adult | Guanshan Station of Farm Irrigation and Engineering Association of Taitung | Guan02C | B | 1 | 23.046861 | 121.169305 |
| *Cycas taitungensis* | Cultivated | Adult | Guanshan Station of Taitung Forest District Office | Guan03C | B | 1 | 23.047280 | 121.160681 |
| *Cycas taitungensis* | Cultivated | Adult | Hongshi, , Haiduan Township, Taitung County | RedStone01C | B | 1 | 23.052083 | 121.169750 |

| | | | | | | | | |
|---|---|---|---|---|---|---|---|---|
| *Cycas taitungensis* | Cultivated | Adult | Taitung Hongshi Elementary School | RedStone02C | B | 1 | 23.052083 | 121.169750 |
| *Cycas taitungensis* | Cultivated | Adult | Taitung Haiduan Elementary School | HaiDuanC | B | 3 | 23.101523 | 121.176367 |
| *Cycas taitungensis* | Cultivated | Adult | Tung-Fong Nursery Garden (Private) | TongFongC | B | 7 | 23.114431 | 121.199563 |
| *Cycas taitungensis* | Cultivated | Adult | Taitung Chulai Elementary School | ChuLai02C | B | 2 | 23.117095 | 121.169028 |
| *Cycas taitungensis* | Cultivated | Adult | Dapu, Chishang Township, Taitung County | ChiShanC | B | 4 | 23.118353 | 121.213199 |
| *Cycas taitungensis* | Cultivated | Adult | Taitung Farm, Chihshang Township, Taitung County | TaiTungFarmC | B | 2 | 23.119285 | 121.210892 |
| *Cycas taitungensis* | Cultivated | Adult | Chulai, Haiduan Township, Taitung County | ChuLai01C | B | 1 | 23.124416 | 121.162591 |
| *Cycas taitungensis* | Cultivated | Adult | Jinping, Haiduan Township, Taitung County | JianPing01C | B | 5 | 23.130533 | 121.175423 |
| *Cycas taitungensis* | Cultivated | Adult | Taitung Jinping Elementary School | JianPing02C | B | 3 | 23.131481 | 121.176667 |
| *Cycas taitungensis* | Cultivated | Progeny | Taitung Nursery Garden (Collected by Taitung Forest District Office) | RLF1 | B | 49 | 22.775817 | 121.139128 |
| *Cycas taitungensis* | Cultivated | Progeny | Taitung Nursery Garden (Collected by Guanshan Station of Taitung Forest District Office) | GuanF1 | B | 16 | 22.775817 | 121.139128 |
| *Cycas taitungensis* | Cultivated | Progeny | Hong-Yeh Village Chiefmayor Mr. Hu's Cycad Orchard | HuF1 | B | 59 | 22.975416 | 121.131056 |
| *Cycas taitungensis* | Cultivated | Progeny | Dianguang Farm at Guanshan Township, Taitung County | DianGuanF1 | B | 10 | 22.987972 | 121.175432 |

* Leaf trait: A, strongly or moderately keeled leaves and revolute leaflet margins; B, moderately keeled (flatter) leaves and plane leaflet margins.

## Conflict of Interest

The authors declare that they have no conflict of interest.

## REFERENCES


1. Mallet, J. Hybridization as an invasion of the genome. *Trends Ecol. Evol* **2005**, *20*, 229–237.
2. Secondi, J.; Faivre, B.; Bensch, S. Spreading introgression in the wake of a moving contact zone. *Mol. Ecol* **2006**, *15*, 2463–2475.
3. Schaefer, J.F.; Duvernell, D.D.; Kreiser, B.R. Ecological and genetic assessment of spatial structure among replicate contact zones between two topminnow species. *Evol. Ecol* **2011**, *25*, 1145–1161.
4. Arnold, M.L. Natural hybridization and the evolution of domesticated, pest and disease organisms. *Mol. Ecol* **2004**, *13*, 997–1007.
5. Ellstrand, N.C.; Prentice, H.C.; Hancock, J.F. Gene flow and introgression from domesticated plants into their wild relatives. *Annu. Rev. Ecol. Syst* **1999**, *30*, 539–563.
6. Nagalingum, N.S.; Marshall, C.R.; Quental, T.B.; Rai, H.S.; Little, D.P.; Mathews, S. Recent synchronous radiation of a living fossil. *Science* **2011**, *334*, 796–799.
7. Chiang, Y.C.; Hung, K.H.; Moore, S.J.; Ge, X.J.; Huang, S.; Hsu, T.W.; Schaal, B.A.; Chiang, T.Y. Paraphyly of organelle DNAs in *Cycas*Sect. *Asiorientales*due to ancient ancestral polymorphisms. *BMC Evol. Biol.* **2009**, *9*.
8. Huang, S.; Chiang, Y.C.; Schaal, B.A.; Chou, C.H.; Chiang, T.Y. Organelle DNA phylogeography of *Cycas taitungensis*, a relict species in Taiwan. *Mol. Ecol* **2001**, *10*, 2669–2681.
9. Huang, S.; Hsieh, H.T.; Fang, K.; Chiang, Y.C. Patterns of genetic variation and demography of *Cycas taitungensis* in Taiwan. *Bot. Rev* **2004**, *70*, 86–92.
10. Zhong, Z.R.; Li, N.; Qian, D.; Jin, J.H.; Chen, T. Maternal inheritance of plastids and mitochondria in *Cycas* L. (Cycadaceae). *Mol. Genet. Genomics* **2011**, *286*, 411–416.
11. Wang, D.; Wu, Y.W.; Shih, A.C.C.; Wu, C.S.; Wang, Y.N.; Chaw, S.M. Transfer of chloroplast genomic DNA to mitochondrial genome occurred at least 300 MYA. *Mol. Biol. Evol* **2007**, *24*, 2040–2048.

12. Anttila, C.K.; Daehler, C.C.; Rank, N.E.; Strong, D.R. Greater male fitness of a rare invader (*Spartina alterniflora*, Poaceae) threatens a common native (*Spartina foliosa*) with hybridization. *Am. J. Bot* **1998**, *85*, 1597–1601.

13. Pellmyr, O.; Tang, W.; Groth, I.; Bergstrom, G.; Thien, L.B. Cycadcone and angiospermfloralvolatiles: Inferences for the evolution of insect pollination. *Biochem. Syst. Ecol* **1991**, *19*, 623–627.

14. Schneider, D.; Wink, M.; Sporer, F.; Lounibos, P. Cycads: Their evolution, toxins, herbivores and insect pollinators.*Naturwissenschaften* **2002**, *89*, 281–294.

15. Kono, M.; Tobe, H. Is *Cycas revoluta* (Cycadaceae) wind- or insect-pollinated? *Am. J. Bot* **2007**, *94*, 847–55.

16. Currat, M.; Ruedi, M.; Petit, R.J.; Excoffier, L. The hidden side of invasions: Massive introgression by local genes.*Evolution* **2008**, *62*, 1908–1920.

17. Kyoda, S.; Setoguchi, H. Phylogeography of *Cycas revoluta* Thunb. (Cycadaceae) on the Ryukyu Islands: Very low genetic diversity and geographical structure. *Plant Syst. Evol* **2010**, *288*, 177–189.

18. Valtuena, F.J.; Preston, C.D.; Kadereit, J.W. Evolutionary significance of the invasion of introduced populations into the native range of *Meconopsis cambrica*. *Mol. Ecol* **2011**, *20*, 4318–4331.

19. Chenault, N.; Arnaud-Haond, S.; Juteau, M.; Valade, R.; Almeida, J.L.; Villar, M.; Bastien, C.; Dowkiw, A. SSR-based analysis of clonality, spatial genetic structure and introgression from the Lombardy poplar into a natural population of *Populus nigra* L. along the Loire River. *Tree Genet. Genomes* **2011**, *7*, 1249–1262.

20. Xia, H.B.; Wang, W.; Xia, H.; Zhao, W.; Lu, B.R. Conspecific crop-weed introgression influences evolution of weedy rice (*Oryza sativa* f. *spontanea*) across a geographical range. *PLoS One* **2011**, *6*, e16189.

21. Liao, P.C.; Tsai, C.C.; Chou, C.H.; Chiang, Y.C. Introgression between cultivars and wild populations of *momordica charantia* L. (Cucurbitaceae) in Taiwan. *Int. J. Mol. Sci* **2012**, *13*, 6469–6491.

22. Via, S.; West, J. The genetic mosaic suggests a new role for hitchhiking in ecological speciation. *Mol. Ecol* **2008**, *17*, 4334–4345.

23. Zhuang, Q.Q.; Zhang, Z.G.; Chen, F.G.; Xia, G.M. Comparative and evolutionary analysis of new variants of ω-gliadin genes from three A-genome diploid wheats. *J. Appl. Genet* **2012**, *53*, 125–131.

24. Cronn, R.; Small, R.L.; Haselkorn, T.; Wendel, J.F. Cryptic repeated genomic recombination during speciation in*Gossypium gossypioides*. *Evolution* **2003**, *57*, 2475–2489.

25. Smits, S.L.; Lavazza, A.; Matiz, K.; Horzinek, M.C.; Koopmans, M.P.; de Groot, R.J. Phylogenetic and evolutionary relationship among torovirus field variants: Evidence for multiple intertypic recombination events. *J. Virol* **2003**, *77*, 9567–9577.

26. Erny, C.; Raoult, P.; Alais, A.; Butterlin, G.; Delobel, P.; Matei-Radoi, F.; Casaregola, S.; Legras, J.L. Ecological success of a group of *Saccharomyces cerevisiae*/*Saccharomyces kudriavzevii* hybrids in the northern European wine-making environment. *Appl. Environ. Microb* **2012**, *78*, 3256–3265.

27. Krutovskii, K.V.; Bergmann, F. Introgressive hybridization and phylogenetic relationships between Norway, *Picea abies*(L.) Karst. and Siberian, *P. obovata* Ledeb. spruce species studied by isozyme loci. *Heredity* **1995**, *74*, 464–480.

28. Schilthuizen, M.; Hoekstra, R.F.; Gittenberger, E. The 'rare allele phenomenon' in a ribosomal spacer. *Mol. Ecol* **2001**,*10*, 1341–1345.

29. Liao, P.C.; Shih, H.C.; Yen, T.B.; Lu, S.Y.; Cheng, Y.P.; Chiang, Y.C. Molecular evaluation of interspecific hybrids between *Acer albopurpurascens* and *A. buergerianum* var. formosanum. *Bot. Stud* **2010**, *51*, 413–420.

30. Hocquigny, S.; Pelsy, F.; Dumas, V.; Kindt, S.; Heloir, M.C.; Merdinoglu, D. Diversification within grapevine cultivars goes through chimeric states. *Genome* **2004**, *47*, 579–589.

31. Moncada, X.; Pelsy, F.; Merdinoglu, D.; Hinrichsen, P. Genetic diversity and geographical dispersal in grapevine clones revealed by microsatellite markers. *Genome* **2006**, *49*, 1459–1472.

32. Orr, M.R.; Smith, T.B. Ecology and speciation. *Trends Ecol. Evol* **1998**, *13*, 502–506.

33. Arteaga, M.C.; McCormack, J.E.; Eguiarte, L.E.; Medellin, R.A. Genetic admixture in multidimensional environmental space: Asymmetrical niche similarity promotes gene flow in armadillos (*Dasypus novemcinctus*). *Evolution* **2011**, *65*, 2470–2480.

34. Johnson, P.A.; Hoppensteadt, F.C.; Smith, J.J.; Bush, G.L. Conditions for sympatric speciation: A diploid model incorporating habitat fidelity and non-habitat assortative mating. *Evol. Ecol* **1996**, *10*, 187–205.

35. Nosil, P.; Vines, T.H.; Funk, D.J. Perspective: Reproductive isolation caused by natural selection against immigrants from divergent habitats. *Evolution* **2005**, *59*, 705–719.

36. Hill, K.D.; Nguyen, H.T.; Loc, P.K. The genus *Cycas* (Cycadaceae) in Vietnam. *Bot Rev* **2004**, *70*, 134–193.

37. Hill, K.D.; Yang, S.L. The genus *Cycas* (Cycadaceae) in Thailand. *Brittonia* **1999**, *51*, 48–73.

38. Hill, K.D. The genus *Cycas* (Cycadaceae) in China. *Telopea* **2008**, *12*, 71–118.

39. Lindstrom, A.J.; Hill, K.D.; Stanberg, L.C. The genus *Cycas* (Cycadaceae) in the Philippines. *Telopea* **2008**, *12*, 119–145.

40. Lindstrom, A.J.; Hill, K.D.; Stanberg, L.C. The genus *Cycas* (Cycadaceae) in Indonesia. *Telopea* **2009**, *12*, 385–418.

41. Ye, Q.G.; Yao, X.H.; Zhang, S.J.; Kang, M.; Huang, H.W. Potential risk of hybridization in *ex situ* collections of two endangered species of *Sinoiackia* Hu (Styracaceae). *J. Integr. Plant. Biol* **2006**, *48*, 867–872.

42. Broome, T. *Hand-pollination of Cycads*, Available online: http://www.plantapalm.com/vce/horticulture/pollination.htm(accessed on 10 April 2013).
43. Bananas.org Forum. Available online: http://www.bananas.org/f8/cycas-taitungensis-x-revoluta-3581.html (accessed on 10 April 2013).
44. Doyle, J.J.; Doyle, J.L. A rapid DNA isolation procedure for small quantities of fresh leaf tissue. *Phytochem. Bull* **1987**,*19*, 11–15.
45. Vos, P.; Hogers, R.; Bleeker, M.; Reijans, M.; van de Lee, T.; Hornes, M.; Frijters, A.; Pot, J.; Peleman, J.; Kuiper, M.; *et al.* AFLP: A new technique for DNA fingerprinting. *Nucleic Acids Res* **1995**, *23*, 4407–4414.
46. Antao, T.; Beaumont, M.A. Mcheza: A workbench to detect selection using dominant markers. *Bioinformatics* **2011**, *27*, 1717–1718.
47. Peakall, R.; Smouse, P.E. GENALEX 6: Genetic analysis in Excel. Population genetic softwarc for teaching and research. *Mol. Ecol. Notes* **2006**, *6*, 288–295.
48. Falush, D.; Stephens, M.; Pritchard, J.K. Inference of population structure using multilocus genotype data: Dominant markers and null alleles. *Mol. Ecol. Notes* **2007**, *7*, 574–578.
49. Falush, D.; Stephens, M.; Pritchard, J.K. Inference of population structure using multilocus genotype data: Linked loci and correlated allele frequencies. *Genetics* **2003**, *164*, 1567–1587.
50. Pritchard, J.K.; Stephens, M.; Donnelly, P. Inference of population structure using multilocus genotype data. *Genetics***2000**, *155*, 945–959.
51. Evanno, G.; Regnaut, S.; Goudet, J. Detecting the number of clusters of individuals using the software structure: A simulation study. *Mol. Ecol* **2005**, *14*, 2611–2620.
52. Earl, D.A.; vonHoldt, B.M. STRUCTURE HARVESTER: A website and program for visualizing STRUCTURE output and implementing the Evanno method. *Conserv. Genet. Resour* **2011**, *4*, 359–361.
53. Stam, P. Construction of integrated genetic-linkage maps by means of a new computer package—JoinMap. *Plant J***1993**, *3*, 739–744.
54. Rhymer, J.M.; Simberloff, D. Extinction by hybridization and introgression. *Annu. Rev. Ecol. Syst* **1996**, *27*, 83–109.
55. Fitzpatrick, B.M.; Johnson, J.R.; Kump, D.K.; Smith, J.J.; Voss, S.R.; Shaffer, H.B. Rapid spread of invasive genes into a threatened native species. *Proc. Natl. Acad. Sci. USA* **2010**, *107*, 3606–3610.
56. Beaumont, M.A.; Nichols, R.A. **1996**, *263*, 1619–1626.
57. Flint, J.; Bond, J.; Rees, D.C.; Boyce, A.J.; Roberts-Thomson, J.M.; Excoffier, L.; Clegg, J.B.; Beaumont, M.A.; Nichols, R.A.; Harding, R.M. Minisatellite mutational processes reduce $F_{ST}$ estimates. *Hum. Gene* **1999**, *105*, 567–576.
58. Antao, T.; Lopes, A.; Lopes, R.J.; Beja-Pereira, A.; Luikart, G. LOSITAN: A workbench to detect molecular adaptation based on a FST-outlier method. *BMC Bioinforma.* **2008**, *9*.
59. Hubisz, M.J.; Falush, D.; Stephens, M.; Pritchard, J.K. Inferring weak population structure with the assistance of sample group information. *Mol. Ecol. Resour* **2009**, *9*, 1322–1332.

# CHAPTER 18 EARLY LIFE ON LAND AND THE FIRST TERRESTRIAL ECOSYSTEMS

Hugo Beraldi-Campesi

## INTRODUCTION

### Definition of "Terrestrial"

Habitable, non-aquatic environments must have existed all throughout the geologic history of Earth unless its surface was entirely under water, which seems unlikely. The definition of a terrestrial environment may not be as trivial as it sounds. "Terrestrial" is defined here as non-aquatic environments. However, even fully aquatic ecosystems, such as lakes and coastal environments, cover a wide spectrum of mixed environments where aquatic and non-aquatic landscapes develop and overlap over time. Habitats above sea level include aquatic (ice-covered and ice-free lakes, ponds and wetlands, peats, rivers and streams, geothermal fields) and non-aquatic environments (especially areas with low rain regimes) that experience drastic changes governed by tectonic activity and climatic conditions, including the rise and fall of sea level, glaciations, and rain regimes (e.g., Romans and Graham 2013). Microbes can be expected in all these environments and, in the long term, they may have strongly influenced the regional topography, sedimentation rates, sedimentary dynamics, and the reworking of previously emplaced materials. This might be difficult to interpret sometimes from the sedimentary record, especially in environments whose configuration and sedimentary dynamics can change in a relatively short time (days to decades), such as coastal areas (deltas, estuaries, lagoons, evaporitic flats, dunes, etc.; e.g., Hamblin and Christiansen 2007), going from aquatic to non-aquatic environments in a few centimeters or meters of rock strata.

Sedimentary deposits originating in fully aquatic environments (fluvial, lacustrine, shallow marine) can be further exposed to the atmosphere for long periods of time and undergo pedogenetic processes, which transform some of the original features of the deposit into a soil (e.g. Paul et al. 2001and references therein). The rocks thus keep gross characteristics of the primary deposit but are overprinted with the in situ, secondary features derived from completely different environmental conditions. Besides, elucidating timescales at the outcrop scale is not always feasible and is frequently overlooked in less resolved regional studies. Ultimately, this has surely contributed to biases in the interpretation of the evolution of the biosphere. In this regard, the study of pedogenetic (e.g., development of horizons, hardpans or duricrusts, peds, and clay compositional and mechanical—e.g., slickensides—features, etc.) and microbial processes in spring and stream microbialites (e.g., travertines, tufas, sinters) and exposed sedimentary and rock surface habitats (endolithic habitats and cryptogamic covers), is of particular importance for the more comprehensive understanding of past life because these represent terrestrial habitats expected on ancient continental surfaces.

### Caution and re-Interpretation of the Rock Record

Through the integrative study of rocks and the understanding of the processes that formed them, including the fossil record and our ability to date materials isotopically, we have built a

**Citation:** Hugo Beraldi-Campesi, Early life on land and the first terrestrial ecosystems, DOI: 10.1186/2192-1709-2-1.

concept of the evolution of the geosphere and the biosphere (see compendiums by Schopf 1983; Canup and Righter 2000; Eriksson et al. 2004; Schieber et al.2007; Van Kranendonk et al. 2007a; Kasting 2009; Taylor et al. 2009; Knoll et al. 2012), despite important and ongoing debate on the details. One key element in this picture is the inception of life, which has been interacting with and changing, maintaining, and recirculating most of the materials existing in the atmosphere and the supracrustal section of the Earth for over ~80% of Earth's history. This scenario has been studied and interpreted over the years, aided by the technology available at the moment, not always correctly and also biased sometimes by the general consensus (e.g., Hallbauer 1975; Gray and Boucot 1994; Windley 2007). Also, our appreciation of the timing of geological phenomena (soil formation, seafloor spreading, mountain building, rock erosion, lake succession, etc.) may be difficult to relate to other global changes (e.g. rapid and profuse volcanism and rapid climatic oscillations coexisting with slow seafloor spreading and continental drift) when interpreting the rock record.

Besides the record being incomplete, the speed at which biology operates compared to geology implies that hiatuses of tens of millions of years (negligible in Pracambrian timescales), represented by only a few centimeters or meters of rock strata, bear enormous opportunities for biological developments and adaptations that may have not been preserved. This conceptualization of the speed at which biotic and geological events occur simultaneously requires a careful examination and re-evaluation of Precambrian geological materials (aided by the advancement of scientific knowledge and technology around it) with a readiness to consider challenging ideas (e.g., Retallack 2013), especially when recognizing fossils or when trying to reconcile them with their paleoenvironments (Xiao and Knauth2013).

The fossil record of microbes is largely related to aquatic environments, and while abundant morphological, chemical, and geochemical evidence of diverse, aquatic Archean life has reached wide acceptance and consensus, the existence of life on Precambrian lands is not always taken for granted. The historical perception of plants as the dominant group on the land, together with the first discoveries of macroscopic fossils only in Phanerozoic rocks and the inability to correctly interpret microbial and algal biosignatures, has perhaps infused the generalized understanding of "colonized" terrestrial ecosystems exclusively for plants (e.g., Bambach 1999). In some instances, even when the existence of Precambrian terrestrial ecosystems may be recognized, they are still treated doubtfully or not given adequate attention (Shear 1991; Behrensmeyer et al. 1992; DiMichele and Hook 1992; Gray and Shear1992; Gray and Boucot 1994) even after previous and important discussions on the topic (e.g., Wright 1985; but see also Labandeira 2005).

The possible misinterpretation of terrestrial paleoenvironments and their relatively poor preservation in the sedimentary record does not necessarily mean that terrestrial life did not exist on the early continents. Today there is growing evidence indicative of non-aquatic environments colonized by microbes early in Earth's history, which is consistent with the extent of modern microbial life on analog "barren" lands (deserts, polar plains, alpine rocks, etc.) their outstanding diversity and metabolic capabilities, and by the great diversity and distribution of Precambrian microfossils, which is a reflection of the microbial ubiquity at that time.

## The Setting for Early Life

The oldest materials yet found in the Solar System occur in meteorites and are ~4,570 Ma (Mega annum, million years) of age (Bouvier and Wadhwa 2010), which may serve as a starting point for the condensation of the first solids in our Solar System. By contrast, the oldest materials on Earth (zircon crystals) go back ~4,400 Ma (Wilde et al. 2001), leaving a hiatus of ~170 Ma in Earth's geological history. Regardless, it is assumed that the Moon was already formed before 4,400 Ma (Canup and Righter 2000; Yu and Jacobsen 2011) and that the Earth's nucleus, mantle, and lithosphere were already differentiated (Nelson 2004; Boyet and Carlson 2005). At least by ~4,200 Ma, but perhaps 200 Ma earlier, large water bodies were in place (Mojzsis et al. 2001; Nutman 2006; Cavosie et al.2007, but see alternative views by Deming 2002), while granitic (continental) and basaltic (oceanic) crusts were constantly growing, resurfacing, and remelting, interacting with water in non-uniform regimes that evolved drastically from the Hadean to the Neoarchean (Komiya et al. 1999; Nutman et al. 2002; Myers 2004; Rino et al. 2004; Van Kranendonk 2004and references therein; Furnes et al. 2007a; Adam et al. 2012), changing from plume-dominated to plate-dominated tectonics toward the late Paleoarchean (Van Kranendonk et al. 2007b). It is plausible then, that by the end of the heavy bombardment (Gomes et al. 2005; Hartmann et al. 2000) some ~3,800 Ma ago, the primitive lands and oceans were open niches ready for the pioneering and rapidly evolving microscopic life forms, for which occasional "sterilizing" perturbations may be irrelevant given the resilience and time scales at which biology operates compared to geology.

Although life may have appeared only a few hundred million years after Earth's accretion (e.g., Lopez-Garcia et al.2006), sedimentary rocks older than ~3,850 Ma (Nutman et al. 1996; Ishizuka 2008; Nutman et al. 2010; O'Neil et al. 2011; Mloszewska et al. 2012), where biotic events are most likely to be imprinted, are uncommon on Earth. Yet, potential traces of life (e.g. biogenically precipitated carbonates) may even be present in this ancient record (Nutman et al. 2010), suggesting that life itself can be several million years older than the oldest known stromatolites and microfossils. Other putative biosignatures older than 3,500 Ma (carbonaceous spherules associated with apatite globules; see McKeegan et al. 2007; Papineau et al. 2010a, 2010b) are also controversial (see Myers 2001; van Zuilen et al. 2002; Fedo and Whitehouse 2002; Papineau et al. 2011) and may not imply a syngenetic timing of formation with the host rock. Biosignatures of particular interest would be those associated with biogenic banded iron formations (e.g., Dauphas et al. 2004; Trendall and Blockley 2004; Kappler et al. 2005; Konhauser et al. 2005; Koehler et al. 2010; Mloszewska et al. 2012) given their potential antiquity of ~4,300 Ma (O'Neil et al. 2009).

Microfossils, microbialites, and isotopic and molecular biomarkers indicate that prokaryotic life was abundant by 3,500–3,400 Ma ago in shallow and deep marine environments (Lowe 1980; Walter et al. 1980; Awramik et al. 1983; Schopf 1983; Walter 1983; Walsh, 1992;

Walsh and Lowe 1985, 1999; Rasmussen 2000; Westall et al. 2001; Furnes et al. 2004; Shen and Buick 2004; Tice and Lowe 2004; Allwood et al. 2006; Banerjee et al. 2006; Westall et al. 2006a, 2006b; Ueno et al. 2006; Schopf et al. 2007and references therein; Shen et al. 2009; Westall 2010; Wacey et al. 2011), which supports the notion that coastal and estuarine areas could have been very productive by that time and that photosynthesis was already operating (Awramik 1992; Rosing and Frei 2004; Tice and Lowe2004; Buick 2008; Hoashi et al. 2009; Kato et al. 2009; Kendall et al. 2010), though perhaps not necessarily oxygenic (Kirschvink and Kopp 2008; Westall et al. 2011; Li et al. 2012).

Many different settings have been proposed as likely or "optimum" for the emergence and prosperity of life, ranging from deep-sea hydrothermal vents and geothermal springs, to land surfaces and mineral-water-air interphases (Baross and Hoffman 1985; Retallack 1986a; Holm 1992; Battistuzzi and Hedges 2009; Aller et al. 2010; Hazen and Sverjensky 2010; Mulkidjanian et al. 2012). However, one preferred environment where many of the oldest signs of life are found is shallow marine continental margins (see references in Schopf and Klein 1992). Whether this is a true fact or a consequence of the incompleteness/selectivity of the record is still to be resolved. However, in these coastal environments microbes were likely to have been periodically exposed and desiccated, as happens in most such environments today, and likely developed adaptations for long-term desiccation regimes (e.g., thick hygroscopic sheaths) and high UV radiation (e.g., living interstitially).

Some of the oldest examples of life activity, which come precisely from aquatic, shallow marine (Klein et al. 1987; Schopf and Klein 1992; Van Kranendonk et al. 2008; Westall 2010; Van Kranendonk 2011; Hickman and Van Kranendonk 2012), shallow lacustrine (Awramik and Buchheim 2009; Hickman and Van Kranendonk 2012), and intertidal environments (e.g., Noffke et al. 2006; Noffke 2010; Noffke et al. 2011; Westall et al. 2011), show signs of evaporation (e.g., Noffke et al. 2008; Westall et al. 2011; Hickman and Van Kranendonk 2012), which suggests that early microbial communities in shallow waters had to deal with periodic desiccation and UV radiation >3,400 Ma ago. This further implies adaptations to resist desiccation, salinity fluctuations, and UV radiation that could be successfully used even after prolonged desiccation. Dry conditions can be expected also for lacustrine and fluviatile environments. Desiccation allows dispersion by wind, which seems like a reasonable means for land colonization. Through this mode of dispersion, communities would tend to be at or near the surface instead of underground, even when migration to aquifers can occur. Perhaps environments with periodic subaerial exposure (especially estuarine and intertidal) were crucial scenarios for a biological transition from water to land.

Apparently not only prokaryotes were abundant in shallow Precambrian environments; the oldest eukaryotic-like fossils (acritarchs; Buick 2010) found so far (that perhaps needed oxygen for advantageous energetic and metabolic capabilities) are ~3,200 Ma old and were also present in estuarine environments (Javaux et al. 2010). This indicates that life achieved a relatively rapid global presence and had diversified enough (Kandler 1994; Altermann and Schopf 1995; Ueno et al. 2006; Blank 2009; David and Alm 2011) to occupy a wide variety of ecological niches by the Paleoarchean, even in places that may have been severely disturbed by asteroid impacts (see Walsh 1992and references therein). Even greater biological diversity, ubiquity, abundance, and habitats are seen in the younger Proterozoic record (e.g., Schopf 1992a; Schopf and Klein 1992), for which rocks are better preserved and more abundant than Archean ones.

## The Fossil Record of Terrestrial Life

The earliest remnants of continental crust may derive from ≥3,500 Ma-old submillimeter zircons (Nutman 2006) and regional rock outcrops (Buick et al. 1995; Iizuka et al. 2006; Stern and Scholl 2010; Adam et al. 2012). Supplementary evidence for exposed lands may consist of thick soils developed on these ancient surfaces (Buick et al. 1995; Hoffman 1995; Ohmoto et al. 2007; Johnson et al. 2009, 2010). The further growth of continents and their sedimentary cover, implying extensive intracratonic terrestrial settings that remained relatively stable (although still affected by erosion, sea level changes, tectonics, and volcanism), is reflected in the ample record of paleosols onward (see study approaches and examples by Jackson 1967; Gay and Grandstaff 1980; Holland 1984; Aspler and Donaldson 1986; Grandstaff et al. 1986; Kimberley and Grandstaff 1986; Reimer 1986; Retallack 1986b; Farrow and Mossman 1988; Zbinden et al. 1988; Palmer et al. 1989; Holland 1992; Gall 1994; Macfarlane et al. 1994; Martini 1994; Retallack and Mindszenty 1994; Driese et al. 1995; Banerjee 1996; Ohmoto 1996; Prasad and Roscoe1996; Gutzmer and Beukes 1998; Thiry and Simon-Coincon 1999; Rye and Holland 2000; Watanabe et al. 2000; Retallack 2001and references therein; Yang and Holland 2003; Driese and Gordon-Medaris 2008; Pandit et al.2008; Bandopadhyay et al. 2010). This record of old paleosols holds indirect proof of the early environmental conditions on Earth and early life on land.

Currently, the oldest and direct evidence of terrestrial life comes from ~2,900–2,700 Ma-old (see age determination of Witwatersrand deposits in Kositcin and Krapez 2004; Zhao et al. 2006), organic matter–rich paleosols (Watanabe et al. 2000), ephemeral ponds (Rye and Holland 2000) and alluvial sequences, some of them bearing microfossils (Hallbauer and van Warmelo 1974; Mossman et al. 2008). Interestingly, their occurrence in such settings coincides with drastic changes in Earth's crustal configuration and the —perhaps abrupt—emplacement of large continental masses in the late Archean (Condie 2004; Eriksson and Martins-Neto 2004; Van Kranendonk 2004and references therein; Hazen et al. 2012), a marked step in the oxygenation of the atmosphere (Kendall et al. 2010), and also with estimations of land colonization by microbes based on phylogenetic relationships (Battistuzzi et al. 2004). Although microbes could have colonized the land before this time, the Meso- to Neoarchean appears to be an important evolutionary time period for terrestrial microbial communities, perhaps linked to supercontinent growth (Santosh2010) and the emergence of potential new habitats.

Later in time, the amount of organic matter–rich and possibly "biologically weathered" paleosols (Ohmoto 1996; Beukes et al. 2002; Driese and Gordon-Medaris 2008), terrestrial sedimentary structures of presumed biotic origin

(Hupe 1952; Lannerbro 1954; Voigt 1972; Eriksson et al. 2000; Prave 2002), and microfossils themselves (Cloud and Germs 1971; McConnell 1974; Horodyski and Knauth 1994: Strother et al. 2011) drastically increased throughout the Proterozoic. Likewise, marine microfossils display increasing biological developments and adaptations (Knoll et al. 2006). Biotic diversity and abundance become even greater from the Neoproterozoic-Phanerozoic transition to the recent (see Zhuravlev and Riding 2001; Xiao and Kaufman 2006; Gaucher et al. 2010). This timeline suggests a rapid and global development of life on Earth, with life forms adapted to live on the land more than 2,000 Ma before the earliest fossil record of land plants (Heckman et al. 2001; Gensel 2008). Important events in this chronology are depicted in Figure 1.

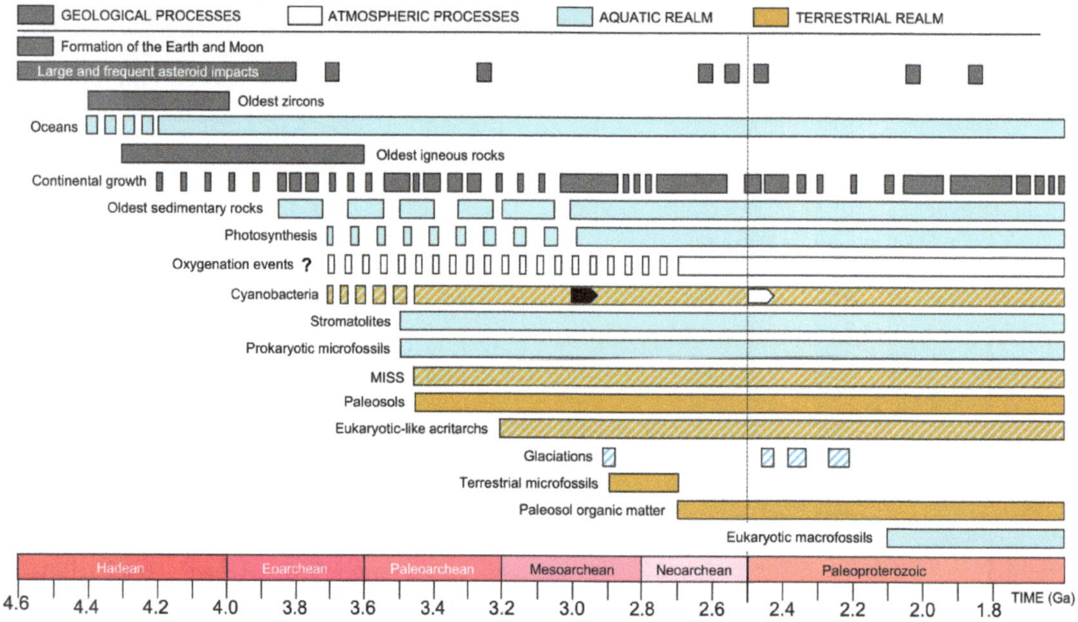


**Figure 1:** Suggested chronology of geological, atmospheric, and biological events during the Hadean, Archean, and Paleoproterozoic eons. Geological events were compiled from references within Canup and Righter (2000), Eriksson et al. (2004), and Van Kranendonk et al. (2007a, 2007b). Asteroidal impact history is by Glikson (2007). Emplacement of the oceans is by Nutman (2006). The dashed line at 2,500 Ma marks an abrupt change into an oxygenated atmosphere (see Kendall et al. 2010), although oxygen may have started accumulating beforehand (Ohmoto 2004). Oxygenation events are considered to have occurred in pulses of unknown magnitude and duration and are correlated with a putative origin of oxygenic photosynthesis by cyanobacteria, based on a theoretical timing estimated by Lazcano and Miller (1994) and considering the emergence of life at 3,800 Ma. On the line of cyanobacteria, alternative origins are indicated with a black (Schirrmeister et al. 2013) or a white mark (Kirschvink and Kopp 2008). The MISS (Microbially Induced Sedimentary Structures; see Noffke et al. 2001), cyanobacteria, and the unicellular eukaryotic-like acritarchs are represented within both aquatic and terrestrial realms. Glaciations are from Hoffman and Schrag (2002), shown in combined colors to represent atmospheric (climatic) and hydrological (ice formation) processes. Other biological evolutionary steps and paleosol-related data are from Hallbauer and van Warmelo (1974), Holland (1984), Schopf (1983), Schopf and Klein (1992), Han and Runnegar (1992), Golubic and Seong-Joo (1999), Rye and Holland (2000), Retallack (2001), Westall et al. (2006a, 2006b), Johnson et al. (2009,2010), El Albani et al. (2010), Javaux et al. (2010), and Noffke (2010).

## Functioning of Primitive Terrestrial Ecosystems and Cyanobacteria

A conceptualization of the functioning of the ancient terrestrial biosphere necessarily requires a general understanding of modern, analog microbial communities to evaluate their living requirements, diversity, physiology, and environmental impact, and to characterize any potential biosignature that could be used to recognize them in the rocks. Modern terrestrial microbial communities are found worldwide and in a great variety of local conditions, in surface (solid rock, regolith) and subsurface (caves, groundwater, deep ground) environments (although the latter could be considered aquatic by some). However, it is unclear which one is more productive in terms of biomass (Pace 1997) and what metabolisms have dominated those systems—and to what extent—over geologic time scales (Sleep and Bird 2007).

An understanding of the biology and distribution of modern microbes, which are ubiquitous in today's Earth's biosphere (Figure 2), seems essential for an understanding of their ancient counterparts and their impact on early terrestrial ecosystems. Estimates of the genetic diversity and biomass distribution in drastically different environments (e.g., Garcia-Pichel et al. 2003; Lozupone and Knight 2007; Nemergut et al. 2011) depict the ample range of strategies that terrestrial organisms, particularly primary producers, have developed for living on the land. Oxygenic photoautotrophy seems to be a particularly important capability of terrestrial organisms, simply because their energy source (light), reductant power (water), and carbon source ($CO_2$) are readily available in these environments. In comparison, other primary producers (e.g., chemolithotrophs) are restricted to aqueous environments because they require soluble sources of reductants (e.g., $H_2$, $Fe^{2+}$, $H_2S$, $HS^-$) and exergonic reactions to maintain their metabolism (White 2000). Besides being restricted to aquatic environments, chemolithotrophs are also less energy-efficient than oxygenic photoautotrophs (DesMarais 2000; Madigan et al.2003; Konhauser 2007), and less likely dominant in subaerial environments.

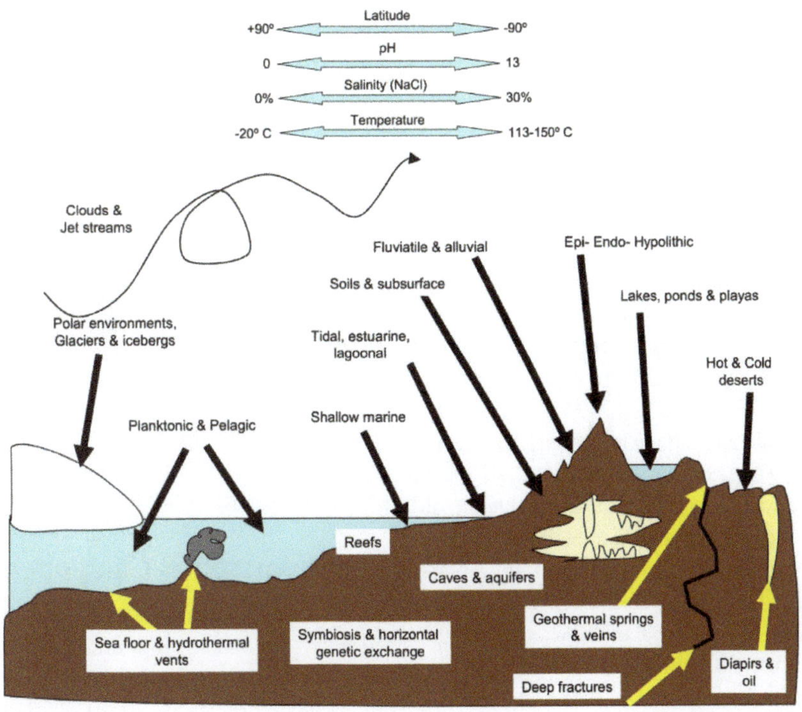


**Figure 2:** Variety of environments known to be inhabited by microbes. Ranges of environmental fluctuations (latitude, pH, salinity, and temperature) where microbes can be found are also shown (see Madigan and Marrs 1997; Madigan et al.2003; Konhauser 2007). Direct and indirect interactions between different microbial communities, as well as symbiotic associations, coevolution, and horizontal genetic exchange (e.g., Davison 1999; Gogarten et al. 2007) are assumed to drive and have driven ecological processes and adaptations to habitat variability.

Cyanobacteria have been the only organisms that developed special pigments and enzymatic capabilities for using water as a source of electrons. This process has allowed them to live outside the water in any suitable environment, even where water might be a limiting factor, such as deserts (e.g., Potts and Friedmann 1981). Oxygenic photosynthesis also contributed to the oxidation of the atmosphere (both by sequestering $CO_2$ and by producing $O_2$), a global and ongoing process with profound geochemical, atmospheric, hydrological, and biological implications (e.g., Rosing et al. 2006; Och and Shields-Zhou 2012; Pufahl and Hiatt 2012). Cyanobacteria and other prokaryotes, can also fix gaseous nitrogen, which seems of great advantage for an independence from dissolved N species, such as $NH_4$ and $NO_3$ (Glass et al. 2009). The appearance of cyanobacterial akinetes (for $N_2$ fixation) in the Paleoproterozoic (Tomitani et al. 2006) attests to this early adaptation. For organisms on the land, a limiting nutrient, such as P, can be supplied by dust deposition (Kennedy et al. 1998; Reynolds et al. 2001; McTainsh and Strong2007), which may be an alternative process for replenishment of nutrient loss by runoff and leaching in such environments (e.g. Beraldi-Campesi et al. 2009); S can also be acquired from minerals, aerosols, and as gaseous sources, likely present in the early atmosphere (Holland 1984). Thus, the nutritional requirements for oxygenic, photoautotrophic, primary producers seem not to have been a limiting factor for the colonization of the land. This idea has also been discussed in light of physiological and genetic characteristics of terrestrial microbes (Battistuzzi et al. 2004; Battistuzzi and Hedges 2009). Yet, an earlier chemotrophic way of life also needs to be considered (Shen and Buick 2004; Sleep and Bird 2007).

Particularly for the early terrestrial biota, a minimum set of adaptations to live subaerially must have included protection against radiation and desiccation effects. Adaptations such as thick polymeric sheaths with hygroscopic capacity against desiccation, efficient DNA repair mechanisms to restore metabolic activities as soon as water is available, and the production of UV-shielding pigments are certainly successful strategies displayed by terrestrial cyanobacteria (Shephard 1987; Garcia-Pichel 1998; Yasui and McCready 1998; Potts 1999; Sinha and Hader 2002; Singh et al. 2010). Refined degrees of adaptation of terrestrial organisms include sunscreens that once placed within the extracellular sheaths, passively protect against UV radiation, even when the cell is dormant or dehydrated (Garcia-Pichel and Castenholz 1991; Gao and Garcia-Pichel 2011). Some of these strategies for living on the land likely evolved early and are partially displayed by microfossils (e.g., thick sheaths), which are sometimes associated with evaporitic sediments, in keeping with subaerial exposure (Schopf 1968; Hofmann 1976; Golubic and Campbell1979; Awramik et al. 1983).

Modern cyanobacteria-driven communities can be found in any terrestrial environment (~30% of modern Earth's surface area). They include endolithic communities (Friedmann 1980; Sun and Friedmann 1999; Büdel et al. 2004) and cryptogamic covers (CGC) on rocks and soils (Belnap and Lange 2001; Elbert et al. 2012). The latter have been shown to be very complex and dynamic and contain many distinct functional groups of prokaryotes and eukaryotes, ranging from primary producers to decomposers of specific materials, and grazers (Fritsch 1922; Fletcher and Martin 1948; Campbell 1979; Bamforth 1984, 2004; Garcia-Pichel et al. 2001; Nagy et al. 2005; Tirkey and Adhikary2005; Chanal et al. 2006; Reddy and Garcia-Pichel 2006; Bates and Garcia-Pichel 2009; Neher et al. 2009; Meadow and Zabinski 2012Bates et al. 2013). This diversity is variable based on local environmental conditions, but all CGC—albeit with few exceptions (e.g., Hoppert et al. 2004)—have in common the presence of cyanobacteria. This speaks for the evolutionary success that cyanobacteria have had over other microbes throughout time.

Although fossil analogs of CGC have been discovered in ancient sediments (Simpson et al. 2010; Beraldi-Campesi et al. 2011; Retallack 2009, 2011; Sheldon 2012), it is unknown what their microbial composition might have been. However, morphological similarities between modern and fossil counterparts are remarkable (Schieber et al. 2007; Noffke 2010). Morphological resemblance between fossils and recent analogs suggests that cyanobacteria are indeed a very old group of bacteria (see Golubic and Seong-Joo 1999) and that at least some morphological traits have been maintained over time (Golubic and Hofmann 1976; Golubic and Campbell 1979; Schopf 1992b). Moreover, as cyanobacteria are such an old group and are so well adapted to colonizing unstable sediments (Booth1941; Campbell et al. 1989; Mazor et al. 1996; Belnap and Gillette 1998; Malam Issa et al. 2001; Hu et al. 2002; Garcia-Pichel and Wojciechowski 2009), even where available water is scarce and there is considerable UV radiation (Fleming and Castenholz 2007; Giordanino et al. 2011), it is likely that they were also primordial components on land surfaces (Campbell 1979) and influenced the formation of sedimentary biostructures and textures represented in fossil examples (e.g., Prave 2002; Schieber et al. 2007; Noffke, 2011). The antiquity of cyanobacteria has been also estimated by molecular clock analyses of genomic distances to be ~3,000 Ma (Battistuzzi and Hedges 2009; Schirrmeister et al. 2013), which more-or-less coincides with the age of the oldest terrestrial microfossils (Mossman et al. 2008). This timing, however, may vary depending on the calibration points used for constructing phylogenies and the extent of horizontal gene transfer. Lastly, the capacity of chlorophyll *a* to absorb higher photonic energies to split water in comparison to other photosynthetic bacteriochlorophylls (Xiong et al. 2000) may be the result of selective pressures to use the shorter wavelengths that reached the Precambrian surface where cyanobacteria had to dwell, a capacity not seen in purple or green phototrophic bacteria that use less energetic wavelengths in submerged/shielded habitats. Thus, from a multi-angular perspective cyanobacteria seem the perfect candidates for the colonization of the earliest lands.

As mentioned above, most CGC have in common the presence of filamentous cyanobacteria. One property of these morphotypes is that they can glide through interstitial spaces using hollow hygroscopic sheaths of mucilage as trails, to shield themselves against radiation, to find their optimum light regime, or to track water (Garcia-Pichel and Pringault 2001). The nature of the filamentous members of these communities also provides more surface area and tension for fastening and binding disaggregated particles (Garcia-Pichel and Wojciechowski 2009). Polysaccharides secreted extracellularly provide additional cementing force to the entire organo-mineral framework, resulting in the formation of a (crust/mat) stable microenvironment. The intrinsic characteristic of filamentous microbes to form cohesive layers at sedimentary surfaces is also known to substantially decrease wind and water erosion in modern arid and semiarid areas of the world (Belnap and Gillette 1998; Belnap and Lange 2001). Although some erosive forces may surpass the tear resistance of CGC in high-energy systems (e.g., Corcoran and Mueller 2004), this property of microbes has been invoked to explain the stability of thick, Precambrian siliciclastic sedimentary sequences (Dott 2003) and the soft deformation properties of microbial mat-like structures (see references in Schieber et al. 2007). This is an important property of microbes for the functioning of siliciclastic ecosystems, and together with the presence of mature and organic-rich soils and microfossils in old Proterozoic strata (see references above), suggests that abundant "cryptogamic" covers were present on Precambrian lands, similar to those covering polar and arid areas of the world today. The addition of new members to these communities over time (most importantly algae and fungi, but also grazers) is expected and may be used to explain increasing weathering rates of the continents (Kennedy et al. 2006) and abrupt changes in the global balance of the C cycle in the Neoproterozoic (Knauth and Kennedy 2009).

## Other Microbial Components

Judging from the rapid achievement of diversity and distribution of early microbial biota and from microbial successions seen in modern "barren" lands (e.g., Sigler et al. 2002; Schmidt et al. 2008; Fierer et al. 2010), it is expected that heterotrophic organisms were also part of land communities, as they seem to be an inevitable component in this type of consortia. Under this perspective, primitive microbial ecosystems cannot be understood as composed only of autotrophic primary producers, but also a myriad of other microbes finding their niche within such pre-existent microenvironments. For example, actinobacteria in modern CGC not only degrade large quantities of organic exudates from cyanobacteria, a process which influences the C cycle, but they also seem to be structural components of these sedimentary biostructures (e.g., Reddy and Garcia-Pichel 2006). The same applies to other bacteria (e.g., Bacteroidetes and Proteobacteria) that secrete large quantities of mucopolysaccharides, which aid in gluing soil particles together and may also have a critical role in the hydraulic conductivity of the surface substrate (Rossi et al. 2012). One of the most important eukaryotic component of of modern CGC are fungi, which must have played a key role in the colonization and weathering of bare rocks in the past (with symbiont cyanobacteria or algae), as well as in the successive development of vascular plants on the land (Smith and Read 2008) and in a radical change toward more "modern" terrestrial ecosystems (Blackwell 2000; Heckman et al. 2001; Taylor et al.2009. See also Gadd 2006).

Although the timing of the origin of these organisms is unknown, terrestrial microbes can certainly drive important chemical transformations in soils (Keller and Wood 1993; Schwartzman and Volk 1989; Chenu and Stotzky 2002; Ehrlich 2002; Chorover et al. 2007) and endolithic habitats (Konhauser et al. 1994; Sun and Friedmann 1999; Büdel et al. 2004; Omelon et al. 2006) that may have operated in the past. These include affecting the reactivity of mineral surfaces with secreted metabolites (Geesey and Jang 1990; Welch et al. 1999), changing the pH and redox potential of the microenvironment (Bennett et al. 2001), or secreting metal ligands and other organic complexes that react with solutes and minerals (Keller and Wood 1993; Schwartzman and Volk 1989; Barker et al. 1998; Welch et al. 1999; Bennett et al. 2001). These mechanisms seem to play a fundamental role in biogeochemistry (weathering, clay formation, nutrient bioavailability, metal concentration and bioavailability, mineral formation or transformation, etc.), and their effects may also be used to trace microbial geochemical biosignatures in the rock record (Beraldi-Campesi et al. 2009). Additionally, the process of soil formation and maturation is usually understood as aided by biology (Keller and Wood 1993; Schwartzman and Volk 1989; Brady and Weil 2008) and differentiated from abiotic regolith development, and involves a critical step prior to plant and animal colonization of the land. All these characteristics displayed by modern CGC could be expected from ancient analog communities, although with variations in the occurrence and magnitude dictated by their limiting factors.

## Dust

The mechanism of dust formation, transport, and deposition reflects one important aspect of the functioning of terrestrial ecosystems because dust can only be formed on the land and because microbes (along with water adhesion and neo-cementation of particles with salts and clays) can stabilize fine dust particles through trapping and binding (e.g., Dong et al. 1987; Liu et al. 1994; Williams et al. 1995; Belnap and Gillette 1998; Hu et al. 2002). Thus, dust production can potentially be regulated by microbes (and other encrusting processes) depending on their degree of development. The more developed, the less dust production.

Dust is an important carrier of nutrients, and its retention on the ground might influence the budget and delivery of those nutrients locally or to other distant ecosystems, such as happens in modern marine environments via deposition of huge loads of dust (Jickells et al. 2005). The capacity of microbes to trap and bind particles has been demonstrated for numerous underwater and subaerial environments (Gunatilaka 1975; Zhang 1992; Takeuchi et al.2001; Altermann 2008; Gradzinski et al. 2010; Williams et al. 2012). If microbes were responsible for much of the global dust capture, retention, and lixiviation on the early continents, recycling effects may have had profound implications for the evolution of global ecosystems through geologic time, as well as for important climatic processes, such as those rooted in atmospheric albedo variations (Harrison et al. 2001; Jickells et al. 2005; Lau et al. 2006).

Finally, dust is also a carrier of microbes and viruses (Abed et al. 2011; Al-Bader et al. 2012), which implies a means for biological dispersion that must have been operating continuously and over long distances in the past, amplifying the potential biogeography of biological entities over vast areas of the oceans and continents. Although the rate of survival and the success of foreign airborne mixed communities on aquatic environments, barren or already colonized surfaces is not known, it is plausible that such a mechanism was vital for the colonization of the early continents and the increase in ecological complexity and genetic exchange (e.g., Gogarten et al. 2007).

## Underground Realm

The underground realm (geothermal veins, aquifers, soil subsurface, all types of caves) should also be considered potential continental habitats for early terrestrial life, as life is abundant there today (e.g., Ghiorse and Wilson 1988; Barton and Northup 2007; Engel 2010). The Precambrian record of caves (e.g., karstic environments) or underground aquifers (detected through nodules and concretions in the rocks) is far less known than the typical shallow marine or lacustrine environments (see examples of karstic and underground environments in Nicholas and Bildgen 1979; Schau and Henderson 1983; Glover and Kah 2006; Skotnicki and Knauth 2007; Rasmussen et al.2009). Nevertheless, these environments must have existed throughout Earth's history, and thus, terrestrial biotas could have adapted to live in those conditions back in the Precambrian (Rasmussen et al. 2009).

In contrast to the most typical subaerial, light-driven primary producer communities, underground microbes require a chemosynthetic metabolism for primary productivity, perhaps relying on the oxidation of sulfur and iron compounds to support growth and continuity, as these are the main energy sources in such environments (Sarbu et al. 1996; Chen et al. 2009; Porter et al. 2009). Because these metabolic pathways are less energetic than photosynthesis, life underground is expected to have been slow-growing, less dynamic in terms of diversity and interactions, and more geographically contained than, for example, subaerial phototrophs. Nevertheless, early underground dwellers may have impacted the subsurface realm (cave formation, buried oil and dissolved organic matter consumption, methane production, etc.) and contributed to the neoformation and dissolution of minerals over the long term, as well as to the generation of gaseous byproducts (e.g., $H_2S$, $CH_4$, $CO_2$) that could be important for geochemical processes on the surface and ultimately for distant communities and global biogeochemical recycling. Moreover, this type of environments could have been better protected from drastic and global crisis than subaerial ones, and thus have functioned as living reservoirs that could later exploit surface environments.

## A Note on Biosignatures

Imprints of life in rocks can be formed in various ways and can be recognized as long as the rocks are preserved and accessible. Although this "fossil" record decreases in outcrop abundance the older the rocks are (basically due to burial, erosion, subduction, and metamorphism), biosignatures can be found in sedimentary rocks (Schopf 1983; Schopf and Klein 1992; Schieber et al. 2007; Noffke 2010), but also in igneous (Banerjee et al. 2006; Furnes et al.2004, 2007b; Fliegel et al. 2010) and metamorphic (Franz et al. 1991; Hanel et al. 1999: Squire et al. 2006; Bernard et al. 2007; Schiffbauer et al. 2007, 2012; Schiffbauer and Xiao 2009; Zang 2007) rocks of all ages.

Preservation will always be favored in underwater settings, especially if biological materials are buried quickly, the sediment is fine grained, and the conditions are overall reducing (anoxic). All of these factors promote rapid mineralization and replacement of biological materials (Farmer 1999; Zonneveld et al. 2010; Allison and Bottjer2011; Lalonde et al. 2012) which can preserve the morphology and organic remains, although this does not mean that preservation always happens (Zonneveld et al. 2010and references therein). Unless protected, organic matter tends to degrade basically by photo-chemical degradation (if exposed to light), chemical bond breaking, biological recycling, mechanical maceration and dissolution. If body fossils are preserved, the lack of diagnostic morphologies for most bacteria and the possible existence of abiotic, microbe-like morphologies (e.g., García-Ruiz et al. 2002, Garcia-Ruiz et al. 2003) make their determination a challenge. Yet, their presence in a suitable geological context and association with sedimentary biostructures may be used as criteria for biogenicity. Molecular biomarkers in hydrocarbons that can be correlated with extant organisms (e.g., Summons et al. 1999) also require careful confirmation of syngenicity for a correct interpretation (Rasmussen et al. 2008; Brocks 2011).

If limiting factors are at play, microbial communities may not develop sufficient biomass to leave behind a biosignature (either chemical, geochemical, mineralogical, or morphological). Water, for example, which is the most basic requirement for survival and reproduction, tends to be a limiting factor on the land compared to a permanent water body. If microbial growth is thus limited, the amount of cells and biomass that can be preserved decreases. Thus, organisms with access to unlimited water resources would be able to grow larger communities and have more possibilities for fossilization, in contrast to terrestrial microbes that depend on dew or rain for their survival and maintenance. For example, the thickness and cohesiveness of a marine-intertidal microbial mat (see Bauld 1981; Bauld et al. 1992) are greater than in a mature biological soil crust (Belnap and Lange 2001), thus the latter will be less prone to fossilization than the marine one. Nevertheless, under favorable climates and landscapes, these too could be preserved (e.g., Prave 2002). Studies on biosignatures left behind by terrestrial microbial communities are needed for comparison against the yet-to-explore rock record.

## CONCLUSIONS

As the Earth was evolving, gradual degassing and accumulation of liquid water at its surface differentiated aquatic and non-aquatic environments. Because terrestrial environments have always existed, it is possible that life evolved on the land (including in lakes, rivers, streams, and flooding areas) as early as aquatic life itself (see Retallack1986aand references therein). In any case, living on the land must have required particular adaptations, such as the capability to acquire nutrients and energy sources outside the aquatic realm, the development of molecular repairing mechanisms, and protection against radiation and desiccation. These adaptations were certainly developed by cyanobacteria, a group with a very old biologic lineage and one of the most conspicuous and successful primary producers on Earth (e.g., Whitton and Potts 2000; Herrero and Flores 2008).

Direct and indirect evidence pointing to inhabited terrestrial environments by the Paleoarchean (Johnson et al. 2009,2010) and the following eras (Stüeken et al. 2012), along with substantive evidence of terrestrialization from the Neoarchean onward (Hallbauer and van Warmelo 1974; McConnell 1974; Horodyski and Knauth 1994; Gutzmer and Beukes 1998; Rye and Holland 2000; Watanabe et al. 2000; Prave 2002; Rasmussen et al. 2009), strongly implies that functional terrestrial ecosystems originated well back in the Precambrian. The implications for such colonization have not been completely understood, but the effects of microbial life on land processes that affect the atmosphere, the lithosphere, and the hydrosphere, are widely diverse and act at all scales. Two main consequences derived from the activities of land biota are the continuous oxygenation of the atmosphere (with consequences for the stratification of the oceans, the formation and maintenance of the ozone layer, and the precipitation of oxides, among others) and the weathering of the continents (Holland 1984; Catling et al. 2001; Stüeken et al. 2012), which indirectly and directly affect marine ecosystems. In contrast to marine biota, which indirectly affect terrestrial ecosystems through atmospheric processes (including gas composition and climate), the establishment of life on the land has an enormous significance for the evolution of the planet through time because gaseous byproducts, such as oxygen produced on the land would be released directly into the atmosphere and not dissolved in the oceans first. Once in the atmosphere oxygen would react with reduced species before starting to accumulate and produce a geochemical signature in marine sediments. Thus, land-based life could have been pivotal for the early oxygenation of the atmosphere, which later affected the oceans as well. A more direct influence of land-based communities over aquatic ones would be the production of dust, clays and leachates on the continents (Kennedy and Wagner 2011and references therein), which would then be carried by rivers and wind into the oceans, and thus increasing the heterogeneity of materials and solutes delivered into oceanic ecosystems and having either beneficial or detrimental consequences for marine life. Yet, an overall retention of sediments on the land via microbial stabilization would be expected for detritic sediments in places with well-developed cryptogamic covers. Finally, it is expected that the time span from the inception of microbial land-based life to the evolution of the first plant ecosystems took long enough (2,000–2,500 Ma) for coastal and inland settings to be transformed into organic- and nutrient-rich substrates that could later be exploited by more evolved communities and organisms toward the Neoproterozoic-Phanerozoic transition.

In general, the logical transition from cyanobacteria (and other bacteria and archea), to algae (and protists and fungi), to non-vascular plants, to vascular plants, may still be valid, but the timing of those evolutionary steps needs to be updated with the latest pertinent information available. The notion that the land was virtually sterile in the Precambrian (e.g., Bambach 1999; Blackwell 2000; Corcoran and Mueller 2004; Nesbitt and Young 2004; Gensel2008) underestimates the impact that microbes could have had on the biosphere. More importantly, the idea the land was first colonized by plants and that they formed the earliest terrestrial ecosystems should be abandoned completely. That is not to say that the advent of plants in the Phanerozoic did not have strikingly enhanced effects on continental weathering, soil formation, and oxygenation of the atmosphere (Labandeira 2005; Taylor et al. 2009), but neglecting the existence of microbial, Paleoarchean-to-recent terrestrial ecosystems would impede a realistic understanding of the evolution of the biosphere and its influence on the geo-atmo-hydrosphere over time.

## DECLARATIONS

### Acknowledgments

I am grateful to Kathleen E. Pigg, Anthony R. Prave, Gregory J. Retallack, Nora Noffke, Fernando Ortega Gutiérrez, Dominic Papineau, Marcela Martínez Millán, and Kelaine Ravdin for their important comments and improvements to this paper. I also thank the editors of Springer and Bettina Weber and Jayne Belnap for organizing and editing this special volume. I also thank the people from SIOV (Seminario Interdisciplinario del Origen de la Vida) at UNAM for fruitful discussions on critical aspects of this topic.

## Competing interests

The author declares that there are no competing interests.

## REFERENCES


1. Abed RMM, Ramette A, Hübner V, De Dekker P, de Beer D: Microbial diversity of eolian dust sources from saline lake sediments and biological soil crusts in arid Southern Australia. *FEMS Microbiol Ecol* 2011, 80(2):294–304.
2. Adam J, Rushmer T, O'Neil J, Francis D: Hadean greenstones from the Nuvvuagittuq fold belt and the origin of the Earth's early continental crust. *Geology* 2012, 40: 363–366.
3. Al-Bader D, Eliyas M, Rayan R, Radwan S: Air-dust-borne associations of phototrophic and hydrocarbon-utilizing microorganisms: promising consortia in volatile hydrocarbon bioremediation. *Environ Sci Pollut R* 2012, 19(9):3997–4005.
4. Aller JY, Aller RC, Kemp PF, Chistoserdov AY, Madrid VM: Fluidized muds: a novel setting for the generation of biosphere diversity through geologic time. *Geobiology* 2010, 8: 169–178.
5. Allison PA, Bottjer DJ: Taphonomy: bias and process through time. In *Taphonomy: process and bias through time. Topics in geobiology, vol. 32*. Edited by: Allison PA, Bottjer DJ. New York: Springer; 2011:1–17.
6. Allwood AC, Walter MR, Kamber BS, Marshall CP, Burch IW: Stromatolite reef from the Early Archaean Era of Australia.*Nature* 2006, 441: 714–718.
7. Altermann W: Accretion, trapping and binding of sediment in Archean stromatolites—morphological expression of the antiquity of life. In *Strategies of life detection. Space Sciences Series of ISSI, vol. 25*. Edited by: Botta O, Bada J, Gómez EJ, Javaux E, Selsis F, Summons R. New York: Springer; 2008:55–79.
8. Altermann W, Schopf JW: Microfossils from the Neoarchean Campbell Group, Griqualand West Sequence of the Transvaal Supergroup, and their paleoenvironmental and evolutionary implications. *Precambrian Res* 1995, 75: 65–90.
9. Aspler LB, Donaldson JA: Paleoclimatology of Nonacho Basin (Early Proterozoic), Northwest Territories, Canada.*Palaeogeogr Palaeoclimateol Palaeoecol* 1986, 56: 17–34.
10. Awramik S: The oldest records of photosynthesis. *Photosynth Res* 1992, 33: 75–89.
11. Awramik SM, Buchheim HP: A giant, Late Archean lake system: the Meentheena Member (Tumbiana Formation; Fortescue Group), Western Australia. *Precambrian Res* 2009, 174: 215–240.
12. Awramik SM, Schopf JW, Walter MR: Filamentous fossil bacteria from the Archean of Western Australia. *Precambrian Res*1983, 20: 357–374.
13. Bambach RK: Energetics in the global marine fauna: a connection between terrestrial diversification and change in the marine biosphere. *Geobios* 1999, 32(2):131–144.
14. Bamforth SS: Microbial distributions in Arizona deserts and woodlands. *Soil Biol Biochem* 1984, 16(2):133–137.
15. Bamforth SS: Water film fauna of microbiotic crusts of a warm desert. *J Arid Environ* 2004, 56: 413–423.
16. Bandopadhyay PC, Eriksson PG, Roberts RJ: A vertic paleosol at the Archean-Proterozoic contact from the Singhbhum-Orissa craton, eastern India. *Precambrian Res* 2010, 177(3–4):277–290.
17. Banerjee DM: A lower Proterozoic paleosol at BGC–Aravalli boundary in south-central Rajasthan. *India J Geol Soc* 1996, 48: 277–288.
18. Banerjee NR, Furnes H, Muehlenbachs K, Staudigel H, de Wit MJ: Preservation of ca. 3.4–3.5 Ga microbial biomarkers in pillow lavas and hyaloclastites from the Barberton Greenstone Belt, South Africa. *Earth Planet Sci Lett* 2006, 241: 707–722.
19. Barker WW, Welch SA, Chu S, Banfield JF: Experimental observations of the effects of bacteria on aluminosilicate weathering. *Am Mineral* 1998, 83: 1551–1563.
20. Baross JA, Hoffman SE: Submarine hydrothermal vents and associated gradient environments as sites for the origin and evolution of life. *Orig Life* 1985, 15: 327–345.
21. Barton HA, Northup DE: Geomicrobiology in cave environments: past, current and future perspectives. *J Cave Karst Stud*2007, 69: 163–178.
22. Bates ST, Garcia-Pichel F: A culture-independent study of free-living fungi in biological soil crusts of the Colorado Plateau: their diversity and relative contribution to microbial biomass. *Environ Microbiol* 2009, 11(1):56–67.
23. Bates ST, Clemente JC, Flores GE, Walters WA, Wegener-Parfrey L, Knight R, Fierer N: Global biogeography of highly diverse protistan communities in soil. *ISME Jour* 2013, 7: 652–659.
24. Battistuzzi FU, Hedges SB: A major clade of prokaryotes with ancient adaptations to life on land. *Mol Biol Evol* 2009, 26: 335–344.
25. Battistuzzi FU, Feijao A, Hedges SB: A genomic timescale of prokaryote evolution: insights into the origin of methanogenesis, phototrophy, and the colonization of land. *BMC Evol Biol* 2004, 4(44):1–14.
26. Bauld J: Geobiological role of cyanobacterial mats in sedimentary environments: production and preservation of organic matter. *BMR J Aust Geol Geophys* 1981, 6: 307–317.
27. Bauld J, D'Amelio E, Farmer JD: Modern microbial mats. In *The Proterozoic biosphere*. Edited by: Schopf JW, Klein C. New York: Cambridge University Press; 1992:261–269.
28. Behrensmeyer AK, Damuth JD, DiMichele WA, Potts R: *Terrestrial ecosystems through time: evolutionary paleoecology of terrestrial plants and animals*. Chicago Il: University of Chicago Press; 1992.

29. Belnap J, Gillette DA: Vulnerability of desert biological soil crusts to wind erosion: the influences of crust development, soil texture, and disturbance. *J Arid Environ* 1998, 39: 133–142.

30. Belnap J, Lange OL: *Biological soil crusts: structure, function, and management. Ecological Studies Series, vol. 150*. Berlin: Springer; 2001.

31. Bennett PC, Rogers JR, Choi WJ: Silicates, silicate weathering, and microbial ecology. *Geomicrobiol J* 2001, 18: 3–19.

32. Beraldi-Campesi H, Hartnett HE, Anbar A, Gordon GW, Garcia-Pichel F: Effect of biological soil crusts on soil elemental concentrations: implications for biogeochemistry and as traceable biosignatures of ancient life on land. *Geobiology* 2009, 7: 348–359.

33. Beraldi-Campesi H, Farmer JD, Garica-Pichel F: *Evidence for Mesoproterozoic life on land and its modern counterpart in arid soils*. Proceedings of the GSA Annual Meeting, Minneapolis, 9–12 October 2011; 2011.

34. Bernard S, Benzerara K, Beyssac O, Menguy N, Guyot F, Brown GE Jr, Goffe B: Exceptional preservation of fossil plant spores in high-pressure metamorphic rocks. *Earth Planet Sc Lett* 2007, 262: 257–272.

35. Beukes NJ, Dorland H, Gutzmer J, Nedachi M, Ohmoto H: Tropical laterites, life on land, and the history of atmospheric oxygen in the Paleoproterozoic. *Geology* 2002, 30: 491–494.

36. Blackwell M: Terrestrial life: fungal from the start? *Science* 2000, 289(5486):1884–1885.

37. Blank CE: Not so old Archaea—the antiquity of biogeochemical processes in the archaeal domain of life. *Geobiology* 2009, 7(5):495–514.

38. Booth WE: Algae as pioneers in plant succession and their importance in erosion control. *Ecology* 1941, 22: 38–46.

39. Bouvier W: The age of the Solar System redefined. *Nat Geosci* 2010, 3: 637–641.

40. Boyet M, Carlson RW: $^{142}$Nd evidence for early (>4.53 Ga) global differentiation of the silicate Earth. *Science* 2005, 309: 576–581.

41. Brady NC, Weil RR: *The nature and properties of soils*. 14th edition. Upper Saddle River, NJ: Pearson-Prentice Hall; 2008.

42. Brocks JJ: Millimeter-scale concentration gradients of hydrocarbons in Archean shales: live-oil escape or fingerprint of contamination? *Geochim Cosmochim Acta* 2011, 75(11):3196–3213.

43. Büdel B, Weber B, KuhlM PH, Sultemeyer D, Wessels D: Reshaping of sandstone surfaces by cryptoendolithic cyanobacteria: bioalkalisation causes chemical weathering in arid landscapes. *Geobiology* 2004, 2: 261–268.

44. Buick R: When did oxygenic photosynthesis evolve? *Philos Trans R Soc Lond B Biol Sci* 2008, 363(1504):2731–2743.

45. Buick R: Early life: ancient acritarchs. *Nature* 2010, 463(7283):885–886.

46. Buick R, Thornett JR, McNaughton NJ, Smith JB, Barley ME, Savage M: Record of emergent continental crust 3.5 billion years ago in the Pilbara Craton of Australia. *Nature* 1995, 375: 574–577.

47. Campbell SE: Soil stabilization by a prokaryotic desert crust: implications for Precambrian land biota. *Orig Life* 1979, 9: 335–348.

48. Campbell SE, Seeler J, Golubic S: Desert crust formation and soil stabilization. In: J. Skujins (ed) Uses of microbiological processes in arid lands for desertification control and increased productivity (UNEP). *Arid Soil Res Rehab* 1989, 3: 217–228.

49. Canup RM, Righter K (Eds): *Origin of the Earth and moon*. Tucson, AZ: University of Arizona Press; 2000.

50. Catling DC, Zahnle KJ, McKay CP: Biogenic methane, hydrogen escape, and the irreversible oxidation of early Earth.*Science* 2001, 293: 839–843.

51. Cavosie AJ, Valley JW, Wilde S: The oldest terrestrial mineral record: a review of 4400 to 3900 Ma detrital zircons from Jack Hills, Western Australia. In *Earth's oldest rocks. developments in Precambrian geology, Series 15*. Edited by: Van Kranendonk MJ, Smithies RH, Bennett V. Amsterdam: Elsevier; 2007:91–111.

52. Chanal A, Chapon V, Benzerara K, Barakat M, Christen R, Achouak W, Barras F, Heulin T: The desert of Tataouine: an extreme environment that hosts a wide diversity of microorganisms and radiotolerant bacteria. *Environ Microbiol* 2006, 8(3):514–525.

53. Chen Y, Wu L, Boden R, Hillebrand A, Kumaresan D, Moussard H, Baciu M, Lu Y, Murrell JC: Life without light: microbial diversity and evidence of sulfur- and ammonium-based chemolithotrophy in Movile Cave. *ISME J* 2009, 3: 1093–1104.

54. Chenu C, Stotzky G: Interactions between microorganisms and soil particles: an overview. In *Interactions between soil particles and microorganisms: impact on the terrestrial ecosystem*. Edited by: Huang PM, Bollag JM, Senesi N. Chichester: John Wiley and Sons; 2002:4–40.

55. Chorover J, Kretzschmar R, Garcia-Pichel F, Sparks DL: Soil biogeochemical processes within the Critical Zone. *Elements*2007, 3: 321–326.

56. Cloud P, Germs A: New pre-paleozoic nannofossils from the Stoer formation (Torridonian), Northwest Scotland. *Geol Soc Am Bull* 1971, 82: 3469–3474.

57. Condie KC: Precambrian superplume events. In *The Precambrian Earth: tempos and events. Developments in Precambrian geology, vol. 12*. Edited by: Eriksson PG, Altermann W, Nelson DR, Mueller WU, Catuneanu O. Amsterdam: Elsevier; 2004:163–173.

58. Corcoran PL, Mueller WU: Aggressive Archaean weathering. In *The Precambrian Earth: tempos and events. Developments in Precambrian geology, vol. 12*. Edited by: Eriksson PG, Altermann W, Nelson DR, Mueller WU, Catuneanu O. Amsterdam: Elsevier; 2004:494–504.

59. Dauphas N, van Zuilen M, Wadhwa M, Davis AM, Martey B, Janney PE: Clues from Fe isotope variations on the origin of early Archean BIFs from

Greenland. *Science* 2004, 302: 2077–2080.

60. David LA, Alm EJ: Rapid evolutionary innovation during an Archaean genetic expansion. *Nature* 2011, 469(7328):93–96.

61. Davison J: Genetic exchange between bacteria in the environment. *Plasmid* 1999, 42(2):73–91.

62. Deming D: Origin of the ocean and continents: a unified theory of the earth. *Int Geol Rev* 2002, 44: 137–152.

63. DesMarais DJ: When did photosynthesis emerge on Earth? *Science* 2000, 289: 1703–1705.

64. DiMichele WA, Hook RW: Paleozoic terrestrial ecosystems. In *Terrestrial ecosystems through time*. Edited by: Behrensmeyer AK, Damuth JD, DiMichele WA, Potts R, Suess HD, Wing SL. Chicago: Chicago University Press; 1992:205–325.

65. Dong GR, Li CZ, Jin T, Gao SY, Wu D: Some results on soil wind-tunnel imitating experiment. *Chinese Sci Bull* 1987, 32: 297–301.

66. Dott RH Jr: The importance of eolian abrasion in supermature quartz sandstones and the paradox of weathering on vegetation-free landscapes. *J Geol* 2003, 111: 387–405.

67. Driese SG, Gordon-Medaris L Jr: Evidence for biological and hydrological controls on the development of a Paleoproterozoic paleoweathering profile in the Baraboo Range, Wisconsin, USA. *J Sediment Res* 2008, 78: 443–457.

68. Driese SG, Simpson EL, Eriksson KA: Redoximorphic paleosols in alluvial and lacustrine deposits, 1.8 Ga lochness formation, Mount Isa, Australia: pedogenic processes and implications for paleoclimate. *J Sediment Res* 1995, A65: 675–689.

69. Ehrlich HL: *Geomicrobiology*. New York, NY: Marcel Dekker; 2002.

70. El Albani A, Bengtson S, Canfield DE, Bekker A, Macchiarelli R, Mazurier A, Hammarlund EU, Boulvais P, Dupuy JJ, Fontaine C, Fürsich FT, Gauthier-Lafaye F: Large colonial organisms with coordinated growth in oxygenated environments 2.1 Gyr ago. *Nature* 2010, 466(7302):100–104.

71. Elbert W, Weber B, Burrows S, Steinkamp J, Büdel B, Andreae MO, Pöschl U: Contribution of cryptogamic covers to the global cycles of carbon and nitrogen. *Nat Geosci* 2012, 5: 459–462.

72. Engel AS: Microbial diversity of cave ecosystems. In *Geomicrobiology: molecular and environmental perspectives*. Edited by: Barton L, Mandl M, Loy A. New York: Springer; 2010:219–238.

73. Eriksson PG, Martins-Neto MA: Commentary. In *The Precambrian Earth: tempos and events. Developments in Precambrian geology, vol. 12*. Amsterdam: Elsevier; 2004:677–680.

74. Eriksson PG, Simpson EL, Eriksson KA, Bumby AJ, Steyn GL, Sarkar S: Muddy roll-up structures in siliciclastic interdune beds of the c. 1.8 Ga Waterberg Group, South Africa. *PALAIOS* 2000, 15: 177–183.

75. Eriksson PG, Altermann W, Nelson DR, Mueller WU, Catuneanu O: *The Precambrian Earth: tempos and events. Developments in Precambrian geology, vol. 12*. Amsterdam: Elsevier; 2004.

76. Farmer J: Taphonomic modes in microbial fossilization. In *Space Studies Board (ed) Size limits of very small microorganisms*. Washington DC: National Research Council, National Academy Press; 1999:94–102.

77. Farrow CE, Mossman DJ: Geology of Precambrian paleosols at the base of the Huronian supergroup, Elliot Lake, Ontario, Canada. *Precambrian Res* 1988, 42: 107–139.

78. Fedo CM, Whitehouse MJ: Metasomatic origin of quartz-pyroxene rock, Akilia, Greenland, and implications for Earth's earliest life. *Science* 2002, 296: 1448–1452.

79. Fierer N, Nemergut D, Knight R, Craine JM: Changes through time: integrating microorganisms into the study of succession. *Res Microbiol* 2010, 161: 635–642.

80. Fleming ED, Castenholz RW: Effects of periodic desiccation on the synthesis of the UV-screening compound, scytonemin, in cyanobacteria. *Environ Microbiol* 2007, 9: 1448–1455.

81. Fletcher JE, Martin WP: Some effects of algae and mold in the rain-crust of desert soils. *Ecology* 1948, 29(1):95–100.

82. Fliegel D, Wirth R, Simonetti A, Furnes H, Staudigel H, Hanski E, Muehlenbachs : Septate-tubular textures in 2.0-Ga pillow lavas from the Pechenga Greenstone Belt: a nano-spectroscopic approach to investigate their biogenicity. *Geobiology* 2010, 8: 372–390.

83. Franz G, Mosbrugger V, Menge R: Carbo-Permian pteridophyll leaf fragments from an amphibolite facies basement, Tauern Window, Austria. *Terra Nova* 1991, 3: 137–141.

84. Friedmann EI: Endolithic microbial life in hot and cold deserts. *Orig Life* 1980, 10: 223–235.

85. Fritsch FE: The terrestrial alga. *J Ecol* 1922, 10(2):220–236.

86. Furnes H, Banerjee NR, Muehlenbachs K, Staudigel H, de Wit M: Early life recorded in Archean pillow lavas. *Science* 2004, 304: 578–81.

87. Furnes H, de Wit M, Staudigel H, Rosing M, Muehlenbachs K: Vestige of Earth's oldest ophiolite. *Science* 2007, 315: 1704–1707.

88. Furnes H, Banerjee NR, Staudigel H, Muehlenbachs K, McLoughlin N, de Wit M, Van Kranendonk M: Comparing petrographic signatures of bioalteration in recent to Mesoarchean pillow lavas: tracing subsurface life in oceanic igneous rocks. *Precambrian Res* 2007, 158: 156–176.

89. Gadd GM (Ed): *Fungi in biogeochemical cycles*. Cambridge, UK: Cambridge University Press; 2006.

90. Gall Q: The Proterozoic Thelon paleosol, Northwest Territories, Canada. *Precambrian Res* 1994, 68: 115–137.

91. Gao Q, Garcia-Pichel F: Microbial ultraviolet sunscreens. *Nat Rev Microbiol* 2011, 9: 791–802.

92. García Ruiz JM, Carnerup A, Christy AG, Welham NJ, Hyde ST: Morphology: an ambiguous indicator of

biogenicity.*Astrobiology* 2002, 2(2):353–369.

93. Garcia-Pichel F: Solar ultraviolet and the evolutionary history of cyanobacteria. Origins Life Evol. *Biosphere* 1998, 28: 321–347.
94. Garcia-Pichel F, Castenholz RW: Characterization and biological implications of scytonemin, a cyanobacterial sheath pigment. *J Phycol* 1991, 27: 395–409.
95. Garcia-Pichel F, Pringault O: Cyanobacteria track water in desert soils. *Nature* 2001, 413: 380–381.
96. Garcia-Pichel F, Wojciechowski MF: The evolution of a capacity to build supra-cellular ropes enabled filamentous cyanobacteria to colonize highly erodible substrates. *PLoS One* 2009, 4(11):e7801.
97. Garcia-Pichel F, Lopez-Cortes A, Nubel U: Phylogenetic and morphological diversity of cyanobacteria in soil desert crusts from the Colorado Plateau. *Appl Environ Microbiol* 2001, 67: 1902–1910.
98. Garcia-Pichel F, Belnap J, Neuer S, Schanz F: Estimates of global cyanobacterial biomass and its distribution. *Algol Stud*2003, 109: 213–227.
99. Garcia-Ruiz JM, Hyde ST, Carnerup AM, Christy AG, Van Kranendonk MJ, Welham NJ: Self-assembled silica-carbonate structures and detection of ancient microfossils. *Science* 2003, 302: 1194–1197.
100. Gaucher C, Sial AN, Halverson GP, Frimmel HE: *Neoproterozoic-Cambrian tectonics, global change and evolution: a focus on Southwestern Gondwana. Developments in Precambrian Geology, vol. 16.* Amsterdam: Elsevier; 2010.
101. Gay AL, Grandstaff DE: Chemistry and mineralogy of Precambrian paleosols at Elliot Lake, Ontario, Canada. *Precambrian Res* 1980, 12: 349–373.
102. Geesey G, Jang L: Extracellular polymers for metal binding. In *Microbial mineral recovery*. Edited by: Ehrlich HL, Brierley CL. New York: McGraw-Hill; 1990:223–249.
103. Gensel PG: The earliest land plants. *Annu Rev Ecol Evol Syst* 2008, 39: 459–477.
104. Ghiorse W, Wilson J: Microbial ecology of the terrestrial subsurface. In *Advances in applied microbiology*. Edited by: Laskin A. New York: Academic; 1988:107–172.
105. Giordanino F, Strauch SM, Villafañe VE, Helbling EW: Influence of temperature and UVR on photosynthesis and morphology of four species of cyanobacteria. *J Photochem Photobiol B: Biol* 2011, 103: 68–77.
106. Glass JB, Wolfe-Simon F, Anbar AD: Coevolution of metal availability and nitrogen assimilation in cyanobacteria and algae.*Geobiology* 2009, 7: 100–123.
107. Glikson A: Early Archean asteroid impacts on Earth: stratigraphic and isotopic age correlations and possible geodynamic consequences. In *Earth's oldest rocks. Developments in Precambrian geology, Series 15.* Edited by: Van Kranendonk MJ, Smithies RH, Bennett V. Amsterdam: Elsevier; 2007:1087–1103.
108. Glover JF, Kah LC: *Speleothem deposits in a Proterozoic paleokarst, Mesoproterozoic dismal lakes group, Arctic Canada*. 2006. Geological Society of America Abstracts with Programs 38(3):36 Geological Society of America Abstracts with Programs 38(3):36
109. Gogarten JP, Fournier G, Zhaxybayeva O: Gene transfer and the reconstruction of life's early history from genomic data. In*Strategies of life detection. Space Science Series of ISSI, vol. 25*. Edited by: Botta O, Bada JL, Gomez-Elvira J, Javaux E, Selsis F, Summons R. New York: Springer; 2007:115–131.
110. Golubic S, Campbell SE: Analagous microbial forms in recent subaerial habitats and in Precambrian cherts: *Gloethece coerulea* Geitler and *Eosynechococcus moorei* Hoffmann. *Precambrian Res* 1979, 8: 201–217.
111. Golubic S, Hofmann HJ: Comparison of modern and mid-Precambrian Entophysalidaceae (Cyanophyta) in stromatolitic algal mats: cell division and degradation. *J Paleont* 1976, 50: 1074–1082.
112. Golubic S, Seong-Joo L: Early Cyanobacterial fossil record: preservation, palaeoenvironments and identification. *Eur J Phycol* 1999, 34: 339–348.
113. Gomes R, Levison HF, Tsiganis K, Morbidelli A: Origin of the cataclysmic late heavy bombardment period of the terrestrial planets. *Nature* 2005, 435(7041):466–469.
114. Gradzinski M, Chmiel MJ, Lewandowska A, Michalska-Kasperkiewicz B: Siliciclastic micro-stromatolites in a sandstone cave: role of trapping and binding of detrital particles in formation of cave deposits. *Ann Soc Geol Polon* 2010, 80(3):303–314.
115. Grandstaff DE, Edelman MJ, Foster RW, Zbinden E, Kimberley MM: Chemistry and mineralogy of Precambrian paleosols at the base of the Dominion and Pongola Groups. *Precambrian Res* 1986, 32: 97–131.
116. Gray J, Boucot AJ: Early Silurian nonmarine animal remains and the nature of the early continental ecosystem. *Acta Palaeontol Polon* 1994, 38(3–4):303–328.
117. Gray J, Shear WA: Early life on land. *Am Sci* 1992, 80: 444–456.
118. Gunatilaka A: Some aspects of the biology and sedimentology of laminated algal mats from Mannar Lagoon, northwest Ceylon. *Sediment Geol* 1975, 14: 275–300.
119. Gutzmer J, Beukes NJ: Earliest laterites and possible evidence for terrestrial vegetation in the Early Proterozoic. *Geology*1998, 26: 263–266.
120. Hallbauer DK: The plant origin of the Witwatersrand carbon. *Minerals Sci Eng* 1975, 7(2):111–131.
121. Hallbauer DK, van Warmelo KT: Fossilized plants in thucholite from Precambrian rocks of the Witwatersrand, South Africa.*Precambrian Res* 1974, 1: 199–212.
122. Hamblin WK, Christiansen EH: *Earth's dynamic systems*. 10th edition. Upper Saddle River, NJ: Prentice Hall; 2007.
123. Han TM, Runnegar B: Megascopic eukaryotic algae from the 2.1-billion-year-old negaunee iron-formation, Michigan.*Science* 1992, 257(5067):232–235.
124. Hanel M, Montenari M, Kalt A: Determining sedimentation ages of high-grade metamorphic gneisses by their palynological record: a case study in the northern Schwarzwald (Variscan Belt, Germany). *Int J Earth*

*Sci* 1999, 88: 49–59.

125. Harrison SP, Kohfeld KE, Roelandt C, Claquin T: The role of dust in climate changes today, at the last glacial maximum and in the future. *Earth Sci Rev* 2001, 54: 43–80.
126. Hartmann WK, Ryder G, Dones L, Grinspoon D: The time-dependent intense bombardment of the primordial Earth/Moon system. In *Origin of the Earth and moon*. Edited by: Canup RM, Righter K. Tucson: University of Arizona Press; 2000:493–512.
127. Hazen RM, Sverjensky DA: Mineral surfaces, geochemical complexities, and the origins of life. In *The origins of life, volume 5*. 2nd edition. Edited by: Deamer D, Szostak JW. Cold Spring Harbor: Cold Spring Harbor Laboratory Press; 2010:1–21.
128. Hazen RM, Golden J, Downs RT, Hystad G, Grew ES, Azzolini D, Sverjensky DA: Mercury (Hg) mineral evolution: a mineralogical record of supercontinent assembly, changing ocean geochemistry, and the emerging terrestrial biosphere. *Am Mineral* 2012, 97: 1013–1042.
129. Heckman DS, Geiser DM, Eidell BR, Stauffer RL, Kardos NL, Hedges SB: Molecular evidence for the early colonization of land by fungi and plants. *Science* 2001, 293(5532):1129–1133.
130. Herrero A, Flores E (Eds): *The cyanobacteria: molecular biology, genomics and evolution*. Norfolk: Caister Academic Press; 2008.
131. Hickman AH, Van Kranendonk MJ: Early Earth evolution: evidence from the 3.5–1.8 Ga geological history of the Pilbara region of Western Australia. *Episodes* 2012, 35(1):283–297.
132. Hoashi M, Bevacqua DC, Otake T, Watanabe Y, Hickman AH, Utsunomiya S, Ohmoto H: Primary haematite formation in an oxygenated sea 3.46 billion years ago. *Nat Geosci* 2009, 2: 301–306.
133. Hoffman PF: The oldest terrestrial landscape. *Nature* 1995, 375(6532):537–538.
134. Hoffman PF, Schrag DP: The snowball Earth hypothesis: testing the limits of global change. *Terra Nova* 2002, 14: 129–155.
135. Hofmann HJ: Precambrian microflora, Belcher Islands, Canada; significance and systematics. *J Paleontol* 1976, 50: 1040–1073.
136. Holland HD: *The chemical evolution of the atmosphere and oceans*. Princeton: Princeton University Press; 1984.
137. Holland HD: Distribution and paleoenvironment interpretation of Proterozoic paleosols. In *The Proterozoic biosphere*. Edited by: Schopf JW, Klein C. Cambridge: Cambridge University Press; 1992:153–155.
138. Holm NG: Marine hydrothermal systems and the origins of life. *Orig Life Evol Biosph* 1992, 22: 181–242.
139. Hoppert M, Reimer R, Kemmling A, Schröder A, Günzl B, Heinken T: Structure and reactivity of a biological soil crust from a xeric sandy soil in central Europe. *Geomicrobiol J* 2004, 21: 183–191.
140. Horodyski RJ, Knauth PL: Life on land in the Precambrian. *Science* 1994, 263(5146):494–498.
141. Hu C, Liu Y, Song L, Zhang D: Effect of desert soil algae on the stabilization of fine sands. *J Appl Phycol* 2002, 14(4):281–292.
142. Hupe P: Sur des problematica du Precambrien III. *Division des Mines et de la Géologie, Service Géologique de Morocco, Notes et Memoires* 1952, 103: 297–383.
143. Iizuka T, Horie K, Komiya T, Maruyama S, Hirata T, Hidaka H, Windley BF: 4.2 Ga zircon xenocryst in an Acasta gneiss from northwestern Canada: evidence for early continental crust. *Geology* 2006, 34: 245–248.
144. Ishizuka H: Protoliths of the Napier Complex in Enderby Land, East Antarctica; an overview and implication for crustal formation of Archean continents. *J Miner Petrol Sci* 2008, 103: 218–225.
145. Jackson TA: Fossil Actinomycetes in Middle Precambrian glacial varves. *Science, New Series* 1967, 155(3765):1003–1005.
146. Javaux EJ, Marshall CP, Bekker A: Organic-walled microfossils in 3.2-billion-year-old shallow-marine siliciclastic deposits.*Nature* 2010, 463: 934–938.
147. Jickells TD, An ZS, Andersen KK, Baker AR, Bergametti G, Brooks N, Cao JJ, Boyd PW, Duce RA, Hunter KA, Kawahata H, Kubilay N, laRoche J, Liss PS, Mahowald N, Prospero JM, Ridgwell AJ, Tegen I, Torres R: Global iron connections between desert dust, ocean biogeochemistry, and climate. *Science* 2005, 308: 67–71.
148. Johnson I, Watanabe Y, Stewart B, Ohmoto H: *Earth's oldest (~3.4 Ga) lateritic paleosol in the Pilbara Craton, Western Australia*. Davos, Switzerland: Proceedings of the Goldschmidt conference; 2009.
149. Johnson I, Watanabe Y, Stewart B, Ohmoto H: *Evidence for terrestrial life and an O2-rich atmosphere in the oldest (~3.4 Ga) paleosol in the east Pilbara craton, Western Australia*. League City, TX: Proceedings of the 6th Astrobiology Science Conference; 2010. 20–26 April 2010 20–26 April 2010
150. Kandler O: The early diversification of life. In *Early life on earth*. Edited by: Bengston S. New York: Columbia University Press; 1994:152–161.
151. Kappler A, Pasquero C, Konhauser KO, Newman DK: Deposition of banded iron formations by anoxygenic phototrophic Fe(II)-oxidizing bacteria. *Geology* 2005, 33(11):865–868.
152. Kasting J: *How to find a habitable planet*. Princeton: Princeton University Press; 2009.
153. Kato Y, Suzuki K, Nakamura K, Hickman AH, Nedachi M, Kusakabe M, Bevacqua DC, Ohmoto H: Hematite formation by oxygenated groundwater more than 2.76 billion years ago. *Earth Planet Sci Lett* 2009, 278: 40–49.
154. Keller CK, Wood BD: Possibility of chemical weathering before the advent of vascular land plants. *Nature* 1993, 364: 223–225.
155. Kendall B, Reinhard CT, Lyons TW, Kaufman AJ, Poulton SW, Anbar AD: Pervasive oxygenation along late Archaean ocean margins. *Nat Geosci* 2010, 3(9):647–652.

156. Kennedy MJ, Wagner T: Clay mineral continental amplifier for marine carbon sequestration in a greenhouse ocean. *Proc Nat Acad Sci USA* 2011, 108: 9776–9781.

157. Kennedy MJ, Chadwick OA, Vitousek PM, Derry LA, Hendricks DM: Replacement of weathering with atmospheric sources of base cations during ecosystem development, Hawaiian Islands. *Geology* 1998, 26: 1015–1018.

158. Kennedy M, Droser M, Mayer LM, Pevear D, Mrofka D: Late Precambrian oxygenation; inception of the clay mineral factory.*Science* 2006, 311: 1446–1449.

159. Kimberley MM, Grandstaff DE: Profiles of elemental concentrations in Precambrian paleosols on basaltic and granitic parent materials. *Precambrian Res* 1986, 32: 133–154.

160. Kirschvink JL, Kopp RE: Palaeoproterozoic ice houses and the evolution of oxygen-mediating enzymes: the case for a late origin of photosystem II. *Phil Trans R Soc B* 2008, 363: 2755–2765.

161. Klein C, Beukes NJ, Schopf JW: Filamentous microfossils in the early Proterozoic Transvaal supergroup: their morphology, significance, and paleoenvironmental setting. *Precambrian Res* 1987, 36: 81–94.

162. Knauth LP, Kennedy MJ: The late Precambrian greening of the Earth. *Nature* 2009, 460: 728–732.

163. Knoll AH, Javaux EJ, Hewitt D, Cohen P: Eukaryotic organisms in Proterozoic oceans. *Philosophical Transactions of the Royal Society B* 2006, 361(1470):1023–1038.

164. Knoll A, Canfield D, Konhauser K (Eds): *Fundamentals of geobiology*. Chichester: Wiley-Blackwell; 2012.

165. Koehler I, Konhauser KO, Kappler A: Role of microorganisms in banded iron formations. In *Geomicrobiology: molecular and environmental perspective*. Edited by: Barton L, Mandl M, Loy A. New York: Springer; 2010:309–324.

166. Komiya T, Maruyama S, Nohda S, Masuda T, Hayashi M, Okamoto S: Plate tectonics at 3.8–3.7 Ga; field evidence from the Isua accretionary complex, southern West Greenland. *J Geol* 1999, 107: 515–554.

167. Konhauser K: *Introduction to geomicrobiology*. Oxford: Blackwell; 2007.

168. Konhauser KO, Schultzelam S, Ferris FG, Fyfe WS, Longstaffe FJ, Beveridge TJ: Mineral precipitation by epilithic biofilms in the Speed River, Ontario, Canada. *Appl Environ Microbiol* 1994, 60: 549–553.

169. Konhauser KO, Newman DK, Kappler A: The potential significance of microbial Fe(III) reduction during deposition of Precambrian banded iron formations. *Geobiology* 2005, 3: 167–177.

170. Kositcin N, Krapez B: SHRIMP U-Pb detrital zircon geochronology of the late Archean Witwatersrand Basin, relation between zircon provenance age spectra and basin evolution. *Precambrian Res* 2004, 129: 141–168.

171. Labandeira CC: Invasion of the continents: cyanobacterial crusts to tree-inhabiting arthropods. *Trends Ecol Evol* 2005, 20(5):253–262.

172. Lalonde K, Mucci A, Ouellet A, Gélinas Y: Preservation of organic matter in sediments promoted by iron. *Nature* 2012, 483(7388):198–200.

173. Lannerbro R: Description of some structures, possibly fossils, in Jotnian sandstone from Mångsbodarna in Dalecarlia.*Geologiska Föreningens i Stockholm Förhandlingar* 1954, 76: 46–50.

174. Lau KM, Kim MK, Kim KM: Asian monsoon anomalies induced by aerosol direct effects. *Clim Dyn* 2006, 26: 855–864.

175. Lazcano A, Miller SL: How long did it take for life to begin and evolve to cyanobacteria? *J Mol Evol* 1994, 39: 546–554.

176. Li W, Johnson CM, Beard BL: U–Th–Pb isotope data indicate Phanerozoic age for oxidation of the 3.4 Ga Apex Basalt.*Earth Planet Sci Lett* 2012, 319–320: 197–206.

177. Liu YZ, Dong GR, Li CZ: A study on the factors influencing soil erosion through wind tunnel experiments. *Chin J Arid Land Res* 1994, 7: 359–367.

178. Lopez-Garcia P, Moreira D, Douzery EJP, Forterre P, van Zuilen M: Ancient fossil record and early evolution (ca. 3.8 to 0.5 Ga). In *From suns to life: a multidisciplinary approach to the history of life on Earth. Earth, moon, and planets*. Edited by: Gargaud M. New York: Springer; 2006:247–290.

179. Lowe DR: Stromatolites 3,400-Myr old from the Archean of Western Australia. *Nature* 1980, 284: 441–443.

180. Lozupone CA, Knight R: Global patterns in bacterial diversity. *Proc Natl Acad Sci USA* 2007, 104: 11436–11440.

181. Macfarlane AW, Danielson A, Holland HD: Geology and major and trace element chemistry of the late Archean weathering profiles in the Fortescue Group, Western Australia: implications for atmospheric $pO_2$. *Precam Res* 1994, 65: 297–317.

182. Madigan MT, Marrs BL: Extremophiles. *Scientific American* 1997, 276(4):66–71.

183. Madigan M, Martinko J, Parker J: *Brock: biology of microorganisms*. 10th edition. Upper Saddle River, NJ: Pearson-Prentice Hall; 2003.

184. Malam Issa O, Le Bissonnais Y, Défarge C, Trichet J: Role of a cyanobacterial cover on structural stability of sandy soils in the Sahalian part of western Niger. *Geoderma* 2001, 101: 15–30.

185. Martini JEJ: A late Archaean–Palaeoproterozoic (2.6 Ga) palaeosol on ultramafics in the eastern Transvaal, South Africa.*Precambrian Res* 1994, 67: 159–180.

186. Mazor G, Kidron GJ, Vonshak A, Abeliovich A: The role of cyanobacterial exopolysaccharides in structuring desert microbial crusts. *FEMS Microbiol Ecol* 1996, 21: 121–130.

187. McConnell RL: Preliminary report of microstructures of probable biologic origin from the Mescal Formation (Proterozoic) of central Arizona. *Precambrian Res* 1974, 1(3):227–234.

188. McKeegan KD, Kudryavtsev AB, Schopf JW: Raman and ion microscopic imagery of graphitic inclusions in apatite from older than 3830 Ma Akilia supracrustal

rocks, west Greenland. *Geology* 2007, 35: 591–594.

189. McTainsh GH, Strong CL: The role of aeolian dust in ecosystems. *Geomorphology* 2007, 89(1–2):39–54.

190. Meadow JF, Zabinski CA: Spatial heterogeneity of eukaryotic microbial communities in an unstudied geothermal diatomaceous biological soil crust: Yellowstone National Park, WY, USA. *FEMS Microbiol Ecol* 2012, 82(1):182–191.

191. Mloszewska AM, Pecoits E, Cates NL, Mojzsis SJ, O'Neil J, Robbins LJ, Konhauser KO: The composition of Earth's oldest iron formations: the Nuvvuagittuq Supracrustal Belt (Quebec, Canada). *Earth Planet Sci Lett* 2012, 317–318: 331–342.

192. Mojzsis SJ, Harrison MT, Pidgeon RT: Oxygen-isotope evidence from ancient zircons for liquid water at the Earth's surface 4,300 Myr ago. *Nature* 2001, 409: 178–181.

193. Mossman DJ, Minter WEL, Dutkiewicz A, Hallbauer DK, George SC, Hennigh Q, Reimer TO, Horscroft FD: The indigenous origin of Witwatersrand "carbon". *Precambrian Res* 2008, 164: 173–186.

194. Mulkidjanian AY, Bychkovc AY, Dibrova DV, Galperin MY, Koonin EV: Origin of the first cells at terrestrial, anoxic geothermal fields. *Proc Natl Acad Sci USA* 2012, 109: E821-E830.

195. Myers JS: Protoliths of the 3.8–3.7 Ga Isua Greenstone Belt, West Greenland. *Precambrian Res* 2001, 105: 129–141.

196. Myers JS: Isua enigmas: illusive tectonic, sedimentary, volcanic and organic features of the >3.7 Ga Isua greenstone belt, southwest Greenland. In *The Precambrian Earth: tempos and events. Developments in Precambrian geology, vol. 12*. Edited by: Eriksson PG, Altermann W, Nelson DR, Mueller WU, Catuneanu O. Amsterdam: Elsevier; 2004:66–73.

197. Nagy M, Perez A, Garcia-Pichel F: The prokaryotic diversity of biological soil crusts in the Sonoran desert (Organ Pipe Cactus National Monument, AZ). *FEMS Microbiol Ecol* 2005, 54: 233–245.

198. Neher DA, Lewins SA, Weicht TR, Darby BJ: Microarthropod communities associated with biological soil crusts in the Colorado Plateau and Chihuahuan deserts. *J Arid Environ* 2009, 73: 672–677.

199. Nelson DR: Earth's formation and first billion years. In *The Precambrian Earth: tempos and events,\. Developments in Precambrian geology, vol. 12*. Edited by: Eriksson PG, Altermann W, Nelson DR, Mueller WU, Catuneanu O. Amsterdam: Elsevier; 2004:3–27.

200. Nemergut DR, Costello EK, Hamady M, Lozupone C, Jiang L, Schmidt SK, Fierer N, Townsend AR, Cleveland CC, Stanish L, Knight R: Global patterns in the biogeography of bacterial taxa. *Environ Microbiol* 2011, 13(1):135–144.

201. Nesbitt HW, Young GM: Aggressive Archaean weathering. In *The Precambrian Earth: tempos and events. Developments in Precambrian geology, vol. 12*. Edited by: Eriksson PG, Altermann W, Nelson DR, Mueller WU, Catuneanu O. Amsterdam: Elsevier; 2004:482–493.

202. Nicholas J, Bildgen P: Relations between the location of the karst bauxites in the northern hemisphere, the global tectonics, and the climatic variations during geological time. *Palaeogeogr Palaeoclimateol Palaeoecol* 1979, 28: 205–239.

203. Noffke N: *Geobiology: microbial mats in sandy deposits from the Archean era to today*. Berlin: Springer; 2010.

204. Noffke N: A Modern Perspective on Ancient Life: Microbial Mats in Sandy Marine Settings from the Archean Era to Today. In*Earliest Life on Earth: Habitats. Environments and Methods of Detection*. Edited by: Golding SD, Glikson M. Dordrecht: Springer; 2011:171–182.

205. Noffke N, Gerdes G, Klenke T, Krumbein WE: Perspectives. Microbially induced sedimentary structures—a new category within the classification of primary sedimentary structures. *J Sediment Res* 2001, 71(5):649–656.

206. Noffke N, Eriksson KA, Hazen RM, Simpson EL: A new window into Early Archean life: microbial mats in Earth's oldest siliciclastic tidal deposits (3.2 Ga Moodies Group, South Africa). *Geology* 2006, 34(4):253–256.

207. Noffke N, Beukes N, Bower D, Hazen RM, Swift DJP: An actualistic perspective into Archean worlds – (cyano-)bacterially induced sedimentary structures in the siliciclastic Nhlazatse Section, 2.9 Ga Pongola Supergroup, South Africa. *Geobiology*2008, 6: 5–20.

208. Noffke N, Christian DR, Hazen RM: A (cyano-)bacterial ecosystem in the Archean 3.49 Ga Dresser Formation, Pilbara, Western Australia. GSA Annual Meeting in Minneapolis. Paper No. 56–11. *Geological Society of America Abstracts with Programs* 2011, 43(5):159.

209. Nutman AP: Antiquity of the oceans and continents. *Elements* 2006, 2: 223–227.

210. Nutman AP, McGregor VR, Friend CRL, Bennett VC, Kinny PD: The Itsaq Gneiss Complex of southern West Greenland; the world's most extensive record of early crustal evolution (3900–3600 Ma). *Precambrian Res* 1996, 78: 1–39.

211. Nutman AP, Friend CRL, Bennett VC: Evidence for 3650–3600 Ma assembly of the northern end of the Itsaq Gneiss Complex, Greenland: implication for early Archaean tectonics. *Tectonics* 2002, 21(1):5–1–5–28.

212. Nutman AP, Friend CRL, Bennett VC, Wright D, Norman MD: ≥3700 Ma pre-metamorphic dolomite formed by microbial mediation in the Isua supracrustal belt (W. Greenland): simple evidence for early life? *Precambrian Res* 2010, 183: 725–737.

213. O'Neil J, Francis D, Carlson RW: Implications of the Nuvvuagittuq greenstone belt for the formation of earth's early crust. *J Petrol* 2011, 52: 985–1009.

214. Och LM, Shields-Zhou GA: The Neoproterozoic oxygenation event. Environmental perturbations and biogeochemical cycling. *Earth Sci Rev* 2012, 110(1–4):26–57.

215. Ohmoto H: Evidence in pre-2.2 Gapaleosols for the early evolution of atmospheric oxygen and terrestrial

biota. *Geology*1996, 24: 1135–1138.

216. Ohmoto H: Archean atmosphere, hydrosphere, and biosphere. In *The Precambrian Earth: tempos and events. Developments in Precambrian geology, vol. 12*. Edited by: Eriksson PG, Altermann W, Nelson DR, Mueller WU, Catuneanu O. Amsterdam: Elsevier; 2004:361–368.

217. Ohmoto H, Watanabe Y, Allwood A, Burch I, Knauth P, Yamaguchi K, Johnson I, Altinok E: Formation of probable lateritic soils ~3.43 Ga in the Pilbara Craton, Western Australia. *Geochimica et Cosmochimica Acta, Supplement* 2007, 71(15):A733.

218. Omelon CR, Pollard WH, Ferris FG: Chemical and ultrastructural characterization of high arctic cryptoendolithic habitats.*Geomicrobiol J* 2006, 23: 189–200.

219. O'Neil J, Carlson RW, Francis D, Stevenson RK: Response to comment on "Neodymium-142 evidence for Hadean mafic crust". *Science* 2009, 325: 267b.

220. Pace N: A molecular view of microbial diversity and the biosphere. *Science* 1997, 276: 734–740.

221. Palmer JA, Phillips GN, McCarthy TS: Paleosols and their relevance to Precambrian atmospheric composition. *J Geol* 1989, 97: 77–92.

222. Pandit MK, de Wall H, Chauhan NK: Paleosol at the Archean-Proterozoic contact in NW India revisited: evidence for oxidizing conditions during paleo-weathering? *J Earth Syst Sci* 2008, 117(3):201–209.

223. Papineau D, DeGregorio BT, Cody GD, Fries MD, Mojzsis SJ, Steele A, Stroud RM, Fogel ML: Ancient graphite in the Eoarchean quartz-pyroxene rock from Akilia in southwest Greenland I: petrographic and spectroscopic characterization.*Geochim Cosmochim Acta* 2010, 74: 5862–5883.

224. Papineau D, DeGregorio BT, Stroud RM, Steele A, Pecoits E, Konhauser K, Wang J, Fogel ML: Ancient graphite in the Eoarchean quartz-pyroxene rock from Akilia in southern West Greenland II: isotopic and chemical compositions and comparison with Paleoproterozoic banded iron formations. *Geochim Cosmochim Acta* 2010, 74: 5884–5905.

225. Papineau D, De Gregorio BT, Cody GD, O'Neil J, Steele A, Stroud RM, Fogel ML: Young poorly crystalline graphite in the 3.8-Gyr-old Nuvvuagittuq banded iron formation. *Nat Geosci* 2011, 4(6):376–379.

226. Paul EA, Collins HP, Leavitt SW: Dynamics of resistant soil carbon of Midwestern agricultural soils measured by naturally occurring $^{14}C$ abundance. *Geoderma* 2001, 104(3–4):239–256.

227. Porter ML, Engel AS, Kinkle B, Kane TC: Productivity-diversity relationships from chemolithoautotrophically based sulfidic karst systems. *Int J Speleol* 2009, 38: 27–40.

228. Potts M: Mechanisms of desiccation tolerance in cyanobacteria. *Eur J Phycol* 1999, 34: 319–328.

229. Potts M, Friedmann EI: Effects of water stress on cryptoendolithic cyanobacteria from hot desert rocks. *Arch Microbiol* 1981, 130: 267–271.

230. Prasad N, Roscoe SM: Evidence of anoxic to oxic atmospheric change during 2.45–2.22 Ga from lower and upper sub-Huronian paleosols, Canada. *Catena* 1996, 27: 105–121.

231. Prave AR: Life on land in the Proterozoic: evidence from the Torridonian rocks of northwest Scotland. *Geology* 2002, 30(9):811–814.

232. Pufahl PK, Hiatt EE: Oxygenation of the Earth's atmosphere–ocean system: a review of physical and chemical sedimentologic responses. *J Mar Petrol Geol* 2012, 32(1):1–20.

233. Rasmussen B: Filamentous microfossils in a 3,235-million-year-old volcanogenic massive sulphide deposit. *Nature* 2000, 405: 676–679.

234. Rasmussen B, Fletcher IR, Brocks JJ, Kilburn MR: Reassessing the first appearance of eukaryotes and cyanobacteria.*Nature* 2008, 455(7216):1101–1104.

235. Rasmussen B, Blake TS, Fletcher IR, Kilburn MR: Evidence for microbial life in synsedimentary cavities from 2.75 Ga terrestrial environments. *Geology* 2009, 37: 423–426.

236. Reddy SG, Garcia-Pichel F: The community and phylogenetic diversity of biological soil crusts in the Colorado Plateau studied by molecular fingerprinting and intensive cultivation. *Microb Ecol* 2006, 52: 345–357.

237. Reimer TO: Alumina-rich rocks from the Early Precambrian of the Kaapvaal craton as indicators of paleosols and as products of other decompositional reactions. *Precambrian Res* 1986, 32: 155–179.

238. Retallack GJ: The fossil record of soils. In *Paleosols: their recognition and interpretation*. Edited by: Wright PV. Oxford: Blackwell; 1986:1–57.

239. Retallack GJ: Reappraisal of a 2200 Ma-old paleosol near Waterval Onder, South Africa. *Precambrian Res* 1986, 32: 195–232.

240. Retallack GJ: *Soils of the past*. 2nd edition. London: Blackwell Science; 2001.

241. Retallack GJ: Cambrian–Ordovician non-marine fossils from South Australia. *Alcheringa* 2009, 33: 355–391.

242. Retallack : Criteria for distinguishing microbial mats and earths. In: Noffke N, Chafetz H (ed) Microbial mats in siliciclastic depositional systems through time. SEPM Special Publication no. 101. Society for Sedimentary Geology, Tulsa, OK, pp. 2001, 139–152.

243. Retallack GJ: Ediacaran life on land. *Nature* 2013, 493(7430):89–92.

244. Retallack GJ, Mindszenty A: Well preserved Late Precambrian paleosols from northwest Scotland. *J Sediment Res* 1994, A64: 264–281.

245. Reynolds R, Belnap J, Reheis M, Lamothe P, Luiszer F: Aeolian dust in Colorado Plateau soils: nutrient inputs and recent change in source. *Proc Natl Acad Sci USA* 2001, 98: 7123–7127.

246. Rino S, Komiya T, Windley BF, Katayama S, Motoki A, Hirata T: Major episodic increases of continental crust growth determined from zircon ages of river sands; implication for mantle overturns in the early

Precambrian. *Phys Earth Planet Inter* 2004, 146: 369–394.

247. Romans BW, Graham SA: A deep-time perspective of land-ocean linkages in the sedimentary record. *Annu Rev Mar Sci*2013, 5: 69–94.

248. Rosing MT, Frei R: U-rich Archaean sea-floor sediments from Greenland: indications of > 3700 Ma oxygenic photosynthesis. *Earth Planet Sc Lett* 2004, 217: 237–244.

249. Rosing MT, Bird DK, Sleep NH, Glassley W, Albarede F: The rise of continents – an essay on the geologic consequences of photosynthesis. *Palaeogeogr Palaeoclimateol Palaeoecol* 2006, 232: 99–113.

250. Rossi F, Potrafka RM, Garcia-Pichel F, De Philippis R: The role of the exopolysaccharides in enhancing hydraulic conductivity of biological soil crusts. *Soil Biol Biochem* 2012, 46: 33–40.

251. Rye R, Holland HD: Life associated with a 2.76 Ga ephemeral pond? Evidence from Mount Roe #2 paleosol. *Geology* 2000, 28: 483–486.

252. Santosh M: A synopsis of recent conceptual models on supercontinent tectonics in relation to mantle dynamics, life evolution and surface environment. *J Geodyn* 2010, 50(3–4):116–133.

253. Sarbu SM, Kane TC, Kinkle BK: A chemoautotrophically based cave ecosystem. *Science* 1996, 272: 1953–1955.

254. Schau M, Henderson JB: Archean chemical weathering at three localities on the Canadian shield. *Dev Precambrian Geol*1983, 7: 81–116.

255. Schieber J, Bose PK, Eriksson PG, Banerjee S, Sarkar S, Altermann W, Catuneau O (Eds): *Atlas of microbial mat features preserved within the siliclastic rock record.* Amsterdam: Elsevier; 2007.

256. Schiffbauer JD, Xiao S: Novel application of focused ion beam-electron microscopy (FIB-EM) in preparation and analysis of microfossils ultrastructures: a new view of complexity in early eukaryotic organisms. *Palaios* 2009, 24: 616–626.

257. Schiffbauer JD, Yin L, Bodnar RJ, Kaufman AJ, Meng F, Hu J, Shen B, Yuan X, Bao H, Xiao S: Ultrastructural and geochemical characterization of Archean-Paleoproterozoic graphite particles: implications for recognizing traces of life in highly metamorphosed rocks. *Astrobiology* 2007, 7: 684–704.

258. Schiffbauer JD, Wallace AF, Hunter JL Jr, Kowalewski M, Bodnar RJ, Xiao S: Thermally-induced structural and chemical alteration of organic-walled microfossils: an experimental approach to understanding fossil preservation in metasediments.*Geobiology* 2012, 10(5):402–423.

259. Schirrmeister BE, de Vosb JM, Antonelli A, Bagheri HC: Evolution of multicellularity coincided with increased diversification of cyanobacteria and the Great Oxidation Event. *PNAS-USA* 2013. 10.1073/pnas.1209927110

260. Schmidt SK, Reed SC, Nemergut DR, Stuart-Grandy A, Cleveland CC, Weintraub MN, Hill AW, Costello EK, Meyer AF, Neff JC, Martin AM: The earliest stages of ecosystem succession in high-elevation (5000 metres above sea level), recently deglaciated soils. *Proc Biol Sci* 2008, 275(1653):2793–2802.

261. Schopf JW: Microflora of the bitter springs formation, late Precambrian, central Australia. *J Paleontol* 1968, 42: 651–688.

262. Schopf JW (Ed): *Earth's earliest biosphere.* Princeton, NJ: Princeton University Press; 1983.

263. Schopf JW: Paleobiology of the Archean. In *The Proterozoic biosphere*. Edited by: Schopf JW, Klein C. New York: Cambridge University Press; 1992.

264. Schopf JW: Proterozoic prokaryotes: affinities, geologic distribution, and evolutionary trends. In *The Proterozoic biosphere.* Edited by: Schopf JW, Klein C. New York: Cambridge University Press; 1992.

265. Schopf JW, Klein C (Eds): *The Proterozoic biosphere.* New York: Cambridge University Press; 1992.

266. Schopf JW, Walter MR, Ruiji C: Earliest evidence of life on earth. *Precambrian Res* 2007, 158: 139–140.

267. Schwartzman DW, Volk T: Biotic enhancement of weathering and the habitability of earth. *Nature* 1989, 340: 457–460.

268. Shear WA: The early development of terrestrial ecosystems. *Nature* 1991, 351: 283–289.

269. Sheldon ND: Microbially induced sedimentary structures in the ca. 1100 Ma terrestrial midcontinent rift of North America. In*Microbial mats in siliciclastic depositional systems through time. SEPM Special Publication no. 101*. Edited by: Noffke N, Chafetz H. Tulsa, OK: Society for Sedimentary Geology; 2012:153–162.

270. Shen Y, Buick R: The antiquity of microbial sulfate reduction. *Earth Sci Rev* 2004, 64: 243–272.

271. Shen Y, Farquhar J, Masterson A, Kaufman AJ, Buick R: Evaluating the role of microbial sulfate reduction in the early Archean using quadruple isotope systematics. *Earth Planet Sc Lett* 2009, 279: 383–391.

272. Shephard KL: Evaporation of water from the mucilage of a gelatinous algal community. *Br Phycol J* 1987, 22: 181–185.

273. Sigler WV, Crivii S, Zeyer J: Bacterial succession in glacial forefield soils characterized by community structure, activity and opportunistic growth dynamics. *Microb Ecol* 2002, 44: 306–316.

274. Simpson WS, Simpson EL, Wizevich MC, Malenda HF, Hilbert-Wolf HL, Tindall SE: A preserved Late Cretaceous biological soil crust in the capping sandstone member, Wahweap Formation, Grand Staircase-Escalante National Monument, Utah, Palaeoclimatic implications. *Sediment Geol* 2010, 230: 139–145.

275. Singh SP, Kumari S, Rastogi RP, Singh KL, Richa SRP: Photoprotective and biotechnological potentials of cyanobacterial sheath pigment, scytonemin. *Afr J Biotechnol* 2010, 9: 580–588.

276. Sinha RP, Häder DP: UV-induced DNA damage and repair: a review. *Photochem Photobiol Sci* 2002, 1: 225–236.

277. Skotnicki SJ, Knauth LP: The Middle Proterozoic Mescal Paleokarst, Central Arizona, USA: karst development,

silicification, and cave deposits. *J Sediment Res* 2007, 77(12):1046–1062.

278. Sleep NH, Bird DK: Niches of the pre-photosynthetic biosphere and geologic preservation of Earth's earliest ecology.*Geobiology* 2007, 5: 101–117.
279. Smith SE, Read D: *Mycorrhizal symbiosis*. 3rd edition. New York: Elsevier; 2008.
280. Squire RJ, Stewart IR, Zang WL: Acritarchs in polydeformed and highly altered Cambrian rocks in western Victoria. *Aust J Earth Sci* 2006, 53: 697–705.
281. Stern RJ, Scholl DW: Yin and yang of continental crust creation and destruction by plate tectonic processes. *Int Geol Rev*2010, 52: 1–31.
282. Strother PK, Battison L, Brasier MD, Wellman CH: Earth's earliest non-marine eukaryotes. *Nature* 2011, 473: 505–509.
283. Stüeken EE, Catling DC, Buick R: Contributions to late Archaean sulphur cycling by life on land. *Nat Geosci* 2012, 5: 722–725.
284. Summons RE, Jahnke LL, Hope JM, Logan GA: 2-Methylhopanoids as biomarkers for cyanobacterial oxygenic photosynthesis. *Nature* 1999, 400: 554–557.
285. Sun HJ, Friedmann EI: Growth on geological time scales in the Antarctic cryptoendolithic microbial community. *Geomicrobiol J* 1999, 16: 193–202.
286. Takeuchi N, Kohshima S, Seko K: Structure, formation, darkening process of albedo reducing material (cryoconite) on a Himalayan glacier: a granular algal mat growing on the glacier. *Arct Antarct Alp Res* 2001, 33: 115–122.
287. Taylor TN, Taylor EL, Krings M: *Paleobotany: the biology and evolution of fossil plants*. 2nd edition. Amsterdam: Elsevier; 2009.
288. Thiry M, Simon-Coincon R (Eds): *Palaeoweathering, palaeosurfaces and related continental deposits. Special publication 27 of the International Association of Sedimentologists*. Oxford: Blackwell Science; 1999.
289. Tice MM, Lowe DR: Photosynthetic microbial mats in the 3,416-Myr-old ocean. *Nature* 2004, 431: 549–552.
290. Tirkey J, Adhikary SP: Cyanobacteria in the biological soil crusts of India. *Curr Sci* 2005, 89: 515–521.
291. Tomitani A, Knoll AH, Cavanaugh CM, Ohno T: The evolutionary diversification of cyanobacteria: molecular-phylogenetic and paleontological perspectives. *Proc Natl Acad Sci USA* 2006, 103: 5442–5447.
292. Trendall AF, Blockley JG: Precambrian iron-formations. In *The Precambrian Earth: tempos and events. Developments in Precambrian geology, vol. 12*. Edited by: Eriksson PG, Altermann W, Nelson DR, Mueller WU, Catuneanu O. Amsterdam: Elsevier; 2004:403–420.
293. Ueno Y, Yamada K, Yoshida N, Maruyama S, Isozaki Y: Evidence from fluid inclusions for microbial methanogenesis in the early Archaean era. *Nature* 2006, 440: 516–519.
294. Van Kranendonk MJ: Archaean tectonics 2004: a review. *Precambrian Res* 2004, 131: 143–151.
295. Van Kranendonk MJ: Stromatolite morphology as an indicator of biogenicity for Earth's oldest fossils from the 3.5–3.4 Ga Pilbara Craton, Western Australia. In *Advances in stromatolite geobiology. Lecture Notes in Earth Sciences, vol. 131*. Edited by: Reitner J, Queric NV, Arp G. Germany: Springer; 2011.
296. Kranendonk MJ Van, Smithies RH, Bennett V (Eds): *Earth's oldest rocks. Developments in Precambrian geology, series 15*. Amsterdam: Elsevier; 2007a.
297. Van Kranendonk MJ, Smithies RH, Hickman AH, Champion DC: Review: secular tectonic evolution of Archean continental crust: interplay between horizontal and vertical processes in the formation of the Pilbara Craton, Australia. *Terra Nova* 2007, 19(1):1–38.
298. Van Kranendonk MJ, Philippot P, Lepot K, Bodorkos S, Pirajno F: Geological setting of Earth's oldest fossils in the c. 3.5 Ga Dresser Formation, Pilbara Craton, Western Australia. *Precambrian Res* 2008, 167: 93–124.
299. van Zuilen MA, Lepland A, Arrhenius G: Reassessing the evidence for the earliest traces of life. *Nature* 2002, 418: 627–630.
300. Voigt E: Tonrollen als potentielle Pseudofossilien. *Nat Mus* 1972, 102(11):401–410.
301. Wacey D, Kilburn MR, Saunders M, Cliff J, Brasier MD: Microfossils of sulphur-metabolizing cells in 3.4-billion-year-old rocks of Western Australia. *Nat Geosci* 2011, 4: 698–702.
302. Walsh MM: Microfossils and possible microfossils from the Early Archean Onverwacht Group, Barberton Mountain Land, South Africa. *Precambrian Res* 1992, 54: 271–292.
303. Walsh MM, Lowe DR: Filamentous microfossils from the 3500 Myr-old Onverwacht Group, Barberton Mountain Land, South Africa. *Nature* 1985, 314: 530–532.
304. Walsh MM, Lowe DR: Modes of accumulation of carbonaceous matter in the early Archaean: A petrographic and geochemical study of the carbonaceous cherts of the Swaziland Supergroup. In *Geologic Evolution of the Barberton Greenstone Belt, South Africa, Geological Society of America Special Paper 329* Edited by: Lowe DR, Byerly GR. 1999, 115–132.
305. Walter MR: Archean stromatolites: evidence of the Earth's earliest benthos. In *Earth's earliest biosphere*. Edited by: Schopf JW. Princeton: Princeton University Press; 1983:187–213.
306. Walter MR, Buick R, Dunlop JSR: Stromatolites 3,400–3,500 Myr old from the North Pole area, Western Australia. *Nature*1980, 284: 443–445.
307. Watanabe Y, Martini JEJ, Ohmoto H: Geochemical evidence for terrestrial ecosystems 2.6 billion years ago. *Nature* 2000, 408: 574–578.
308. Welch SA, Barker WW, Banfield JF: Microbial extracellular polysaccharides and plagioclase dissolution. *Geochim Cosmochim Acta* 1999, 63: 1405–1419.
309. Westall F: Early life: nature, distribution and evolution. In *Origins and evolution of life*. Edited by: Gargaud M, López-Garcìa P, Martin H. Cambridge: An

astrobiological perspective. Cambridge University Press; 2010:391–413.

310. Westall F, De Wit MJ, Dann J, Van Der Gaast S, De Ronde C, Gerneke D: Early Archaean fossil bacteria and biofilms in hydrothermally influenced, shallow water sediments, Barberton Greenstone Belt, South Africa. *Precambrian Res* 2001, 106: 91–112.

311. Westall F, de Vries ST, Nijman W, Rouchon V, Orberger B, Pearson V, Watson J, Verchovsky A, Wright I, Rouzaud JN, Marchesini D, Anne S: The 3.466 Ga Kitty's Gap Chert, an Early Archaean microbial ecosystem. In *Processes on the early Earth. Geological Society of America special paper 405*. Edited by: Reimold WU, Gibson R. Boulder, CO: Geological Society of America; 2006a:105–131.

312. Westall F, de Ronde CEJ, Southam G, Grassineau N, Colas M, Cockell C, Lammer H: Implications of a 3.472–3.333 Ga-old subaerial microbial mat from the Barberton Greenstone Belt, South Africa for the UV environmental conditions on the early Earth. *Phil Trans R Soc B* 2006, 361: 1857–1875.

313. Westall F, Cavalazzi B, Lemelle L, Marrocchi Y, Rouzaud JN, Simionovici A, Salomé M, Mostefaoui S, Andreazza C, Foucher F, Toporski J, Jauss A, Thiel V, Southam G, MacLean L, Wirick S, Hofmann A, Meibom A, Robert F, Défarge C: Implications of in situ calcification for photosynthesis in a 3.3 Ga-old microbial biofilm from the Barberton greenstone belt, South Africa. *Earth Planet Sci Lett* 2011, 310(3–4):468–479.

314. White D: *The physiology and biochemistry of prokaryotes*. 2nd edition. Oxford: Oxford University Press; 2000.

315. Whitton BA, Potts M: *The ecology of cyanobacteria: their diversity in time and space*. Dordrecht: Kluwer; 2000.

316. Wilde SA, Valley JW, Peck WH, Graham CM: Evidence from detrital zircons for the existence of continental crust and oceans on the Earth 4.4 Gyr ago. *Nature* 2001, 409: 175–178.

317. Williams JD, Dobrowolsk JP, West NE, Gillette DA: Microphytic crust influence on wind erosion. *Trans ASAE* 1995, 38: 131–137.

318. Williams AJ, Buck BJ, Beyene MA: Biological soil crusts in the Mojave Desert, USA: micromorphology and pedogenesis. *Soil Sci Soc Am J* 2012, 76(5):1685–1695.

319. Windley B: Overview and history of investigation of early Earth rocks. In *Earth's oldest rocks. Developments in Precambrian geology, series 15*. Edited by: Van Kranendonk MJ, Smithies RH, Bennett V. Amsterdam: Elsevier; 2007:3–7.

320. Wright VP: The precursor environment for vascular plant colonization. *Phil Trans R Soc London B* 1985, 309: 143–145.

321. Xiao S, Kaufman AJ (Eds): *Neoproterozoic geobiology and paleobiology. Topics in Geobiology, vol. 27*. Dordrecht: Springer; 2006.

322. Xiao S, Knauth LP: Palaeontology: fossils come in to land. *Nature* 2013, 493(7430):28–29.

323. Xiong J, Fischer WM, Inoue K, Nakahara M, Bauer CE: Molecular evidence for the early evolution of photosynthesis.*Science* 2000, 289: 1724–1730.

324. Yang W, Holland HD: The Hekpoort paleosol in strata 1 at Garborone, Botswana: soil formation during the Great Oxidation Event. *Am J Sci* 2003, 303: 187–220.

325. Yasui A, McCready SJ: Alternative repair pathways for UV-induced DNA damage. *Bioessays* 1998, 20(4):291–297.

326. Yu G, Jacobsen SB: Fast accretion of the Earth with a late Moon-forming giant impact. *PNAS USA* 2011, 108(43):17604–9.

327. Zang WL: Deposition and deformation of late Archean sediments and preservation of microfossils in the Harris Greenstone Domain, Gawler Craton, South Australia. *Precambrian Res* 2007, 156: 107–124.

328. Zbinden EA, Holland HD, Feakes CR, Dobos SK: The Sturgeon Falls paleosol and the composition of the atmosphere 1.1 Ga BP. *Precambrian Res* 1988, 42: 141–163.

329. Zhang J: Observation on algal effects on subaerial karst sedimentation. *Geogr Res* 1992, 11(2):26–33.

330. Zhao B, Robb LJ, Harris C, Jordaan LJ: Origin of hydrothermal fluids and gold mineralization associated with the Ventersdorp contact reef, Witwatersrand Basin, South Africa: constraints from S, O, and H isotopes. In *Processes on the early Earth. Geological Society of America special paper 405*. Edited by: Reimold WU, Gibson RL. Boulder, CO: Geological Society of America; 2006:333–352.

331. Zhuravlev AY, Riding R (Eds): *The ecology of the Cambrian radiation*. New York: Columbia University Press; 2001.

332. Zonneveld KAF, Versteegh GJM, Kasten S, Eglinton TI, Emeis KC, Huguet C, Koch BP, De Lange GJ, De Leeuw JW, Middelburg JJ, Mollenhauer G, Prahl FG, Rethemeyer J, Wakeham SG: Selective preservation of organic matter in marine environments; processes and impact on the sedimentary record. *Biogeosciences* 2010, 7: 483–511.

CHAPTER 19

# A NEW LATE DEVONIAN GENUS WITH SEED PLANT AFFINITIES

Deming Wang and Le Liu

**Keywords:** China, Early seed plant, Famennian, Late devonian, Leaf trace, Petiole, Protostele, Tizikou formation, *Yiduxylon trilobum*

## BACKGROUND

Seed plants (spermatophytes) began to diversify in the Famennian (374–359 million years ago) of the Late Devonian and many fossils have been documented since the Carboniferous [1]. Till now, about twenty Famennian genera with ovules and/or pollen organs have been studied in detail [2]. Most of these earliest seed plants were collected from Laurussia (Europe and North America) and only two genera from China, i.e., *Kongshania* [3] and *Cosmosperma* [2]. However, few plants have been recognized on the basis of stem and/or petiole anatomy. In contrast, there are more than forty Early Carboniferous stem morphotaxa throughout Euramerica [4]. Such taxa may be the primary line of evidence for phylogenetic analyses of early spermatophytes [5].

Famennian seed plants known for their stem vascular system include *Buteoxylon* [6], *Kerryoxylon* [7], *Laceya*[8,9], *Tristichia* [7] and cf. *Tetrastichia bupatides* [7] from Ireland, and *Elkinsia* from USA [10]. They are (tentatively) attributed to the Buteoxylonales (*Buteoxylon*), Elkinsiales (*Elkinsia*) and Lyginopteridales (*Kerryoxylon*, *Laceya*, *Tristichia* and *Tetrastichia*). Anatomically preserved spermatophytes from the New Albany Shale of USA have raised great controversy over their Latest Devonian or earliest Carboniferous age [11-19].

Buteoxylonales, Calamopityales and Lyginopteridales (or Buteoxylonaceae, Calamopityaceae and Lyginopteridaceae) are represented by numerous stem morphotaxa and/or fertile organs from the Late Devonian and especially Carboniferous. However, as noted by many workers such as [4,20], the definition and relationship of these three orders need reinterpretation because of shared characters and the lack of whole plant knowledge. These early spermatophytes have been interpreted as lianas or upright plants [20,21], and their possession of sclerotic nests in the stem inner cortex has been interpreted as a response to insect infections [15,20]. The derivation of the early spermatophyte eustele from protostele has been widely accepted [1,22,23], but the evolution of the spermatophyte protostele from the ancestral aneurophyte progymnosperm protostele is poorly understood.

South China was an isolated plate in the Late Devonian and showed great diversification of many plant lineages excluding seed plants. This paper now reports a new genus with seed plant affinities, *Yiduxylon trilobum* gen. et sp. nov., having both vegetative morphology and stelar structure including petiolar vascular supply preserved from the Upper Devonian (Famennian) Tizikou Formation of Hubei Province, South China. The age of this formation is determined mainly by the stratigraphic sequence and comparisons of the spore assemblages. Based on morphological and anatomical studies of stem/frond as well as on detailed comparisons, this paper discusses the taxonomy and growth habit of *Yiduxylon*, and the evolution of seed plant protostelic architecture during the Late Devonian and earliest Carboniferous. The stelar feature and leaf trace production of *Yiduxylon* indicate transitional anatomical traits in the evolution of the seed plants from the Middle to Late Devonian aneurophytes, although the latter group is currently unknown in South China.

**Citation:** Deming Wang and Le Liu, A new Late Devonian genus with seed plant affinities, DOI: 10.1186/s12862-015-0292-6.

## MATERIAL AND METHODS

Fossils were collected from the Upper Devonian Xiejingsi (Hsiehchingssu) Formation sensu lato, on a hill named Tizikou, which is to the south of Maohutang village, Wangjiafan Town, Yidu City, western Hubei Province, China. In the Three Gorges area of the Yangtze River including western Hubei, the Upper Devonian strata are referred to the Huangjiadeng Formation and overlying Xiejingsi Formation sensu lato (see summary in [24]). The latter formation includes the Frasnian-Famennian Xiejingsi Formation sensu stricto, Famennian Tizikou Formation and Changyang Formation in ascending order [25]. The Changyang Formation containing marine invertebrates is Famennian [25,26] or Lower Carboniferous (basal Tournaisian) [27,28]. The present plant occurs together with a lycopsid *Sublepidodendron songziense*. The latter species has been discovered and studied in detail from the Famennian Tizikou Formation of Songzi City, western Hubei and from the Famennian Wutong (Wutung) Formation of Chaohu City, Anhui Province [24,29]. Miospore assemblages indicate that the Tizikou Formation in western Hubei and northwestern Hunan is similar in age to the Famennian (Fa2d-Tn1a) Xikuangshan Formation (upper part) and Shaodong Formation (lower part) in central Hunan, the Zhewang Formation and Gelaohe Formation (lower part) in southeastern Guizhou, Wutong Formation (upper part: Leigutai Member) in the lower reaches of the Yangtze River, and the Wangjiadian Formation in southern Gansu [30-33]. Some spores in the PN assemblage, e.g., *Vallatisporites pusillites*,*Aneurospora greggsii* and *Rugospora flexuosa* occur only in Famennian deposits [33]. These data suggest that the Tizikou Formation is middle to late Famennian but not latest Famennian.

At the Tizikou section, the Tizikou Formation is about 20 m thick and characterized by black-gray mudstone and siltstone, and gray or gray-white fine sandstone [27]. There are some lycopsids and the sphenopsids*Hamatophyton verticillatum* and *H. yiduense* in the mid-lower part of this formation [27]. *H. verticillatum* and*H. yiduense* have been emended as *Rotafolia songziensis* [34], which was later anatomically studied [35]. Near the horizon of *R. songziensis*, *Yiduxylon trilobum* gen. et sp. nov. occurs as impressions and compressions in black-gray silty mudstone.

The poorly preserved stems, branches and leaves show little contrast with the rock matrix. Under a microscope, clastic material was removed by steel needles to expose the plant morphology. Coalified material of *Yiduxylon trilobum* has been well preserved, which was embedded, sectioned and ground to show the anatomy. Altogether nine stems were transversely sectioned to produce 62 slides, and six stems were longitudinally sectioned to produce 16 slides. These sections were observed with light microscope (LM). Some stems were examined with a scanning electron microscope (SEM).

We have not carried out our field works in private areas, since in China all the land belongs to the country. And we have not caused any destruction to National Nature Reserves during fossil excavation. All the fossil collecting works have been done obeying the Chinese fossil collection and mining laws, and have been allowed by local government.

## RESULTS

### Vegetative Morphology

Two stems are up to 120 mm long and measure 12–20 mm in diameter (Figure 1a,b). They decrease in diameter acropetally. The proximal bifurcation of the fronds is sometimes visible at an angle of 60° (Figure 1b, middle frond). Rachises of fronds depart from the stem at 45-60°, are up to 60 mm long, and are separately 5.2-10 mm and 5.0-6.3 mm in diameter before and after proximal bifurcation. One stem has three fronds appearing to be alternately arranged (Figure 1a). In another stem (Figure 1b), the upper frond departs at 50°, the middle frond emerges at 20°, while the basal frond (Figure 1b, arrow) plunges into the rock matrix. In this case, the three fronds on this stem are borne helically, approaching a 1/3 phyllotaxis. The interval between the attachment of two adjacent frond rachises is 20–70 mm. Narrow and longitudinally parallel ridges 120–350 μm in width occur on the surface of stems and frond rachises and alternate with narrow grooves. These ridges correspond to the bands of sclerenchyma cells in the outer cortex (see below).

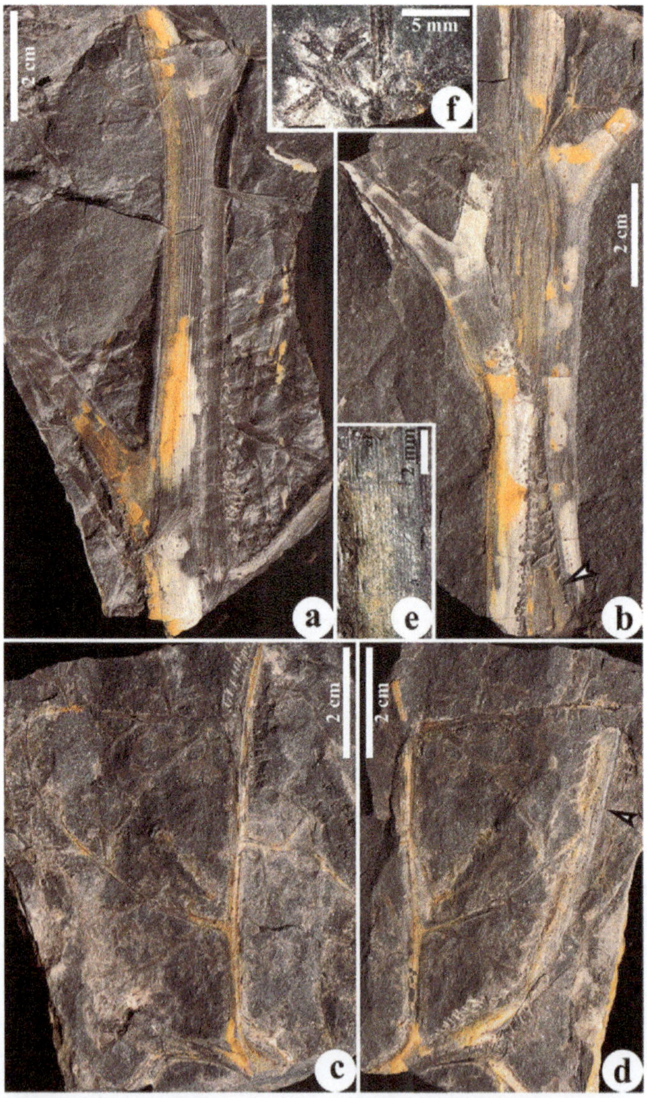


**Figure 1: Vegetative morphology of *Yiduxylon trilobum* gen. et sp. nov. from Hubei. (a)** Stem attached by three fronds. PKUB14401. **(b)** Stem with three fronds in helix, with middle frond bifurcating and lower frond extending into rock (arrow). PKUB144403. **(c-d)** Part and counterpart of specimen showing connection of frond, pri-

mary and secondary pinnae and planate pinnules in one plane. Arrow in **(d)** indicating frond rachis enlarged in **(e)**. PKUB14402a, PKUB14402b. **(e)** Enlargement of **(d, arrow)** displaying longitudinal ridges and grooves. **(f)** Secondary pinna rachis with planate pinnules. PKUB14407.

The part and counterpart of a specimen display an axis with lateral branches bearing planate pinnules (Figure 1c, d, Additional file 1: Figure S1a, b). This axis is about 85 mm long and 4.8-5.2 mm in diameter, and possesses longitudinal ridges (Figure 1d, arrow, e). The diameter and striations suggest that this axis may represent the upper part of a frond rachis after initial bifurcation. In this case, there are two orders of pinnae within the frond. Since plant shows little contrast with the rock matrix, the configuration of the pinnules is very difficult to recognize. However, narrow longitudinal ridges are visible on the rachises of the primary pinnae. The rachises of two primary pinnae are 1.8-2.6 mm in diameter, and attached separately at 70° and 30° to the frond rachis. One primary pinna is only basally preserved, whereas another is 76 mm long and bears four secondary pinnae. The rachises of secondary pinnae, up to 60 mm long and 0.8-1.3 mm in width, are alternately arranged. Except for the basal one, the other three secondary pinnae depart at about 90°. The distance between the attachment of two neighboring secondary pinnae is 16–21 mm. Nonlaminate pinnules are borne at 45-90° and alternately on the secondary pinna rachis, and are planate and highly dissected. A pinna rachis possesses at least seven pinnules. The distance between two adjacent pinnules is 4.6-13 mm. Pinnules are up to 21.7 mm long and about 19.4 mm wide. Each pinnule unequally dichotomizes to produce four or more units. These units, which are up to 11 mm long and about 12 mm wide, arise at 20-90°, and divide unequally twice or thrice to form narrow segments which are 0.5-1.2 mm wide (Figure 1f, Additional file1: Figure S1).

## Anatomy

### *Stems*

Transverse sections of stems show a small three-ribbed protostele (actinostele), broad secondary xylem, narrow cortex and circular-oval leaf traces (Figures 2 and 3, Additional file 2: Figure S2 and Additional file 3: Figure S3). Longitudinal sections of stems demonstrate the secondary xylem and cortex (Figure 4a-e), with the former tissue being also observed with SEM (Figure 4f-k). The stems are 7.2-10.5 mm in diameter and the primary xylem is 1.8-2.0 mm in diameter (Figure 2a, b, Additional file 2: Figure S2a-d and Additional file 3: Figure S3a, b, e-g). The ratio of stelar primary xylem diameter to total stem diameter is around 1/5 (calculated from transverse section illustrated in Additional file 2: Figure S2c). The three ribs of primary xylem, which differ in length (Additional file 3: Figure S3a, arrows 1-3), are 0.5-1.2 mm long and 0.2-0.3 mm wide at their tips. The ribs become narrower toward the stelar center, being no more than 0.1-0.2 mm. The angles between two adjacent ribs of a stele are approximately 90°, 125° and 145°. The mesarch primary xylem only with tracheids (Additional file 3: Figure S3 and Additional file 4: Figure S4) possesses a single protoxylem strand located near the periphery of each rib (Additional file 3: Figures S3c, arrow, d, left arrow, h, i, and S4). The xylem rib is longitudinally continuous (Additional file 2: Figure S2a-d, Additional file 3: Figure S3e-g, Additional file 5: Figure S5a-g and Additional file 6: Figure S6a-g). Protoxylem and metaxylem tracheids are 3.0-10 μm and 13–85 μm in diameter, respectively.

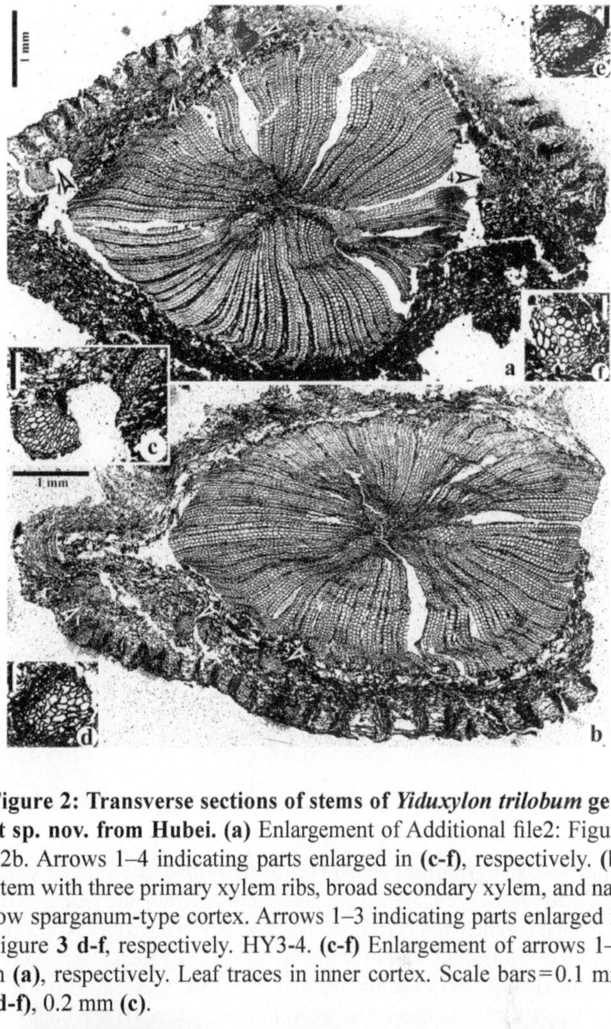


**Figure 2: Transverse sections of stems of *Yiduxylon trilobum* gen. et sp. nov. from Hubei. (a)** Enlargement of Additional file2: Figure S2b. Arrows 1–4 indicating parts enlarged in **(c-f)**, respectively. **(b)** Stem with three primary xylem ribs, broad secondary xylem, and narrow sparganum-type cortex. Arrows 1–3 indicating parts enlarged in Figure **3 d-f**, respectively. HY3-4. **(c-f)** Enlargement of arrows 1–4 in **(a)**, respectively. Leaf traces in inner cortex. Scale bars = 0.1 mm **(d-f)**, 0.2 mm **(c)**.

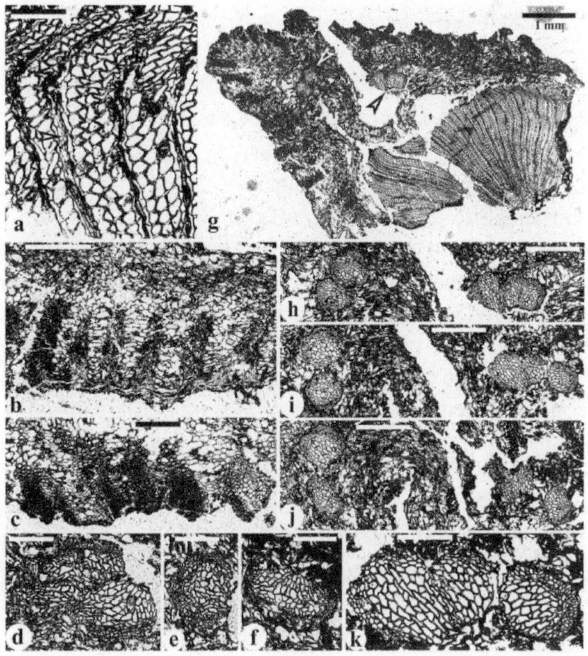


**Figure 3: Transverse sections of stems of *Yiduxylon trilobum* gen. et sp. nov. from Hubei. (a)** Secondary xylem with ray cells (arrow) and tracheids. HY6-3. **(b-c)** Inner cortex with parenchyma cells, and outer cortex with bands of sclerenchyma cells and intervening paren-

chyma cells. HY8-5, HY1-7. **(d-f)** Enlargement of Figure 2b (arrows 1–3), respectively, showing tangentially dividing leaf trace **(d)** and individual leaf traces in inner cortex. **(g)** Partial stem showing two dividing leaf traces in inner cortex (arrows), which are enlarged in **(h)**. HY6-7. **(h-j)** Serial transverse sections of two leaf traces showing their gradual tangential division to produce four traces. **(h)** Enlargement of **(g, arrows)**. HY6-7, HBY6-5, HBY6-3. **(k)** Enlargement of right leaf traces in **(i)**. Scale bars=0.2 mm **(a, d-f, k)**, 0.5 mm **(b, c, h-j)**, 1 mm **(g)**.

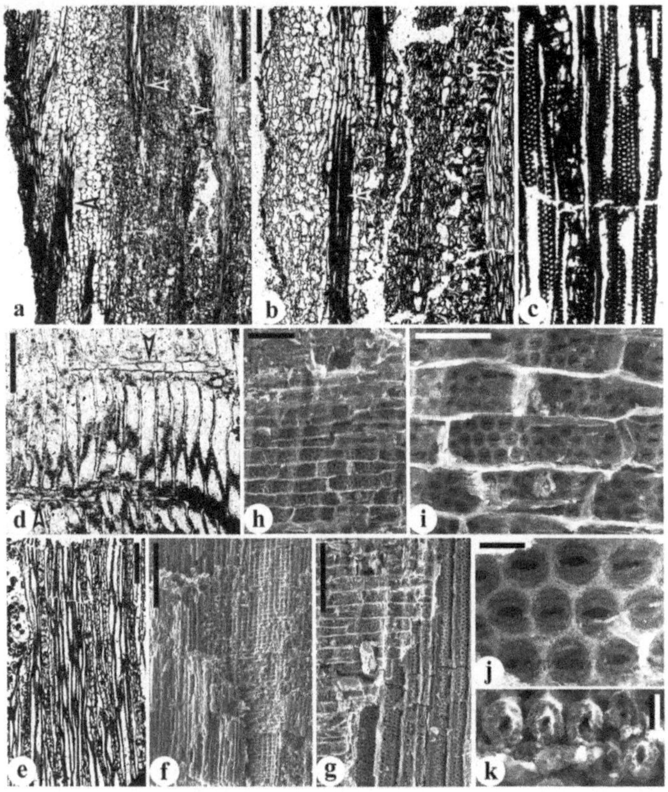


**Figure 4: Longitudinal view of stems of *Yiduxylon trilobum* gen. et sp. nov. from Hubei. (a)** Tangential section showing leaf traces (upper two arrows), sclerenchyma cells (lowermost arrow) and parenchyma cells of cortex. HY5-2. **(b)**Tangential section showing sclerenchyma (arrow) and parenchyma cells of cortex, and secondary xylem (far right). HY2-3. **(c)** Tangential section of secondary xylem showing multiseriate pitting on tangential wall of tracheids. HY7-2.**(d)** Radial section of secondary xylem showing rays (arrows) and pitting on radial walls of tracheids. HY4-3. **(e)**Tangential section of secondary xylem showing high and mainly biseriate rays. Note pitting on tangential walls of tracheids. HY9-2. **(f-k)** SEM observation of secondary xylem. **(f)** Secondary xylem in radial fracture showing very high rays. **(g)** Radial fracture showing rays and tracheids. Note pitting on radial walls of tracheids. **(h)** Radial fracture showing rays and underlying tracheids. **(i)** Enlargement of **(h)** showing ray cells and tracheids with pits. **(j)**Enlargement of **(i)** showing bordered pits of a tracheid. **(k)** Bordered pits of a tracheid. Scale bars=1 mm **(a)**, 0.5 mm**(f)**, 0.2 mm **(b, d, e, g)**, 0.1 mm **(c, h)**, 50 μm **(i)**, 10 μm **(j, k)**.

Arising close to the stelar center, the secondary xylem extends outward up to a distance of 2.7 mm (Figure 2a, b, Additional file 2: Figure S2a-d). In transverse section, the secondary xylem tracheids are of various shapes (circular, oval, square, rectangular and irregular) as in the metaxylem tracheids, and are 12–75 μm in diameter and arranged in radial files. Multiseriate pits of circular and oval shape occur on both tangential (Figure 4c, e) and radial (Figure 4d, f-i) walls of secondary xylem tracheids. These pits are bordered (Figure 4j, k) and 3.0-13 μm in diameter. Rays occur between every two to about ten rows of secondary xylem tracheids (Figures 2a, b, 3a, g and 4e). Rays are usually 2 and sometimes up to 6 cells wide (Figure 4e, Additional file 3: Figure S3a, arrow), and more than 120 cells high (Figure 4e, f). Ray parenchyma cells are 65–128 μm in radial and 10–32 μm in tangential dimensions, and 22–48 μm high (Figures 3a and 4d, arrow, e-i).

The cortex sometimes appears to have been compressed to present a wing-like outline (Figure 2b, Additional file 2: Figure S2a, b). It is 0.8-1.1 mm thick (Figure 2b, Additional file 2: Figure S2a-d) and lacks sclerotic nests (Figure 3b, c). The inner cortex consists of circular and oval parenchyma cells which are 25–160 μm in diameter. Longitudinal sections show that these cells are up to 180 μm high (Figure 4a, b). Sparganum-type outer cortex contains many radial bands of sclerenchyma cells with thick walls (Figures 2a, b and 3b, c, g). These bands at the periphery of the cortex are 300–700 μm in radial and 100–520 μm in tangential dimensions, and the circular to square sclerenchyma cells range from 14–68 μm in diameter. In longitudinal sections, the sclerenchyma cells are continuous, up to 1.0 mm high and do not anastomose (Figure 4a, lowermost arrow, b, arrow). Intervening parenchyma cells are tangentially elongate between two adjacent bands with intervals of 130–270 μm.

Individual leaf traces are circular and slightly oval in outline, and their diameter is 0.3-0.4 mm near the tip of the primary xylem rib (Additional file 2: Figure S2i, k and Additional file 5: Figure S5d), but becomes 0.4-0.5 mm in the secondary xylem (Additional file 2: Figure S2c, right arrow, d, lower arrow, h, j and S5a-f), and reaches 0.3-0.7 mm in the inner cortex of the stem (Figures 2a, arrows 1–4, b, arrows 1–3, c-f and 3d-f, g, arrows, h-k, Additional file 2: Figure S2a, arrow, c, left arrow, d, upper arrow, e-g). The tracheids of the leaf traces are up to 56 μm in diameter.

Serial transverse sections of a stem indicate that each rib of the primary xylem successively initiates two leaf traces. The first trace is located at the periphery of the inner cortex (Additional file 2: Figure S2a, arrow, e and Additional file 6: Figure S6h). The tip of a xylem rib obliquely and radially expands to form a protoxylem strand representing the incipience of another leaf trace (Additional file 3: Figure S3a, arrow 3, b, d, right arrow). Then, this trace appears to tangentially diverge from the rib tip (Additional file 2: Figure S2i) and afterwards becomes detached within the secondary xylem (Additional file 2: Figure S2c, right arrow, h and Additional file 6: Figure S6i). More distally, a leaf trace diverges from the tip of another rib with protoxylem strand (Additional file 2: Figure S2k, arrow) and is then surrounded by the secondary xylem (Additional file 2: Figure S2d, lower arrow, j). This trace contains two protoxylem strands (Additional file 2: Figure S2j, arrows, k). Another trace thought to depart from the same rib tip is however invisible. It is suggested that two traces may subsequently diverge from the tip of the third rib, thus indicating a 1/3 phyllotaxy.

Serial transverse sections of another stem clearly show the successive origination of two leaf traces. One trace in the secondary xylem has been tangentially produced from a tip of the primary xylem rib (Additional file 5: Figure S5a and Additional file 6: Figure S6a), then gradually moves toward (Additional file 5: Figure S5b, c and Additional file 6: Figure S6b, c) and reaches the periphery of the inner cortex (Additional file 5: Figure S5d-g and Additional file 6: Figure S6d-g). Here (Additional file 5: Figure S5g, arrow) the trace with two protoxylem strands is tangentially elongate (Additional file 5: Figure S5i, arrows), thus indicating further division. The same rib tip obliquely and radially expands to produce a protoxylem strand representing the incipience of another trace (Additional file 5: Figure S5a-c and Additional file 6: S6a-c), which radially diverges, then passes through the secondary xylem (Additional file 5: Figure S5d-f and Additional file 6: Figure S6d-f) and reaches the inner cortex (Additional file 5: Figure S5g and Additional file 6: S6g). Because of the irregular configuration of ribbed primary xylem, the apparent radial emission of this leaf trace may result from preservation and distortion of the specimen due to compression. This trace in the secondary xylem (Additional file 5: Figure S5f, arrow) displays one protoxylem strand (Additional file 5: Figure S5h, arrow).

In the inner cortex of one stem, two leaf traces derived from a primary xylem rib begin to tangentially divide once (Figure 3g, arrows). The subsequent gradual divisions result in four traces (Figure 3h-k). Four leaf traces from a primary xylem rib occur in the inner cortex of the second stem (Figure 2a, arrows 1–3, c-e), with two traces having just tangentially separated (Figure 2a, arrow 1, c). Four leaf traces from a primary xylem rib are present in the inner cortex of the third stem (Figures 2b, arrows 1–3 and 3d-f), with two traces beginning to separate (Figures 2b, arrow 1 and 3d). Leaf traces in the cortex of these three stems have mesarch primary xylem (two protoxylem strands, see Figures 2f and 3e, k, Additional file 2: Figure S2e).

## Frond Base with Trace

Two successive transverse sections of a stem demonstrate the attachment of a petiole base (Figure 5a-d). This stem has the same structure as the other stems, which includes a three-ribbed primary xylem of mesarch maturation, thick secondary xylem and narrow cortex with sclerenchyma cells (such cells are present only in the first section as shown by Figure 5a, b). The petiole base is ca. 5.6 mm tangentially (Figure 5b). The vascular supply of the petiole consists of four bundles, with two shown in oblique longitudinal section (Figure 5b, arrow, e) and two in oblique transverse section (Figure 5f). The latter two bundles are tangentially oriented, asymmetrical and C- or broadly U-shaped (adaxially concave). They are ca. 2.0 mm tangentially and 0.2-0.4 mm radially, with the concave surface facing the stem center. The inner bundle tends to divide into three portions, with the abaxial surface of two bundles possessing protoxylem strands (Figure 5f, arrows). The two bundles in oblique longitudinal section may present the same shape and structure when they are in transverse view. Therefore, in the upward and outward course through the cortex of the stem, a leaf trace dichotomizes to form four bundles that assume two C-configurations in the petiole base. The tracheids of the vascular bundles are 10–52 μm in diameter and embedded in a ground tissue of parenchyma cells which are 30–110 μm in diameter. Bands of sclerenchyma cells occur in partially preserved sparganum-type outer cortex, and are 450–720 μm and 130–190 μm in radial and tangential dimensions, respectively (Figure 5b). Intervals between adjacent bands are ca. 260 μm.

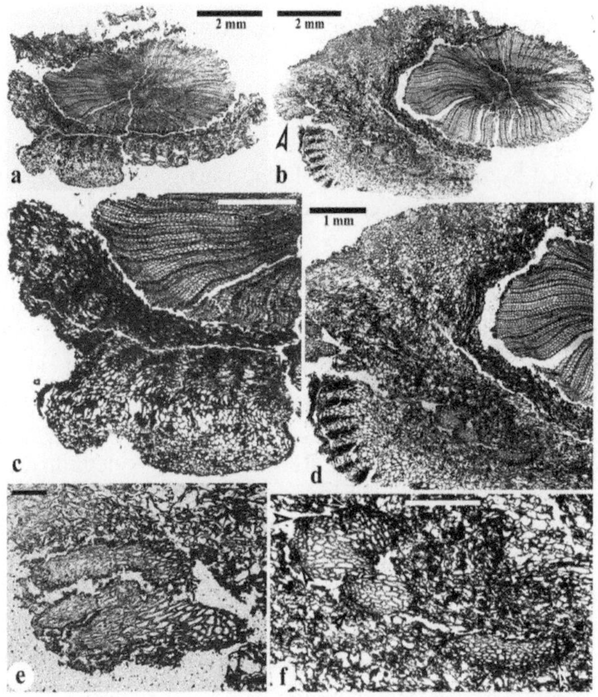


**Figure 5: *Yiduxylon trilobum* gen. et sp. nov. from Hubeiv. (a-b)** Successive transverse sections of a stem bearing a petiole base. Stem with three-ribbed primary xylem, secondary xylem and cortex. Sparganum-type outer cortex in **(a)**. **(b)**Petiole base with vascular bundles, ground tissue and sparganum-type outer cortex. Arrow indicating part enlarged in **(e)**. Slides 4–1 and 4–6. **(c-d)** Enlargement of **(a)** and **(b)**, respectively. **(e)** Enlargement of **(b, arrow)** showing two vascular bundles in oblique longitudinal view. **(f)** Enlargement of lower left part of **(b)** showing two vascular bundles in oblique transverse view. Arrows indicating protoxylem strands. Scale bars = 1 mm **(c)**, 0.2 mm **(e)**, 0.5 mm **(f)**.

## Systematics

? Class Spermatopsida

Order Incertae sedis

Genus *Yiduxylon* Wang et Liu gen. nov.

## Generic Diagnosis

Stems with fronds borne helically in a 1/3 phyllotaxy. Frond rachises bifurcate proximally. Two orders of pinnae with alternately arranged secondary pinna rachises bearing alternate pinnules which are planate and highly dissected. Stems protostelic with small three-ribbed primary xylem, broad secondary xylem, and narrow cortex. Primary xylem mesarch with each rib possessing a single peripheral protoxylem strand. Secondary xylem intermediate between manoxylic and pycnoxylic types. Multiseriate bordered pits on tangential and radial walls of secondary xylem tracheids. Rays usually biseriate and very high. Cortex consisting of an outer sparganum zone and an inner parenchymatous ground tissue. Two leaf traces successively derived tangentially from tip of each primary xylem rib and once divided to produce four traces in stem cortex. Leaf traces possessing mesarch primary xylem. Petiole base containing vascular supply of four bundles in two C-shaped groups (each group with concave surface facing stem center and protoxylem strands on abaxial surface), peripheral and parallel bands of sclerenchyma cells, and ground tissue of parenchyma cells.

Type species *Yiduxylon trilobum* Wang et Liu sp. nov.

## Specific Diagnosis

As in generic diagnosis. Stems 7.2-20 mm in diameter and attached by fronds at 45-60°. Frond rachises 5.2-10 mm and 4.8-6.3 mm in diameter before and after proximal bifurcation at about 60°, respectively. Narrow and longitudinally parallel superficial ridges and grooves alternating on stems and rachises of fronds and primary pinnae. Primary pinna rachises 1.8-2.6 mm in diameter. Secondary pinna borne at about 90°, with rachises 0.8-1.3 mm in diameter, and bearing pinnules at 45-90°. Pinnules up to 21.7 mm long and about 19.4 mm wide, with each unequally dividing into four or more units that are anisotomous twice or thrice. Ratio of primary xylem diameter to stem diameter 1/5. Stem primary xylem 1.8-2.0 mm in diameter, with three undivided ribs varying in length and becoming narrower towards stelar center. Secondary xylem up to 5.2 mm thick. Rays 2–6 cells wide and more than 120 cells high. Cortex 0.8-1.1 mm thick and inner cortex lacking sclerotic clusters. Individual leaf traces circular or slightly oval with two protoxylem strands. At petiole base, each C-shaped group of two vascular bundles asymmetrical with inner bundle tending to divide into three portions.

## Holotype Designated Here

Slide HBY-02 (Figure 2a, c-f, Additional file 2: Figure S2b and Additional file 3: Figure S3b).

## Paratypes Designated Here

Specimens PKUB14401 (Figure 1a), PKUB14403 (Figure 1b), PKUB14402a (Figure 1c), PKUB14402b (Figure 1d, e); Slides HBY-04 (Additional file 2: Figure S2c, f, h), HY3-4 (Figures 2b, 3d-f, Additional file 3: Figure S3g), HY9-2 (Figure 4e), 4–6 (Figure 5b, d-f).

## Etymology

Generic name from Yidu (where the plant was discovered) and the Greek xylon (xylem). Specific epithet from the Greek treis (tri) and lobos (lobes), meaning three-lobed shape of stem primary vascular system in transverse section.

## Locality and Horizon

Tizikou, Maohutang village, Wangjiafan Town, Yidu City, Hubei Province, China; Upper Devonian (Famennian) Tizikou Formation.

## Repository

All specimens and slides are housed in the Department of Geology, Peking University, Beijing, China.

## Comparisons

Although the aneurophyte progymnosperms possess protostele with ribbed primary xylem of mesarch maturation, and secondary xylem with bordered pits on all walls of tracheids [1], they are distinct from *Yiduxylon* mainly in the geological age, other stelar architectures and vegetative morphology. Aneurophytes existed in the Middle Devonian-Late Devonian (Frasnian) and was characterized by protoxylem strands near the tips and along the midplanes of the primary xylem ribs and in the stelar center, pycnoxylic secondary xylem and radial divergence of branch traces [1,36,37]. Furthermore, except for *Triloboxylon ashlandicum* and *Proteokalon petryi* [38,39], the aneurophytes lack planation of the lateral axes (alternate or opposite arrangement) and vegetative ultimate appendages (planate leaves); a decussate pattern occurs in *Tetraxylopteris* with most orders of branches containing a four-ribbed primary vascular system [36]. Protoxylem strands along the midplanes of the rib of the Aneurophytes and Stenokoleales usually accompany and probably have been derived from a single strand at the stelar center [36,40,41]. Concerning *Yiduxylon*, the primary xylem is poorly preserved, but many sections indicate the absence of protoxylem strands along stelar ribs and two sections perhaps lack a central protoxylem strand (Additional file 4: Figure S4a, b). The secondary xylem is intermediate between pycnoxylic and manoxylic type (see below). Two leaf traces depart successively from the tip of a stem xylem rib (Additional file 6: Figure S6). One trace diverges and extends up at the boundary between secondary xylem and cortex, and then another trace is initiated. Two leaf traces probably divide tangentially from the rib tip.

Stenokoleales of the Middle Devonian to Early Carboniferous represents a distinct order within the "radiate protoxylem" group comprising plants such as aneurophytes and some early spermatophytes [41]. This order may have a close affinity with the seed plants [42]. Stenokoleales includes *Stenokoleos* and *Crossia* known only from anatomy [41,42] and is characterized by a three-ribbed protostele with a single protoxylem strand at the stelar center and numerous protoxylem strands along the midplanes of the primary xylem ribs. Secondary xylem is sometimes present but only in small amounts. In *Stenokoleos*, successive pairs of traces are produced from the opposite xylem ribs of stem, suggesting that paired fronds are borne alternately [43]. As to*Crossia*, three stelar ribs of the lower orders of axes divide to form six ribs where branch traces are emitted in pairs [41]. However, *Yiduxylon* has protoxylem strands restricted near the periphery of three xylem ribs of stem, large amounts of secondary xylem, two leaf traces successively arising from a xylem rib, and helically arranged petioles.

The stem steles of early seed plants (Late Devonian to Carboniferous Buteoxylonales, Calamopityales and Lyginopteridales of seed ferns) may be divided into three types, i.e. ribbed protostele or actinostele, parenchymatized protostele and eustele with pith [37]. *Yiduxylon* with protostele differs from the spermatophytes with parenchymatized protosteles, e.g., *Buteoxylon* [44], *Galtiera* [15], *Heterangium* [45,46],*Rhetinangium* [47-49], *Schopfiastrum* [50], *Microspermopteris* [51,52], some species of *Calamopitys*, e.g., *C. saturni* and *C. annularis* [53], and most species of *Stenomyelon*, e.g., *S. muratum* [11], *S. bifasciculare* [54], *S. tuedianum* [55,56]. *Yiduxylon* also differs from those with eusteles, e.g., *Diichnia* [11,12,16], *Faironia* [4],*Lyginopitys* [57], *Lyginopteris* [58], *Trivena* [20], *Triichnia* [59], and some species of *Calamopitys*, e.g., *C. solmsii*[53].

The spermatophytes with ribbed protosteles include *Bostonia* [60,61], *Elkinsia* [10], *Kerryoxylon* [7], *Laceya*[8,9], *Stenomyelon primaevum* [55,56], *Tetrastichia* [7,62,63], *Triradioxylon* [64], and *Tristichia* [7,65,66]. However, *Tetrastichia* and *Kerryoxylon* differ from *Yiduxylon* in having four- and six-ribbed primary xylem, respectively. The primary xylem of *Tetrastichia* is sometimes three- or five-ribbed, and a single protoxylem strand is present near the stelar center, or along the midplane of each primary xylem rib; the phyllotaxy of fronds is opposite decussate, 1/3 and 2/5. Leaf traces of *Kerryoxylon* are three-ribbed in the stem cortex.

*Yiduxylon* resembles *Bostonia*, *Elkinsia*, *Laceya*, *Stenomyelon primaevum*, *Triradioxylon* and *Tristichia* in the protosteles with three primary xylem ribs possessing peripheral protoxylem strands. Nevertheless, *Yiduxylon*is characterized by two leaf traces successively derived in tangential pattern from a single rib of stem primary xylem. Specifically, the primary xylem of *Bostonia* comprises three semi-discrete portions and a central portion; the protoxylem strands occur also in the stelar center; neither xylem ribs nor peripheral protoxylem strands are longitudinally continuous; the vascular supply of petioles is three-ribbed and derived from a pair of leaf traces [61]. Protoxylem strands of *Elkinsia* are located also in the stelar center and sometimes along the midplane of xylem ribs; the peripheral protoxylem strand diverges into different numbers and shapes of leaf traces (one papillionoid, one C-shaped, two bilobed, four elliptical ones), although the frond rachis is vascularized by two C-shaped bundles [10]. The leaf trace initiation of *Laceya* is caused by repeated divisions of primary xylem rib resulting in a tip with 6–8 united lobes; the U-shaped leaf trace remains undivided and its abaxial surface shows 6–8 protoxylem strands [8]; vascular supply at the petiole base is a U-/V-shaped bundle with 8–12 protoxylem strands on the convex abaxial surface [9]. The primary xylem of *Stenomyelon primaevum* is separated from the secondary xylem by parenchymatous cells; protoxylem strands occur along the midplanes of xylem ribs and are often associated with a lacuna; pits appear only on radial walls of secondary xylem tracheids; there are eight vascular bundles in each petiole base [56]. One or more protoxylem strands are also located in the stelar center of *Triradioxylon*, and the leaf traces are undivided and the vascular supply in the petiole is T-shaped [64]. Concerning *Tristichia*, protoxylem strands sometimes occur also in the stelar center and along the xylem ribs, and the leaf traces are tetrarch and papillionoid [66]. Sclerotic nests characteristic of the inner cortex of *Elkinsia*, *Triradioxylon* and *Tristichia* are absent in *Yiduxylon*.

As in *Yiduxylon*, the pinnae and pinnules of Late Devonian *Kongshania* [3] are alternately arranged. In contrast, the pinnules of *Kongshania* are laminate and wedge-tongue shaped; the poorly preserved frond rachis in description and diagram consists mainly of two bean-shaped vascular bundles surrounded by secondary xylem, which are however difficult to see in the photographs.

## DISCUSSION

### Taxonomy

It is widely agreed that the seed plants originated from the aneurophyte progymnosperms [1]. As Galtier [37] summarized, the early spermatophytes are characterized by both primitive aneurophyte and derived morphological and anatomical features. Primitive anatomical features include the protostelic primary vascular system, secondary xylem/phloem produced by a bifacial cambium, multiseriate and elliptical bordered pitting of tracheids, and outer cortex with longitudinally oriented plates of sclerenchyma cells. Derived features refer to such as bipartite fronds with dichotomous rachis, dorsiventral (adaxial/abaxial) identity of petioles, nonlaminate to entire pinnules, loss of protoxylem strands in stelar center, pits restricted to the radial walls of secondary xylem tracheids, and tangential divergence of leaf traces. In addition, the early seed plants often have their fronds borne helically on the stem [67]. *Yiduxylon* possesses these characteristic vegetative morphology and stelar architecture except for unknown secondary phloem. Manoxylic secondary xylem is another important feature of the early seed plants but it is difficult to define [37] or may be described as having large tracheids (more than 150 μm in diameter) and rays (over 5 cells wide and up to 200 cells high), which contrasts to pycnoxylic woods with small tracheids (less than 50 μm in diameter) and rays (uniseriate and short) [68,69]. In this case, the size of secondary xylem tracheids (12–75 μm in diameter) and rays (2–6 cells wide and more than 120 cells high) of *Yiduxylon* is intermediate between pycnoxylic and manoxylic wood anatomy.

The Buteoxylonales has only two genera, i.e., Early Carboniferous *Buteoxylon* [44] and *Triradioxylon* [64], with the former also reported from the Latest Devonian [6]. Calamopityales diversified in the Early Carboniferous and Lyginopteridales in the Carboniferous, but they (probably) occurred in the Late Devonian [1,20]. As noted by many workers [4,15,20,37,67], the Calamopityales and Lyginopteridales are difficult to define in that they share many characters such as helical phyllotaxis, protostele to eustele, cortex with sclerotic clusters and sclerenchyma cells (*sparganum* and *dictyoxylon* type). Their differentiation may depend on a combination of morphological (e.g., adventitious roots) and anatomical characters such as the ratio of primary vasculature diameter to stem diameter, leaf trace origination and vascular supply of frond rachises. Among them, the type of petiole vascular supply is crucial to define the Calamopityales and Lyginopteridales [64], although this character has been questioned (especially when it is independently used) because of considerable overlap of petiole morphology between the two orders [4,15,20].

According to Galtier [37] and Dunn et al. [20], adventitious roots are generally absent in the Calamopityales, the ratio of primary vasculature diameter to stem diameter is 1/5-1/10, the leaf trace origination involves multiple divisions and sometimes is double, the petiole is *Kalymma* type with four or more vascular bundles in the base of a frond rachis (multifascicular petiole), the total volume of these bundles exceeds the volume of cauline primary vasculature at the same level. The cortex of the calamopityaleans has been emphasized as remarkably broad [1,70]. In contrast, the Lyginopteridales is characterized by adventitious roots, the ratio of primary vasculature diameter to stem diameter is 1/2-1/4, the leaf trace usually remains undivided to produce a single petiolar vascular supply with several protoxylem strands, or it divides once into two closely associated bundles of U-, or V-, or W-shape (*Lyginorachis*-type petiole), the volume of the petiolar vascular supply never exceeds the stelar volume. There are however some exceptions to the characters of Lyginopteridales. For example, *Laceya* probably has no adventitious roots [9] and *Heterangium* may possess *Kalymma*-type petioles [71].

*Yiduxylon* cannot be assigned to the Buteoxylonales with their undivided leaf trace and T-shaped petiolar vascular supply. It differs from the Elkinsiales represented by *Elkinsia* mainly in the stelar architecture and leaf trace number/configurations (see Comparisons). *Yiduxylon* lacks adventitious roots, has a ratio of stelar diameter to stem diameter of 1/5, double leaf traces in origin, four leaf traces in the inner cortex, and a multifascicular petiole base with four discrete bundles that tend to further divide and exceed the volume of stem primary vasculature. These characters suggest anatomical similarities with the Calamopityales. However, the ratio of 1/5 is not a very low value and it results mainly from the small primary xylem and thick secondary xylem (i.e., the cortex is thin). Individual leaf traces are undivided in the cauline secondary tissue. Furthermore, as in some lyginopteridaleans, the four petiolar bundles of *Yiduxylon* are in two C-shaped traces that possess protoxylem strands on the abaxial surface. The origin of the leaf traces is unique because the two traces were produced from a stem primary xylem rib at different stages. *Yiduxylon* further exemplifies the anatomical similarities among early seed plants and indicates that their taxonomy depends primarily on fertile organs with attachment or association of stem/frond anatomy.

## Growth habit

Stems of the Lyginopteridales and Calamopityales are generally less and more than 20 mm in diameter, respectively [37]. As summarized by Dunn et al. [20], due to the presence of adventitious roots, narrow stems and large fronds, the lyginopteridaleans are probably more similar to lianas or vines and possess a scrambling or climbing habit; the calamopityaleans lacking adventitious roots are considered to be primarily upright plants. In addition, the wide angle of attachment of the fronds and *Dictyoxylon*-type outer cortex of the lyginopteridaleans are thought to be adaptations to the liana habit [58]. Adventitious roots are absent in the specimens and stem sections of *Yiduxylon*, the outer cortex is *Sparganum*-type, and the angle of frond attachment is not so wide, whereas the stem diameter is generally less than 20 mm and the complex fronds consist of two orders of long pinnae with many large pinnules. Therefore, it is difficult to decide the growth habit of *Yiduxylon*, although it appears to have had an upright stem. Future biomechanical analyses such as those by Masselter et al. [21] may provide more accurate information on this subject.

## Evolution of Vegetative Organs and Protostele in Seed Plants

The aneurophyte progymnosperms of Middle to Late Devonian (Givetian-Frasnian) were widespread on the Laurussia continent and Gondwana continents. As the descendant of the aneurophytes, the spermatophytes were widespread in the Late Devonian (Famennian) of Laurussia. However, surprisingly, the aneurophytes have not been reported from South China [72,73]. There are now three genera of seed plants known from the Famennian of China or eastern Asia, i.e., *Cosmosperma* [2], *Kongshania* [3] and *Yiduxylon* (this paper). In *Yiduxylon*, the secondary pinnae and highly dissected pinnules are borne alternately. These vegetative characters support the hypothesis that Late Devonian seed plants generally possess pinnae in one plane and planate rather than laminate pinnules, in contrast to the three-dimensional branches and vegetative ultimate appendages of the aneurophytes [2].

Various stelar configurations occur in the early seed plants, ranging from simple protostele to eustele with sympodia and pith. The eustele is suggested to have evolved from the protostele (e.g., *Stenomyelon primaevum*) by gradual loss of the tracheidal tissue in the center of protostele; the transitional stelar form refers to parenchymatized protostele (e.g., *Stenomyelon tuedianum* and some species of *Calamopitys*), in which the stele has both parenchyma cells and tracheids [1,22,23,74]. Parenchymatized protostele and eustele were widespread in the Carboniferous Buteoxylonales, Calamopityales and Lyginopteridales.

Except for *Buteoxylon gordonianum* [6], where the stem primary vascular system with pith is unclear in detail, Late Devonian (Famennian) spermatophytes are primitive in having a protostele with three to six ribs. These plants are *Elkinsia polymorpha* [10], *Kerryoxylon hexalobatum* [7], *Laceya hibernica* [8,9], *Tristichia* sp. [7], cf. *Tetrastichia bupatides* [7] and *Yiduxylon trilobum* (this paper). Among them, *Elkinsia polymorpha* possesses a three-ribbed primary xylem with a single central protoxylem strand producing other strands along the ribs, which is very similar to the condition of the ancestral aneurophytes. Its frond rachises contain a *Lyginorachis*-like trace with abaxial protoxylem strands, which characterizes the early spermatophytes. This combination of stelar and petiolar architectures indicates the close evolutionary relationship between the aneurophytes and spermatophytes [75]. Therefore, middle Famennian *Elkinsia* with its protostele is more primitive than other Famennian taxa. In the Latest Devonian seed plants from Ireland, the protoxylem strand occurs in/near the stelar center of *Tristichia* sp. and cf. *Tetrastichia bupatides*, or along the midplane of stelar ribs of *Laceya hibernica*, or only at the periphery of xylem rib of *Kerryoxylon hexalobatum*. *Yiduxylon* is probably younger than *Elkinsia* but older than the Irish taxa. If *Yiduxylon* had only peripheral protoxylem strands, it suggests that the protostele of early seed plants may have lost the protoxylem strands in the stelar center and along the stelar ribs by the Famennian. Ribbed protostele extends to the earliest Carboniferous (late Tournaisian)*Triradioxylon primaevum* [64], species of *Tristichia* [66] and *Bostonia perplexa* [61] with central protoxylem strands, *Stenomyelon primaevum* [55,56] and *Tetrastichia bupatides* [62,63] without such strands.

It is clear that the protosteles of *Elkinsia* [10], *Laceya* [8] and *Yiduxylon* are not fully evolved, as they still show the aneurophyte character of bordered pits on both radial and tangential walls of secondary xylem tracheids. This feature also occurs in *Tetrastichia* [62], *Triradioxylon* [64] and the earliest Carboniferous (Tournaisian)*Stenomyelon bifasciculare* with parenchymatized protosteles [54]. Secondary xylem of *Elkinsia* [10] may be pycnoxylic in that the tracheids are small (15–88 μm in diameter) and rays are 1–2 cells wide. That of*Buteoxylon* [6] and *Laceya* [8] is manoxylic. Secondary xylem of *Yiduxylon* is probably intermediate between pycnoxylic and manoxylic types separately characterizing the aneurophytes and spermatophytes (see above). However, *Elkinsia*, *Laceya* and *Yiduxylon* demonstrate the derived features of bipartite fronds with dorsiventral identity and tangential divergence of leaf traces. In general, the Late Devonian and some of the earliest Carboniferous seed plants possessed the more or less protostelic architectures of ancestral progymnosperms. Among these taxa, *Yiduxylon* obviously demonstrates the transitional stelar traits between these two plant groups. Such traits include primary vascular system, type of secondary xylem, pit distribution on secondary xylem tracheid walls, and mode of leaf trace production (Figure 6).

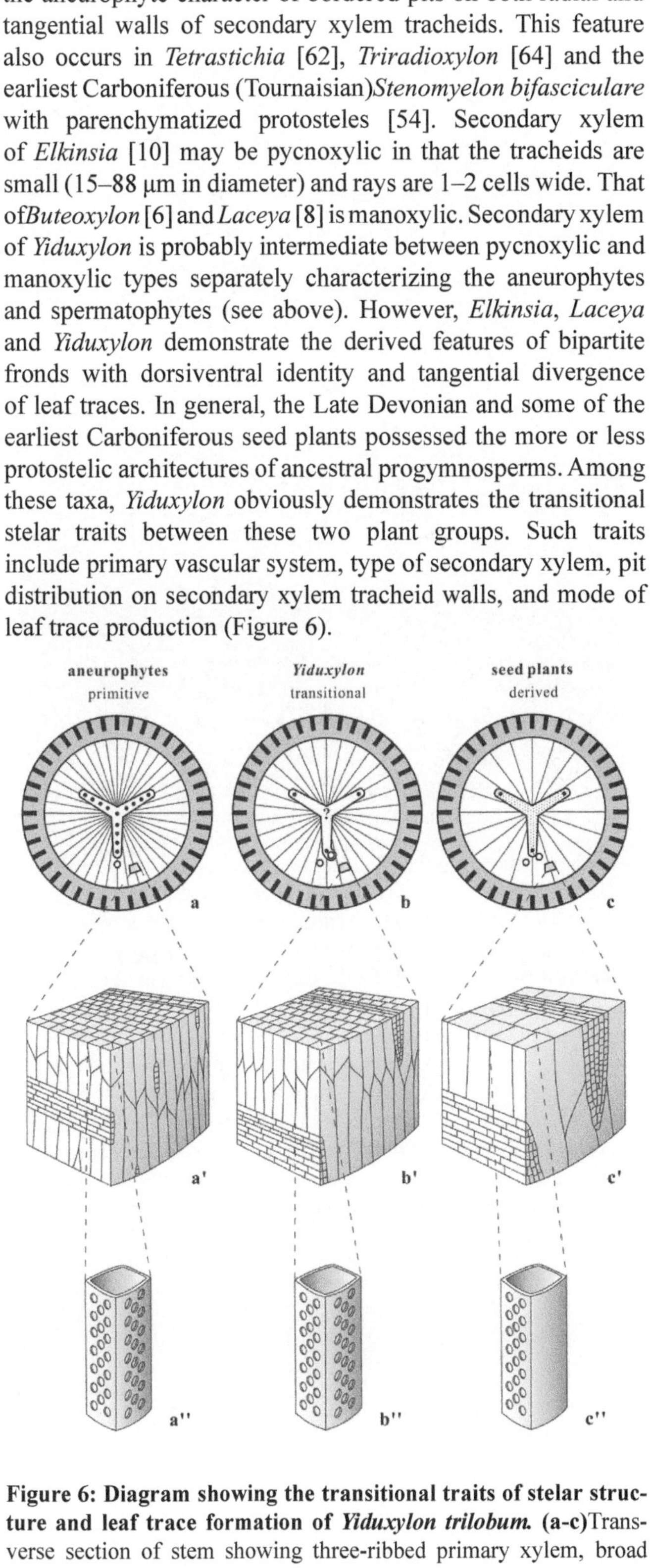


**Figure 6: Diagram showing the transitional traits of stelar structure and leaf trace formation of *Yiduxylon trilobum.* (a-c)**Transverse section of stem showing three-ribbed primary xylem, broad secondary xylem (represented by thin lines directed radially), narrow cortex with outer cortex containing peripheral bands of sclerenchyma cells (thick lines directed radially). **(a'-c')** Three-dimensional view of secondary xylem with tracheids (vertically aligned) and ray parenchyma cells (horizontally aligned). **(a"-c")** Secondary xylem tracheid with bordered pits. **(a-a')** Aneurophyte progymnosperms showing protostele with a central and several protoxylem strands (black dots) along primary xylem rib, radial divergence of branch trace (small circle) **(a)**, pycnoxylic secondary xylem with small tracheids and rays **(a')**, and pits on both radial and tangential walls of tracheid **(a")**. **(b-b")** *Yiduxylon* with protostele possessing a protoxylem strand near the periphery of primary xylem rib but lacking strands along rib and ? a central strand, tangential divergence of leaf traces (small circles) **(b)**, pycnoxylic-manoxylic secondary xylem with moderately sized tracheids and rays **(b')**, and pits on both radial and tangential walls of tracheid **(b")**. **(c-c")** Early seed plants with parenchymatized protostele possessing a protoxylem strand near periphery of primary xylem rib, tangential divergence of leaf traces **(c)**, manoxylic secondary xylem with large tracheids and rays **(c')**, and pits restricted on radial walls of tracheid **(c")**.

## Palaeogeographic Consideration

It is commonly accepted that the origin of seed plants is closely related to the aneurophyte gymnosperms. Accordingly, *Yiduxylon* provides transitional anatomical evidence linking these two plant groups. However, as stated before in this paper, typical aneurophytes have not yet been recognized from South China and thus need to be sought after in the Middle Devonian (Givetian) flora of this block. Alternatively, the seed plants may be nested within earlier representatives, e.g., *Celatheca beckii* with fertile structures putatively resembling cupulate ovules or pollen organs, which occurs in the Lower Devonian (Pragian) Posongchong Formation, Yunnan Province in South China [76-78].

At present, most of the earliest seed plants have been discovered from the Late Devonian of the Laurussian continent. Other coeval spermatophyte taxa come from South China which was an isolated block near the equator during the Devonian [79,80]. The Late Devonian flora of South China is well known for other lineages of vascular plants including many cosmopolitan and/or endemic elements. These lineages include the lycopsids, preferns, sphenophyllalean sphenopsids and archaeopterid progymnosperms. The rare record of Famennian spermatophytes in China may be related to the palaeogeographic position, or palaeoclimatic influence, limited dispersal, or insufficient collection in the past.

## CONCLUSIONS

*Yiduxylon* is the third genus of seed plants or a taxon closely related to seed plants to have been reported from the Late Devonian of China or East Asia. *Yiduxylon* provides evidence for stem/petiole anatomy in this period. Its vegetative organs support the hypothesis that Late Devonian spermatophytes had evolved planate pinnae and pinnules. *Yiduxylon*, probably upright in habit, differs from other early seed plants mainly in the stem protostele with three-ribbed primary xylem of mesarch maturation, successive divergence of two leaf traces in tangential pattern from a primary xylem rib, four leaf traces in the cortex, and a petiolar base with four vascular bundles in two C-shaped groups. The difficulty in differentiating stelar architecture of the early seed plants indicates that they are anatomically similar and their taxonomy depends largely on fertile organs. The stems of all Late Devonian and some of the earliest Carboniferous seed plants possessed a ribbed protostele, which lacks or contains a central protoxylem strand as a character of the ancestral aneurophyte progymnosperms. Protostelic architecture and leaf trace formation of *Yiduxylon* suggest transitional traits in the evolution of early seed plants from aneurophytes.

## DECLARATIONS

### Acknowledgements

The authors thank Professor David K. Ferguson, University of Vienna, for important discussion, and Jia Qiuyue and Ni Debao, Peking University, Beijing, for laboratory assistance. This work is supported by the National Natural Science Foundation of China (Nos. 41172007) and the National Basic Research Program of China (2012CB821900) of the Ministry of Science and Technology of China.

## ADDITIONAL FILES

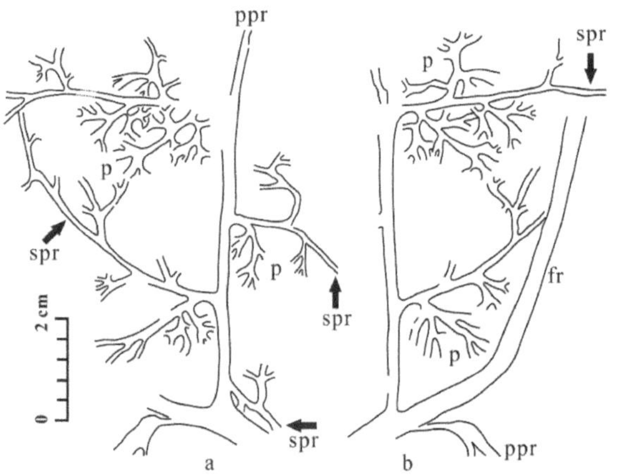


**Additional file 1. Figure S1.** *Yiduxylon trilobum* gen. et sp. nov. from Hubei. **(a-b)** Line-drawings of part and counterpart of specimen in Figure 1c,d, respectively. fr: frond rachis, ppr: primary pinna rachis, spr: secondary pinna rachis, p: pinnule. PKUB14402a, PKUB14402b.

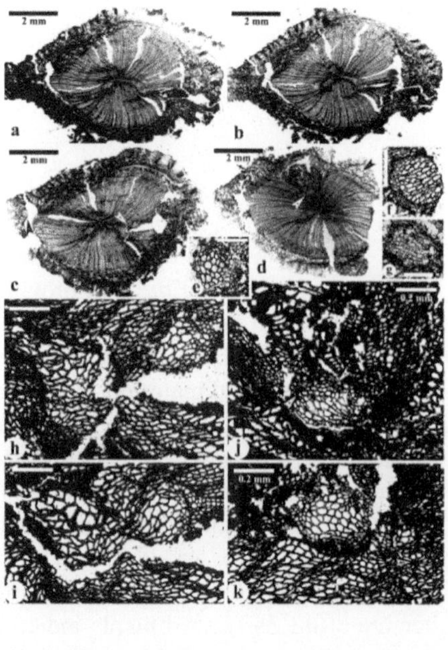

**Additional file 2. Figure S2.** Transverse sections of a stem of *Yiduxylon trilobum* gen. et sp. nov. from Hubei. **(a-d)** Successive transverse sections of a stem showing small three-ribbed primary xylem, broad secondary xylem, and narrow sparganum-type cortex. HBY-01, HBY-02, HBY-04, HBY-16. **(a)** Arrow indicating part enlarged in **(e)**. **(c)** Left and right arrows indicating parts enlarged in **(f)** and **(h)**, respectively. **(d)** Upper and lower arrows indicating parts enlarged in **(g)** and **(i)**, respectively. **(e-g)**Individual leaf traces in inner cortex. Enlargement of (**a**, arrow), (**c**, left arrow) and (**d**, upper arrow), respectively. **(h)**Enlargement of (**c**, right arrow) showing leaf trace diverging from tip of a primary xylem rib. **(i)** Earlier stage of leaf trace divergence in **(h)**. HBY-03. **(j)** Enlargement of (**d**, lower arrow) showing leaf trace in secondary xylem. Arrows indicating two protoxylem strands. HBY-16. **(k)** Earlier stage of leaf trace divergence in **(j)**. Arrow showing a protoxylem strand near rib tip of stem primary xylem. HBY-11. Scale bars=0.2 mm **(e-k)**.

**Additional file 3. Figure S3.** Transverse sections of stems of *Yiduxylon trilobum* gen. et sp. nov. from Hubei. **(a-b)** Enlargement of Additional file 2: Figure S2a,b, respectively. Three-ribbed primary xylem surrounded by secondary xylem. Arrows 1–3 in **(a)**showing three xylem ribs, with arrows 2–3 also indicating parts enlarged in **(c-d)**, respectively. **(c)** Enlargement of (**a**, arrow 2) showing protoxylem strand (arrow). **(d)** Enlargement of (**a**, arrow 3) showing protoxylem strand (left arrow) and incipient leaf trace (right arrow). **(e-g)** Successive transverse sections of a stem showing three-ribbed primary xylem of mesarch maturation. HY3-1, HY3-3, HY3-4. Arrows in **(e-f)** indicating parts enlarged in **(h)** and **(i)**, respectively. **(g)** Enlargement of Figure 2b. **(h-i)**Enlargement of (e, arrow) and (f, arrow), respectively, showing peripheral protoxylem strand. Scale bars=0.2 mm **(c, d, h, i)** , 0.5 mm **(a, b, e-g)**.

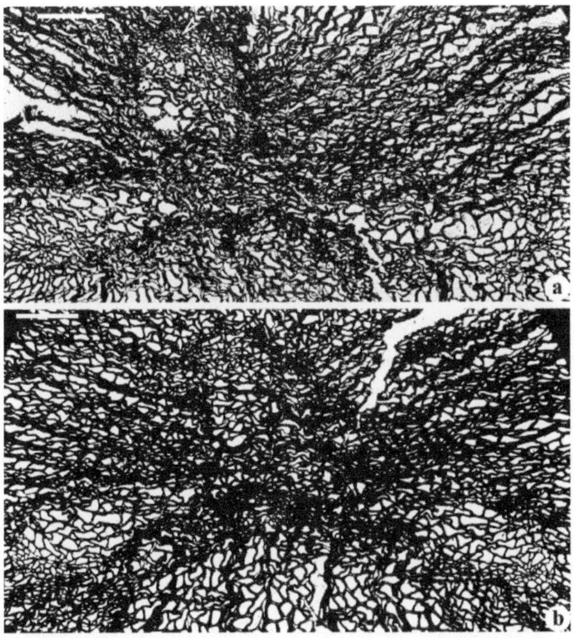

**Additional file 4. Figure S4.** Transverse sections of stems of *Yiduxylon trilobum* gen. et sp. nov. from Hubei. **(a)** Enlargement of Additional file 3: Figure S3e showing primary xylem probably without central protoxylem strand. Arrow indicating a protoxylem strand near apex of xylem tip. **(b)** Three-ribbed stem primary xylem probably lacking central protoxylem strand but with peripheral strands near rib tip. 2–1. Scale bars = 0.2 mm.

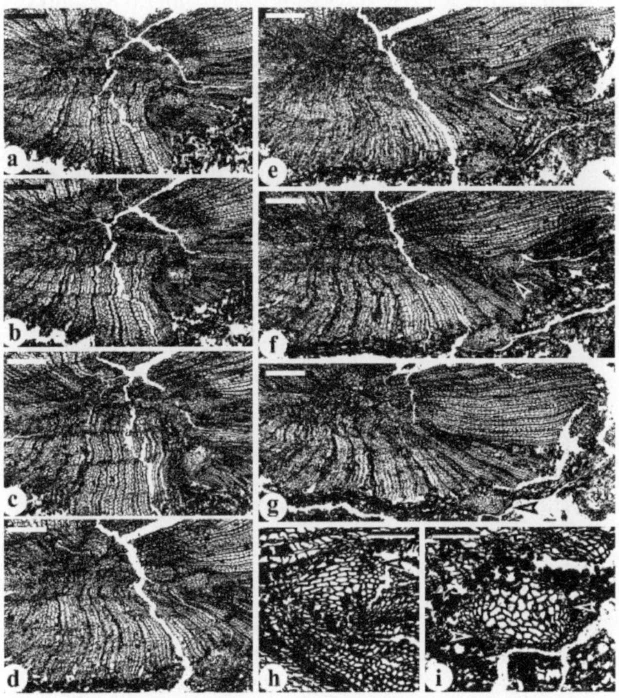

**Additional file 5. Figure S5.** Transverse sections of a stem of *Yiduxylon trilobum* gen. et sp. nov. from Hubei. **(a-g)** Serial sections showing stages of two leaf traces. 3–1 to 3–7. Arrows in **(f-g)** indicating parts enlarged in **(h)** and **(i)**, respectively. **(h)**Enlargement of (**f**, arrow) showing leaf trace in secondary xylem and its protoxylem strand (arrow). **(i)** Enlargement of (**g**, arrow) showing leaf trace at periphery of inner cortex and two protoxylem strands (arrows). Scale bars = 0.2 mm **(h, i)**, 0.5 mm**(a-g)**.

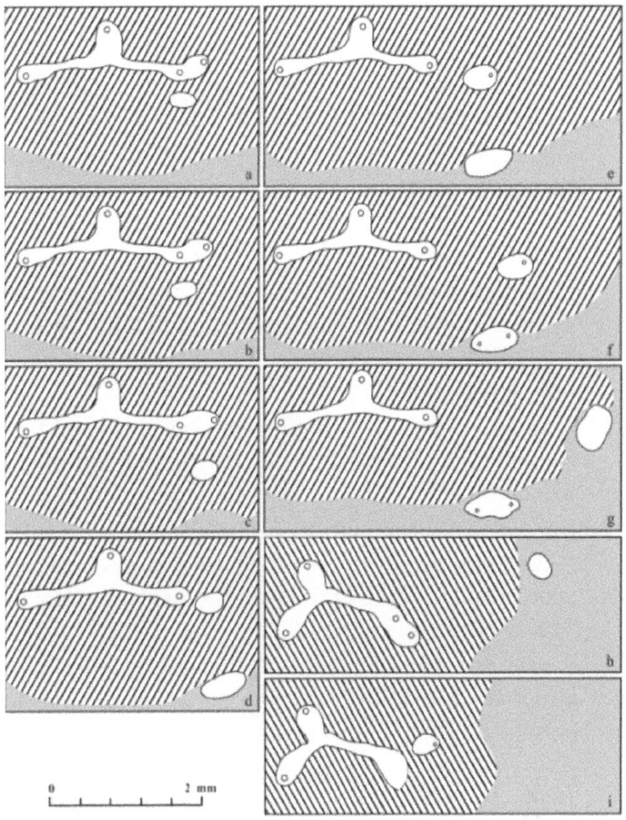

**Additional file 6. Figure S6.** *Yiduxylon trilobum* gen. et sp. nov. from Hubei. **(a-g)** Line-drawings of serial transverse sections of a stem in Additional file 5: Figure S5a-g, respectively. 3–1 to 3–7. **(h-i)** Line-drawings of portions of two transverse sections of a stem in Additional file 2: Figure S2a,c, respectively. HBY-01, HBY-04. white area: primary xylem of stem or leaf trace, small circle: protoxylem strand, oblique lines: secondary xylem, gray area: cortex.

## Competing interests

The authors declare that there have no competing interests.

## Authors' contributions

DMW collected the fossils, conceived the research, conducted the experiments and observations of specimens and wrote the manuscript. DMW and LL prepared the figures and performed the evolutionary interpretations. Both authors read and approved the final manuscript.

## REFERENCES

1. Taylor TN, Taylor EL, Krings M. Paleobotany: the biology and evolution of fossil plants. 2nd ed. Burlington: Academic Press; 2009.
2. Wang DM, Liu L, Meng MC, Xue JZ, Liu T, Guo Y. *Cosmosperma polyloba* gen. et sp. nov., a seed plant from the Upper Devonian of South China. Naturwissenschaften. 2014;101:615–22.
3. Wang Y. *Kongshania* gen. nov. A new plant from the Wutung Formation (Upper Devonian) of Jiangning County, Jiangsu, China. Acta Palaeontol Sin. 2000;39:42–56.
4. Decombeix AL, Galtier J, Meyer-Berthaud B. *Faironia difasciculata*, a new gymnosperm from the Early

Carboniferous (Mississippian) of Montagne Noire, France. Rev Palaeobot Palynol. 2006;142:79–92.

5. Dunn MT. A review of permineralized seed fern stems of the Upper Paleozoic. J Torrey Bot Soc. 2006;133:20–32.

6. Matten LC, Lacey WS, May BI, Lucas RC. A megafossil flora from the Uppermost Devonian near Ballyheigue, Co. Kerry, Ireland. Rev Palaeobot Palynol. 1980;29:241–51.

7. Matten LC, Tanner WR, Lacey WS. Additions to the silicified Upper Devonian/Lower Carboniferous flora from Ballyheigue, Ireland. Rev Palaeobot Palynol. 1984;43:303–20.

8. May BI, Matten LC. A probable pteridosperm from the uppermost Devonian near Ballyheigue, Co. Kerry, Ireland. Bot J Linn Soc. 1983;86:103–23.

9. Klavins SD, Matten LC. Reconstruction of the frond of *Laceya hibernica*, a lyginopterid pteridosperm from the uppermost Devonian of Ireland. Rev Palaeobot Palynol. 1996;93:253–68.

10. Serbet R, Rothwell GW. Characterizing the most primitive seed ferns. I. A reconstruction of *Elkinsia polymorpha*. Int J Plant Sci. 1992;153:602–21.

11. Read CB. A Devonian flora from Kentucky. J Paleontol. 1936;10:213–27.

12. Read CB. The flora of the New Albany Shale. Part 1. *Diichnia kentuckiensis*, a new representative of the Calamopityeae. US Geol Surv Prof Pap. 1936;185H:149–55.

13. Read CB, Campbell G. Preliminary account of the New Albany Shale flora. Amer Midl Nat. 1939;21:435–53.

14. Beck CB, Bailey RE. Plants of the New Albany Shale. III: *Chapelia campbellii* gen. n. Amer J Bot. 1967;54:998–1007.

15. Beck CB, Stein WE. *Galtiera bostonensis*, gen. et sp. nov., a protostelic calamopityacean from the New Albany shale of Kentucky. Can J Bot. 1987;65:348–61.

16. Beck CB, Galtier J, Stein WE. A reinvestigation of *Diichnia* Read from the New Albany Shale of Kentucky. Rev Palaeobot Palynol. 1992;75:1–32.

17. Sebby WS, Matten LC. *Kalymma minuta* (Read) comb. nov. from the New Albany Shale. Bull Torrey Bot Club. 1969;96:79–88.

18. Scott AC, Galtier J, Clayton G. Distribution of anatomically-preserved floras in the Lower Carboniferous in Western Europe. Trans Roy Soc Edinb. 1984;75:311–40.

19. Masselter T, Rowe N, Galtier J, Speck T. Secondary growth and deformation of stem tissues in the Lower Carboniferous seed fern *Calamopitys*. Int J Plant Sci. 2009;170:1228–39.

20. Dunn MT, Rothwell GW, Mapes G. On Paleozoic plants from marine strata: *Trivena arkansana* (Lyginopteridaceae) gen. et sp. nov., a lyginopterid from the Fayetteville Formation (middle Chesterian/Upper Mississippian) of Arkansas, USA. Amer J Bot. 2003;90:1239–52.

21. Masselter T, Rowe NP, Speck T. Biomechanical reconstruction of the Carboniferous seed fern *Lyginopteris oldhamia*: implications for growth form reconstruction and habit. Int J Plant Sci. 2007;168:1177–89.

22. Namboodiri KK, Beck CB. A comparative study of the primary vascular system of conifers. III. Stelar evolution in gymnosperms. Amer J Bot. 1968;55:464–72.

23. Beck CB. The appearance of gymnospermous structure. Biol Rev. 1970;45:379–400.

24. Wang Q, Hao SG, Wang DM, Wang Y, Denk T. A Late Devonian arborescent lycopsid *Sublepidodendron songziense* Chen emend. (Sublepidodendraceae Kräusel et Weyland 1949) from China, with a revision of the genus Sublepidodendron (Nathorst) Hirmer 1927. Rev Palaeobot Palynol. 2003;127:269–305.

25. Zhang RJ, Feng SN, Duan QF. Devonian. In: Wang XF, Chen XH, Zhang RJ, editors. Protection of precise geological remains in the Yangtze gorges area, China with the study of the Archean-Mesozoic multiple stratigraphic subdivision and sea-level change. Beijing: Geological Publishing House; 2002. p. 131–72.

26. Wu JH, Niu ZJ, Wang JX, Yao HZ, Zhang RJ. On the geological age of Changyang Formation, western Hubei Province. Sci Tech Eng. 2011;11:1411–5.

27. Devonian system and plant kingdom. In Biostratigraphy of the Yangtze Gorge area. Volume 3. Late Paleozoic era. Edited by Institute of Geology and Mineral Resources of Yichang. Beijing: Geological Publishing House; 1984: 6–31, 293–317

28. Tan L, Li X, Sun YL. On age of the Changyang Formation of western Hubei. J Stratigr. 2011;35:8–18.

29. Meng MC, Wang DM, Tian T. New insights on the megasporangiate strobilus of *Sublepidodendron songziense* from the Late Devonian of Hubei Province. Acta Palaeontol Sin. 2014;53:180–90.

30. Wang GX, Jing YJ, Zhuang JL, Zhang CF, Hu WQ. Stratigraphical system of Devonian-Lower Carboniferous epoch of Xikuangshan area in the central region of Hunan. Hunan Geol. 1986;5:36–50.

31. Gao LD. Miospore zones in the Devonian-Carboniferous boundary beds in Hunan and their stratigraphical significance. Geol Rev. 1990;36:58–68.

32. Gao LD. Spore assemblages from the Wangjiadian Formation in Zhangxian County, Gansu and their age. Gansu Geol. 1990;11:1–12.

33. Gao LD. Late Devonian and Early Carboniferous miospore zones from southeastern Guizhou and the boundary of the Devonian and Carboniferous. Geol Guizhou. 1991;8:59–69.

34. Wang DM, Hao SG, Wang Q. *Rotafolia songziensis* gen. et sp. comb. nov., a sphenopsid from the Late Devonian of Hubei, China. Bot J Linn Soc. 2005;148:21–37.

35. Wang DM, Hao SG, Wang Q, Xue JZ. Anatomy of the Late Devonian sphenopsid *Rotafolia songziensis*, with a discussion of stelar architecture of the Sphenophyllales. Int J Plant Sci. 2006;167:373–83.

36. Beck CB, Wight DC. Progymnosperms. In: Beck CB, editor. Origin and evolution of gymnosperms. New

York: Columbia University Press; 1988. p. 1–84.

37. Galtier J. Morphology and phylogenetic relationships of early pteridosperms. In: Beck CB, editor. Origin and evolution of gymnosperms. New York: Columbia University Press; 1988. p. 135–76.
38. Scheckler SE, Banks HP. Anatomy and relationships of some Devonian progymnosperms from New York. Amer J Bot. 1971;58:737–51.
39. Scheckler SE, Banks HP. *Proteokalon*, a new genus of progymnosperms from the Devonian of New York State and its bearing on phylogenetic trends in the group. Amer J Bot. 1971;58:874–84.
40. Wight DC. Non-adaptive change in early land plant evolution. Paleobiology. 1987;13:208–14.
41. Beck CB, Stein WE. *Crossia virginiana* gen. et sp. nov., a new member of the Stenokoleales from the Middle Devonian of southwestern Virginia. Palaeontogr Abt B. 1993;229:115–34.
42. Matten LC. Studies on Devonian plants from New York State: *Stenokoleos holmesii* n. sp. from the Cairo flora (Givetian) with an alternative model for lyginopterid seed fern evolution. Cour Forsch Inst Senck. 1992;147:75–85.
43. Matten LC, Banks HP. *Stenokoleos bifidus* sp. n. in the Upper Devonian of New York State. Amer J Bot. 1969;56:880–91.
44. Barnard PDW, Long AG. On the structure of a petrified stem and some associated seeds from the Lower Carboniferous rocks of East Lothian, Scotland. Trans Roy Soc Edinb. 1973;69:91–108.
45. Stidd BM. A new species of *Heterangium* from the Illinois basin of North America. Rev Palaeobot Palynol. 1979;28:249–57.
46. Pigg KB, Taylor TN, Stockey RA. Paleozoic seed ferns: *Heterangium kentuckyensis* sp. nov., from the Upper Carboniferous of North America. Amer J Bot. 1987;74:1184–204.
47. Gordon WT. On *Rhetinangium arberi*, a new genus of Cycadofilices from the Calciferous Sandstone series. Trans Roy Soc Edinb. 1912;48:813–25.
48. Long AG. On the structure of some petioles associated with *Rhetinangium* Gordon. Trans Roy Soc Edinb. 1964;66:1–7.
49. Dunn MT. The Fayetteville flora I: Upper Mississippian (middle Chesterian/lower Namurian A) plant assemblages of permineralized and compression remains from Arkansas, USA. Rev Palaeobot Palynol. 2004;132:79–102.
50. Stidd BM, Phillips TL. The vegetative anatomy of *Schopfiastrum decussatum* from Middle Pennsylvanian of the Illinois Basin. Amer J Bot. 1973;60:463–74.
51. Taylor TN, Stockey RA. Studies of Paleozoic Seed Ferns: Anatomy and Morphology of *Microspermopteris aphyllum*. Amer J Bot. 1976;63:1302–10.
52. Pigg KB, Stockey RA, Taylor TN. Studies of Paleozoic Seed Ferns: Additional Studies of *Microspermopteris aphyllum*Baxter. Bot Gaz. 1986;147:126–36.
53. Galtier J, Meyer-Berthaud B. Studies of the Early Carboniferous pteridosperm *Calamopitys*—A redescription of the type material from Saalfeld (GDR). Palaeontogr Abt B. 1989;213:1–36.
54. Meyer-Berthaud B. *Stenomyelon* from the Upper Tournaisian of the Montagne Noire (France). Can J Bot. 1984;62:2297–307.
55. Long AG. Some specimens of *Stenomyelon* and *Kalymma* from the Calciferous Sandstone Series of Berwickshire. Trans Roy Soc Edinb. 1964;65:435–47.
56. Meyer-Berthaud B, Stein WE. A Reinvestigation of *Stenomyelon* from the Late Tournaisian of Scotland. Int J Plant Sci. 1995;156:863–95.
57. Galtier J. Recherches sur les végétaux à structure conservée du Carbonifère inférieur français. Paléobiol Continent. 1970;1:1–221.
58. Tomescu AMF, Rothwell GW, Mapes G. *Lyginopteris royalii* sp. nov. from the Upper Mississippian of North America. Rev Palaeobot Palynol. 2001;116:159–73.
59. Galtier J, Beck CB. *Triichnia*, a new eustelic calamopityacean from the Lower Carboniferous of France. Palaeontogr Abt B. 1992;224:1–16.
60. Stein WE, Beck CB. *Bostonia perplexa* gen. et sp. nov., a calamopityan axis from New Albany Shale of Kentucky. Amer J Bot. 1978;65:459–65.
61. Stein WE, Beck CB. New information on *Bostonia perplexa*-an unusual member of the Calamopityaceae from North America. Rev Palaeobot Palynol. 1992;72:73–102.
62. Gordon WT. XII.—On *Tetrastichia bupatides*: A Carboniferous Pteridosperm from East Lothian. Trans Roy Soc Edinb. 1938;59:351–70.
63. Dunn MT, Rothwell GW. Phenotypic plasticity of the hydrasperman seed fern *Tetrastichia bupatides* Gordon (Lyginopteridaceae). Int J Plant Sci. 2012;173:823–34.
64. Barnard PDW, Long AG. *Triradioxylon*—a new genus of Lower Carboniferous petrified stems and petioles together with a review of the classification of early Pterophytina. Trans Roy Soc Edinb. 1975;69:231–49.
65. Long AG. *Tristichia ovensi* gen. sp. nov., a protostelic Lower Carboniferous pteridosperm from Berwickshire and East Lothian, with an account of some associated seeds and cupules. Trans Roy Soc Edinb. 1961;64:477–89.
66. Galtier J, Meyer-Berthaud B. The early seed plant *Tristichia tiipos* (Unger) *comb. nov.* from the Lower Carboniferous of Saalfeld, Thuringia. Rev Palaeobot Palynol. 1996;93:299–315.
67. Stewart WN, Rothwell GW. Paleobotany and the Evolution of Plants. 2nd ed. New York: Cambridge University Press; 1993.
68. Galtier J. On the Earliest Arborescent Gymnosperms. Cour Forsch Inst Senckenb. 1992;147:119–25.
69. Galtier J, Meyer-Berthaud B. The diversification of early arborescent seed ferns. J Torrey Bot Soc. 2006;133:7–19.
70. Hotton CL, Stein WE. An ontogenetic model for the Mississippian seed plant family Calamopityaceae. Int J Plant Sci. 1994;155:119–42.

71. Shadle GL, Stidd BM. The frond of *Heterangium*. Amer J Bot. 1975;62:67–75.

72. Berry CM, Wang Y. *Eocladoxylon* (*Protopteridium*) *minutum* (Halle) Koidzumi from the Middle Devonian of Yunnan, China: an early Rhacophyton-like plant? Int J Plant Sci. 2006;167:551–66.

73. Wang Y, Berry CM, Hao SG, Xu HH, Fu Q. The Xichong flora of Yunnan, China: diversity in late Mid Devonian plant assemblages. Geol J. 2007;42:339–50.

74. Beck CB, Schmid R, Rothwell GW. Stelar morphology and the primary vascular system of seed plants. Bot Rev. 1982;48:691–815.

75. Rothwell GW, Erwin DM. Origin of Seed Plants: An aneurophyte/seed-fern link elaborated. Amer J Bot. 1987;74:970–3.

76. Hao SG, Gensel PG. A new genus and species, *Celatheca beckii*, from the Siegenian (Early Devonian) of southeastern Yunnan, China. Int J Plant Sci. 1995;156:896–909.

77. Hao SG, Xue JZ. The Early Devonian Posongchong flora of Yunnan—a contribution to an understanding of the evolution and early diversification of vascular plants. Beijing: Science Press; 2013.

78. Hao SG, Xue JZ. Earliest record of megaphylls and leafy structures, and their initial diversification. Chin Sci Bull. 2013;58:2784–93.

79. Hao SG, Gensel PG. The Posongchong floral assemblages of Southeast Yunnan, China-diversity and disparity in Early Devonian plant assemblages. In: Gensel PG, Edwards D, editors. Plants invade the land: evolutionary and environmental considerations. New York: Columbia University Press; 2001. p. 103–19.

80. Chen X, Ruan YP, Boucot AJ. Paleozoic Climatological Evolution of China. Beijing: Science Press; 2001.

# CHAPTER 20 MECHANISMS OF STOMATAL DEVELOPMENT

Anne Vatén and Dominique C Bergmann

**Keywords:** Stomata, Plant evolution, bHLH transcription factors, Arabidopsis, Maize, Physcomitrella, Rice, Ligand receptor signaling, Cell polarity, Asymmetric cell division

## REVIEW

### Introduction to Stomata and Stomatal Pattern

Plants conquered land more than 400 million years ago. In the fossil record, the appearance of these pioneer species is contemporaneous with the appearance of structures on their surfaces called stomata. Each stoma (plural, stomata) consists of paired epidermal guard cells, a pore between them and an airspace in the photosynthetic mesophyll tissue subtending it. The function of stomata is to regulate gas exchange between the plant and its surroundings. On short timescales (minutes to hours), the opening and closing of the stomatal pore by turgor-driven changes in guard cell shape is a key regulatory step in maintaining water and carbon dioxide balance. Work from many laboratories has defined the intracellular signal transduction cascades that mediate changes in pore size in response to hormone and environmental signals [1].

The current view is that stomata arose only once during evolution [2]. In early land plants, stomatal density was low [3]. During intervening millennia, the stomatal density (SD, number of stomata/unit leaf area) increased, probably in response to reduced aerial $CO_2$ concentration [4]. The stomatal complex has been fine-tuned by several innovations including recruitment of neighboring subsidiary cells to facilitate stomatal opening/closing, relocation of stomatal complexes under protective epidermal cells and incorporation of multiple asymmetric cell divisions in precursors to create a variety of stomatal distributions. Despite the variation, the basic core structure has remained unchanged: two guard cells flank the stomal pore. In nearly all species, two stomata are separated at least by one non-stomatal cell, an arrangement thought to be essential for efficient opening and closing. Stomata are located on aerial organs including leaves, stems, flowers, fruits and seeds and they develop gradually during organ growth such that young organs have fewer total stomata than mature organs, though SD often decreases as the neighboring epidermal cells expand during maturation. The frequency and positioning of stomata are organ and species-specific characters, but are also affected by environmental factors.

### Innovations

Paleobotanical analyses utilizing the fossils of the early land plants and their currently living descendants have been combined with phylogenetic analyses to address the origins of stomata (Figures 1 and 2A). Liverworts, mosses and hornworts comprise the bryophytes, a basal land plant group. Liverworts do not have stomata; gas exchange is facilitated by epidermal air pores, structures whose development and morphology differ from stomata. Stomata are found in mosses and hornworts, making it likely that liverworts diverged from other bryophytes before the origin of stomata. Intriguingly, in extant bryophytes, both guard cell morphology and regulation of pore aperture can closely resemble higher plant stomata.

**Citation:** Vatén and Bergmann: Mechanisms of stomatal development. doi:10.1186/2041-9139-3-11.

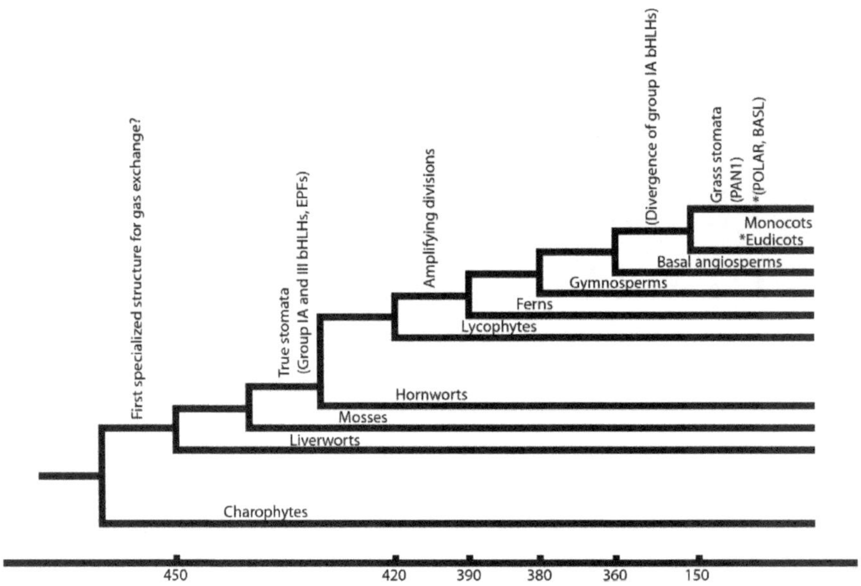


**Figure 1: Divergence of major land plant lineages and appearance of stomatal characteristics.** Phylogenetic tree of extant land plants indicating positions of major innovations in the evolution of stomata, following Ruszala *et al.*[5]. Those in brackets indicate predicted appearance of stomatal development regulatory genes. Numbers on the x-axis refer to multiples of millions of years.

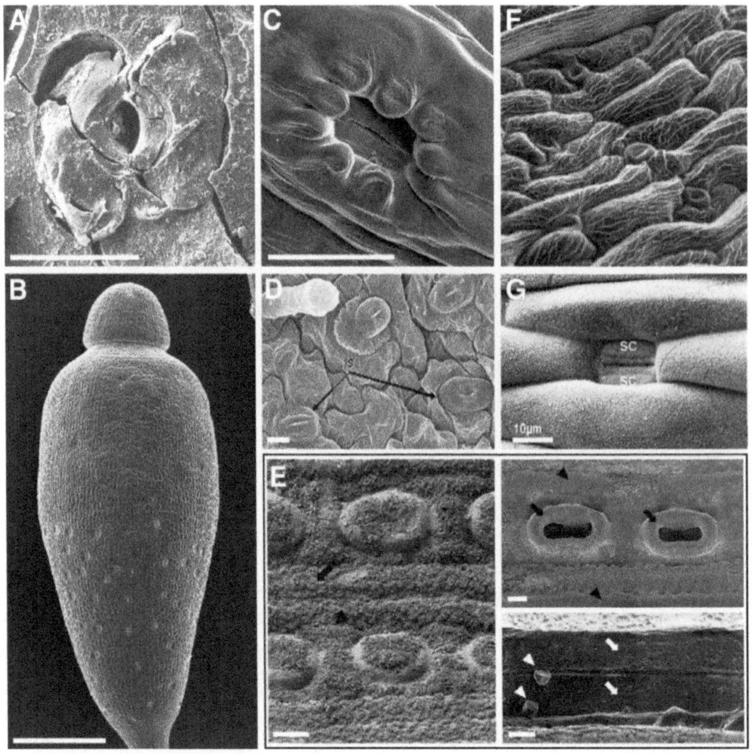

**Figure 2: Representative stomatal complexes and patterns from different species.** (**A**) Scanning electron micrograph (SEM) of Silurian fossil stoma displaying common morphology. Scale bar, 20 μm [3]. (**B**) SEM of moss *Bryum capillare*sporangium with stomata visible on the lower half. Scale bar, 600 μm [2]. (**C**) SEM of moss *Bryum capillare*sporangium stoma sunken below epidermal cells. Scale bar, 50 μm [2]. (**D**) SEM of fern *Thelypteris ovata* var.*lindheimeri* (sporophyte) leaf with stomata separated by pavement cells. Scale bar, 10 μm; s, stomata [6]. (**E**) Left panel, field emission SEM of *Pinus koraiensis* (gymnosperm) stomata arranged in rows on needle surface; granular material is surface wax. Scale bar, 10 μm. Upper right, dewaxed stomata. Scale bar, 10 μm. Lower right, dewaxed guard cells (arrows) within an epistomatal chamber. Scale bar, 2 μm [7]. (**F**) SEM of dicot *Arabidopsis thaliana*stomatal pattern in the sepal. (**G**) SEM of monocot *Poa annua* stoma, with subsidiary cells (sc) flanking the narrow guard cells. Scale bar, 10 μm [8].

In the most recently derived plant group (angiosperms, or flowering plants) there is a dedicated epidermal lineage that produces stomata. In dicot plants, such as the research model *Arabidopsis*[9], these lineages are initiated from various sites on the leaf (Figure 3). In each lineage, a committed protodermal cell called the meristemoid mother cell (MMC) divides asymmetrically to give rise to a larger stomatal lineage ground cell (SLGC) and smaller meristemoid. The meristemoid undergoes one to three asymmetric divisions (amplifying divisions) before it differentiates into a guard mother cell (GMC). Later, SLGCs can also divide asymmetrically and produce more meristemoids (spacing divisions). The GMC divides symmetrically to create two guard cells, and in some species the GMCs recruit neighboring subsidiary cells. These subsidiary cells can provide mechanical assistance and a source of ions required for guard cell movement. Amplifying and spacing divisions and subsidiary recruitment all require cell to cell communication and together they contribute to pattern. The frequency with which cells participate in these division types can be modified to yield the extraordinary diversity of stomatal patterns seen in nature [10] (Figure 2B2D and 2F).

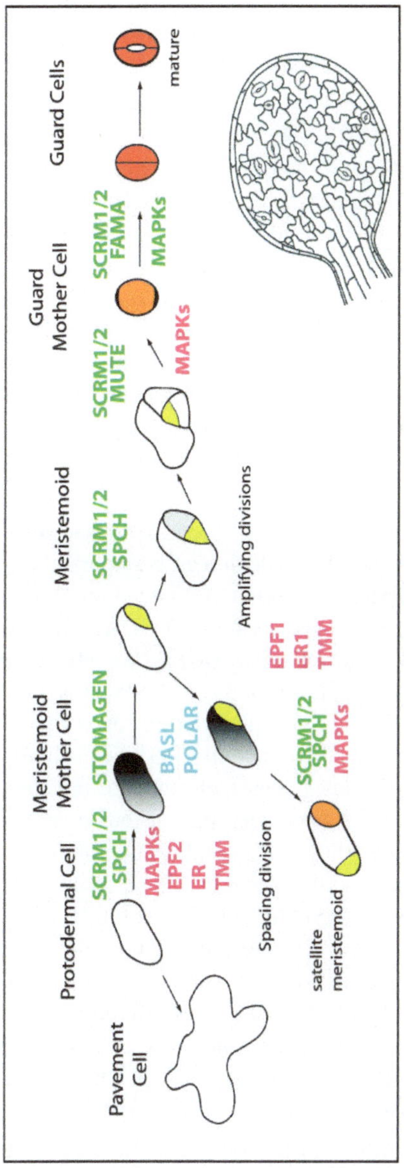


**Figure 3: Stomatal development in *Arabidopsis*.** Diagram of major stages in stomatal development with place of action of the subset of regulatory genes discussed in this review noted. Positive regulators are written in green, negative regulators in red and polarity regulators in blue. Not all genes known to regulate stomata are presented. The image of the young leaf in the lower right corner is to represent the dispersed nature of stomatal lineage initiation. Color code: yellow, meristemoid; orange, guard mother cell; red, guard cell; grey, meristemoid mother cell (MMC).

Monocots exhibit a strong base to tip gradient of leaf differentiation with stomata-producing cell lineages forming at the base of the leaf. Asymmetric cell divisions produce GMCs without prior transit through a self-renewing meristemoid stage (Figure 4). Protodermal cells in files flanking the GMC polarize towards the GMC and divide asymmetrically giving rise to subsidiary cells. After this, the GMC divides to produce guard cells that exhibit a novel flattened or dumbbell-shaped morphology (Figure 2G). In monocots like *Tradescantia,* overall stomatal pattern can be refined when GMCs change fate and differentiate into epidermal cells [11]. This fate change is dependent on distance from neighboring stomata, suggesting an inhibitory communication mechanism.

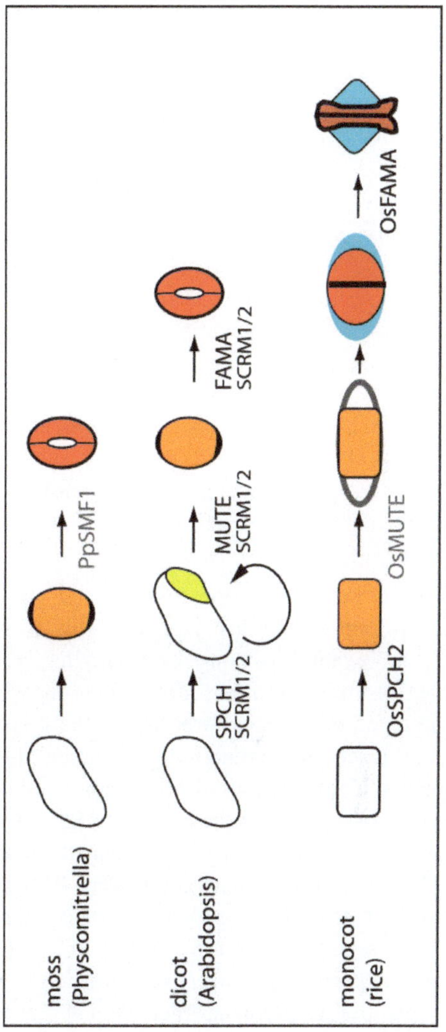


**Figure 4: Comparison of the molecular and morphological features of stomatal development in *Arabidopsis* and representatives of the grasses and mosses for which molecular data exist.** Presentation of a simplified stomatal lineage displaying only cell identities (in the same color codes as Figure 3), with the addition of blue to mark the subsidiary cells in monocots. Genetic regulators of the processes are included at their points of action, with black text indicating that there is direct functional evidence supporting the placement and grey text representing inferences from cross-species complementation tests. The curved arrow in the dicot lineage represents the continued asymmetric amplifying divisions made by meristemoids.

Given the current patterns and developmental processes associated with stomata of flowering plants, what were their origins? In the simplified ontogeny seen in some mosses, stomatal development involves a single asymmetric cell division giving rise directly to a GMC (Figure 4). This GMC may not divide completely, as seen in*Funaria hygrometrica* where two guard cell nuclei are separated by an incomplete cell wall [12], or may break the spacing rule as in *Polytrichastrum formosum* where stomata sometimes form next to each other. Already in mosses, diverse stomatal morphologies are seen [2]. Several (both early- and late-divergent) moss species also lack stomata [13]. The function of stomata in these plants may also be unusual, for example the basal genus *Sphagnum* displays pseudostomata which might function in spore desiccation rather than typical $CO_2$acquisition [14].

The appearance of amplifying divisions in ferns provided novel mechanisms to control cell number as well stomatal density and to produce specialized subsidiary cells [15, 16]. Here an epidermal cell may go through one or two asymmetric cell divisions before it differentiates into a GMC. Subsidiary cells in gymnosperms (for example, pines) can arise from meristemoid divisions or division of protodermal cells next to stomata, or both. In *Pinus strobus* and in *Pinus banksiana*, meristemoids divide once symmetrically to generate a GMC and a subsidiary cell [17]. The subsidiary cell, as well as neighboring epidermal cells, expands in a polar fashion over the GMC. As a result the GMC, and later the guard cell pair, is overlaid by a group of epidermal subsidiary cells of mixed origin and in addition, is closely connected to hypodermal subsidiary cells. Also in gymnosperms, we begin to see stomata incorporating some of the biochemical innovations of this group (Figure 2E). For example, subsidiary cells display thick, waterproof cuticles, and the guard cells become reinforced with lignin, a cell wall polymer that is not present in bryophytes [18].

Despite the benefits of stomata-mediated gas-exchange, some plant lineages have lost stomata. This is sometimes facultative; for example among heterophyllic species, two alternate leaf forms are made, depending on whether the leaf is submerged in water or airborne. In these species, leaf submergence leads to elimination of stomata [19]. Some parasitic plants, whose sources of fixed carbon are their hosts, may also lose or inactivate their stomata [16]. Other plant groups, like the small, predominantly aquatic isoetes, have members that have lost stomata completely. The astomatous isoetes gain $CO_2$ from the sediment via their extensive root system [20]. Isoetes perform a variant of photosynthesis common among cacti (crassulacean acid metabolism (CAM)), in which separation of particular biochemical reactions allow (stomatous) plants to only open stomata during the night to decrease water loss. Use of a root-derived carbon source enabled astomatous isoetes to fix carbon continuously without a threat of stomata-related water loss. In general, astomatous species are small and only exist in a narrow growth environment. It has been suggested that functional stomata allow plants to develop to larger sizes and to adapt to a wider range of growth conditions [21].

## Pathways for Stomatal Development in *Arabidopsis*

The regulation of stomatal development is best understood at a molecular level in *Arabidopsis.* Here, individual cell fate transitions in the stomatal lineage are promoted by three closely related basic helix-loop-helix (bHLH) transcription factors, SPEECHLESS (SPCH), MUTE and FAMA [22, 23, 24] (Figure 3). These bHLHs are expressed in the stomatal lineage, each in a specific developmental window, and each of them is absolutely required for stomata formation. SPCH is expressed in subset of young epidermal cells, often in two adjacent cells [24] and SPCH expression is dynamic. After an asymmetric cell division, SPCH disappears from the SLGC, but remains in the meristemoid, which continues asymmetric cell divisions [25]. Loss of *SPCH* leads to a complete loss of the stomatal lineage whereas overexpression of *SPCH* leads to ectopic asymmetric cell divisions [23, 24, 26]; thus it is required for entry into the stomatal lineage. MUTE is expressed in late meristemoids and is required for exit from the amplifying division stage, and it promotes the meristemoid to GMC transition [23, 24, 26]. FAMA is expressed in the GMC and in immature guard cells. Overexpression of FAMA leads to ectopic formation of unpaired guard cells indicating that FAMA promotes stomatal cell fate while restricting (symmetric) divisions [22].

Proteins encoded by the paralogous bHLHs, *INDUCER OF CBF EXPRESSION1/SCREAM (ICE1/SCRM)* and *SCRM2*, form heterodimers with SPCH, MUTE and FAMA and promote all three stomatal fate transitions [27]. A semidominant *scrm-D* mutant converts the epidermis into stomata, a phenotype identical to MUTE overexpression, whereas double mutants of *ICE1/SCRM* and *SCRM2* resemble *spch*[27]. Interestingly, ICE1/SCRM has been shown to be involved in cold stress response [28]. Since stomatal development is regulated by both environmental [21] and developmental factors [29], it is possible that ICE1/SCRM is a cross-regulatory node where several signaling pathways are integrated to direct stomatal development.

More signal integration occurs via mitogen-activated protein kinases (MAPKs) which regulate stomatal development and stress responses through a three-step phosphorylation cascade. MAPK kinase kinase YODA, MAPK kinases (MKK4/5/7/9) and MAPKs (MPK3/6) are essential for normal stomatal spacing [30, 31, 32]. SPCH is a direct target of MAPK-mediated phosphorylation and this serves to negatively regulate SPCH activity [33]. The MAPK pathway also regulates the later stages of stomatal development, but the targets have not been identified. More complexity arises from the recent finding that signaling intermediates from the steroid hormone brassinosteroid (BR) pathway phosphorylate both YODA [34] and SPCH [35]. Interestingly, BR-modulated phosphorylation as mediated through YODA and SPCH actually produces opposite stomatal phenotypes. Combined with other evidence that SPCH is differentially methylated under certain environmental conditions [36], we are seeing just hints of the complex interactions and precise tuning to which the early parts of the stomatal pathway may be subjected.

Upstream of the intracellular signalling cascades, genetic studies have revealed that stomatal spacing is regulated by secreted peptides of the EPIDERMAL PATTERNING FACTOR-LIKE (EPFL) family [37, 38, 39, 40, 41], by three leucine-rich repeat receptor kinases (LRR-RLKs), ERECTA (ER), ERECTA-LIKE 1 (ERL1) and ERL2 [42] and one LRR-receptor-like protein, TOO MANY MOUTHS (TMM) [43, 44]. Members of the ER-family (ERf) are broadly expressed and their absence leads to severe stomatal overproliferation and mispatterning, as well as pleiotropic growth phenotypes, indicating that they regulate multiple developmental processes [42, 45]. ER acts predominantly as a negative regulator of entry divisions whereas ERL1 and ERL2 control later stages [42].

TMM is expressed in the early stomatal lineage and, thus far, only roles in stomatal development have been described [44].

EPF1 and EPF2 peptides are stomatal lineage-expressed and regulate the number and orientation of asymmetric divisions [37, 38, 39, 40, 41]. Loss of either *EPF1* or *EPF2* results in more stomata, but mutant and overexpression phenotypes indicate that EPF2 prevents entry into the stomatal lineage whereas EPF1 acts later. Their paralogue, *STOMAGEN/EPFL9*, by contrast, is expressed in the underlying cell layer (mesophyll) and travels to the epidermis to promote stomatal differentiation [46, 47]. EPF1, EPF2 and STOMAGEN require the receptor TMM for full activity [37, 38, 39, 40, 41, 46, 47]. Surprisingly, the function of three other EPFLs, EPFL6/CHALLAH (CHAL), EPFL4 and EPFL5, is inhibited by the presence of TMM [40]. Although CHAL was originally identified by its stomatal phenotype in a *tmm* background [41], *CHAL/EPFL4/EPFL5* are expressed in internal tissues and their loss leads to a compromised growth phenotype resembling loss of *ER*[40]. Thus, they likely represent ligands for ER's non-stomatal roles. The identification of EPF family members with distinct developmental roles has led to interesting models of how signaling specificity is achieved by using the non-kinase receptor TMM to modulate ligand interactions with the ERf kinases in specific tissues [40]. Recently, elegant biosensor approaches demonstrated ER and ERL1 primarily bind EPF2 and EPF1, respectively, *in vitro*[48] and that *in planta*, TMM can heterodimerize with ER and ERL1. Clarifying the physical interactions and*in vivo* activities of the four receptors with the 11 members of the EPF family looks to be an exciting future area of research in *Arabidopsis*. Homologues of ERf, TMM and EPFLs are found in diverse species, including monocots and mosses, indicating that the potential for conserved signaling systems exist. To date, however, no experimental information is available outside of *Arabidopsis*.

Stomatal lineages in *Arabidopsis* are established by asymmetric cell divisions, and these unusual and unequal divisions involve several other novel, plant-specific, proteins: BREAKING OF ASYMMETRY IN THE STOMATAL LINEAGE (BASL) [49] and POLAR LOCALIZATION DURING ASYMMETRIC DIVISION AND REDISTRIBUTION (POLAR) [50]. BASL displays dynamic spatiotemporal localization in the stomatal lineage. Before asymmetric cell division, BASL is detected in both the nucleus and at the cell periphery distal to the cell division plane. After the division, daughter cells inherit BASL in a manner that defines their fate: nuclear localization (differentiation to guard cells), peripheral localization (differentiation to pavement cell), or both (continued asymmetric cell divisions) [49]. *BASL* mutants display misoriented asymmetric cell division and overexpression of BASL leads to ectopic outgrowths in the positions where BASL is peripherally concentrated [49]. Hence, it seems possible that BASL controls or mediates cell polarity during asymmetric cell division in the stomatal lineage. POLAR shares some features of the BASL localization pattern; it is peripheral and distal to the cell division site before asymmetric cell division and shows unequal behaviors in the daughters, disappearing from the larger daughter and being upregulated in the smaller, meristematic, daughter [50]. Although no phenotypes have been ascribed to loss of *POLAR*, its localization is dependent on *BASL* suggesting that they act in the same pathway [50].

## Additional Rules for Monocot Stomata

One of the major differences between dicot and monocot (specifically, grass) stomatal pathways is that, in the latter, subsidiary cells are recruited from cell files flanking the stomatal lineage. This process requires the generation of a highly polarized cell division that is specifically oriented toward the GMC. After formation of a GMC (itself formed by asymmetric division within the stomatal lineage), neighboring subsidiary mother cells (SMCs) divide asymmetrically to produce small subsidiary cells next to the GMC (Figures 2G and 4). SMC polarization involves localization of F-actin patches along the cell wall flanking the GMC, and nuclear migration towards the actin patches [51]. In maize, actin patches co-localize with an LRR-RLK protein, PANGLOSS 1 (PAN1) [52]. Despite shared roles in stomatal development, PAN1 is not in the same LRR-kinase family as ERf and there are no polarly localized LRR-kinases implicated yet in *Arabidopsis* stomatal development. Loss of*PAN1* leads to mislocalization of actin and the nucleus. This disrupts asymmetric cell divisions and results in abnormal subsidiary cells [52, 53]. Recently, the actin regulators Rho of plants 2 (ROP2) and ROP9 were shown to localize polarly in SMCs and to promote SMC polarization [54]. PAN1, ROP2 and ROP9 interact and localization of ROP2 and ROP9 is dependent on PAN1, but PAN1 localization is independent of ROPs. It is attractive to speculate that these proteins work in a common pathway to receive polarity cues and translate them into the cellular reorganization necessary for SMC polarization [54].

## Evolution of Stomatal Regulators

### *Arabidopsis stomatal bHLH genes are in stomatal-producing plant lineages*

As described above, in *Arabidopsis*, five bHLH genes are major determinants of the identities and behaviors of different stomatal lineage precursors. SPCH, MUTE and FAMA are fairly restricted in their expression pattern to subsets of the stomatal lineage whereas ICE1/SCRM and SCRM2 are expressed throughout the lineage and in additional non-stomatal lineage cells. When considering the diversity of stomatal pattern in nature, it is interesting to think about how the expression, regulation and function (and existence) of this class of regulators may change. Moreover, one might ask whether the heterodimeric partnership between SPCH, MUTE and FAMA with ICE1/SCRM and SCRM2 could be ancient or whether this is a new innovation.

The bHLH family is characterized by a conserved DNA binding region, but there are easily recognizable sub-families within. *SPCH MUTE* and *FAMA* belong to the group IA bHLHs [55] whose genic intron/exon structure and protein C-termini are distinctive enough to serve as high confidence group characters throughout the flowering plants and out to *Selaginella* (a model lycophyte) and *Physcomitrella* (a model moss) [56]. Distinction among individual 1A members is only clear within the flowering plants. ICE1/SCRM and SCRM2 are members of group III and representatives of this group are found in many clades back to the mosses (http://www.phytozome.net/). In the incomplete transcriptome and genome sequences from plant lineages predating the emergence of stomata, neither group IA nor group III bHLH genes are obvious [56] (C MacAlister,*personal communication*). Based on sequences currently available, in all cases where group 1A members can be distinguished, there is also a group III bHLH, suggesting that their partnership can be ancient. A group 1A homologue from *Physcomitrella* can partially complement *Arabidopsis mute* and *fama*, but not *spch* mutants [56]. These cross species complementation results are interesting in light of the shortened pathway for development of stomata in *Physcomitrella;* in this moss, no early asymmetric divisions are evident and instead a single GMC is specified and undergoes incomplete cytokinesis to form two connected guard cells (Figure 4). This pathway would require MUTE-and FAMA-like fate promoting activities, but not the division-promoting activity of SPCH [56].

In grasses, the positions of stomata are determined and fixed at early stages of leaf development and amplifying divisions are not present. In fact, only the differentiation (GMC to guard cells) step is similar between grasses and *Arabidopsis* (Figure 4). Nonetheless, *SPCH MUTE* and *FAMA* genes can be identified in the genomes of maize, rice and *Brachypodium*. There has also been a duplication of *SPCH* in these plants [57]. Perhaps due to the different stomatal ontogenies, however, the rice homologues OsSPCH1/2 are expressed very early during plant development, possibly before the production of stomatal lineage [57]. *OsSPCH2*mutants in rice, do, however, have reduced stomatal numbers and resemble weak mutant alleles of*AtSPCH*[57]. Overexpression of *OsMUTE* and *OsFAMA* recapitulates overexpression phenotypes of the*Arabidopsis* genes, indicating that their GMC and guard cell identity-promoting functions are conserved. Of the three genes, the only one acting at a stage common to stomatal development in both plant groups (the GMC to stomatal guard cell transition), *FAMA*, is most highly conserved in terms of expression pattern and loss of function phenotypes in both rice and *Arabidopsis*[57].

## New Appearance of Polarity Regulators

Homologues of the cell fate regulators and many of the signaling components discussed above appear in many plant species [10, 58, 59]. In contrast, the two proteins shown to exhibit polarized localization in stomatal lineage cells of *Arabidopsis*, POLAR and BASL, do not. BASL does not resemble any other *Arabidopsis*proteins and only in the congeneric *A. lyrata* is there a significantly similar sequence. In *Arabidopsis POLAR* is moderately similar to another gene (*POLAR-LIKE1*) and homologues of *POLAR* and *POLAR-LIKE1* can be found in closely related (dicot) species such as poplar (POPTR B9IL54). Already in rice, however, the sequence similarity becomes restricted to a very small domain of the proteins. It is interesting to consider whether this inability to find such homologues is because the function of BASL and POLAR is required only in the dicots, or because similar functions are carried out by different genes and the apparent uniqueness of these proteins represents either fast substitution rates or that their roles can be served by other proteins. For example, scaffold proteins that bind others together into complexes play important roles in polarity generation in yeast and animals, yet these scaffolds are often not well conserved at the sequence level and consist primarily of multiple interaction surfaces.

## *Evolution of Regulated Stomatal Pore Opening*

A developmental approach concerns itself with the correct specification and pattern of stomata. From a physiological point of view, however, the behavior of these final products is key. Modulation of the stomatal pore aperture depends on coordinated morphologies of the guard cell pair and, particularly in the case of the grasses, on the coordination of guard cells and the specialized subsidiary cells that are obligate parts of the stomatal complex [60]. For stomatal pore aperture to be optimized for daily and seasonal fluctuations in light, temperature, humidity and $CO_2$ availability, the guard cells must be able to sense such environmental factors. Guard cells in angiosperms appear to sense many of these factors autonomously, and key kinases (OST1), phosphatases (PP2C) and receptors for the "drought stress" hormone, ABA (PYR1) have been identified in*Arabidopsis*[1]. Recent studies of $CO_2$ and ABA responsiveness in non-vascular plants have come to different conclusions about when the sensing of these different environmental cues arose. Monitoring stomatal pore closure in response to ABA, [61] concluded that responsiveness to this hormone was a new feature and was absent in fern and lycophyte species. Other studies, however, provide evidence that ABA sensing may have arisen quite early. By cross-species complementation, Ruszala [5] and Chater [62] showed that the OST1 homologues from *Selaginella moellendorffii* and *Physcomitrella patens* could partially restore the ability of*Arabidopsis ost1* stomata to respond to ABA. Moreover, knockout of *PpOST1-1* significantly attenuated ABA response in *P. patens* stomata [62]. The differing conclusions from these studies could be due to the different "representative" species chosen, a general caution in evolutionary studies of this system that is also echoed in the behavior of maize and rice bHLHs [57].

# CONCLUSIONS

Stomatal development in *Arabidopsis* has been used as a model genetic system for the analysis of cell fate, cell polarity and cell to cell communication. The nature of the gene products identified in such analysis, coupled with the long tradition of evaluating the numbers and patterns of stomata in diverse plants for taxonomic purposes makes this system a useful natural laboratory to look at the parallel evolution of genes and developmental trajectories. As the number of completed plant genomes increases and tools for experimental manipulation of non-model species develop, we believe there will be an excellent opportunity to test the roles of candidate cell fate- and cell signaling factor-encoding genes in creating developmental diversity.

# DECLARATIONS

## Acknowledgments

Funding for work on stomata in the author's laboratory is provided by grants from the US National Science Foundation (IOS-0845521) and the National Institutes of Health (R01GM086632). DCB is a Gordon and Betty Moore Foundation investigator of the Howard Hughes Medical Institute.

## Competing interests

The authors declare that they have no competing interests.

## Authors' contributions

AV and DCB and designed the study, wrote the manuscript and prepared the figures. All authors read and approved the final manuscript.

## REFERENCES


1. Sirichandra C, Wasilewska A, Vlad F, Valon C, Leung J: The guard cell as a single-cell model towards understanding drought tolerance and abscisic acid action. J Exp Bot. 2009, 60: 1439-1463. 10.1093/jxb/ern340.
2. Ligrone R, Duckett JG, Renzaglia KS: Major transitions in the evolution of early land plants: a bryological perspective. Ann Bot. 2012, 109: 851-871. 10.1093/aob/mcs017.
3. Edwards D, Kerp H, Hass H: Stomata in early land plants: an anatomical and ecophysiological approach. J Exp Bot. 1998, 49 (Suppl 1): 255-278.
4. Franks PJ, Beerling DJ: Maximum leaf conductance driven by CO2 effects on stomatal size and density over geologic time. Proc Natl Acad Sci USA. 2009, 106: 10343-10347. 10.1073/pnas.0904209106.
5. Ruszala EM, Beerling DJ, Franks PJ, Chater C, Casson SA, Gray JE, Hetherington AM: Land plants acquired active stomatal control early in their evolutionary history. Curr Biol. 2011, 21: 1030-1035. 10.1016/j.cub.2011.04.044.
6. de León MEM, Pérez-García B, Márquez-Guzmán J, Mendoza-Ruiz A: Developmental gametophyte morphology of seven species of Thelypteris subg. Cyclosorus (Thelypteridaceae). Micron. 2008, 39: 1351-1362. 10.1016/j.micron.2008.02.001.
7. Kim KW, Lee S, Bae S, Kim P: 3d surface profiling and high resolution imaging for refining the florin rings and epicuticular wax crystals of Pinus koraiensis needles. Microsc Res Tech. 2011, 74: 1166-1173. 10.1002/jemt.21010.
8. Koch K, Bhushan B, Barthlott W: Multifunctional surface structures of plants: an inspiration for biomimetics. Progress in Materials Science. 2009, 54: 137-178. 10.1016/j.pmatsci.2008.07.003.
9. Zhao L, Sack FD: Ultrastructure of stomatal development in Arabidopsis (Brassicaceae) leaves. Am J Bot. 1999, 86: 929-10.2307/2656609.
10. Peterson KM, Rychel AL, Torii KU: Out of the mouths of plants: the molecular basis of the evolution and diversity of stomatal development. Plant Cell. 2010, 22: 296-306. 10.1105/tpc.109.072777.
11. Boetsch J, Chin J, Croxdale J: Arrest of stomatal initials in Tradescantia is linked to the proximity of neighboring stomata and results in the arrested initials acquiring properties of epidermal cells. Dev Biol. 1995, 168: 28-38. 10.1006/dbio.1995.1058.
12. Sack FD, Paolillo DJ: Incomplete cytokinesis in Funaria stomata. Am J Bot. 1985, 72: 1325-1333. 10.2307/2443504.
13. Egunyomi A: On the stomata of some tropical African mosses. Lindbergia. 1982, 8: 121-124.
14. Duckett JG, Pressel S, P'ng KM, Renzaglia KS: Exploding a myth: the capsule dehiscence mechanism and the function of pseudostomata in Sphagnum. New Phytol. 2009, 183: 1053-1063. 10.1111/j.1469-8137.2009.02905.x.
15. Cotthem WV: Comparative morphological study of the stomata in the Filicopsida. Bulletin du Jardin botanique national de Belgique/Bulletin van de National Plantentuin van België. 1970, 40: 81-151. 10.2307/3667713. +1-88
16. Ziegler H: The evolution of stomata. Stomatal Function. Edited by: Zeiger E, Farquhar GD, Cowan IR. 1987, Stanford University Press, Stanford, CA, 29-57.
17. Johnson RW, Riding RT: Structure and ontogeny of the stomatal complex in Pinus strobus L. and Pinus banksiana lamb. Am J Bot. 1981, 68: 260-268. 10.2307/2442858.
18. Chabot JF, Chabot BF: Ultrastructure of the epidermis and stomatal complex of balsam fir (Abies balsamea). Can J Bot. 1977, 55: 1064-1075. 10.1139/b77-125.
19. Pedersen O, Sand-Jensen K: Adaptations of submerged Lobelia dortmanna to aerial life form: morphology, carbon sources and oxygen dynamics. Oikos. 1992, 65: 89-96. 10.2307/3544890.
20. Keeley JE, Busch G: Carbon assimilation characteristics of the aquatic CAM plant, isoetes howellii. Plant Physiol. 1984, 76: 525-530. 10.1104/pp.76.2.525.
21. Hetherington AM, Woodward FI: The role of stomata in sensing and driving environmental change. Nature. 2003, 424: 901-908. 10.1038/nature01843.
22. Ohashi-Ito K, Bergmann D: Arabidopsis FAMA controls the final proliferation/differentiation switch during stomatal development. Plant Cell. 2006, 18: 2493-2505. 10.1105/tpc.106.046136.
23. MacAlister CA, Ohashi-Ito K, Bergmann DC: Transcription factor control of asymmetric cell divisions that establish the stomatal lineage. Nature. 2007, 445: 537-540. 10.1038/nature05491.
24. Pillitteri LJ, Sloan DB, Bogenschutz NL, Torii KU: Termination of asymmetric cell division and differentiation of stomata. Nature. 2007, 445: 501-505. 10.1038/nature05467.
25. Robinson S, de Barbier Reuille P, Chan J, Bergmann D, Prusinkiewicz P, Coen E: Generation of spatial patterns through cell polarity switching. Science. 2011, 333: 1436-1440. 10.1126/science.1202185.
26. Macalister CA, Bergmann DC: Stomatal Patterning. eLS. Edited by: Anonymous. 2001, John Wiley & Sons, Ltd
27. Kanaoka MM, Pillitteri LJ, Fujii H, Yoshida Y, Bogenschutz NL, Takabayashi J, Zhu JK, Torii KU: SCREAM/ICE1 and SCREAM2 specify three cell-state transitional steps leading to arabidopsis stomatal differentiation. Plant Cell. 2008, 20: 1775-1785. 10.1105/tpc.108.060848.
28. Chinnusamy V, Ohta M, Kanrar S, Lee BH, Hong X, Agarwal M, Zhu JK: ICE1: a regulator of cold-induced

transcriptome and freezing tolerance in Arabidopsis. Genes Dev. 2003, 17: 1043-1054. 10.1101/gad.1077503.

29. Dong J, Bergmann DC: Stomatal patterning and development. Curr Top Dev Biol. 2010, 91: 267-297.

30. Lampard GR, Lukowitz W, Ellis BE, Bergmann DC: Novel and expanded roles for MAPK signaling in Arabidopsis stomatal cell fate revealed by cell type-specific manipulations. Plant Cell. 2009, 21: 3506-3517. 10.1105/tpc.109.070110.

31. Bergmann DC, Lukowitz W, Somerville CR: Stomatal development and pattern controlled by a MAPKK kinase. Science. 2004, 304: 1494-1497. 10.1126/science.1096014.

32. Wang H, Ngwenyama N, Liu Y, Walker JC, Zhang S: Stomatal development and patterning are regulated by environmentally responsive mitogen-activated protein kinases in Arabidopsis. Plant Cell. 2007, 19: 63-73. 10.1105/tpc.106.048298.

33. Lampard GR, Macalister CA, Bergmann DC: Arabidopsis stomatal initiation is controlled by MAPK-mediated regulation of the bHLH SPEECHLESS. Science. 2008, 322: 1113-1116. 10.1126/science.1162263.

34. Kim TW, Michniewicz M, Bergmann DC, Wang ZY: Brassinosteroid regulates stomatal development by GSK3-mediated inhibition of a MAPK pathway. Nature. 2012, 482: 419-422. 10.1038/nature10794.

35. Gudesblat GE, Schneider-Pizon J, Betti C, Mayerhofer J, Vanhoutte I, van Dongen W, Boeren S, Zhiponova M, de Vries S, Jonak C, Russinova E: SPEECHLESS integrates brassinosteroid and stomata signalling pathways. Nat Cell Biol. 2012, 14: 548-554. 10.1038/ncb2471.

36. Tricker PJ, Gibbings JG, Rodríguez López CM, Hadley P, Wilkinson MJ: Low relative humidity triggers RNA-directed de novo DNA methylation and suppression of genes controlling stomatal development. J Exp Bot. 2012, in press.

37. Hara K, Yokoo T, Kajita R, Onishi T, Yahata S, Peterson KM, Torii KU, Kakimoto T: Epidermal cell density is autoregulated via a secretory peptide, EPIDERMAL PATTERNING FACTOR 2 in Arabidopsis leaves. Plant Cell Physiol. 2009, 50: 1019-1031. 10.1093/pcp/pcp068.

38. Hunt L, Gray JE: The signaling peptide EPF2 controls asymmetric cell divisions during stomatal development. Curr Biol. 2009, 19: 864-869. 10.1016/j.cub.2009.03.069.

39. Hara K, Kajita R, Torii KU, Bergmann DC, Kakimoto T: The secretory peptide gene EPF1 enforces the stomatal one-cell-spacing rule. Genes Dev. 2007, 21: 1720-1725. 10.1101/gad.1550707.

40. Abrash EB, Davies KA, Bergmann DC: Generation of signaling specificity in Arabidopsis by spatially restricted buffering of ligand-receptor interactions. Plant Cell. 2011, 23: 2864-2879. 10.1105/tpc.111.086637.

41. Abrash EB, Bergmann DC: Regional specification of stomatal production by the putative ligand CHALLAH. Development. 2010, 137: 447-455. 10.1242/dev.040931.

42. Shpak ED, McAbee JM, Pillitteri LJ, Torii KU: Stomatal patterning and differentiation by synergistic interactions of receptor kinases. Science. 2005, 309: 290-293. 10.1126/science.1109710.

43. Yang M, Sack FD: The too many mouths and four lips mutations affect stomatal production in Arabidopsis. Plant Cell. 1995, 7: 2227-2239.

44. Nadeau JA, Sack FD: Control of stomatal distribution on the Arabidopsis leaf surface. Science. 2002, 296: 1697-1700. 10.1126/science.1069596.

45. Shpak ED, Berthiaume CT, Hill EJ, Torii KU: Synergistic interaction of three ERECTA-family receptor-like kinases controls Arabidopsis organ growth and flower development by promoting cell proliferation. Development. 2004, 131: 1491-1501. 10.1242/dev.01028.

46. Kondo T, Kajita R, Miyazaki A, Hokoyama M, Nakamura-Miura T, Mizuno S, Masuda Y, Irie K, Tanaka Y, Takada S, Kakimoto T, Sakagami Y: Stomatal density is controlled by a mesophyll-derived signaling molecule. Plant Cell Physiol. 2009

47. Sugano SS, Shimada T, Imai Y, Okawa K, Tamai A, Mori M, Hara-Nishimura I: Stomagen positively regulates stomatal density in Arabidopsis. Nature. 2010, 463: 241-244. 10.1038/nature08682.

48. Lee JS, Kuroha T, Hnilova M, Khatayevich D, Kanaoka MM, McAbee JM, Sarikaya M, Tamerler C, Torii KU: Direct interaction of ligand-receptor pairs specifying stomatal patterning. Genes Dev. 2012, 26: 126-136. 10.1101/gad.179895.111.

49. Dong J, MacAlister CA, Bergmann DC: BASL controls asymmetric cell division in Arabidopsis. Cell. 2009, 137: 1320-1330. 10.1016/j.cell.2009.04.018.

50. Pillitteri LJ, Peterson KM, Horst RJ, Torii KU: Molecular profiling of stomatal meristemoids reveals new component of asymmetric cell division and commonalities among stem cell populations in Arabidopsis. Plant Cell. 2011, 23: 3260-3275. 10.1105/tpc.111.088583.

51. Gallagher K, Smith L: Discordia mutations specifically misorient asymmetric cell divisions during development of the maize leaf epidermis. Development. 1999, 126: 4623-4633.

52. Cartwright HN, Humphries JA, Smith LG: PAN1: a receptor-like protein that promotes polarization of an asymmetric cell division in maize. Science. 2009, 323: 649-651. 10.1126/science.1161686.

53. Gallagher K, Smith LG: Roles for polarity and nuclear determinants in specifying daughter cell fates after an asymmetric cell division in the maize leaf. Curr Biol. 2000, 10: 1229-1232. 10.1016/S0960-9822(00)00730-2.

54. Humphries JA, Vejlupkova Z, Luo A, Meeley RB, Sylvester AW, Fowler JE, Smith LG: ROP GTPases act with the receptor-like protein PAN1 to polarize asymmetric cell division in maize. Plant Cell. 2011, 23: 2273-2284. 10.1105/tpc.111.085597.

55. Heim MA, Jakoby M, Werber M, Martin C, Weisshaar B, Bailey PC: The basic helix-loop-helix transcription factor family in plants: a genome-wide study of protein structure and functional diversity. Mol Biol Evol. 2003, 20: 735-747. 10.1093/molbev/msg088.

56. MacAlister CA, Bergmann DC: Sequence and function of bHLHs required for stomatal development in Arabidopsis are deeply conserved in land plants. Evol Dev. 2011, 13: 182-192. 10.1111/j.1525-142X.2011.00468.x.

57. Liu T, Ohashi-Ito K, Bergmann DC: Orthologs of Arabidopsis thaliana stomatal bHLH genes and regulation of stomatal development in grasses. Development. 2009, 136: 2265-2276. 10.1242/dev.032938.

58. Rychel AL, Peterson KM, Torii KU: Plant twitter: ligands under 140 amino acids enforcing stomatal patterning. J Plant Res. 2010, 123: 275-280. 10.1007/s10265-010-0330-9.

59. Katsir L, Davies KA, Bergmann DC, Laux T: Peptide signaling in plant development. Curr Biol. 2011, 21: R356-R364. 10.1016/j.cub.2011.03.012.

60. Franks PJ, Farquhar GD: The mechanical diversity of stomata and its significance in gas-exchange control. Plant Physiol. 2007, 143: 78-87.

61. Brodribb TJ, McAdam SA: Passive origins of stomatal control in vascular plants. Science. 2011, 331: 582-585. 10.1126/science.1197985.

62. Chater C, Kamisugi Y, Movahedi M, Fleming A, Cuming AC, Gray JE, Beerling DJ: Regulatory mechanism controlling stomatal behavior conserved across 400 million years of land plant evolution. Curr Biol. 2011, 21: 1025-1029. 10.1016/j.cub.2011.04.032.

# INDEX

## M



## P



## R



## S



## T



## U



## X